Nelson

Chemistry 12

College Preparation

Authors

Lucille Davies
Limestone District School Board

Maurice Di Giuseppe
Toronto Catholic District School Board

Ted Gibb
Formerly of Thames Valley District School Board

Milan Sanader
Dufferin-Peel Catholic District School Board

Angela Vavitsas
Toronto District School Board

Program Consultant

Maurice Di Giuseppe
Toronto Catholic District School Board

THOMSON
NELSON

Australia   Canada   Mexico   Singapore   Spain   United Kingdom   United States

THOMSON

NELSON

Nelson Chemistry 12: College Preparation
By Lucille Davies, Maurice Di Giuseppe,
Ted Gibb, Milan Sanader, and Angela
Vavitsas

Director of Publishing
David Steele

Publisher
Kevin Martindale

**Executive Managing Editor,
Development and Testing**
Cheryl Turner

Program Managers
Lina Mockus
John Yip-Chuck

Developmental Editors
Julie Bedford
Lisa Kafun
Julia Lee
Lina Mockus

Editorial Assistant
Jennie Worden

**Executive Managing Editor,
Production**
Nicola Balfour

Senior Production Editor
Debbie Davies-Wright

Copy Editor/Proofreader
Paula Pettitt-Townsend

Senior Production Coordinator
Sharon Latta Paterson

Creative Director
Angela Cluer

Art Director
Ken Phipps

Art Management
Suzanne Peden

Illustrators
Andrew Breithaupt
Steven Corrigan
Deborah Crowle
Margo Davies LeClair
John Fraser
Irma Ikonen
Norman Lanting
Dave Mazierski
Dave McKay
Linda Neale
Frank Netter
Peter Papayanakis
Ken Phipps

Marie Price
Myra Rudakewich
Gabriel Sierra
Katherine Strain
Bart Vallecoccia
Jane Whitney

Composition
Nelson Gonzalez

Interior Design
Kyle Gell
Allan Moon

Cover Design
Peter Papayanakis

Cover Image
James King-Holmes/
Science Photo Library

Photo Research and Permissions
Mary Rose MacLachlan

Set-up Photos
Dave Starrett
Media Services

Printer
Transcontinental Printing Inc.

**National Library of Canada
Cataloguing in Publication Data**

Main entry under title:
Chemistry 12 : college preparation/
Maurice Di Giuseppe ... [et al.].

Includes index.
ISBN 0-17-626533-3

1. Chemistry—Textbooks. I. Di
Giuseppe, Maurice II. Title:
Chemistry twelve.

QD33.C44 2003   540
C2003-902499-7

Acknowledgments
Nelson and the authors of *Nelson
Chemistry 12: College Preparation*
thank the staff and students of
Mary Ward Catholic Secondary
School for the use of their
facilities, and for the grace and
generosity of their help.

Reviewers

▸ CONTENTS

▸ Unit 1
Matter and Qualitative Analysis

▸ Unit 2
Quantities in Chemistry

▸ **Unit 3**
Organic Chemistry

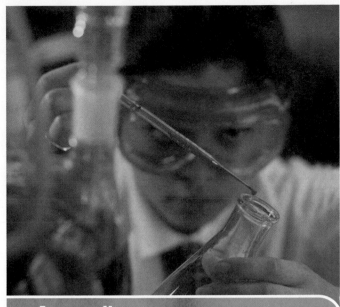

▶ Appendices

Matter and Qualitative Analysis

Many of us have experienced the brilliant sights and sounds of a fireworks display. Yet, while we watch and enjoy the artistry lighting the night sky with a range of colours in different patterns, we rarely think about the science involved. The production of different colours of light by fireworks is a wonderful display of chemistry. Fireworks contain different substances that produce different colours of light (**Table 1**). For example, lithium carbonate produces red light, and sodium nitrate produces yellow light. Scientists believe that the colours are caused by excited electrons releasing energy. The next time you are at a fireworks display, marvelling at the beautiful colours, try to identify the substances that may be responsible for the different colours you are seeing, based on what you will learn in this unit. Identifying substances by their properties is what chemists call *qualitative analysis*.

Table 1 Colours Produced by the Compounds Used in Fireworks

Colour	Compound
red	strontium carbonate, lithium carbonate
orange	calcium chloride
yellow	sodium nitrate
blue	copper(II) chloride
purple	mixture of strontium and copper compounds

▶ Overall Expectations

In this unit, you will be able to

- demonstrate an understanding of the basic principles of qualitative analysis and underlying theories;
- carry out qualitative analyses, using flow charts and appropriate laboratory equipment and instruments;
- describe the role and importance, in society, of some of the applications of qualitative analysis.

ARE YOU READY?

▶ **Prerequisites**

Concepts

- balanced chemical equations
- the states of matter
- physical and chemical changes and properties
- the symbols for common elements and the formulas for common compounds
- the formulas of common ionic and molecular compounds, based on the periodic table and an IUPAC table of ions
- Bohr–Rutherford diagrams

Skills

- design and conduct controlled experiments
- select and use laboratory materials accurately and safely
- organize and display experimental observations and data in suitable tabular and graphical formats

Knowledge and Understanding

1. Calcium chloride (**Figure 1**) is a salt that is added to roads and sidewalks in the winter to melt ice and snow. State whether each of the following properties of calcium chloride is physical or chemical, and whether each physical property is qualitative or quantitative.
   (a) Calcium chloride has a melting point of 782°C.
   (b) Calcium chloride has a density of 2150 kg/m^3.
   (c) Calcium chloride is soluble in water.
   (d) In the presence of sulfuric acid, calcium chloride reacts to form hydrogen chloride.
   (e) When heated to decomposition, calcium chloride produces chlorine gas.

2. Classify each of the following changes as physical or chemical. Explain your classification.
   (a) Water boils in a kettle.
   (b) Propane burns in a barbecue.
   (c) Acid rain corrodes iron.
   (d) An apple rots.
   (e) Sugar dissolves in hot tea.
   (f) An egg is boiled.
   (g) Butter melts on hot toast.
   (h) Wood burns.
   (i) Copper wire bends.
   (j) A candle burns.
   (k) Snow melts.

3. What physical property is described by each of the following statements?
   (a) Aluminum can be hammered into thin sheets.
   (b) Copper wire is used for electrical circuitry in homes.
   (c) One millilitre of water has a mass of one gram.
   (d) Ice melts at 0°C.
   (e) Diamond can scratch glass.

4. Distinguish between the two terms in each of the following pairs of terms. Provide examples where possible.
   (a) element and compound
   (b) solute and solvent
   (c) mechanical mixture and solution
   (d) homogeneous mixture and heterogeneous mixture
   (e) proton and neutron
   (f) metal and nonmetal
   (g) atom and molecule
   (h) atomic number and mass number
   (i) pure substance and mixture

Figure 1
Calcium chloride is used to melt ice on walkways and roads in the winter.

5. Look carefully at the following diagrams (**Figure 2**). Decide whether each diagram represents an element, a compound, or a mixture. If the diagram represents a mixture, state how many elements and how many compounds are present in the mixture. Note that each different circle represents a different atom.

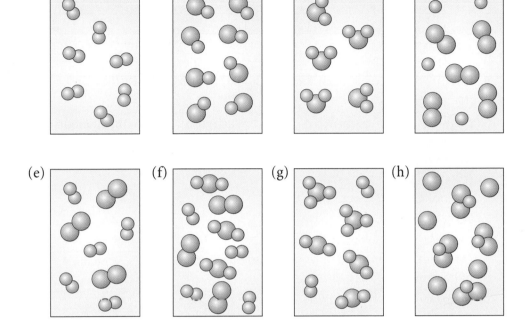

Figure 2
Elements, compounds, and mixtures

6. Copy and complete **Table 1**.

Table 1   Subatomic Particles

Particle	Relative mass	Relative charge	Location within atom
proton			
electron			
neutron			

7. List the elements that belong to each of the following groups in the periodic table:
   (a) halogens
   (b) alkali metals
   (c) noble gases
   (d) alkaline earth metals

(a)

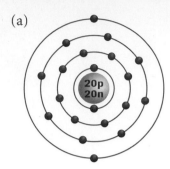

(b)

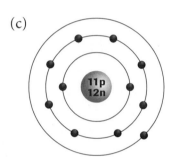

(c)

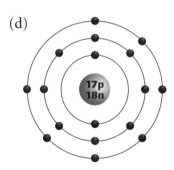

(d)

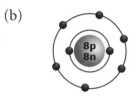

Figure 3
Bohr–Rutherford diagrams for four elements

8. Copy and complete **Table 2**.

Table 2 Identifying Elements

Chemical name	Chemical symbol	Atomic number	Number of protons	Number of neutrons	Mass number	Number of electrons
helium						
		16				
			8			
						10
	K					

9. Identify the elements that are represented by the Bohr–Rutherford diagrams in **Figure 3**.

10. Draw a Bohr–Rutherford diagram for each of the following elements:
    (a) magnesium
    (b) boron
    (c) helium
    (d) potassium
    (e) oxygen

11. In **Table 3**, each element in column A can form a compound with one element in column B, in a 1:1 ratio.
    (a) What type of element is found in column A?
    (b) What type of element is found in column B?
    (c) Name the three chemical compounds that can form between the elements in column A and the elements in column B, in a 1:1 ratio.

Table 3 Forming Compounds

Column A	Column B
Ca	N
K	O
Al	F

Inquiry and Communication

12. Imagine that you are given a mixture that contains iron filings, sand, water, and rubbing alcohol.
    (a) Design an experiment that will allow you to separate this mixture into its components.
    (b) Make a list of the equipment you will need in order to carry out your experiment.
    (c) Write a procedure for your experiment, including any safety precautions.

13. Imagine that you are given an unidentified gas that is hydrogen, oxygen, or carbon dioxide. Design an experiment that will allow you to determine, conclusively, the identity of the gas. (*Hint*: You must use diagnostic tests to confirm or eliminate each of the three gases.)

Technical Skills and Safety

14. State whether each of the following actions is safe or unsafe. If an action is unsafe, explain why it is unsafe and then describe the corresponding safe action.

    (a) A student breaks a beaker due to poor laboratory equipment management. She picks up the glass with a paper towel and places it in the classroom garbage can.

    (b) During a lab, a student who is not wearing safety glasses gets an unidentified liquid in one of his eyes. The eye starts to sting. The student rubs the eye with his hand and hopes that the stinging will go away.

    (c) A student performs a lab that requires heating a solution in a test tube. She lights the Bunsen burner correctly and uses a test-tube holder to pick up the test tube that contains the solution. She then places the bottom of the test tube in the hottest part of the flame, pointing the mouth of the test tube away from her classmates.

    (d) A student is asked to identify a substance in the laboratory. He observes that the substance is a white, crystalline solid, so he tastes the substance to determine whether it is salt or sugar.

15. The following WHMIS symbols (**Figure 4**) appear on chemicals in the laboratory. Identify each symbol.

    (a)   (b)   (c)

    Figure 4
    The Workplace Hazardous Materials Information System (WHMIS) provides information regarding hazardous products.

Figure 1
The Mars Global Surveyor entered the Martian orbit approximately one year after it was launched, on September 11, 1997.

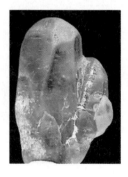

Figure 2
The mineral olivine is high in magnesium and iron.

Figure 3

Rover 1 and *Rover 2* travel 100 m in a Martian day, collecting data to be transmitted back to Earth by computer downloads.

Humans have always been interested in the world around them. In the last century, scientists even started to explore other planets. In 1960, the Russians attempted to launch the first Martian probe, *Marsnik 1*. Since then, scientists have used orbiters, landers, rovers, and probes to learn more about Mars. Scientists have collected a vast amount of data about the atmosphere, composition, and climate of Mars, without ever having stepped on the planet's surface!

On November 7, 1996, the U.S. National Aeronautics and Space Administration (NASA) launched the Mars Global Surveyor (MGS) (**Figure 1**). The MGS is equipped with a Thermal Emission Spectrophotometer (TES), which collects information about the composition of Martian rock and soil and, indirectly, the presence of water on Mars.

The TES measures infrared (IR) energy as a result of heat transfer. IR energy is a form of electromagnetic energy that you can detect with your skin but not with your eyes. Since all matter absorbs or reflects IR energy differently, scientists can use IR radiation that is emitted by objects to distinguish between different types of matter.

As the MGS orbits Mars, the TES detector senses and records the different amounts of IR energy that are released by the surface of Mars. Scientists on Earth receive the data and compare the values with known IR emission values for minerals. In this way, scientists can determine the soil composition of Mars. Determining the composition of a sample of matter from its *physical and chemical properties* and comparing it with the composition of known substances is called *qualitative analysis.*

Scientists have discovered that the rocks on Mars contain large quantities of the mineral olivine (**Figure 2**). Olivine is sometimes used in jewellery, and it is easily eroded by water. Since large quantities of olivine are found on Mars, some scientists believe that Mars has always been very cold and dry.

NASA launched the Mars Exploration Rovers in 2003 (**Figure 3**). The rovers landed on Mars in early 2004. They were equipped with instruments for further detecting and analyzing mineral and water samples. The data collected

will provide scientists with a more detailed picture of the planet Mars. Using these data, scientists hope to develop technologies that will assist in the colonization of Mars by humans sometime in the future.

▶ *TRY THIS* activity   *An Introduction to Qualitative Analysis*

Chromatography is the separation of a mixture (liquid or gas) into its components by passing it through a material (such as chalk) that absorbs the components at different rates. In this activity, you will use chalk chromatography to separate and identify the components of black ink.

Materials: black marker (water-soluble ink), 250-mL beaker, 2 full pieces of white chalk, water, pencil

1. Use the black marker to draw a single ring of ink around one piece of white chalk, approximately 2 cm from one end. Trace over the ink ring two or three times.

2. Stand the chalk in the centre of a 250-mL beaker, with the ink ring near the bottom of the beaker. The chalk must remain freestanding.

3. Using a finger to hold the chalk in place, gently pour water into the beaker until the water level is just below the ink ring on the chalk (**Figure 4**). Let the chalk stand in the water for 20 to 30 min.

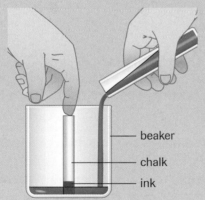
          — beaker
          — chalk
          — ink

Figure 4
Laboratory setup for separating "black" ink

4. Remove the chalk from the water when the level of water in the chalk is just below the top of the chalk. Stand the chalk on a bench, and let it dry.

(a) Observe the banding pattern on the chalk. Predict the result you would obtain if you repeated steps 1 to 4 with a similar piece of chalk and the same black ink.

5. Repeat steps 1 to 4 with the other piece of chalk. Remove the chalk from the beaker when the water level reaches the same height as it did in the first piece of chalk you used.

(b) What happened to the black ink as it moved up the pieces of chalk?

(c) Describe the separation distances between the coloured bands. Are the separation distances all the same, or do they vary?

(d) State a hypothesis for the banding pattern you observed. Suggest an experiment to test your hypothesis.

(e) Evaluate the prediction you made in (a).

(f) Describe several advantages and several disadvantages of this separation method.

(g) Suggest changes in the procedure that may improve the band separation.

💡 *REFLECT* on your learning

1. Make a list of physical and chemical properties you are familiar with and you think can be used to identify matter.

2. Why is it important for scientists to develop models in order to present their scientific research?

3. Describe some careers in which people need to identify matter.

Identifying a Mystery Powder

Forensic science is the application of science to criminal investigations. Forensic scientists collect, analyze, and evaluate evidence from a crime scene. The evidence is then used in law enforcement. Qualitative analysis (that is, identifying a sample of matter from its physical and chemical properties) is an integral part of forensic science.

In this activity, imagine that you are a forensic scientist. You suspect that a death was caused by the ingestion of a mysterious white powder. Traces of the powder were found near the victim. The victim's partner claims that a severe heart attack was the cause of death. It is your responsibility to identify the mystery powder.

Question

What is the identity of the mystery powder?

Materials

eye protection
lab apron
6 different white powders, labelled A to F, obtained
    from your teacher
unidentified white powder, obtained from your
    teacher
microtray
5 eyedroppers
scoopula
toothpicks
distilled water
universal indicator
dilute hydrochloric acid, $HCl_{(aq)}$
dilute iron(III) (ferric) nitrate solution, $Fe(NO_3)_{3(aq)}$
iodine solution, $I_{2(aq)}$

LEARNING TIP

State Subscripts
The following subscripts are used to indicate the physical state of a substance:
(s)   solid
(l)   liquid
(g)   gas
(aq) dissolved in water

Procedure

1. Label the microtray using a grease pencil, as shown in **Figure 1**.

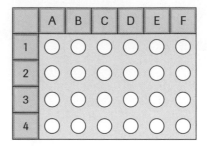

Figure 1
The distribution of the six powders in a microtray

2. Using a scoopula, add a small sample of the first white powder to four wells in column A of the microtray (**Figure 1**). Place samples of the other five powders in columns B, C, D, E, and F, respectively. Record the identity and appearance of each powder.

3. Using an eyedropper, add five drops of water to each of the six samples in the *first row* of the microtray. Mix the contents of each well using a different toothpick. Record your observations in a table.

 Wear eye protection and a lab apron. Hydrochloric acid, iron(III) nitrate, iodine solution, and universal indicator are corrosive if they contact the skin or eyes.

4. Using an eyedropper, add one drop of universal indicator to each well containing a sample and water in the *first row* of the microtray. Mix the contents of each well with a different toothpick. Record your observations.

5. Using an eyedropper, add five drops of dilute hydrochloric acid to each sample in the *second row* of the microtray. Mix the contents of each well with a different toothpick. Record your observations.

6. Using an eyedropper, add five drops of dilute iron(III) nitrate solution to each of the six samples in the *third row* of the microtray. Mix the contents of each well with a different toothpick. Record your observations.

7. Using an eyedropper, add five drops of iodine solution to each of the six samples in the *fourth row* of the microtray. Mix the contents of each well with a different toothpick. Record your observations.

8. Obtain an unidentified labelled powder from your teacher. This powder is one of the six substances you have just tested. Repeat steps 2 to 7 with the mystery powder. Record your observations.

Analysis

(a) Analyze your observations, and then answer the Question.

(b) List the physical properties of the six samples that helped you identify the mystery powder.

(c) List the chemical properties of the six samples that helped you identify the mystery powder.

(d) Why was it important to test the six samples before identifying the mystery powder?

(e) What kind of analysis did you perform in this activity? Justify your answer.

Evaluation

(f) How confident are you in your answer to the Question? Justify your level of confidence using evidence from this activity.

(g) Describe additional tests that you could perform to increase your level of confidence in your answer.

(h) Suggest sources of error and changes to the Procedure that would help to reduce these sources of error.

(i) In your own words, explain why the qualitative analysis of matter plays an important role in forensic science.

Observation and Inference

From an early age, you have relied on experience in order to understand your environment. Babies and young children are "natural scientists" (**Figure 1**). They acquire understanding of their world through exploration and experimentation. For example, if you place a baby in a highchair and put different objects on the highchair tray, the baby will likely drop the objects onto the floor in rapid succession, focusing intently on where the objects fall. If you pick up the objects from the floor and put them on the highchair tray again, the baby will probably repeat the same actions.

observation a statement that is based on what you see, hear, taste, touch, and smell

The baby is not trying to annoy you, but to learn more about the objects through experimentation. Do the objects fall in the same spot? Do they make the same noise when they fall? Through **observation** and experience, the baby mentally collects information about the objects and then tries to make sense of the observations. The baby may collect additional information about the objects by tasting, touching, or shaking the objects. Scientists take a similar approach. They gather data through repeated experimentation and observation in order to test theories and hypotheses.

inference a judgment or opinion that is based on observations and/or conclusions from testing

To understand the universe, scientists need to do more than simply observe it. Scientists need to go one step further and *infer* knowledge from their observations. An **inference** is a judgment or opinion that is based on direct observation. As you learned in "Getting Started," scientists know that large quantities of the mineral olivine exist on Mars. Given that olivine is easily eroded by water and that it is present in large quantities on Mars, some scientists infer that the frozen water on Mars may have always existed in a solid state.

Figure 1
Babies are natural experimenters. They test the parameters of the environment around them in order to gain understanding.

Observation and inference are integral components of qualitative analysis. When trying to identify a sample of matter, scientists first *observe* the sample. Then they *infer* its identity based on their observations, by comparing their observations to characteristics of known matter. When you identified the mystery powder in section 1.1, you made an inference based on your observations of the physical and chemical properties of the other powders you tested.

▶ **TRY THIS** activity   ***The Burning Candle***

In many situations, it is important to understand the difference between an observation and an inference. For example, a police officer who is collecting data from a crime scene must provide statements based on observations only.

Materials: candle, match, pencil, paper, watch glass, ruler

1. Obtain a candle, match, and watch glass from your teacher.

2. Light the candle, and drip some of the wax onto the watch glass. Place the candle upright in the hot wax in order to secure it.

3. Observe the candle for 3 to 5 min.

 A flickering candle can trigger a migraine or an epileptic seizure. Do not do this activity if you have either condition.

(a) Write down as many statements about the burning candle as you can think of.

(b) Divide the statements into two categories: inferences and observations.

(c) Compare the number of observations you made with the number of inferences.

(d) Provide one example of an observation and one example of an inference that could be stated by each of the following people:

   (i) a nurse examining a patient with a high fever

   (ii) a firefighter sifting through the debris of a recently extinguished fire

   (iii) a chef tasting a new recipe

Empirical Knowledge and Theoretical Knowledge

Scientific knowledge is acquired through observation and inference. Information that is gathered by the *senses* or by the *extended senses* (using scientific equipment) is called **empirical knowledge**. Empirical knowledge *describes* what is being observed. For example, scientists have observed that when a thermometer placed in boiling water reads 100°C, the water changes state from liquid to gas (vapour). **Theoretical knowledge** attempts to *explain* how or why something occurs. To explain what happens when water boils, scientists use the *kinetic molecular theory* (**Figure 2**, on the next page). According to this theory, water molecules, clinging to each other by weak bonds in the liquid state, absorb energy when heated. When the temperature reaches 100°C, the water molecules have absorbed enough energy to overcome the weak bonds, and they enter the gaseous state. A **theory** is an explanation of a large number of related observations.

empirical knowledge knowledge coming directly from observations

theoretical knowledge knowledge based on ideas that are created to explain observations

theory an explanation of a large number of related observations

Models

Once scientists have developed a theory, they must communicate it. A model is an effective way to communicate a theory or idea. An architect may construct a model of a building to present the design to a client. The globe in geography class is a physical model of the planet Earth. In science, a **model** is a restricted representation of a theory. A model can change over time as new information is added. For example, in the second century A.D., an Alexandrian astonomer and geographer named Ptolemy put forth the theory that Earth is at the centre of the universe (the geocentric model). In the sixteenth century, based on many years of extensive observations by himself and others, Nicolaus Copernicus, a Polish astronomer, presented his theory that Earth and the other planets revolve around the Sun (the heliocentric model).

Models are used not only to visualize a theory, but also to suggest ways to test the theory. If a theory fails to correctly predict new observations, it may need to be changed, along with the model.

In the next few sections, you will examine the evolution of the model of the *atom*.

Figure 2
A vibrating box containing marbles is a physical model that represents the motion of particles, as described by the kinetic molecular theory of gases. (You will learn more about this theory in Unit 4.)

model a representation of a theoretical concept

▶ Section 1.2 Questions

Understanding Concepts

1. Classify each of the following statements as either an inference or an observation:
   (a) The wood does not burn because it is wet.
   (b) The boiling point of methanol is 67.5°C.
   (c) The light that a glow stick emits can be prolonged if the glow stick is placed in a fridge.
   (d) The temperature of a metal increased due to an increase in the vibrations of the atoms within it.
   (e) There is 125 mL of water in the flask.

2. "Theories cannot be proven; they can only be supported with experimental evidence." Comment on this statement.

3. Explain how models are useful for conveying a theory or idea.

Making Connections

4. Using a concept map, illustrate how the following terms are interconnected: empirical knowledge, theoretical knowledge, inference, observation, theory.

5. In a court of law, responses from witnesses who state opinions are often struck from the record. Give reasons why a lawyer may want a witness to make statements based on observations rather than inferences.

6. Qualitative chemical analysis involves identifying a substance through diagnostic tests. Diagnostic tests are based on physical and chemical properties of substances. Is qualitative analysis empirical or theoretical? Give reasons for your answer.

Ancient Models of Matter

Since ancient times, humans have wondered what matter is made of. Scientists have developed models of what was too small for them to see. If you were a student in ancient Greece, in the fifth century B.C., your model of matter would differ greatly from today's model of matter. You would have no trouble remembering the ancient periodic table. Empedocles, a philosopher, proposed that all substances in the world are made of combinations of four fundamental elements: earth, air, fire, and water (**Figure 1**).

Empedocles' model was not the only model at the time. Democritus, another Greek philosopher, proposed that matter could be divided into smaller and smaller pieces until a single indivisible particle was reached. He called this indivisible particle an **atom**. He proposed that all matter was made up of atoms and the void (empty space).

Both Empedocles and Democritus came up with their models of matter without any experimentation. Experimentation gained importance during the Middle Ages, when alchemists were trying to turn base metals, such as copper and nickel, into precious metals, such as silver and gold. Methodical observation and experimentation were crucial parts of their quest. Through experimentation, they accumulated invaluable empirical knowledge, which led to more theoretical knowledge and more sophisticated equipment for conducting experiments.

Dalton's Atomic Theory

Experimentation became fashionable in England during the reign of Elizabeth I (1533–1603), when many people chose to pursue science as both a career and a hobby. In 1803, an English schoolteacher named John Dalton came up with his atomic theory. Dalton's theory was based on many years of experimentation by many scientists. His theory included the following ideas (**Figure 2**):

- Matter consists of definite particles called atoms.
- Each element is made up of its own type of atom.
- Atoms of different elements have different properties.
- Atoms of two or more elements can combine in constant ratios to form new substances.
- Atoms cannot be created, destroyed, or subdivided in a chemical change.

Dalton's theory supported Democritus' 2000-year-old model.

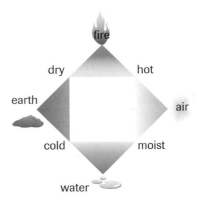

Figure 1
Empedocles suggested that all matter is composed of four elements: earth, air, fire, and water. Different types of matter are combinations of these basic elements, in varying degrees. For example, a dry substance is probably a combination of fire and earth.

atom the smallest particle of an element that has all the properties of this element

Model

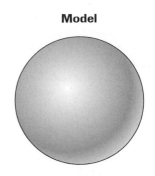

(featureless sphere)

Analogy

(billiard balls)

Figure 2
Dalton's atomic theory states that an atom is a featureless sphere. An analogy to his model is a group of identical billiard balls, representing atoms of the same element, with spaces between them.

Model

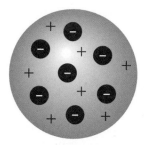

(uniform charge distribution)

Analogy

(raisin bun)

Figure 3
Thomson's model is analogous to a raisin bun. The atom is a positively charged sphere (the dough) embedded with negatively charged electrons (the raisins). Overall, the atom is neutral in charge.

electron a negatively charged subatomic particle

nucleus the positively charged centre of an atom

proton a positively charged subatomic particle found in the nucleus of an atom

> **▶ TRY THIS** activity

An Analogy to Probing the Atom

Dalton's model of the atom became widely accepted during the early part of the nineteenth century. Experiments conducted at the time led scientists to suppose that the atom contained subatomic particles. Scientists were faced with the challenge of determining the nature of these particles, which were too tiny to see.

When we do not know how the internal parts of a device or a model work, we may refer to it as a *black box*. In this activity, you will be provided with an object encased in Plasticine that represents a black box. Imagine that it is an atom. You will probe your black box and then write a detailed description of it based on your observations.

Materials: Plasticine "black box," pin

1. Obtain a "black box" and pin from your teacher.
2. Probe the "black box" with the pin. Record your observations in point form as you proceed. You are not allowed to break open the Plasticine.

(a) In paragraph form, write a detailed description of the object encased in the Plasticine.

(b) How could you improve your description? Remember that you cannot physically open the black box.

(c) How is investigating the contents of the black box similar to investigating the contents of an atom? How is it different?

(d) What does the pin represent?

(e) Make a list of other black boxes you have encountered, or you are aware of, in your daily life.

The Subatomic Particles

As more technology became available during the Industrial Revolution (eighteenth and nineteenth centuries), scientists had access to increasingly sophisticated equipment that allowed them to probe matter further. In 1897, J.J. Thomson conducted experiments using a cathode ray tube. His results suggested that the atom was not the smallest unit of matter: there were particles *within* the atom. In Thomson's model, the atom was a positively charged sphere with negatively charged particles embedded within it (**Figure 3**). The negatively charged particles were later named **electrons**. Thomson's atomic model is like a raisin bun. The dough represents the positively charged sphere, and the raisins represent the negatively charged electrons. The atom is neutral overall.

Thomson's model of the atom was tested in 1911, when Ernest Rutherford conducted his famous gold foil experiment. From the results of his experiment, he concluded that an atom contains a positively charged **nucleus** surrounded by mostly empty space. Some of this space is occupied by negatively charged electrons (**Figure 4**). A few years later, Rutherford further concluded that the atom's nucleus consists of positively charged subatomic particles, which he named **protons**. In 1932, Rutherford's model was modified by James Chadwick. Through experimentation, Chadwick suggested the existence of **neutrons** within the nucleus of the atom. Neutrons are similar in mass to protons but carry no charge.

The term **isotope** is used to distinguish between atoms of an element that contain the same number of protons but different numbers of neutrons in their nuclei. Isotopes may be represented by the element's name followed by its mass number. Thus, chlorine-35 and chlorine-37 symbolize the two natural isotopes of chlorine.

Niels Bohr

If the atom's nucleus is positively charged and the electrons surrounding it are negatively charged, and opposite charges attract, why do electrons not collapse into the nucleus? Even though Rutherford's atomic model was an improvement over previous atomic models, it could not explain this phenomenon. Niels Bohr (1885–1962), a Danish physicist, developed an atomic model that explained the behaviour of electrons in atoms. In the next two sections, you will learn about the phenomena that led Bohr to come up with his model of the atom.

neutron an uncharged subatomic particle found in the nucleus of an atom

isotope an atom of an element that has the same number of protons as the element, but different numbers of neutrons

Model

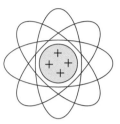

(nuclear model)

Analogy

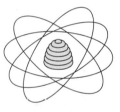

(beehive)

Figure 4
Rutherford's model is analogous to a beehive. The centre or nucleus of the atom is composed of a dense positive core (the beehive) and the electrons (bees) orbit the centre.

> ### Section 1.3 Questions

Understanding Concepts

1. Explain why Empedocles' and Democritus' atomic models were not scientific models like the models proposed by Dalton and Rutherford.

2. List the ideas in Dalton's atomic model.

3. Which philosopher or scientist first suggested each of the following ideas?
   (a) Atoms contain a dense positively charged core.
   (b) Atoms contain electrons and protons.
   (c) Atoms cannot be divided.
   (d) Electrons surround a central positive core.

4. (a) In Thomson's raisin bun model, what electrical charge was represented by the raisins?
   (b) What electrical charge was represented by the bun?

5. (a) How did Democritus' model differ from Dalton's model?
   (b) Suggest reasons why the model of the atom took so long to evolve.

6. By 1932, Chadwick had modified Rutherford's model of the atom to include neutrons. Define each of the following terms according to Chadwick's modified model:
   (a) nucleus
   (b) proton
   (c) electron
   (d) neutron

Making Connections

7. Using the development of atomic theory as your example, explain each of the following statements:
   (a) Scientific knowledge is tentative.
   (b) The progress of science is highly dependent on concurrent advances in technology.
   (c) Scientific knowledge is cumulative.

DID YOU KNOW?

Rutherford at McGill
Ernest Rutherford was professor of physics at McGill University in Montreal, Canada, from 1898 to 1907.

electromagnetic energy light energy that travels in the form of waves

frequency the number of cycles per second

wavelength the distance between successive crests or troughs in a wave

nanometre 10^{-9} m; unit nm

Electromagnetic energy is commonly known as light energy. Visible light, infrared light, ultraviolet light, and X rays are examples of light energy. Light energy is thought to move in the form of waves.

One of the ways in which light waves differ from each other is frequency. The concept of frequency can be applied to any event that occurs at regular intervals. For example, the frequency of a person's paycheque may be once every two weeks or once a month. Frequency is defined as cycles per unit time. The **frequency** of a light wave is the number of cycles that pass a point in one second.

A wave has maximum and minimum values called crests and troughs, respectively. The distance between successive crests or successive troughs is known as the **wavelength** (**Figure 1**). The wavelength of visible light is usually measured in **nanometres** (nm). The symbol for wavelength is the Greek letter *lamda* (λ).

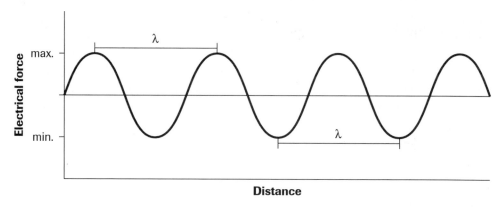

Figure 1
The distance between two crests or two troughs (maxima or minima) represents the wavelength.

Another characteristic of light is speed. All forms of light travel at the same speed: 3.0×10^8 m/s, or the speed of light. When the distance between the successive peaks of a light wave is short (the wave has a short wavelength), the time that the wave takes to pass a point is short, which means that the wave has a high frequency. Alternatively, when a light wave has a long wavelength, fewer cycles pass a point in a given time and the wave has a low frequency. A higher-frequency wave has more energy than a lower-frequency wave.

The electromagnetic spectrum consists of light waves of different frequencies (**Figure 2**). For example, radio waves have low frequencies. Therefore, they have long wavelengths and low energy. Radio waves do not pose a risk to humans. X rays are high-frequency, high-energy waves. Exposure to high-energy waves over a prolonged period of time can be harmful to human health. For this reason, an X-ray technician leaves the room when X-ray images are taken.

Humans can detect certain types of light waves but not other types. For example, you can detect infrared light waves as heat. You cannot detect radio waves without a radio receiver, even though they pass above, below, around, and through you. The special band of light waves that the *human eye* can

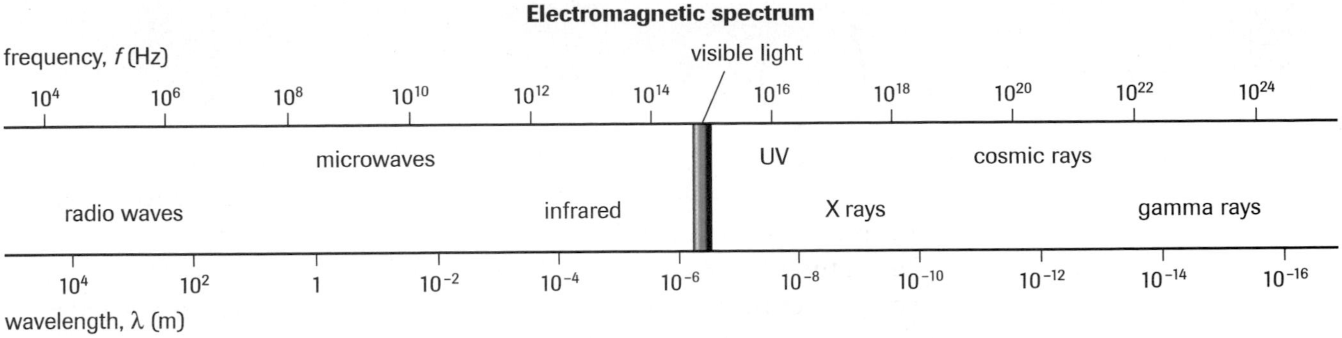

Electromagnetic spectrum
frequency, f (Hz)
visible light
10^4 10^6 10^8 10^10 10^12 10^14 10^16 10^18 10^20 10^22 10^24
microwaves
UV
cosmic rays
radio waves
infrared
X rays
gamma rays
10^4 10^2 1 10^-2 10^-4 10^-6 10^-8 10^-10 10^-12 10^-14 10^-16
wavelength, λ (m)

Figure 2
The electromagnetic spectrum encompasses an enormous range of wavelengths and energies. Waves in the X-ray range are 10^{11} times more energetic than waves in the radio range.

detect is known as the **visible spectrum**. The visible spectrum is composed of light waves with wavelengths in the range of 400 nm to 700 nm. Each wavelength of visible light is seen as a different colour. A rainbow contains all the colours in the visible spectrum and is an example of a **continuous spectrum**: a spectrum in which all the wavelengths of light are represented as an uninterrupted sequence (**Figure 3**).

As you learned in "Getting Started," different types of matter emit different wavelengths of light. Different types of matter also emit characteristic line spectra. Unlike a continuous spectrum, a **line spectrum** consists of *distinct coloured lines* rather than a rainbow. Analyzing the line spectrum of a sample is one way to distinguish between different types of matter. An element's line spectrum is analogous to a human's fingerprint. **Figure 4** shows the characteristic line spectra of sodium and helium. Line spectra can be observed

visible spectrum the region of the electromagnetic spectrum that the human eye can detect

continuous spectrum an uninterrupted pattern of colours that is observed when a narrow beam of white light passes through a prism

line spectrum a discontinuous spectrum that is produced when light emitted by an element is directed through a prism or a diffraction grating; unique to an element

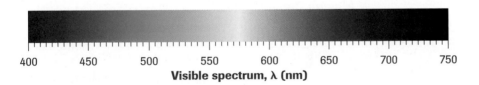

400
450
500
550
600
650
700
750
Visible spectrum, λ (nm)

Figure 3
A continuous spectrum is produced when light from an incandescent lamp is directed through a prism. The continuous spectrum contains all the colours of light in the visible spectrum.

DID YOU KNOW?

Infrared Energy and Fever
Doctors and nurses can measure a patient's temperature without touching the patient. They insert a hand-held device into the patient's ear and measure infrared (heat) energy. A person with a fever emits more infrared energy than a healthy person.

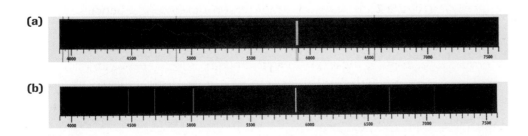

(a)
(b)

Figure 4
The line spectra produced by sodium **(a)** and helium **(b)**

Two Analogies for Continuous and Line Spectra

A line spectrum can be compared to playing a scale on a piano. A pianist can play tones and semitones, but no notes in between. A continuous spectrum is like playing a scale on a violin. A violinist can play all the intermediate notes between tones. A continuous spectrum is also analogous to a ramp, which provides a smooth transition from one level to the next. A line spectrum can be compared to a staircase, where the transition between levels is broken into steps.

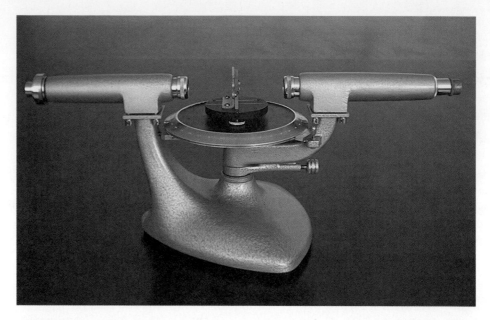

Figure 5
A simple prism spectroscope separates light from a source into its component colours.

spectroscope an optical instrument that separates light energy into its component wavelengths; used in qualitative analysis

using a **spectroscope**: an instrument that separates light into its component colours using a prism or a diffraction grating (**Figure 5**). In the next section, you will examine the line spectra of several gases.

▶ Section 1.4 Questions

Understanding Concepts

1. Describe the difference between radio waves and X rays using the concepts of wavelength, frequency, and energy.

2. What range of wavelengths of electromagnetic radiation can the human eye detect?

3. White light is composed of many different colours of light. Explain.

4. Distinguish between a continuous spectrum and a line spectrum.

Making Connections

5. When an X ray is taken, the X-ray technician leaves the room, but the patient is exposed to the X ray.

   (a) Why is it not safe for the technician to be in the room when the X ray is taken?
   (b) What precautions are followed to minimize the patient's and the technician's exposure to X rays?

    GO   www.science.nelson.com

Identifying Gases Using Line Spectra

The light emitted by elements consists of many different wavelengths. If this light is directed through a spectroscope, a bright line spectrum is seen (**Figure 1**). Since each element's line spectrum is unique, line spectra can be used to identify known elements and to predict the existence of new elements. *Spectroscopy* is a branch of science in which light is used to identify and/or quantify substances.

In this activity, you will perform qualitative analysis using a hand-held spectroscope. You will observe a continuous spectrum, as well as the line spectrum of hydrogen gas. When you have completed your observations, you will identify various gases on the basis of their line spectra.

Question

What gases are in the discharge tubes?

Materials

eye protection
incandescent light source (at least 40 W)
fluorescent light source
spectroscope
power supply
hydrogen gas discharge tube
other gas discharge tubes

Procedure

1. Use the spectroscope to observe the spectrum of sunlight.

(a) Draw and label the spectrum.

2. Use the spectroscope to observe the fluorescent lights in the classroom.

(b) Draw and label the spectrum.

3. Turn off all the lights in the classroom. Then turn on the incandescent light.

4. Use the spectroscope to observe the incandescent light.

(c) Draw and label the spectrum that you observe.

5. Connect the hydrogen gas discharge tube to the power supply. Turn off all other lights.

 Discharge tubes operate at high voltage. Do not touch them during operation.

6. Observe the spectrum that is produced when the light passes from the hydrogen gas discharge tube through the spectroscope (**Figure 2**, on the next page).

(d) Draw the spectrum, indicating the colour of each spectral line you observed.

7. Repeat steps 5 and 6 using the gas discharge tubes of unidentified elements.

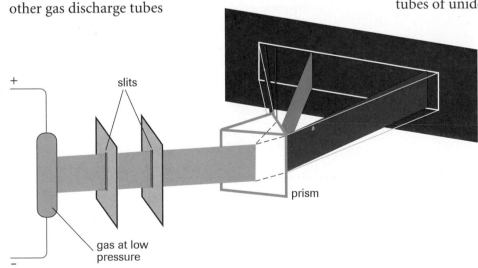

Figure 1
When electricity is passed through a gaseous element, the atoms emit certain wavelengths of light. A spectroscope separates the mixture of wavelengths to produce a line spectrum. The spectroscope may not show all the lines near the edges of the visible spectrum.

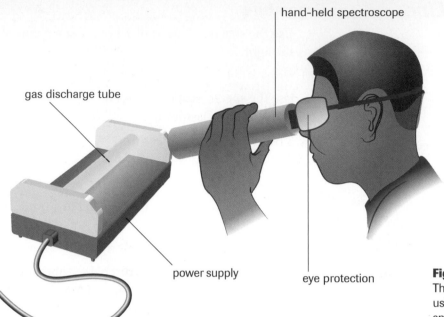

gas discharge tube

hand-held spectroscope

power supply

eye protection

Figure 2
The line spectrum of an element is being viewed using a hand-held spectroscope. This kind of spectroscope uses a diffraction grating rather than a prism to separate the spectrum into lines.

(e) If the patterns are complex, do the best you can to note and record the differences in the number, colour, and spacing of the spectral lines.

Analysis

(f) Compare the line spectra of the unidentified elements in the gas discharge tubes with the line spectra in Appendix C4, and then answer the Question.

Synthesis

(g) How do the spectra of sunlight, fluorescent light, and incandescent light sources compare? What do the differences suggest about the light that is produced by each of the three sources?

(h) Which element produced the largest number of spectral lines?

(i) Describe how line spectroscopy can be used as a qualitative analysis technique. Provide examples of situations in which this technique may be useful.

(j) The second most abundant element in the universe is helium. Helium was discovered in 1868 using line spectroscopy. Describe how helium was discovered. Write a short paragraph, outlining your findings.

 www.science.nelson.com

The Bohr Model of the Hydrogen Atom 1.6

As you learned at the end of section 1.3, Rutherford's model of the atom did not explain why negatively charged electrons orbiting a positively charged nucleus do not collapse into the nucleus. The law of moving charges states that as an electron orbits the nucleus, it should emit energy in the form of electromagnetic radiation. As the electron runs out of energy, it should collapse into the atom's nucleus. Matter is very stable, however. Therefore, scientists have inferred that atoms are stable and that electrons do not collapse into the nucleus. Why?

As you saw in section 1.5, a spectroscope separates light into its component wavelengths, revealing a line spectrum that is unique to each element. Line spectra had been known and observed for many years, but no one could explain this phenomenon. Finally, in 1913, Niels Bohr not only explained how line spectra were produced, but also why electrons do not collapse into the atom's nucleus.

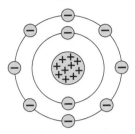

Model

(nuclear model plus orbiting electrons)

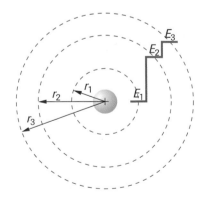

Analogy

(planets orbiting the Sun)

Figure 2
Niels Bohr proposed that electrons circle the nucleus in orbits of fixed energy. For his work, he was awarded the Nobel Prize in physics in 1922.

Figure 1
The line spectrum of hydrogen

Bohr's theory was based on the line spectrum of hydrogen (**Figure 1**) because of the simplicity of hydrogen: the hydrogen atom possesses only one electron. Bohr suggested that electrons revolve around the atom's nucleus in *orbits of fixed energy* (**Figure 2**). These orbits are similar to the fixed orbits of the planets as they revolve around the Sun. The electrons are restricted to certain energy levels, and the energy of electrons is **quantized**. In other words, electrons must possess *specific amounts* of energy at each energy level (**Figure 3**). Quantization is analogous to a ball sitting on a flight of steps. Since the ball cannot sit between steps, it is restricted to specific levels: the first step, the second step, the third step, and so on. As well, the ball can possess only a specific amount of energy, depending on which step it is sitting on. Since the ball cannot exist between steps, it cannot possess energies that are in between specific levels.

When the hydrogen atom's electron absorbs light that has a *specific energy*, the electron jumps to a higher energy level. An electron that occupies a higher energy level is said to be in an **excited state**. Using the staircase analogy, if you want to kick a ball from the second step to the fourth step, you must kick it with precisely the right amount of energy. If you put too much energy into the kick, the ball may reach a height between the fourth and fifth steps and fall back down to the landing. If you put too little energy into the kick, the ball may go up only half a step and fall back down to its original position. Similarly, the orbiting electron must absorb electromagnetic radiation with just the right amount of energy in order to jump to a higher energy level.

quantized possessing a specific value or amount (quantity)

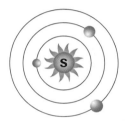

Figure 3
In the Bohr model of the atom, electrons orbit the nucleus as the planets orbit the Sun. Only certain orbits are allowed, however, and an electron in each orbit has a specific energy.

Matter and Qualitative Analysis **21**

excited state an electron's state when it absorbs energy and jumps to a higher energy level

ground state the energy level in which an electron is most stable; the electron does not release any electromagnetic radiation in this state

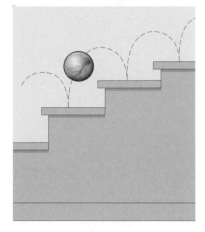

Figure 4
An electron dropping from a higher energy level to a lower energy level is like a ball bouncing down a staircase. The bottom of the staircase is analogous to an electron's ground state in an atom.

In order for the ball to fall back down to its original position on the staircase, it must *release* the same amount of energy that was required to raise it to the higher step (**Figure 4**). Similarly, when the hydrogen atom's electron falls back to its original position, it releases the same amount of energy it absorbed in order to reach the higher energy level. According to Bohr, the lines in the line spectrum for hydrogen are a result of the *energy released* when the hydrogen atom's electron falls from a higher energy level to a lower energy level (**Figure 5**). If the electron is found in the lowest possible energy level, it is said to be in its **ground state**. Electrons in their ground state do not emit any light energy.

Bohr's model of the atom was both a success and a failure. It worked very well for explaining the line spectrum of hydrogen, but it did not work as well for the line spectra of other elements. Nevertheless, it was a breakthrough model because it introduced the idea that electrons orbit the nucleus in fixed energy levels. Furthermore, using evidence from the periodic table, Bohr predicted that each energy level could hold a maximum number of electrons. He predicted that the first energy level could hold two electrons, and the second energy level could hold eight electrons. The third energy level can hold 18 electrons, but it is stable with eight electrons. You have used these numbers in previous courses to draw Bohr–Rutherford diagrams of atoms.

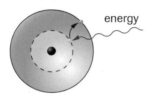

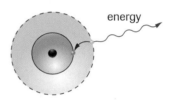

(a) An electron gains a quantum of energy and jumps to a higher energy level.

(b) An electron loses a quantum of energy and drops to a lower energy level.

Figure 5

▶ **Section 1.6 Questions**

Understanding Concepts

1. State the reason why Rutherford's model of the atom failed to describe the observed behaviour of matter.

2. Describe Bohr's model of the atom. How is it different from Rutherford's model? How is it similar to Rutherford's model?

3. What role did spectroscopy play in helping Bohr come up with his atomic model?

4. Electrons can be found in the ground state or in the excited state. What is different about an electron in each state?

5. Even though Bohr's model was an improvement on Rutherford's model of the atom, there were problems with it. What was the major problem with Bohr's atomic model?

6. Why do electrons emit light energy when they drop from a higher energy level to a lower energy level?

7. "Energy levels are quantized." What does this statement mean?

Applying Inquiry Skills

8. Imagine that you have been given two unidentified gases and a spectroscope. Design a simple procedure that you could use to identify the gases.

Making Connections

9. Using your knowledge of Bohr's model of the atom, explain why each element possesses a characteristic line spectrum.

Flame Tests

When electrons in an atom are excited by heat or electricity, they absorb energy and jump from lower energy levels to higher energy levels. When an electron drops from an excited state to its ground state, the atom releases energy in the form of light. This light has a characteristic colour.

Different types of matter produce different colours when subjected to high temperatures. As you learned at the beginning of this unit, fireworks contain different compounds, which produce the different colours at a fireworks display.

In this activity, you will subject different metallic compounds to high heat and observe the colours that are produced. You will use a qualitative analysis technique known as a *flame test*. You will compile your observations into a table, listing the compounds and their respective flame test colours. You will then identify the metal in a compound using a flame test and your table.

Question

What is the unidentified metal in a metallic compound?

Materials

eye protection
lab apron
nichrome test wire
2 test tubes
test-tube rack
cobalt glass squares
Bunsen burner
5 mL dilute hydrochloric acid, $HCl_{(aq)}$
3 mL sodium chloride solution, $NaCl_{(aq)}$
samples of the following solids:
    sodium nitrate, $NaNO_{3(s)}$
    sodium chloride, $NaCl_{(s)}$
    calcium chloride, $CaCl_{2(s)}$
    strontium chloride, $SrCl_{2(s)}$
    lithium chloride, $LiCl_{(s)}$
    potassium chloride, $KCl_{(s)}$
    copper(II) chloride, $CuCl_{2(s)}$
unidentified metallic compound

Procedure

1. Obtain your samples (the seven solids and 3 mL of the sodium chloride solution in a small test tube) and 5 mL of hydrochloric acid (also in a small test tube). Place the test tubes in a test-tube rack.

2. Light the Bunsen burner. Adjust it until it is burning with a blue flame.

3. To clean the nichrome test wire, dip it in the hydrochloric acid and then hold the wire in the flame of the Bunsen burner. Repeat the procedure until the wire adds no colour to the flame.

 Hydrochloric acid is highly corrosive and can burn the skin. Use caution when handling hydrochloric acid, and use it only in a well-ventilated area. Wear eye protection and a lab apron throughout this activity.

 A flickering light, particularly red, can trigger a migraine or an epileptic seizure. Do not do this activity if you have either condition.

4. Pick up a small amount of solid sodium nitrate with the loop of the nichrome test wire. Hold the end of the wire in the flame. Record your observations.

5. Clean the wire as explained in step 3.

6. Pick up a small sample of solid sodium chloride with the loop of the nichrome test wire. Hold the end of the wire in the flame for several seconds. Record your observations.

7. Clean the wire as explained in step 3.

8. Dip the nichrome test wire in the sodium chloride solution. Hold the end of the wire in the flame, and record your observations.

9. Repeat step 4 with each of the remaining solid samples. Remember to clean the wire after each test. Record your observations.

10. Repeat the flame tests for potassium chloride and sodium chloride. This time, look through the cobalt glass to observe each wire in the flame. Record your observations.

11. Obtain an unidentified metallic compound from your teacher. Repeat the flame test for the unidentified compound, and record your results.

Analysis

(a) Assemble your observations for the identified samples into a flame test identification key, which you can use as a quick reference.

(b) Analyze your observations, and then answer the Question.

Evaluation

(c) Suggest possible sources of error in this activity, and describe their possible effects on your results. What changes could you make to the Procedure to reduce these sources of error?

(d) Compare the results of your flame tests for solid sodium nitrate, solid sodium chloride, and sodium chloride solution. What do these results indicate?

(e) Compare the results of your flame tests for potassium chloride and sodium chloride, with and without the cobalt glass. What was the purpose of the cobalt glass?

Synthesis

(f) **Figure 1** shows the results of four flame tests. Using your flame test identification key, identify the metal in each compound in **Figure 1**.

(g) Explain why flame tests are a qualitative analysis technique. Use the Procedure and observations in this activity to support your answer.

(h) Flame emission spectroscopy is a technique that is used to identify the components of different types of matter. Using electronic and print resources, compile a list of situations in which flame emission spectroscopy is used.

GO www.science.nelson.com

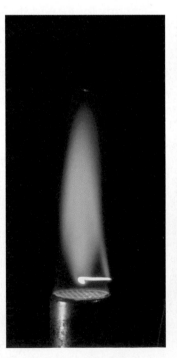

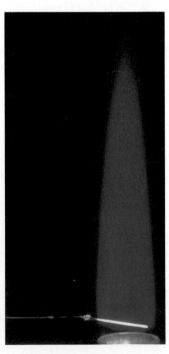

Figure 1
Results of four flame tests, for four solutions of unidentified metallic compounds

Each year, billions of dollars are spent worldwide repairing roads, dams, and other concrete structures. Many factors contribute to concrete degradation. Two of these factors are alkali-aggregate reaction and carbonation.

Alkali-Aggregate Reaction

One factor that can contribute to the accelerated degradation of a concrete structure is a chemical process called *alkali-aggregate reaction*. Concrete contains trace amounts of **alkali metals**, such as sodium, along with other minerals. If moisture is present, the sodium reacts with the minerals in concrete, forming a gel. The gel may absorb water and expand, causing cracks in the concrete. If the concrete is cracked, the strength of a structure is compromised (**Figure 1**).

alkali metal a Group I metal

flame emission spectroscopy a qualitative analysis technique that is used to determine the presence of a substance by exciting the substance's electrons using a flame and then detecting the electromagnetic energy emitted when the excited electrons return to their ground state

Figure 1
Cracks in concrete may make a structure unsafe.

Sodium oxide, $Na_2O_{(s)}$, is an alkali metal compound that is found in concrete. One method that can be used to determine the presence of sodium oxide in a concrete sample is **flame emission spectroscopy.** Flame emission spectroscopy is based on the theory that electrons are promoted to higher energy levels when they absorb energy. When the excited electrons return to their ground state, they release this energy in the form of light.

To test concrete for sodium oxide content using flame emission spectroscopy, a small quantity (about a milligram) of concrete is dissolved in a strong acid. The solution is then placed in a nebulizer, which converts the solution into a fine spray of droplets. A flame vaporizes the strong acid, leaving behind the sodium oxide compound. The energy from the flame excites sodium's outermost electron to a higher energy level. When the excited electron returns to its ground state, it emits electromagnetic energy with a wavelength of 589 nm. If the colour that corresponds to 589 nm is detected, it indicates the presence of sodium oxide.

$$Na + \text{heat energy} \rightarrow Na^* \quad (Na^* \text{ represents a sodium atom with an excited electron.})$$

$$Na^* \rightarrow Na + \text{electromagnetic energy (589 nm)}$$

CAREER CONNECTION

Geological technicians help geologists, engineers, and geophysicists find and assess possible mineral and fuel resources. Geological technicians make observations, collect and analyze rock samples, and record information.

They use a wide variety of instruments (such as UV, visible, atomic absorption, and flame emission spectroscopes) to test for the presence and composition of a mineral in a rock sample.

(i) Why is spectroscopy a good analysis technique for determining the type of mineral that is found in a sample?

(ii) Research where geological technician programs are offered in Ontario. Find out about prerequisite courses, required courses, and the duration of the geological technician program.

 www.science.nelson.com

Carbonation

Carbon dioxide gas, $CO_{2(g)}$, in the atmosphere can also cause concrete structures to weaken. Carbon dioxide reacts with the calcium hydroxide, $Ca(OH)_{2(aq)}$, in concrete, in the presence of moisture (water), to form calcium carbonate (limestone), $CaCO_{3(s)}$:

$$CO_{2(g)} + Ca(OH)_{2(aq)} \rightarrow CaCO_{3(s)} + H_2O_{(l)}$$

base a substance that neutralizes acids, conducts electricity in solution, and has a pH that is greater than 7 (neutral) in solution

pH a numerical scale that is used to measure how acidic or basic a solution is

indicator a chemical that changes colour when the acidity of a solution (its pH) changes

acid a substance that neutralizes bases, conducts electricity in solution, and has a pH that is less than 7 (neutral) in solution

Calcium carbonate is less **basic** than calcium hydroxide. Therefore, the production of calcium carbonate causes concrete to become less basic. The decrease in **pH** increases the potential for any steel reinforcements in the concrete to corrode, resulting in a weakened structure. (You will learn more about acids and bases in Unit 4 and more about corrosion in Unit 5.)

Phenolphthalein, an indicator, is used to test for carbonation. An **indicator** changes colour when it is placed into an **acidic** or a basic solution. Phenolphthalein is pink in a basic solution and colourless in a neutral or acidic solution (**Figure 2**). If phenolphthalein remains colourless when it is applied to a bore obtained from a concrete structure, then carbonation is occurring and the structure must be repaired. Testing for carbonation in concrete is a common qualitative analysis technique used by structural and materials testing engineers.

Figure 2
In a neutral or acidic solution, phenolphthalein is colourless. In a basic solution, phenolphthalein is pink.

▶ Case Study 1.8 Questions

Understanding Concepts

1. In your own words, describe how flame emission spectroscopy works.

2. Why is it important to minimize the alkali metal content of concrete?

3. Describe how atmospheric carbon dioxide can lead to the degradation of concrete.

4. How is the carbonation of concrete detected? Why can this test be considered a qualitative analysis technique?

Making Connections

5. One of the most important uses of flame emission spectroscopy is to detect the presence of sodium and potassium, particularly in biological fluids such as urine. Sodium and potassium ions in your body affect your electrolyte balance.
   (a) Research the role of these electrolytes in your body.
   (b) What do low and high levels of each electrolyte mean for your health?

 www.science.nelson.com

1.9 Detecting Counterfeit Canadian Currency

The Bank of Canada has approximately 1.4 billion bills in circulation, with a total value of $38 billion. Each year, in Canada, approximately 200 000 counterfeit bills are passed, with an approximate value of $3.7 million. In the late 1990s, some retailers began rejecting $100 bills due to the circulation of a large number of counterfeit bills. Counterfeiters switched to making counterfeit $10 bills, hoping that retailers would be less concerned with checking the validity of smaller-denomination bills. With advances in technology, the art of counterfeiting has reached new heights. Colour printers and copiers, along with desktop publishing software, enable counterfeiters to produce high-quality fraudulent currency that is easily mistaken for the real thing. The Bank of Canada is continually trying to develop currency with security features that make it more difficult to counterfeit. As a result, the chance of a Canadian coming across a counterfeit bill is only about 1 in 10 000.

Deterring counterfeiters is an old practice. In the tenth century A.D., the Chinese printed a warning directly on their currency. The warning stated that anyone who was caught making counterfeit money would be boiled alive! The anti-counterfeiting measures adopted by the Bank of Canada today are more subtle and sophisticated. At some retailers, when you use a bill to pay for something, the cashier takes your bill and places it under a UV light (**Figure 1**). The cashier is checking to see whether the bill fluoresces. Higher-denomination bills ($20s, $50s, and $100s) have planchettes that are placed randomly on and within them. A planchette is a small, oval, removable green paper disk, coated with UV fluorescent ink that is added to the pulp while making banknote paper. When the ink is subjected to UV light, its electrons absorb the energy and enter the excited state. When the electrons return to the ground state, they release the energy as fluorescence. When placed under UV light, a genuine banknote fluoresces at random points on its surface. The planchettes of counterfeit bills are not removable because these banknotes are surface printed. Under UV light, counterfeit bills do not fluoresce.

Instead of planchettes, the new $5 and $10 bills of the Canadian Journey series contain blue and white fibres. When placed under UV light, the blue fibres appear blue, but the white fibres appear red. The section that fluoresces includes the denomination of the bill, a coat of arms, and the words "Bank of Canada" in both English and French (**Figure 2**).

Figure 1
A portable UV light source

Figure 2
When a $5 or $10 bill is placed under UV light, the left side fluoresces.

Tech Connect 1.9 Questions

Understanding Concepts

1. Why do some parts of Canadian currency fluoresce when placed under UV light, while other parts do not? Describe the chemistry that is involved.
2. List some of the security features of Canadian currency.
3. Why are small denominations of Canadian currency just as likely to be counterfeited as larger denominations?
4. Why can the detection of counterfeit bills be considered qualitative analysis?

Making Connections

5. Obtain a $20 bill. Identify as many security features as possible. What other qualities distinguish genuine currency from counterfeit currency?

GO www.science.nelson.com

6. What steps are retailers instructed to follow when they receive money that they suspect is counterfeit?

GO www.science.nelson.com

Tech Connect

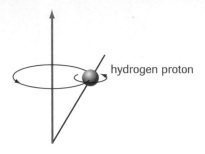

Figure 1
A hydrogen proton spins about an axis within a magnetic field, similar to a toy top spinning on its axis.

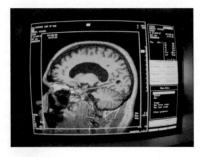

Figure 2
An MRI scan

Magnetic resonance imaging (MRI) is a sophisticated diagnostic medical procedure that uses magnets and radio waves. The human body contains billions of hydrogen atoms. When the protons in the hydrogen nuclei are subjected to a strong magnetic field, they align themselves with the north or south pole of the magnet of the MRI machine. When radio-wave energy is applied to the hydrogen atoms, they absorb this energy and spin (**Figure 1**). When the application of radio-wave energy ceases, the hydrogen atoms return to their original magnetized state, releasing the excess energy they absorbed. This energy is detected and transformed into a detailed image (**Figure 2**) that doctors use to detect cancer and other abnormalities.

Due to the high cost and limited availability of MRI machines, some Ontarians have had to wait for one year before gaining access to one of the 43 MRI clinics in the province. In 2002, in an effort to address the MRI waiting list in Ontario, the provincial government decided to award private companies licences to open 20 new MRI facilities in Ontario in 2003. Under the Canadian Health Act, medically necessary services must be paid for by government health plans, ensuring equal access to basic health services. The new clinics will provide diagnostic services that are paid for by OHIP, but they will also charge a fee for supplementary services not covered by OHIP. In 2002, the Ontario government acknowledged that it will have to monitor the new clinics closely to ensure that they are not offering wealthy people a better standard of health care than the rest of Ontarians. Wealthy individuals could jump the queue by stating that the MRI requested is not medically necessary but a supplementary service. Supporters of private MRI clinics argue that the average Ontarian may choose to spend money on health care rather than on a vacation.

▶ Understanding the Issue

1. Describe how MRI works. Why is it considered to be a form of spectroscopy?

2. What type of energy is absorbed by the hydrogen atom's proton during an MRI scan?

3. Describe the Ontario government's solution to the current MRI shortage in the province.

4. Why must the Ontario government keep close tabs on private MRI clinics? Explain your answer.

▶ Take a Stand

Fair Access

Decision-Making Skills

○ Define the Issue ● Analyze the Issue ● Research
● Defend a Decision ● Identify Alternatives ○ Evaluate

Statement: Ontarians should have the right to choose faster access to MRI and better treatment through personal funding.

In groups, research the issue of fair access to MRIs. Using up-to-date information, decide whether you agree or disagree with the statement. Prepare a position paper, in point form, to support your position. If your group disagrees with the statement, suggest some alternative strategies to meet the health-care needs of Ontarians. Search for information in periodicals and newspapers, and on the Internet. As a group, present your view to your classmates.

 www.science.nelson.com

At a crime scene, detectives collect many clues and pieces of evidence in order to pinpoint the perpetrator of the crime. Suspects are rarely convicted in a court of law based on only one piece of evidence. It is the supporting evidence that usually leads to a conviction. Similarly, scientists do not usually depend on a single method in order to identify matter. If possible, they collect additional information about a substance's physical and chemical properties.

So far in this unit, you have been introduced to three methods for identifying different types of matter using qualitative analysis:

1. thermal emission spectroscopy (TES), in which substances are identified based on the amount of heat they emit ("Getting Started");

2. light spectroscopy, in which substances are identified by their line spectra, since line spectra are unique to each type of substance (section 1.5);

3. flame tests, in which substances are identified by the colours they emit when placed in a flame (section 1.7).

In this section, you will study a fourth qualitative analysis technique that can be used to identify matter: conductivity.

Conductivity

One physical property that can be used to identify a substance is the ability of the substance to conduct electricity. **Conductivity** is a physical property of metals. Few *compounds* are able to conduct electricity in the solid state at room temperature, however. Many compounds are able to conduct electricity when they are heated to become a liquid or when they are dissolved in water to produce a solution. If the solution is able to conduct electricity when a compound is dissolved in water, the compound is called an **electrolyte**. If the solution does not conduct electricity, the compound is called a **nonelectrolyte** (**Figure 1**). Why are some compounds electrolytes, but not others?

conductivity the ability of a substance to conduct electricity; a physical property of matter

electrolyte a compound that, when dissolved in water, produces a solution that conducts electricity

nonelectrolyte a compound that, when dissolved in water, does not produce a solution that conducts electricity

(a)

(b)

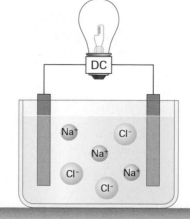
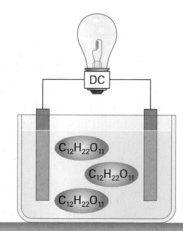

Figure 1
Why do you think the sodium chloride solution can conduct electricity but the sugar solution cannot? To answer this question, look closely at the two diagrams.
(a) Sodium chloride, $NaCl_{(s)}$, is an electrolyte. Scientists know that a sodium chloride solution conducts electricity because the light bulb lights up.
(b) Sugar, $C_{12}H_{22}O_{11(s)}$, is a nonelectrolyte. A sugar solution does not conduct electricity because the bulb remains unlit.

The Formation of Ions

Scientists know that most of the matter on Earth is found as compounds or molecules rather than single atoms. From this evidence, scientists have inferred that most matter is more stable in compound form.

Atoms gain or lose the electrons found in their outermost shells in order to become more stable. Atoms are most stable when their outer shells are full. A full outer shell (except the first shell) contains eight electrons. With the exception of helium, which has two electrons, all noble gases have a full outer shell of eight electrons (**Figure 2**).

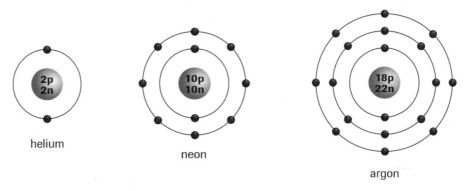

helium

neon

argon

Figure 2
Examine the Bohr–Rutherford diagrams of three of the noble gases. Notice that the outer shells are full and cannot hold any more electrons.

octet rule Atoms gain or lose electrons in their outermost shells in order to attain a noble gas configuration.

ion an atom (or group of atoms) that has lost or gained one or more electrons

anion an atom that carries an overall negative charge because it has more electrons than protons

cation an atom that carries an overall positive charge because it has fewer electrons than protons

valence electrons electrons that are found in the outermost shell of an atom

In previous courses, you learned that the noble gases (Group VIII) are the least reactive elements and, therefore, the most stable of all the elements. An atom that loses or gains electrons in order to attain a noble gas configuration is following the **octet rule**.

Atoms that have gained or lost electrons are known as **ions**. If an atom gains electrons, it has more electrons than protons and becomes a negative ion or **anion**. If an atom loses electrons, it has fewer electrons than protons and becomes a positive ion or **cation**.

Whether an atom gains or loses electrons in order to become a cation or anion depends on the number of electrons it possesses in its outer shell. The electrons in an atom's outermost shell are called **valence electrons**. For example, sodium, Na, is a metal in Group I. It has one electron in its third shell. Losing this electron is easier and energetically more favourable than gaining seven electrons in order to fill its outer shell. As a result, sodium loses its valence electron, revealing the underlying second shell, which is full. Since sodium has lost one electron, the number of electrons it has is now one less than the number of protons. Therefore, the sodium ion carries a charge of $1+$ (Na^+). The sodium ion is stable because it has the same electron configuration as the noble gas neon, Ne (**Figure 3(a)**).

(a)

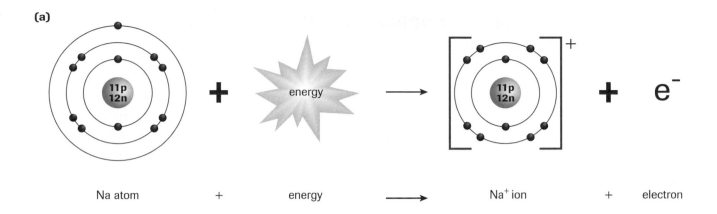

(b)

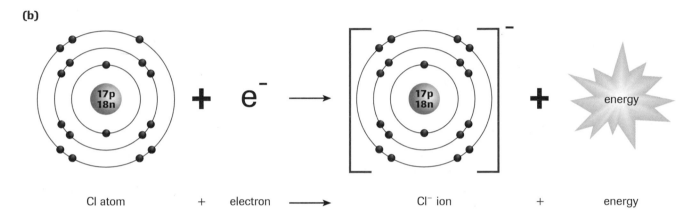

Figure 3
(a) The sodium atom has one lone electron in its third shell. When it loses this electron, it is left with a full second shell. The sodium ion has a net charge of 1+ since it now has 11 protons and 10 electrons. The sodium ion's electron configuration is now identical to neon's electron configuration.
(b) The chlorine atom has 7 electrons in its third shell. When chlorine gains an electron, it attains a net charge of 1– and becomes a chloride ion, with 17 protons and 18 electrons. The chloride ion's electron configuration is identical to argon's electron configuration.

Chlorine is found in Group VII. It has seven valence electrons. Gaining one electron is easier for chlorine than losing seven electrons. Chlorine, therefore, accepts one electron. Since the number of electrons it has is now one more than the number of protons, the chloride ion has a charge of 1– (Cl^-). The chloride ion is stable because it has the same electron configuration as the noble gas argon, Ar (**Figure 3(b)**).

In general, in order to become stable, metals lose electrons and become cations while nonmetals gain electrons and become anions. Metals attain the electron configuration of the noble gas that precedes them, while nonmetals attain the electron configuration of the noble gas that follows them.

Lewis Symbols

In previous courses, you learned how to represent atoms using Bohr–Rutherford diagrams. A Bohr–Rutherford diagram depicts the electron structure of the whole atom. Since only the valence electrons take part in chemical reactions, a **Lewis symbol** is used to depict the electrons in an atom's outermost shell. In a Lewis symbol, the valence electrons are represented by dots and are distributed around the element's symbol. Each quadrant surrounding the symbol can hold up to two valence electrons, for a maximum of eight electrons around the symbol. When drawing Lewis symbols, first place electrons, one at a time, at the 12, 3, 6, and 9 o'clock positions, as required. Then, add more electrons by pairing them, one at a time, in the same order as you placed the first four electrons (**Figure 4**).

Lewis symbol a diagram composed of a chemical symbol and dots, depicting the valence electrons of an atom or ion

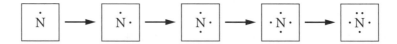

Figure 4
The progressive placement of electrons in a Lewis symbol

The Lewis symbols of elements in the same group of the periodic table have the same number of dots surrounding the symbol (**Figure 5**). For example, the Lewis symbol for magnesium has the same number of dots as the Lewis symbols for calcium and beryllium. Magnesium, calcium, and beryllium, therefore, all have the same number of valence electrons.

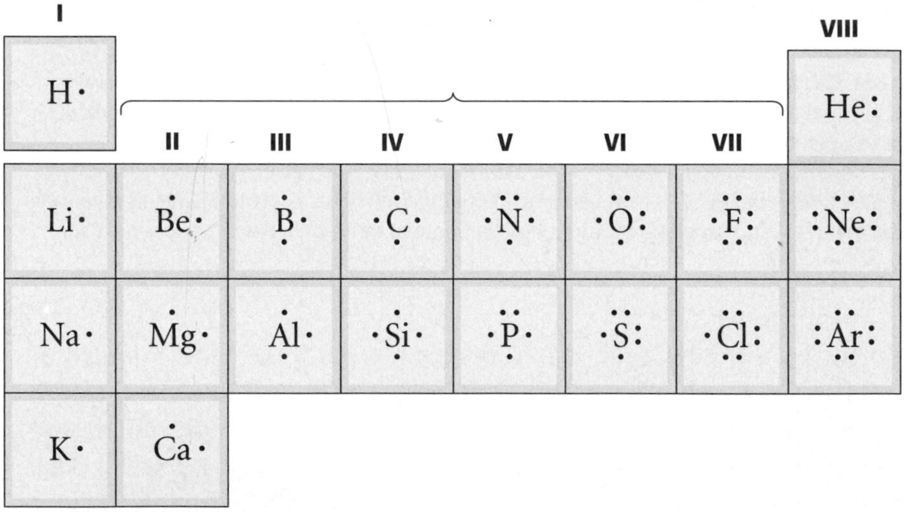

Figure 5
Lewis symbols for the first 20 elements in the periodic table. What do you notice about the Lewis symbols for elements in the same group? What happens to the number of electrons as you go across a period?

I							VIII
H·							He:
	II	III	IV	V	VI	VII	
Li·	Be·	B·	·C·	·N·	·O:	·F:	:Ne:
Na·	Mg·	Al·	·Si·	·P·	·S:	·Cl:	:Ar:
K·	Ca·						

Ions can also be represented by Lewis symbols. **Figure 6** shows the Lewis symbol for the chloride ion. The Lewis symbol is enclosed in square brackets. The charge of the ion is placed outside the brackets.

$$[:\ddot{C}\!l:]^-$$

Figure 6
Lewis symbol for Cl⁻

Polyatomic Ions

Ions that are composed of one atom are called **monatomic ions**. All the ions that have been mentioned in this section, so far, have been monatomic. When an ion is composed of more than one atom, it is called a **polyatomic ion**. **Table 1** lists some common polyatomic ions and their charges.

monatomic ion an ion that is composed of only one atom

polyatomic ion an ion that is composed of two or more atoms

Ionic Compounds

You have just learned that monatomic ions form in order to become more stable. Now you will learn *how* monatomic ions form. Consider sodium and chlorine. As you know, in order to satisfy the octet rule, chlorine needs to *gain* one electron and sodium needs to *lose* one electron. If a sodium atom and a chlorine atom come in contact under the right chemical conditions, sodium will lose its lone valence electron to chlorine, which has seven valence electrons. The transfer of one electron from sodium to chlorine results in both chlorine and sodium attaining a noble gas configuration. It also results in sodium gaining a positive charge and chlorine gaining a negative charge. Since opposite charges attract, the positive sodium ion is now attracted to the negative chloride ion, and vice versa. The bond that arises from the attraction between the two oppositely charged ions is called an **ionic bond** (**Figure 7**).

Ionic bonds form between metals and nonmetals. Metals lose their electrons and become positive ions (cations) more easily than other types of elements. Nonmetals easily gain electrons and become negative ions (anions). The combination of cations and anions forms an **ionic compound**. At a temperature of 25°C and a pressure of 101.3 kPa, ionic compounds form solid **ionic crystals** in which large numbers of cations and anions are arranged in a repeating three-dimensional pattern. Ionic crystals are also known as salts or salt crystals (**Figure 8**).

Table 1 Some Common Polyatomic Ions

Ion	Name
NO_3^-	nitrate
SO_4^{2-}	sulfate
PO_4^{3-}	phosphate
OH^-	hydroxide
CO_3^{2-}	carbonate
BrO_3^-	bromate
ClO_3^-	chlorate
IO_3^-	iodate
$C_2H_3O_2^-$	acetate
HCO_3^-	hydrogen carbonate

ionic bond the bond that results from the electrostatic force of attraction that holds positive and negative ions together

ionic compound a compound that consists of cations and anions held together by ionic bonds

ionic crystal a solid that consists of large numbers of cations and anions arranged in a repeating three-dimensional pattern

$$Na\cdot + \cdot \ddot{\underset{\cdot\cdot}{Cl}}: \rightarrow [Na]^+ \, [:\ddot{\underset{\cdot\cdot}{Cl}}:]^- + energy$$

Figure 7
Formation of an ionic bond

(a)

(b)

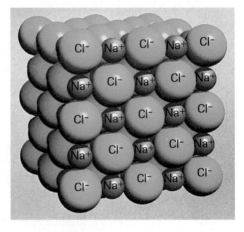

Figure 8
(a) Sodium chloride (table salt) crystals appear as cubes when viewed under a microscope.
(b) A sodium chloride ionic crystal is composed of large numbers of Na^+ ions and Cl^- ions arranged in a repeating three-dimensional pattern. There is one Na^+ ion for every Cl^- ion, thus the ratio of ions is 1:1.

Molecular compounds are made up of individual molecules that contain the types and numbers of atoms given in their molecular formulas. Thus, a single molecule of water, $H_2O_{(l)}$, contains two hydrogen atoms and one oxygen atom. The crystals of ionic compounds, on the other hand, are composed of large numbers of anions and cations. Thus, their chemical formulas do not indicate the types and numbers of ions in the crystal, but rather the types and *ratio* of ions in the crystal. Since a crystal of sodium chloride contains many sodium ions and chloride ions in a 1:1 ratio, the chemical formula for sodium chloride is $NaCl_{(s)}$. The smallest amount of a substance that has the composition given by its chemical formula is called a **formula unit**. Thus, $NaCl_{(s)}$ is the formula unit of sodium chloride because it indicates that sodium chloride contains sodium and chloride ions in a 1:1 ratio (**Figure 9**). Similarly, the formula unit of the ionic compound magnesium bromide is $MgBr_{2(s)}$, indicating that a magnesium bromide crystal contains magnesium ions and bromide ions in a 1:2 ratio.

formula unit the smallest amount of a substance having the composition given by its chemical formula

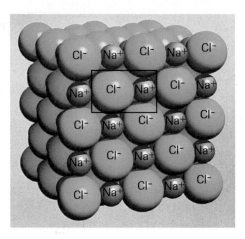

Figure 9
Although a sodium chloride crystal contains large numbers of sodium ions and chloride ions, sodium chloride contains sodium and chloride ions in a 1:1 ratio. Thus, the chemical formula for sodium chloride is $NaCl_{(s)}$.

When soluble ionic compounds are placed in aqueous solution, they conduct electricity: that is, they are electrolytes, as discussed at the beginning of this section. A sodium chloride solution (**Figure 1(a)**) conducts electricity. When placed in water, the positive sodium ions separate from the negative chloride ions. In other words, the sodium chloride compound **dissociates** into its respective ions. To test a solution for its ability to conduct electricity, electrodes, which are attached to the positive and negative terminals of a battery, are placed in the solution. The electrode that is positively charged by the battery attracts the negatively charged ions. The electrode that is negatively charged by the battery attracts the positively charged ions. The movement of ions causes current to flow through the solution. (You will learn more about electrochemistry in Unit 5.)

dissociate separate into positive and negative ions

As mentioned earlier, sugar molecules (**Figure 1(b)**) are not ionic and do not dissociate when they dissolve in water. Instead, the sugar molecules remain neutral, whole molecules in aqueous solution. Since an aqueous solution of sugar molecules does not conduct electric current, sugar is a nonelectrolyte.

Chemical Nomenclature: Naming Elements and Compounds

In the early days of chemistry, when few compounds were known, chemists used common names—such as sugar of lead, quicklime, and milk of magnesia—to identify compounds. To date, nearly five million different compounds have been invented or discovered. Assigning each compound a unique common name, and then keeping track of them all, is almost impossible. The solution is to develop a system for naming compounds in which the name reveals something about the composition of the compound. An international organization of scientists, called the International Union of Pure and Applied Chemistry, or **IUPAC**, develops and sets international rules for chemical names and symbols. See Appendix C6 to review IUPAC rules for naming inorganic compounds. You will learn IUPAC rules for naming organic compounds in Unit 3.

IUPAC International Union of Pure and Applied Chemistry, the organization that establishes the conventions used by chemists

---

▶ **Section 1.11 Questions**

Understanding Concepts

1. Distinguish between an electrolyte and a nonelectrolyte.

2. Explain why some atoms have a tendency to form ions. Which combinations of atoms tend to form ions? Why?

3. Which types of elements form ionic bonds? How?

4. Draw a Lewis symbol for each of the following atoms:
   (a) potassium
   (b) cesium
   (c) iodine
   (d) silicon
   (e) antimony
   (f) krypton
   (g) barium

5. Predict the charge on the most stable ion that is formed by each of the following elements. Write the ion's symbol, including its charge.
   (a) hydrogen
   (b) potassium
   (c) fluorine
   (d) magnesium
   (e) sulfur

6. Draw Lewis symbols for the element ions in question 5. What similarity do you notice? State the rule that is being followed.

7. Using Lewis symbols and the octet rule, illustrate how each of the following pairs of atoms bond:
   (a) potassium and chlorine
   (b) magnesium and sulfur

Applying Inquiry Skills

8. (a) Design an experiment to investigate the conductivity of an ionic solid. With your teacher's approval, perform your experiment.
   (b) Research the conductivity of molten (liquid) ionic compounds.

    www.science.nelson.com

   (c) Use your findings to write a report about the conductivity of ionic compounds in the different states: solid, liquid, and aqueous.
   (d) State a hypothesis for the properties you observe.

1.12 Covalent Bonding

In the previous section, you learned about ionic compounds. When a metal and a nonmetal come in contact, the result is a *transfer of electrons*, the production of positive and negative ions, and the formation of an *ionic compound*. When two nonmetals, such as two chlorine atoms, come in contact, a transfer of electrons would result in less stable atoms. If an electron were transferred from one chlorine atom to another, one chlorine atom would attain eight electrons in its outermost shell, but the other atom would be left with six electrons—a less stable situation. Instead of transferring electrons, the chlorine atoms *share* electrons in order to become more stable. The bond that is formed as a result of sharing electrons is called a **covalent bond**. Atoms that are covalently bonded form chemical entities called **molecules** (**Figure 1**). A **chemical entity** is a single chemical unit, such as an atom, an ion, or a molecule. When two atoms form a covalent bond, the sharing of valence electrons allows both atoms to satisfy the octet rule.

In a hydrogen molecule, two hydrogen atoms share a pair of electrons (one from each atom) and, therefore, form a single covalent bond. Some molecules contain double and triple covalent bonds. For example, oxygen molecules, $O_{2(g)}$, are composed of two oxygen atoms held together by a double covalent bond (two shared electron pairs). Nitrogen molecules, $N_{2(g)}$, are composed of two nitrogen atoms held together by a triple covalent bond (three shared electron pairs).

Molecules can be represented by **Lewis structures**. A Lewis structure shows the electrons that are shared between the atoms as a dash, and the remaining surrounding valence electrons as dots. The pairs of valence electrons that surround each atom are referred to as **lone pairs**. The Lewis structure of a chlorine molecule is shown below.

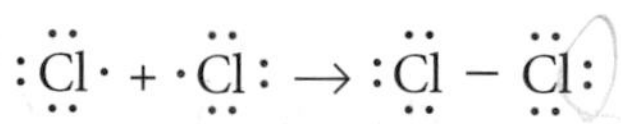

$$:\ddot{\text{Cl}}\cdot + \cdot\ddot{\text{Cl}}: \rightarrow :\ddot{\text{Cl}} - \ddot{\text{Cl}}:$$

covalent bond a bond that arises when two atoms share one or more pairs of electrons between them. The shared electron pairs are attracted to the nuclei of both atoms.

molecule two or more atoms that are joined by covalent bonds

chemical entity a chemical unit, such as an atom, an ion, or a molecule

Figure 1
The pair of shared electrons between the nuclei of two chlorine atoms results in a single covalent bond.

Lewis structure a representation of covalent bonding based on Lewis symbols, with shared electron pairs shown as lines and lone pairs shown as dots

lone pair a pair of valence electrons that is not involved in bonding

> ▶ **SAMPLE** problem

Drawing Lewis Structures

(a) Draw Lewis symbols for the reaction between two iodine atoms, and draw a Lewis structure for the resulting iodine molecule.

Step 1: Draw Lewis Symbols for Atoms
Draw Lewis symbols for the two iodine atoms. Arrange the iodine atoms as you would expect them to be arranged in an iodine molecule. Since there are only two atoms, they are side by side.

$$:\ddot{\text{I}}\cdot \quad \cdot\ddot{\text{I}}:$$

Step 2: Arrange Shared Electron Pairs According to Octet Rule
Arrange the electrons so that at least one pair of electrons is shared by the two atoms and each atom is surrounded by a total of eight electrons, thus obeying the octet rule. Since iodine is found in Group VII, it has seven valence electrons. Each iodine atom contributes one electron to form a shared electron pair.

$$:\ddot{I}:\ddot{I}:$$

Step 3: Represent Bonds as Lines
The shared electron pair (covalent bond) is represented by a line.

$$:\ddot{I} - \ddot{I}:$$

Step 4: Rewrite Chemical Equation Using Lewis Symbols and Lewis Structures

$$:\ddot{I}\cdot + \cdot\ddot{I}: \longrightarrow :\ddot{I} - \ddot{I}:$$

(b) Draw Lewis symbols for the reaction between two oxygen atoms, and draw a Lewis structure for the resulting oxygen molecule.

Step 1: Draw Lewis Symbols for Atoms
Draw Lewis symbols for the two oxygen atoms. Arrange the oxygen atoms as you would expect them to be arranged in an oxygen molecule. Since there are only two atoms, they are side by side.

$$:\ddot{O}\cdot \ \cdot\ddot{O}:$$

Step 2: Arrange Shared Electron Pairs According to Octet Rule
Arrange the electrons so that at least one pair of electrons is shared by the two atoms and each atom is surrounded by a total of eight electrons, thus obeying the octet rule. Since oxygen is found in Group VI, it has six valence electrons. Each oxygen atom contributes two electrons to form two shared electron pairs. Lone pairs repel each other and move as far apart as possible.

$$:\ddot{O}: :\ddot{O}:$$

Step 3: Represent Bonds as Lines
A shared electron pair is represented by a line. Since an oxygen molecule has two shared electron pairs, there are two lines, representing a double bond.

$$:\ddot{O} = \ddot{O}:$$

Step 4: Rewrite Chemical Equation Using Lewis Symbols and Lewis Structures

$$:\ddot{O}: + :\ddot{O}: \longrightarrow :\ddot{O} = \ddot{O}:$$

(c) Draw Lewis symbols for the reaction between two nitrogen atoms, and draw a Lewis structure for the resulting nitrogen molecule.

Step 1: Draw Lewis Symbols for Atoms

Draw Lewis symbols for the two nitrogen atoms. Arrange the nitrogen atoms as you would expect them to be arranged in a nitrogen molecule. Since there are only two atoms, they are side by side.

$$\cdot \ddot{\text{N}} \cdot \quad \cdot \ddot{\text{N}} \cdot$$

Step 2: Arrange Shared Electron Pairs According to Octet Rule

Arrange the electrons so that at least one pair of electrons is shared by the two atoms and each atom is surrounded by a total of eight electrons, thus obeying the octet rule. Since nitrogen is found in Group V, it has five valence electrons. Each nitrogen atom contributes three electrons to form three shared electron pairs. Lone pairs repel each other and move as far apart as possible.

$$:\text{N}::\text{N}:$$

Step 3: Represent Bonds as Lines

A shared electron pair is represented by a line. Since a nitrogen molecule has three shared electron pairs, there are three lines, representing a triple bond.

$$:\text{N} \equiv \text{N}:$$

Step 4: Rewrite Chemical Equation Using Lewis Symbols and Lewis Structures

$$:\dot{\text{N}}\cdot + \cdot\dot{\text{N}}: \rightarrow :\text{N} \equiv \text{N}:$$

(d) Draw Lewis symbols for the reaction between two hydrogen atoms and one oxygen atom. Draw a Lewis structure for the resulting water molecule.

Step 1: Draw Lewis Symbols for Atoms

Draw Lewis symbols for the two hydrogen atoms and one oxygen atom. Arrange the atoms as you would expect them to be arranged in a water molecule. In general, molecules are symmetrical. Here the oxygen atom is placed in the middle, with the hydrogen atoms on each side.

$$\text{H} \cdot \quad \ddot{\text{O}} \quad \cdot \text{H}$$

Step 2: Arrange Shared Electron Pairs According to Octet Rule

Arrange the electrons so that at least one pair of electrons is shared by the two atoms and each atom is surrounded by a total of eight electrons, thus obeying the octet rule. Since oxygen is found in Group VI, it has six electrons in its outer shell. Therefore, it requires two more electrons in order to satisfy the octet rule. Hydrogen can donate one electron to a shared pair and needs one electron in order to become stable. (It is an exception to the octet rule.) Therefore, each hydrogen atom donates one electron to oxygen, forming a shared pair. The result is a single bond between each hydrogen atom and the central oxygen atom. Hydrogen is stable since it now has two electrons in its outermost shell, achieving the noble gas configuration of helium.

$$H \quad \overset{\cdot\cdot}{\underset{\cdot\cdot}{O}} \quad H$$

Step 3: Represent Bonds as Lines

A shared electron pair is represented by a line. Since an oxygen molecule has two shared pairs of electrons, there are two lines, one going to each hydrogen atom.

$$\overset{\cdot\cdot}{O} \diagdown$$
$$H \qquad H$$

Step 4: Rewrite Chemical Equation Using Lewis Symbols and Lewis Structures

$$H\cdot + H\cdot + \overset{\cdot\cdot}{\underset{\cdot\cdot}{O}} \cdot \longrightarrow \quad \overset{\cdot\cdot}{O}\diagdown$$
$$H \qquad H$$

Example

Draw Lewis symbols for the reaction between two oxygen atoms and a carbon atom. Draw the Lewis structure for the resulting carbon dioxide molecule.

Solution

$$\overset{\cdot\cdot}{\underset{\cdot\cdot}{O}}: \quad \cdot \overset{\cdot}{C} \cdot \quad : \overset{\cdot\cdot}{\underset{\cdot\cdot}{O}}$$

$$\overset{\cdot\cdot}{\underset{\cdot\cdot}{O}} :: C :: \overset{\cdot\cdot}{\underset{\cdot\cdot}{O}}$$

$$\overset{\cdot\cdot}{\underset{\cdot\cdot}{O}} = C = \overset{\cdot\cdot}{\underset{\cdot\cdot}{O}}$$

$$\overset{\cdot\cdot}{\underset{\cdot\cdot}{O}}: + \cdot \overset{\cdot}{C} \cdot + : \overset{\cdot\cdot}{\underset{\cdot\cdot}{O}} \rightarrow \overset{\cdot\cdot}{\underset{\cdot\cdot}{O}} = C = \overset{\cdot\cdot}{\underset{\cdot\cdot}{O}}$$

> ## ▶ Practice

Understanding Concepts

1. Draw Lewis symbols for the reaction between two bromine atoms.
2. Draw Lewis symbols for the reaction between one nitrogen atom and three hydrogen atoms. Draw the Lewis structure for the resulting ammonia molecule.
3. Draw Lewis symbols for the reaction between one carbon atom and four hydrogen atoms. Draw the Lewis structure for the resulting methane molecule.
4. Draw Lewis symbols for the reaction between one silicon atom and two oxygen atoms. Draw the Lewis structure for the resulting silicon dioxide molecule.

Electronegativity

electronegativity a measure of an atom's ability to attract a shared pair of electrons within a covalent bond

Electronegativity is a measure of an atom's ability to attract the pair of electrons it shares with another atom within a covalent bond. **Table 1** lists the electronegativities of some elements. In general, metals have lower electronegativities than nonmetals. Fluorine has the highest electronegativity: 4.0. Therefore, it has the greatest ability to attract shared electrons.

Table 1 The Electronegativities of Some Elements

	Element	Electronegativity
	H	2.1
Metals	Li	1.0
	Be	1.5
	Na	0.9
	Mg	1.2
	K	0.8
	Ca	1.0
Nonmetals	C	2.5
	N	3.0
	O	3.5
	F	4.0
	P	2.1
	S	2.5
	Cl	3.0

One factor that plays a role in determining an atom's electronegativity is the atom's atomic radius. The larger an atom is, the weaker its attraction for shared electron pairs. The many layers of electron shells that separate the shared electrons from the positively charged nucleus act as a shield between the positive charge of the nucleus and the negative charge of the shared electrons. An atom with a small atomic radius has a stronger attraction for a shared pair of electrons than a larger atom. Electronegativity *decreases* as you descend a family in the periodic table because atomic radii *increase*. Electronegativity *increases* as you move from left to right across the periodic table because atomic radii *decrease*—there are more protons in the nucleus, causing a stronger attraction for electrons (**Figure 2**). Metals generally have lower electronegativities than nonmetals.

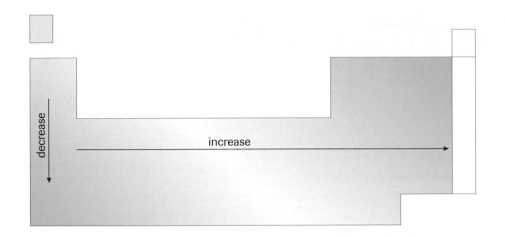

Figure 2
In general, electronegativity decreases as you descend a family and increases as you travel across a period from left to right. This trend is related to atomic radius, which increases as you move down the periodic table and decreases as you move from left to right across the periodic table.

Polar and Nonpolar Covalent Bonds

If an electron pair is shared equally, then the bond is a **nonpolar covalent bond**. The two hydrogen atoms in a hydrogen molecule, $H_{2(g)}$, are held together by a nonpolar covalent bond because the two identical hydrogen nuclei (one proton each) attract electrons with equal force (**Figure 3**). Think of a nonpolar covalent bond as a tug of war in which both people who are tugging have exactly the same strength of pull.

Even though a covalent bond involves the sharing of electrons, the sharing is not always equal. The shared electron pair may spend more time around one atom than around the other atom. When an electron pair is not shared equally, there is a localized negative charge around one atom, represented by the symbol δ^-. The other atom is more positively charged. It has a localized positive charge, represented by δ^+. The bond between the two atoms is called a **polar covalent bond**. A polar covalent bond has a slightly negative end and a slightly positive end.

Whether or not a bond is polar covalent depends on the difference between the electronegativities of the bonded atoms. Hydrogen chloride, $HCl_{(g)}$, is an example of a molecule that has a polar covalent bond (**Figure 4**). In a hydrogen chloride molecule, the electron pair is unequally shared because chlorine has a higher electronegativity than hydrogen. The electron pair of the covalent bond therefore spends a greater amount of time in the space surrounding the chlorine nucleus than in the space surrounding the hydrogen nucleus. As a result, the hydrogen end of the molecule has a slightly positive charge, δ^+, and the chlorine end has a slightly negative charge, δ^-.

nonpolar covalent bond a bond in which an electron pair is shared equally between a pair of atoms having the same electronegativity

Figure 3
Two identical atoms are bonded together by a nonpolar covalent bond.

polar covalent bond a bond in which an electron pair is shared unequally between a pair of atoms that have different electronegativities

$$\delta^+ \qquad \delta^-$$
$$H \qquad :\overset{..}{\underset{..}{C}l}:$$

Figure 4
In hydrogen chloride, $HCl_{(g)}$, the shared electron pair spends more time near the chlorine atom. Therefore, the chlorine end of hydrogen chloride possesses a slightly negative charge, represented by δ^-. Since the hydrogen atom's electron spends more time around the chlorine atom, the hydrogen end has a slightly positive charge, represented by δ^+.

Polar and Nonpolar Molecules

polar molecule a molecule that has a slightly positive charge on one end and a slightly negative charge on the other end

Polar molecules are molecules that have a positively charged end and a negatively charged end. **Nonpolar molecules** do not have charged ends. The polarity of a molecule depends on two characteristics of the molecule:

1. the presence of polar covalent bonds;

2. the three-dimensional shape (geometry) of the molecule.

nonpolar molecule a molecule that has no charged ends

Ammonia, $NH_{3(g)}$, is a polar molecule because it contains polar covalent bonds and a pyramidal shape (**Figure 5(a)**). Nitrogen forms covalent bonds with three hydrogen atoms. Because nitrogen has a higher electronegativity than hydrogen, it has a stronger attraction for each of the three shared electron pairs. Therefore, ammonia has a positively charged end and a negatively charged end.

Methane, $CH_{4(g)}$, contains slightly polar covalent bonds between the carbon atom and each of the four hydrogen atoms. Because carbon is slightly more electronegative than hydrogen, it has a stronger attraction for each of the shared electron pairs. Methane is a nonpolar molecule, however, because its polar covalent bonds are all arranged symmetrically about the central carbon atom (**Figure 5(b)**). Molecules made up of identical atoms, such as nitrogen, $N_{2(g)}$, contain only nonpolar covalent bonds. They are always linear in shape and nonpolar (**Figure 5(c)**).

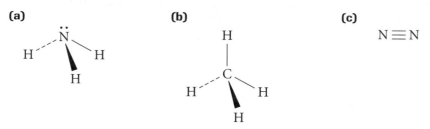

(a)   **(b)**   **(c)**

Figure 5
(a) Ammonia, $NH_{3(g)}$, has polar covalent bonds and a pyramidal shape. The lone pair of electrons in the nitrogen atom repels the electrons in the hydrogen atoms, causing them to move closer to each other. The result is a polar molecule.
(b) Methane, $CH_{4(g)}$, has slightly polar covalent bonds and a symmetrical tetrahedral shape, resulting in a nonpolar molecule.
(c) Nitrogen, $N_{2(g)}$, has nonpolar bonds. Both nitrogen nuclei have equal pull on the shared electron pairs.

Intermolecular Bonds

intermolecular bonds bonds between molecules; forces of attraction that form between a molecule and its neighbouring molecules

Covalent bonds are strong forces of attraction that hold the atoms of a molecule together. Molecules are attracted to other molecules by a group of much weaker forces of attraction, known as **intermolecular bonds** (bonds between molecules). The strength of intermolecular bonds determines the physical state (solid, liquid, or gas) of a molecular compound at a particular temperature and pressure, and also determines the melting point and boiling point of the compound. Intermolecular bonds are broken when a molecular compound melts and boils.

There are two different types of intermolecular bonds that we will consider. One type, called the **dipole–dipole force (DDF)**, occurs between polar molecules, such as hydrogen chloride, HCl. The slightly positive end of one hydrogen chloride molecule is attracted to the slightly negative end of a neighbouring hydrogen chloride molecule (**Figure 6**). Another type of intermolecular bond, called the London dispersion force, occurs between all molecules, polar and nonpolar. It is the most important intermolecular bond that occurs between nonpolar molecules. **London dispersion forces (LDF)** are formed when the electrons in the atoms of a molecule happen to be located on one side of the atoms, leaving a deficiency of electrons on the other side. The side of the atoms with more electrons develops a temporary negative charge, and the side with fewer electrons develops a temporary positive charge. If the same thing happens to the atoms of a neighbouring molecule at the same time, the atoms of the two molecules, and thus the molecules themselves, experience a force of attraction: the positive side of one molecule attracts the negative side of the other molecule. Since electrons move quickly, the dipole lasts for only a fraction of a second.

If the molecules are large and have lots of atoms in them, then the chance of creating London dispersion forces between atoms of adjacent molecules is much higher. London dispersion forces, therefore, arise more often (and are more effective) between molecules composed of large numbers of atoms, where the chance of electron imbalances is high. Dipole–dipole forces and London dispersion forces are referred to as **van der Waals forces**, after the nineteenth-century chemist who first suggested the existence of intermolecular bonds.

dipole–dipole force (DDF) an intermolecular force of attraction that forms between the slightly positive end of one polar molecule and the slightly negative end of an adjacent polar molecule

London dispersion force (LDF) an intermolecular force of attraction that forms between atoms of neighbouring molecules as a result of a temporary imbalance in the position of the atoms' electrons; forms between all molecules, polar and nonpolar

van der Waals forces forces of attraction between molecules, such as the dipole–dipole force and the London dispersion force

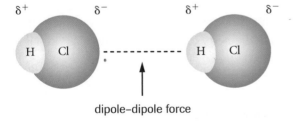

dipole–dipole force

Figure 6
The slightly positive end of one hydrogen chloride molecule is attracted to the slightly negative end of the other hydrogen chloride molecule, creating a dipole–dipole force between them.

Building Molecular Models

In section 1.2, you learned about the importance of models for understanding how matter behaves. Lewis structures are models that are used to represent molecules. You can also represent molecules using three-dimensional molecular models. In this activity, you will build molecular models of different compounds using a molecular model kit.

Materials: molecular model kit

1. In your molecular model kit, identify the pieces that represent the following elements: chlorine, oxygen, hydrogen, nitrogen, and carbon.

(a) Copy and complete **Table 2** by drawing the Lewis structures for the following molecules: O_2, CO_2, H_2O, CH_4, H_2, HCl, C_2H_2, and CCl_4.

2. Build models of the molecules in **Table 2**.

(b) Describe the shape of each molecule you built. Is it linear, bent, pyramidal, or tetrahedral?

(c) What do the holes in the round pieces represent? What do the pegs or springs represent?

(d) Which molecules contain single bonds? Which molecules contain double bonds, and which contain triple bonds?

(e) Identify the polar and nonpolar molecules.

Table 2 Lewis Structures

Molecule	O_2	CO_2	H_2O	CH_4	H_2	HCl	C_2H_2	CCl_4
Lewis structure								

Water

Water is an excellent example of a polar molecule. The difference between the electronegativities of the oxygen and hydrogen atoms is 1.4. Since oxygen has the higher electronegativity, the electrons spend more time around oxygen than they do around hydrogen. Therefore, the oxygen end of a water molecule has a slightly negative charge. (Since hydrogen has the lower electronegativity and therefore less attraction for the electron pair, the hydrogen end of the molecule is slightly positive.) The arrangement of atoms and covalent bonds forms a bent molecule that is highly polar (**Figure 7**).

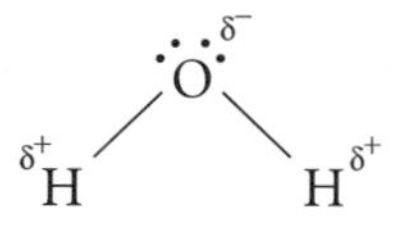

Figure 7

Water is a polar molecule due to the difference in the electronegativities of oxygen and hydrogen, and its bent shape.

> **Section 1.12 Questions**

Understanding Concepts

1. Which of the following pairs of atoms form covalent bonds, and which form ionic bonds? How do you know?
   (a) sulfur and oxygen
   (b) sodium and iodine
   (c) bromine and bromine

2. Draw a Lewis structure for each of the following molecules:
   (a) F_2 (c) O_2 (e) CO_2
   (b) H_2 (d) H_2S

3. Which of the molecules in question 2 are polar molecules? Which are nonpolar?

4. How are a covalent bond and an ionic bond different? How are they similar?

5. Using the periodic table and your knowledge of electronegativity, predict which of the atoms in the following pairs has a higher electronegativity. Justify your prediction.
   • beryllium and strontium
   • sodium and chlorine

6. Identify the more polar bond in each of the following pairs.
   (a) H–F and H–Cl   (e) C–H and N–H
   (b) N–O and C–O   (f) S–O and P–O
   (c) S–H and O–H   (g) C–N and N–N
   (d) P–Cl and S–Cl

7. Explain how some molecules that contain polar covalent bonds can be nonpolar.

8. (a) What is an intermolecular bond?
   (b) What type(s) of intermolecular forces of attraction may form between nonpolar molecules? What type(s) may form between polar molecules? Give reasons in both cases.

9. Copy and complete **Table 3**.

Table 3   Intermolecular Forces

Molecule	Intermolecular force(s) (LDF, DDF, or both)
hydrogen, H_2	
carbon tetrachloride, CCl_4	
hydrogen sulfide, H_2S	

10. How do intermolecular bonds help to explain why the boiling point of methane, CH_4, is much lower than the boiling point of hydrogen bromide, HBr?

Applying Inquiry Skills

11. A student was given four sample liquids and asked to determine whether the liquids were affected by a positively charged object or a negatively charged object. The student tested a thin stream of each liquid by holding a positively charged object near the liquid stream. The student then repeated the procedure using a negatively charged object. Complete the Prediction, Observations, Analysis, and Synthesis in the following lab report:

Question

How does a charged object affect a thin stream of each of the following liquids: NCl_3, H_2O, Br_2, and CCl_4?

Prediction

(a) Using what you know about polar and nonpolar molecules, predict an answer to the Question.

Observations

Table 4   Effects of Charged Objects on Four Liquids

Sample	Positive charge	Negative charge
1	There was no effect.	There was no effect.
2	There was no effect.	There was no effect.
3	The stream moved toward the charged object.	The stream moved toward the charged object.
4	The stream moved toward the charged object.	The stream moved toward the charged object.

(b) Which of the four substances could be samples 1 and 2? Which could be samples 3 and 4?

Analysis

(c) Use your Observations to answer the Question.

Synthesis

(d) Provide a theoretical explanation to justify your answer for (b).
(e) Speculate as to why the liquids were affected by both positive and negative charges.

1.13 Investigation

Classifying Solids Using Physical Properties

Scientists classify matter according to physical and chemical properties. To do so, they conduct many qualitative analysis tests on different substances and then group the substances into categories based on properties. In this investigation, you will explore and identify the physical properties of two different classes of solids: ionic and molecular. Then you will use the established criteria to classify an unidentified solid provided by your teacher.

The physical properties that you will explore are conductivity (ability to conduct electricity), hardness (resistance to being scratched), and solubility (ability to dissolve in water). You will be given the melting point of each solid (another physical property).

Question

How do ionic and molecular solids compare on the basis of conductivity, solubility, hardness, and melting point?

Prediction

(a) From its chemical formula, predict whether each solid (potassium iodide, sodium chloride, sucrose, camphor) is ionic or molecular.

Hypothesis

(b) Explain your Prediction.

Materials

eye protection
lab apron
potassium iodide, $KI_{(s)}$
sodium chloride, $NaCl_{(s)}$
sucrose, $C_{12}H_{22}O_{11(s)}$
camphor, $C_{10}H_{16}O_{(s)}$
low-voltage conductivity apparatus
6 large test tubes
test-tube rack

distilled water
wax pencil
scoopula
stirring rod
medicine dropper
10-mL graduated cylinder
4 watch glasses
unidentified solid sample

Procedure

Part 1: Testing for Solubility

1. Place the test tubes in the test-tube rack.

2. Using the wax pencil, label the first four test tubes with the following labels: sucrose, camphor, KI, NaCl.

3. Pour 10 mL of distilled water into each of the four test tubes.

4. Add a small amount (fill the tip of the scoopula) of each solid to its respective test tube.

5. Using a stirring rod, stir the mixtures. Observe how completely each solid dissolves. Record your observations in a table.

6. Keep the solutions for Part 2 of the investigation.

Part 2: Testing for Conductivity

7. Fill a test tube with 10 mL of distilled water. Test for conductivity using the low-voltage conductivity apparatus demonstrated by your teacher.

8. Test the conductivity of each solution you prepared in Part 1. Record your observations in your table.

9. Follow your teacher's instructions for disposing of the solutions.

Part 3: Testing for Hardness

10. Using the scoopula, place a few small crystals of each solid on individual watch glasses.

11. Using the scoopula, attempt to crush a few of the small crystals between the scoopula and the watch glass.

12. Rank the relative hardness of each sample. Record your rankings in your table.

Part 4: Melting Point

13. Potassium iodide melts at 686°C, sodium chloride melts at 801°C, camphor melts at 177°C, and sucrose melts at 185°C. Record these values in your table.

Part 5: Classifying an Unidentified Sample

14. Obtain an unidentified sample from your teacher. Your teacher will tell you the melting point of the sample. Test the sample for solubility in water, conductivity, and hardness, as you did in steps 1 to 12. Record your observations in your table.

Analysis

(c) Use your observations to group the first four solids into two categories.

(d) Briefly summarize the physical properties of each category.

(e) Answer the Question.

(f) Based on your summary of the physical properties of the two categories, decide to which category your unidentified sample belongs. Give detailed reasons for your choice, using the different properties of matter that you investigated.

Evaluation

(g) Are the physical properties that you studied in this investigation sufficient for classifying the solids into two categories? What other physical properties could you investigate?

(h) Suggest possible sources of error in the Procedure, and explain their effects on your results. How could you modify the Procedure to reduce these sources of error?

(i) Evaluate your Prediction based on your Analysis.

chemical reaction a change that forms one or more new substances with different physical and chemical properties

So far, you have focused on qualitative analysis techniques that are based on physical properties, such as conductivity, line spectra, and flame tests. Observing how a substance reacts with another substance in a chemical reaction can also help you identify a compound. A **chemical reaction** has occurred when substances change into new substances with new physical and chemical properties.

Many diagnostic tests have been devised based on the chemical properties of matter. You may recall, from previous courses, that the presence of hydrogen gas can be detected using a burning splint and the presence of oxygen can be detected using a glowing splint. If a burning splint is placed in the presence of hydrogen gas, it will make a popping noise. If a glowing splint is placed in the presence of oxygen gas, it will re-ignite.

Another example of a diagnostic test based on the chemical properties of matter is the Kastle-Meyer colour test. Forensic scientists may use the Kastle-Meyer colour test to determine whether or not blood is present on a piece of evidence. When phenolphthalein reagent (an acid–base indicator) in combination with hydrogen peroxide is added to blood, the blood turns a deep pink colour. Once forensic scientists have established that blood is indeed on the evidence, they can conduct additional tests to identify the blood type or they can perform DNA analysis.

Types of Reactions

To conduct qualitative chemical analysis, you first need to know the types of chemical reactions. You may recall, from a previous chemistry course, that there are four types of chemical reactions: synthesis, decomposition, single displacement, and double displacement. In this section, you will review each type of reaction.

Synthesis Reactions

synthesis reaction a chemical reaction in which two or more simple substances combine to form a more complex substance

A **synthesis reaction** involves the combination of two or more simple substances to form a more complex substance (**Figure 1**). A synthesis reaction (sometimes referred to as a *combination reaction* or an *addition reaction*) is represented by the following equation:

$$A + B \rightarrow AB$$

Figure 1
A synthesis reaction results in the production of a complex molecule from simple molecules.

An example of a synthesis reaction is the very powerful reaction that is used to propel a space shuttle into orbit: the synthesis of water. Liquid hydrogen and liquid oxygen, under high pressure, react to form water.

$$2\,H_{2(l)} + O_{2(l)} \rightarrow 2\,H_2O_{(g)}$$

Engineers at NASA have designed a system in which the energy that is released from this violent reaction helps to launch a space shuttle (**Figure 2**).

Synthesis reactions can also involve the combination of simple compounds into more complex compounds. For example, sulfuric acid, $H_2SO_{4(aq)}$, is produced in two synthesis reactions involving sulfur dioxide, sulfur trioxide, oxygen, and water. Sulfur dioxide, $SO_{2(g)}$, is a byproduct of the combustion of gasoline in automobiles. Sulfur dioxide from the cars reacts with oxygen in the atmosphere, producing sulfur trioxide, $SO_{3(g)}$, in the first synthesis reaction:

$$2\,SO_{2(g)} + O_{2(g)} \rightarrow 2\,SO_{3(g)}$$

In the second synthesis reaction, sulfur trioxide reacts with water to produce sulfuric acid, a chemical that also contributes to acid rain:

$$SO_{3(g)} + H_2O_{(l)} \rightarrow H_2SO_{4(aq)}$$

Some synthesis reactions are also classified as combustion reactions if oxygen is a reactant. When an element combines rapidly with oxygen gas to form more complex molecules by burning, the reaction is classified as a **combustion reaction**. For example, the reaction between hydrogen and oxygen to produce water is a combustion reaction as well as a synthesis reaction. Some other examples of combustion reactions include the synthesis of carbon and oxygen to form carbon dioxide, and the synthesis of sulfur and oxygen to form sulfur dioxide:

$$C_{(s)} + O_{2(g)} \rightarrow CO_{2(g)}$$
$$S_{(s)} + O_{2(g)} \rightarrow SO_{2(g)}$$

A special type of combustion reaction occurs when *hydrocarbons* (compounds composed of hydrogen and carbon atoms only) are burned. You will learn about this type of reaction in Unit 3.

Decomposition Reactions

Decomposition reactions involve the breakdown of large, complex molecules or ionic compounds into smaller, simpler entities (**Figure 3**). The general equation for a decomposition reaction is

$$AB \rightarrow A + B$$

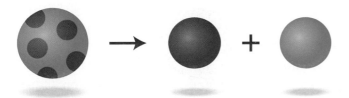

Figure 3
A decomposition reaction involves the breakdown of a compound into simpler substances.

DID YOU KNOW?

Why *Challenger* Exploded
In 1986, the space shuttle *Challenger* exploded seconds after takeoff due to a damaged O-ring in its solid rocket motor. The damaged O-ring allowed hot gases to come in contact with the liquid hydrogen and liquid oxygen fuel, causing the explosion. Seven astronauts were killed.

Figure 2
The external tank of the space shuttle contains the liquid hydrogen and liquid oxygen fuel. The fuel is supplied, under pressure, to the three main engines of the space shuttle during liftoff. The synthesis reaction between hydrogen and oxygen results in the production of water and the release of a large amount of energy, which propels the shuttle into orbit. The external tank does not go into orbit but is jettisoned into a remote part of the ocean.

combustion reaction a chemical reaction that occurs when a substance reacts rapidly with oxygen, releasing energy

decomposition reaction a chemical reaction in which a molecule or ionic compound is broken down into simpler entities

enzyme a molecule, found within a biological system, that increases the speed of a chemical reaction

solution a homogeneous mixture of two or more pure substances

solute a pure substance in a solution that is dissolved by a solvent; usually the substance in lesser quantity

solvent the pure substance in a solution that dissolves other components; usually the substance in greater quantity

homogeneous mixture a mixture that has uniform chemical and physical properties throughout because the components are uniformly distributed

aqueous solution a solution in which water is the solvent; denoted by the subscript (aq)

Figure 4
The decomposition of hydrogen peroxide results in the production of water and oxygen gas. The oxygen gas causes the soap that is added to the hydrogen peroxide to bubble and foam.

Have you ever dabbed hydrogen peroxide, $H_2O_{2(l)}$, on a cut? If so, you may have noticed that bubbling and fizzing occur when the hydrogen peroxide makes contact with blood. The bubbling is caused by the production of oxygen gas (**Figure 4**). An **enzyme** that is present in blood is responsible for speeding up the breakdown of hydrogen peroxide (a compound) into oxygen gas (a simple molecule) and water (a simpler compound). The equation for this chemical reaction is

$$2\ H_2O_{2(l)} \rightarrow O_{2(g)} + 2\ H_2O_{(l)}$$

Chemical Reactions in Solution

A **solution** is a homogeneous mixture in which a pure substance, called the **solute** (usually the substance in lesser quantity), is dissolved in another pure substance, called the **solvent** (usually the substance in greater quantity). Thus, if 10 mL of an alcohol is dissolved in 100 mL of water, the alcohol is the solute and the water is the solvent. Some solutions contain several different solutes dissolved in a solvent. Seawater is an example of a solution in which different salts (the solutes) are dissolved in water (the solvent). As a **homogeneous mixture**, a solution possesses uniform properties. In other words, all samples of a solution have the same physical and chemical characteristics because solute entities are evenly distributed within the solution, and all samples have the same composition (**Figure 5**).

An **aqueous solution** is a solution in which water is the solvent. (*Aqua* is Latin for "water.") Thus, seawater is an example of an aqueous solution. Other examples include vinegar (acetic acid dissolved in water), antifreeze (ethylene glycol dissolved in water), and rubbing alcohol (isopropyl alcohol dissolved in

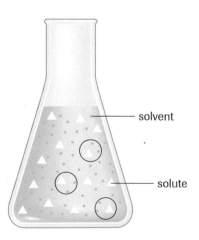

solvent

solute

Figure 5
A solution possesses uniform chemical and physical properties throughout because solute and solvent entities are uniformly distributed.

water). The subscript (aq) following the solute's chemical formula indicates that the solute is dissolved in water to form an aqueous solution. Thus, $HC_2H_3O_{2(aq)}$ denotes an aqueous solution of acetic acid.

Chemical reactions that commonly occur in aqueous solution are single displacement reactions and double displacement reactions.

Single Displacement Reactions

A **single displacement reaction** involves the reaction of an element with a compound to produce a new element and a new compound (**Figure 6**). The general equation for a single displacement reaction is

$$A + BC \rightarrow AC + B$$

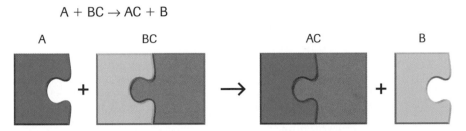

Figure 6
In a single displacement reaction, an element reacts with a compound to replace another element within the compound.

Single displacement reactions usually involve the displacement of one metal (or hydrogen) by another metal. For example, if a zinc strip is placed in a solution of lead(II) nitrate, $Pb(NO_3)_{2(aq)}$, and left to stand (**Figure 7**), solid lead, $Pb_{(s)}$, and aqueous zinc nitrate, $Zn(NO_3)_{2(aq)}$, will form:

$$Zn_{(s)} + Pb(NO_3)_{2(aq)} \rightarrow Pb_{(s)} + Zn(NO_3)_{2(aq)}$$

You will find out more about single displacement reactions in Unit 5.

Double Displacement Reactions

A **double displacement reaction** is the reaction of two compounds in aqueous solution. In a double displacement reaction, the positive and negative ions in their respective compounds switch places to form two new compounds (**Figure 8**). The general equation for a double displacement reaction is

$$AB + CD \rightarrow AD + CB$$

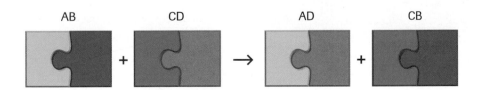

Figure 8
In a double displacement reaction, the anions and cations from the two compounds switch places, forming two new compounds.

single displacement reaction the reaction of an element and a compound to produce a new element and a new compound

Figure 7
Zinc placed in a solution of lead(II) nitrate displaces the lead in lead(II) nitrate, forming lead metal and zinc nitrate.

double displacement reaction a chemical reaction in which two compounds in aqueous solution react to form two new compounds

Two examples of double displacement reactions are given below:

aqueous silver nitrate + aqueous calcium chloride → solid silver chloride + aqueous calcium nitrate

$$2\,AgNO_{3(aq)} \quad + \quad CaCl_{2(aq)} \quad \rightarrow \quad 2\,AgCl_{(s)} \quad + \quad Ca(NO_3)_{2(aq)}$$

aqueous sodium hydroxide + aqueous calcium nitrate → solid calcium hydroxide + aqueous sodium nitrate

$$2\,NaOH_{(aq)} \quad + \quad Ca(NO_3)_{2(aq)} \quad \rightarrow \quad Ca(OH)_{2(s)} \quad + \quad 2\,NaNO_{3(aq)}$$

precipitate a solid formed in a reaction that takes place in aqueous solution

Notice two things about these two reactions:

1. The *reactants* are commonly in aqueous state, (aq): that is, the compounds have been dissolved in water and have dissociated into their respective positive and negative ions. For example, aqueous silver nitrate, $AgNO_{3(aq)}$, is a solution in which silver nitrate, $AgNO_{3(s)}$, has been dissolved, yielding $Ag^+_{(aq)}$ ions and $NO_3{}^-_{(aq)}$ ions.

2. One of the *products* is in aqueous solution $(Ca(NO_3)_{2(aq)})$, and the other product is a solid $(AgCl_{(s)})$. A solid that is formed as a result of the chemical reaction of two aqueous solutions is called a **precipitate** (**Figure 9**). Because a precipitate forms, double displacement reactions are sometimes called precipitation reactions.

To predict whether or not a reaction will occur between two aqueous compounds, you must first understand the behaviour of the compounds in water. In the next section, you will learn how various substances react when placed in water.

Figure 9
A precipitate is formed when aqueous solutions of calcium chloride, $CaCl_{2(aq)}$, and sodium carbonate, $Na_2CO_{3(aq)}$, are mixed.

▶ *Section 1.14* *Questions*

Understanding Concepts

1. Classify each of the following reactions as a synthesis reaction or a decomposition reaction. Give reasons for your classification.
   (a) $2 KI_{(s)} \rightarrow 2 K_{(s)} + I_{2(s)}$
   (b) $H_{2(g)} + Cl_{2(g)} \rightarrow 2 HCl_{(g)}$
   (c) $ZnCO_{3(s)} \rightarrow ZnO_{(s)} + CO_{2(g)}$

2. Water can be broken down into its elements by a technique called electrolysis. Write a balanced chemical equation that represents the electrolysis of water. Include the states of the reactants and products.

3. Why can some synthesis reactions be considered combustion reactions? Give an example to support your answer.

4. Complete each of the following single displacement reactions. Make sure that your final equation is balanced.
   (a) $Zn_{(s)} + CuCl_{2(aq)} \rightarrow$
   (b) $Ca_{(s)} + HCl_{(aq)} \rightarrow$
   (c) $Na_{(s)} + H_2O_{(l)} \rightarrow$

5. What types of reactants are usually involved in synthesis, decomposition, single displacement, and double displacement reactions?

6. Synthesis and decomposition reactions are sometimes referred to as opposite reactions. Explain why.

7. Classify each of the following reactions as a single displacement reaction or a double displacement reaction. Predict the products of each reaction, and write the balanced chemical equation.
   (a) Silver metal is recovered in a laboratory by placing aluminum foil into aqueous silver nitrate:

   $Al_{(s)} + AgNO_{3(aq)} \rightarrow$

   (b) Hydrogen gas can be produced in the lab from the reaction of zinc and sulfuric acid:

   zinc + sulfuric acid $\rightarrow$

   (c) The presence of chloride ions in a water sample is indicated by the formation of a white precipitate when aqueous silver nitrate is added to the sample:

   aqueous magnesium chloride + aqueous silver nitrate $\rightarrow$

   (d) Sodium metal reacts vigorously with water to produce a flammable gas and a basic (hydroxide) solution:

   sodium + water $\rightarrow$

   (e) Aqueous potassium hydroxide is added to a sample of well water. The formation of a rusty brown precipitate indicates the presence of an iron(III) compound in the water:

   $KOH_{(aq)} + FeCl_{3(aq)} \rightarrow$

Applying Inquiry Skills

8. An investigation is planned and carried out to test the general formula for decomposition reactions:

   $AB \rightarrow A + B$

   Complete the Prediction for the following lab report:

   Question

   What are the products of the decomposition of lithium oxide, $Li_2O_{(s)}$, magnesium oxide, $MgO_{(s)}$, and zinc chloride, $ZnCl_{2(s)}$?

   Prediction

   (a) Predict the answer to the Question.

Making Connections

9. Catalytic converters in automobiles are used to decompose nitrogen monoxide, $NO_{(g)}$.
   (a) Predict the products of this decomposition reaction.
   (b) Research how a catalytic converter functions in an automobile and why it is effective for reducing air pollution.

 www.science.nelson.com

10. Most metals are found in nature as compounds and ores rather than as pure elements. The most common types of metal ores are oxides, sulfides, halides, carbonates, sulfates, and silicates. Choose a metal, and answer the following questions:
    (a) In what form is this metal most often found in nature?
    (b) How is this metal extracted and purified?
    (c) What types of reactions are used to purify it?
    (d) What environmental concerns are associated with its purification?

 www.science.nelson.com

solubility a measure of the extent to which a solute dissolves in a solvent at a given temperature and pressure

Double displacement reactions occur in aqueous solutions. Every substance dissolves in water to a certain degree. **Solubility** is a measure of the amount of a substance that dissolves in water at a given temperature and pressure. A substance that remains solid in water is said to have low solubility. An example of a substance with low solubility is chalk (calcium carbonate, $CaCO_{3(s)}$). An example of a soluble substance is table salt (sodium chloride, $NaCl_{(s)}$). In this section, you will learn how to predict the products of double displacement reactions using solubility rules.

The Solubility Rules

Being able to predict the products of a chemical reaction is very useful to chemists and other professionals. For example, water quality engineers can test water samples for the presence of different ions. If they know the product of a chemical reaction, then they can determine which reactants they need to add to the water in order to determine the presence of certain ions. A vast amount of information has been collected on the solubility of various compounds in water. This information helps chemists and chemical technicians identify the contents of solutions.

Table 1 Solubility Rules for Ionic Compounds in Water

Anions	+	Cations	→	Solubility of compounds
most		alkali ions (Li^+, Na^+, K^+, Rb^+, Cs^+, Fr^+)		soluble
most		hydrogen ion, $H^+_{(aq)}$		soluble
most		ammonium ion, NH_4^+		soluble
nitrate, NO_3^-		most		soluble
acetate, $C_2H_3O_2^-$		Ag^+		low solubility
		most others		soluble
chloride, Cl^- bromide, Br^- iodide, I^-		Ag^+, Pb^{2+}, Hg_2^{2+}, Cu^+, Tl^+		low solubility
		all others		soluble
sulfate, SO_4^{2-}		Ca^{2+}, Sr^{2+}, Ba^{2+}, Pb^{2+}, Ra^{2+}		low solubility
		all others		soluble
sulfide, S^{2-}		alkali ions, $H^+_{(aq)}$, NH_4^+, Be^{2+}, Mg^{2+}, Ca^{2+}, Sr^{2+}, Ba^{2+}, Ra^{2+}		soluble
		all others		low solubility
hydroxide, OH^-		alkali ions, $H^+_{(aq)}$, NH_4^+, Sr^{2+}, Ba^{2+}, Ra^{2+}, Tl^+		soluble
		all others		low solubility
phosphate, PO_4^{3-} carbonate, CO_3^{2-} sulfite, SO_3^{2-}		alkali ions, $H^+_{(aq)}$, NH_4^+		soluble
		all others		low solubility

The solubility of different ionic compounds in water is summarized in **Table 1**. You can use this table to predict whether or not an ionic compound formed during a double displacement reaction is soluble in water (no precipitate forms) or has low solubility in water (a precipitate forms).

Predicting the Formation of a Precipitate

In the previous section, you learned that some double displacement reactions in aqueous solution result in the formation of a precipitate and an ionic compound in aqueous state (**Figure 1**). You looked at the following two examples:

aqueous silver nitrate + aqueous calcium chloride → solid silver chloride + aqueous calcium nitrate

$$2\,AgNO_{3(aq)} \quad + \quad CaCl_{2(aq)} \quad \rightarrow \quad 2\,AgCl_{(s)} \quad + \quad Ca(NO_3)_{2(aq)}$$

aqueous sodium hydroxide + aqueous calcium nitrate → solid calcium hydroxide + aqueous sodium nitrate

$$2\,NaOH_{(aq)} \quad + \quad Ca(NO_3)_{2(aq)} \quad \rightarrow \quad Ca(OH)_{2(s)} \quad + \quad 2\,NaNO_{3(aq)}$$

When two ionic compounds in aqueous state are mixed together, you can use the solubility rules to predict whether or not a chemical reaction will result in the formation of a precipitate.

▶ **SAMPLE** problem **1**

Predicting Precipitate Formation

(a) **Determine the products (if any) when a solution of sodium chloride is mixed with a solution of silver nitrate. If a reaction occurs, summarize the reaction as a balanced chemical equation.**

Step 1: Identify Type of Reaction and Possible Products
Sodium chloride and silver nitrate are both ionic compounds. Therefore, if a reaction occurs, it will be a double displacement reaction. Write the possible double displacement reaction as a word equation:

sodium chloride + silver nitrate → silver chloride + sodium nitrate

Step 2: Look Up Solubility of Both Products
silver chloride: low solubility
sodium nitrate: soluble

Step 3: Indicate States of Reactants and Products

aqueous sodium chloride + aqueous silver nitrate →
solid silver chloride + aqueous sodium nitrate

Since silver chloride has low solubility, it will precipitate out of solution. Sodium chloride, silver nitrate, and sodium nitrate are all soluble.

Step 4: Write Chemical Equation for Reaction

$$NaCl_{(aq)} + AgNO_{3(aq)} \rightarrow AgCl_{(s)} + NaNO_{3(aq)}$$

Figure 1
In this double displacement reaction, two aqueous ionic compounds react to form a new solid compound and a new aqueous ionic compound. A solution of sodium iodide, $NaI_{(aq)}$, reacts immediately when mixed with a solution of lead(II) nitrate, $Pb(NO_3)_{2(aq)}$, to produce a bright yellow precipitate.

Step 5: Balance Equation

The equation is already balanced.

$$NaCl_{(aq)} + AgNO_{3(aq)} \rightarrow AgCl_{(s)} + NaNO_{3(aq)}$$

(b) Determine the products (if any) when a solution of sodium chloride is mixed with a solution of potassium bromide. If a reaction occurs, summarize the reaction as a balanced chemical equation.

Step 1: Identify Type of Reaction and Possible Products

Sodium chloride and potassium bromide are both ionic compounds. If a reaction occurs, it will be a double displacement reaction. Write the possible double displacement reaction as a word equation:

sodium chloride + potassium bromide →
sodium bromide + potassium chloride

Step 2: Look Up Solubility of Both Products

sodium bromide: soluble
potassium chloride: soluble

Step 3: Indicate States of Reactants and Products

Since both sodium bromide and potassium chloride are soluble in water, all ions stay in solution and no reaction takes place.

Step 4: Write Chemical Equation for Reaction

sodium chloride + potassium bromide → no reaction

Example 1

Determine the products (if any) when a solution of sodium sulfate is mixed with a solution of lead(II) nitrate. If a reaction occurs, summarize the reaction as a balanced chemical equation.

Solution

Double displacement reaction:

sodium sulfate + lead(II) nitrate → lead(II) sulfate + sodium nitrate

lead(II) sulfate: low solubility
sodium nitrate: soluble

aqueous sodium sulfate + aqueous lead(II) nitrate →
solid lead(II) sulfate + aqueous sodium nitrate

$$Na_2SO_{4(aq)} + Pb(NO_3)_{2(aq)} \rightarrow PbSO_{4(s)} + NaNO_{3(aq)} \qquad \text{(unbalanced)}$$
$$Na_2SO_{4(aq)} + Pb(NO_3)_{2(aq)} \rightarrow PbSO_{4(s)} + 2\,NaNO_{3(aq)} \qquad \text{(balanced)}$$

Example 2

Determine the products (if any) when a solution of sodium nitrate is mixed with a solution of potassium chloride. If a reaction occurs, summarize the reaction as a balanced chemical equation.

Solution

Double displacement reaction:

sodium nitrate + potassium chloride → sodium chloride + potassium nitrate

sodium chloride: soluble
potassium nitrate: soluble

sodium nitrate + potassium chloride → no reaction

▶ **Practice**

Understanding Concepts

1. Determine the products (if any). If a reaction occurs, summarize the reaction as a balanced chemical equation.
   (a) a solution of lead(II) nitrate mixed with a solution of sodium chloride
   (b) a solution of sodium sulfate mixed with a solution of calcium chloride
   (c) a solution of magnesium acetate mixed with a solution of silver nitrate
   (d) a solution of sodium acetate mixed with a solution of potassium chloride

Total Ionic and Net Ionic Equations

When a solution of sodium sulfate is combined with a solution of calcium chloride, calcium sulfate precipitate forms and sodium chloride remains in solution. The chemical reaction can be represented by the following equation:

$$Na_2SO_{4(aq)} + CaCl_{2(aq)} \rightarrow CaSO_{4(s)} + 2\,NaCl_{(aq)}$$

What this equation does not tell you is which ions are actually involved in the chemical reaction. A **net ionic equation** depicts only the ions that are involved in the chemical reaction. To write the net ionic equation for a given reaction, you must first determine the total ionic equation. A **total ionic equation** illustrates the separation of soluble ionic compounds into their respective ions.

In the sodium sulfate and calcium chloride reaction, all the compounds dissociated into anions and cations. The *total ionic equation* is

$$2\,Na^+_{(aq)} + SO^{2-}_{4(aq)} + Ca^{2+}_{(aq)} + 2\,Cl^-_{(aq)} \rightarrow CaSO_{4(s)} + 2\,Na^+_{(aq)} + 2\,Cl^-_{(aq)}$$

This equation indicates which ions are in solution and which ions are found in precipitate form. Notice that the sodium and chloride ions are found as both reactants and products. They have not changed their state during the reaction and appear exactly the same on both sides of the equation arrow. They are called **spectator ions** and can be crossed out:

$$2\,\cancel{Na^+_{(aq)}} + SO^{2-}_{4(aq)} + Ca^{2+}_{(aq)} + 2\,\cancel{Cl^-_{(aq)}} \rightarrow CaSO_{4(s)} + 2\,\cancel{Na^+_{(aq)}} + 2\,\cancel{Cl^-_{(aq)}}$$

net ionic equation an equation that depicts only the ions that are involved in a chemical reaction

total ionic equation a chemical equation that illustrates all soluble ionic compounds in their ionic form

spectator ion an ion that is present during a chemical reaction but does not participate in the reaction

LEARNING *TIP*

Subscripts and Coefficients
When a compound dissociates into ions, the number of ions is denoted by the subscript. When writing the total ionic equation, the subscript becomes the coefficient. For example, in $Na_2SO_{4(aq)}$, there are two sodium ions. Therefore, when this compound dissociates, it is represented as

$$2\,Na^+_{(aq)} + SO^{2-}_{4(aq)}$$

The remaining ions are the ions that participate in the chemical reaction and are directly responsible for the formation of the precipitate. For the reaction of sodium sulfate with calcium chloride, the *net ionic equation* is

$$Ca^{2+}_{(aq)} + SO^{2-}_{4(aq)} \rightarrow CaSO_{4(s)}$$

> **SAMPLE** problem 2

Writing Total and Net Ionic Equations

(a) Write the total ionic equation and net ionic equation for the reaction between barium sulfide and sodium sulfate.

Step 1: Identify Type of Reaction and Possible Products
Barium sulfide and sodium sulfate are both ionic compounds. Therefore, if a reaction takes place, it will be a double displacement reaction. Write the possible double displacement reaction as a word equation:

barium sulfide + sodium sulfate → barium sulfate + sodium sulfide

Step 2: Look Up Solubility of Both Products
barium sulfate: low solubility
sodium sulfide: soluble

Step 3: Indicate States of Reactants and Products

aqueous barium sulfide + aqueous sodium sulfate →
solid barium sulfate + aqueous sodium sulfide

Step 4: Write Chemical Equation for Reaction

$$BaS_{(aq)} + Na_2SO_{4(aq)} \rightarrow BaSO_{4(s)} + Na_2S_{(aq)}$$

Step 5: Balance Equation
The equation is already balanced.

$$BaS_{(aq)} + Na_2SO_{4(aq)} \rightarrow BaSO_{4(s)} + Na_2S_{(aq)}$$

Step 6: Write Total Ionic Equation
Using your knowledge of nomenclature and solubility, rewrite the equation with all the ionic compounds that are soluble in water in ionic form. The total ionic equation is

$$Ba^{2+}_{(aq)} + S^{2-}_{(aq)} + 2\,Na^{+}_{(aq)} + SO^{2-}_{4(aq)} \rightarrow BaSO_{4(s)} + 2\,Na^{+}_{(aq)} + S^{2-}_{(aq)}$$

Step 7: Write Net Ionic Equation
First cancel identical amounts of identical ions that appear on both sides of the equation:

$$Ba^{2+}_{(aq)} + \cancel{S^{2-}_{(aq)}} + \cancel{2\,Na^{+}_{(aq)}} + SO^{2-}_{4(aq)} \rightarrow BaSO_{4(s)} + \cancel{2\,Na^{+}_{(aq)}} + \cancel{S^{2-}_{(aq)}}$$

Then write the net ionic equation with the remaining ions and precipitate. Ensure that any coefficients are reduced, if applicable.

$$Ba^{2+}_{(aq)} + SO^{2-}_{4(aq)} \rightarrow BaSO_{4(s)}$$

(b) **Write the total ionic equation and net ionic equation for the reaction between ammonium hydroxide and sodium nitrate.**

Step 1: Identify Type of Reaction and Possible Products

Ammonium hydroxide and sodium nitrate are both ionic compounds. If a reaction takes place, it will be a double displacement reaction. Write the possible double displacement reaction as a word equation:

ammonium hydroxide + sodium nitrate →
ammonium nitrate + sodium hydroxide

Step 2: Look Up Solubility of Both Products

ammonium nitrate: soluble
sodium hydroxide: soluble

Step 3: Indicate States of Reactants and Products

aqueous ammonium hydroxide + aqueous sodium nitrate →
aqueous ammonium nitrate + aqueous sodium hydroxide

Step 4: Write Chemical Equation for Reaction

Using information from the solubility rules, write a balanced chemical equation. Indicate the states of the reactants and products in the reaction.

$$NH_4OH_{(aq)} + NaNO_{3(aq)} \rightarrow NH_4NO_{3(aq)} + NaOH_{(aq)}$$

Step 5: Balance Equation

Since the products of this reaction are aqueous, you can conclude that there is no reaction.

Step 6: Write Total Ionic Equation

Since there is no reaction, all entities in the equation dissociate into ions. Therefore, the total ionic equation is

$$NH_{4(aq)}^{+} + OH_{(aq)}^{-} + Na_{(aq)}^{+} + NO_{3(aq)}^{-} \rightarrow$$
$$NH_{4(aq)}^{+} + NO_{3(aq)}^{-} + Na_{(aq)}^{+} + OH_{(aq)}^{-}$$

Step 7: Write Net Ionic Equation

Cancel identical amounts of identical ions that appear on both sides of the equation:

$$\cancel{NH_{4(aq)}^{+}} + \cancel{OH_{(aq)}^{-}} + \cancel{Na_{(aq)}^{+}} + \cancel{NO_{3(aq)}^{-}} \rightarrow$$
$$\cancel{NH_{4(aq)}^{+}} + \cancel{NO_{3(aq)}^{-}} + \cancel{Na_{(aq)}^{+}} + \cancel{OH_{(aq)}^{-}}$$

Since there are no ions remaining (no reaction took place), there is no net ionic equation.

Example

Write the total ionic equation and net ionic equation for the reaction between aqueous sodium chloride and aqueous lead(II) nitrate.

Solution

Double displacement reaction:

sodium chloride + lead(II) nitrate → lead(II) chloride + sodium nitrate

lead(II) chloride: low solubility
sodium nitrate: soluble

aqueous sodium chloride + aqueous lead(II) nitrate →
solid lead(II) chloride + aqueous sodium nitrate

$NaCl_{(aq)} + Pb(NO_3)_{2(aq)} → PbCl_{2(s)} + NaNO_{3(aq)}$ (unbalanced)

$2\,NaCl_{(aq)} + Pb(NO_3)_{2(aq)} → PbCl_{2(s)} + 2\,NaNO_{3(aq)}$ (balanced)

$2\,Na^+_{(aq)} + 2\,Cl^-_{(aq)} + Pb^{2+}_{(aq)} + 2\,NO^-_{3(aq)} → PbCl_{2(s)} + 2\,Na^+_{(aq)} + 2\,NO^-_{3(aq)}$
(total ionic equation)

$2\,\cancel{Na^+_{(aq)}} + 2\,Cl^-_{(aq)} + Pb^{2+}_{(aq)} + 2\,\cancel{NO^-_{3(aq)}} → PbCl_{2(s)} + 2\,\cancel{Na^+_{(aq)}} + 2\,\cancel{NO^-_{3(aq)}}$

$Pb^{2+}_{(aq)} + 2\,Cl^-_{(aq)} → PbCl_{2(s)}$ (net ionic equation)

SUMMARY   Writing Net Ionic Equations

1. Write the double displacement reaction as a word equation.

2. Using the solubility rules, determine whether the products of the reaction have high or low solubility in water.

3. Using this information, indicate the states of the reactants and products.

4. Write a chemical equation for the reaction.

5. Balance the equation.

6. Rewrite the equation, with all the ionic compounds that are soluble in water separated into their respective ions. This equation is the total ionic equation. Cancel identical amounts of identical ions that appear on both the reactant and product sides of the equation.

7. Write the net ionic equation, reducing coefficients if necessary.

▶ Practice

Understanding Concepts

2. Write the total ionic equation and the net ionic equation for each of the following reactions:
   (a) the reaction between aqueous barium chloride and aqueous silver nitrate
   (b) the reaction between aqueous zinc chloride and aqueous lead(II) nitrate

Qualitative Chemical Analysis

The information in the solubility rules can be applied to create diagnostic tests for the presence of specific ions in aqueous solution. Chemists can use the solubility rules to determine the presence of certain ions in a solution by conducting double displacement reactions: they use solutions that contain ions that form a precipitate with the ions they wish to detect. For example, a chemist may suspect that a solution contains acetate ions, $C_2H_3O_{2(aq)}^-$. The solubility rules indicate that silver acetate, $AgC_2H_3O_{2(s)}$, has low solubility. If the chemist adds a source of silver ions, $Ag_{(aq)}^+$, to the solution and a precipitate forms, then the chemist might infer that acetate ions are present. The chemist must be aware, however, that other ions, such as $Cl_{(aq)}^-$, may be present in solution, which may also cause the acetate ions to precipitate. To rid the solution completely of acetate ions, the chemist can continue adding silver ions until no more precipitate forms.

Most solutions contain more than one type of ion. Therefore, chemists must design procedures to identify and remove any suspected ions one at a time. For example, a chemist may suspect that a solution contains both strontium ions, $Sr_{(aq)}^{2+}$, and iron(II) ions, $Fe_{(aq)}^{2+}$. In order to determine whether one or both types of ions are present, the chemist must design and carry out a procedure that precipitates each type of suspected ion out of solution one at a time. The solubility rules indicate that if hydroxide ions, $OH_{(aq)}^-$, are added to a solution that contains both strontium and iron(II) ions, the hydroxide ions will react with only the iron(II) ions. Therefore, the chemist may add a solution of sodium hydroxide, $NaOH_{(aq)}$. If iron(II) ions are present, they will react with the hydroxide ions, producing a precipitate according to the following net ionic equation:

$$Fe_{(aq)}^{2+} + 2\,OH_{(aq)}^- \rightarrow Fe(OH)_{2(s)}$$

If no precipitate forms, the chemist can infer that no iron(II) ions are present in the solution.

To ensure that all the iron(II) ions are removed from the solution so that they do not interfere with any future reactions, the chemist may continue adding sodium hydroxide solution until no more precipitate forms. The chemist must then remove the iron(II) hydroxide precipitate, $Fe(OH)_{2(s)}$, from the solution. One method is to use a centrifuge. A **centrifuge** is a piece of laboratory equipment that separates different substances in a solution based on their densities. A centrifuge spins the solution at very high speeds, causing higher-density (heavier) particles to sink to the bottom of the test tube faster than lower-density particles. During centrifugation, the heavier particles of iron(II) hydroxide precipitate settle to the bottom of the container while the **supernate**, which contains all the other aqueous ions, is left at the top of the container. After centrifugation, the iron(II) hydroxide precipitate can be separated from the remaining supernate by pouring the supernate into another container.

Once all the iron(II) ions have been removed from the solution as iron(II) hydroxide precipitate, the chemist can test for the presence of strontium ions. From the solubility rules, the chemist knows that sulfate ions, $SO_{4(aq)}^{2-}$, will react with strontium ions to form a precipitate. This reaction is represented by the following net ionic equation:

$$Sr_{(aq)}^{2+} + SO_{4(aq)}^{2-} \rightarrow SrSO_{4(s)}$$

If a precipitate forms, the chemist can infer the presence of strontium ions. The strontium ions can be removed from the solution using aqueous sodium sulfate, $Na_2SO_{4(aq)}$. The sulfate ions will react with the strontium ions.

centrifuge a piece of laboratory equipment that spins solutions at very high speeds, to separate the different particles from each other based on their densities

supernate the part of a centrifuged solution that does not settle to the bottom of the centrifuge tube

Chemists use solubility to test for the presence of specific ions, and also to remove ions from solution (**Figure 2**). In the next section, you will use the solubility rules to design and perform an experiment that allows you to determine which ions are present in a solution.

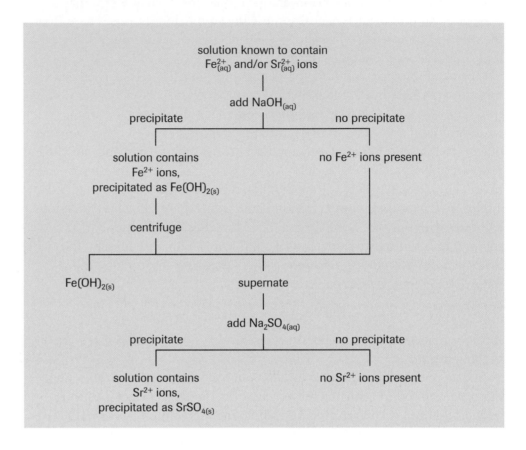

Figure 2
This flow chart summarizes the steps taken to determine whether $Sr^{2+}_{(aq)}$ and/or $Fe^{2+}_{(aq)}$ ions are present in solution.

▶ *Section 1.15 Questions*

Understanding Concepts

1. Using the solubility rules, determine whether each of the following compounds is soluble in water:
   (a) lead(II) sulfate
   (b) ammonium sulfide
   (c) silver nitrate
   (d) silver chloride
   (e) calcium carbonate
   (f) ammonium hydroxide
   (g) barium hydroxide

2. Using the solubility rules, determine whether a precipitate will form if each of the following pairs of compounds is mixed. If a precipitate forms, identify the precipitate and write its chemical formula.

   (a) strontium nitrate and sodium sulfate
   (b) sodium acetate and silver nitrate
   (c) barium nitrate and ammonium phosphate
   (d) sodium hydroxide and calcium nitrate

3. Write the total ionic equation and net ionic equation for each reaction in question 2.

Making Connections

4. Some pollutants in natural waters, such as heavy metals and organic compounds, are classified as having low solubility. What are the origins of these pollutants in natural waters? If these pollutants are low-solubility compounds, why are they a problem?

 www.science.nelson.com

Analyzing a Household Cleaning Product

Different household cleaning products (**Figure 1**) contain different ions that contribute to their cleaning ability. Two examples of ions that are present in common household cleaning products are ammonium ions and hydroxide ions. In this activity, you will use precipitation reactions to test for the presence of different ions in a household cleaning product.

Question

Which ions are present in the cleaning product?

Materials

microtray
eyedroppers
ionic (testing) solutions (chosen using the solubility rules)
cleaning product

(a) Using the solubility rules, determine which ions you will need to use, to test for the presence of the ions you wish to detect. (In other words, determine which ions will cause the ions in the cleaning product to precipitate.)

Procedure

1. Place three drops of the cleaning product in wells of the microtray. Use one well for each ion you are testing for. For example, if you are testing for three different ions, then set up three different wells.

2. From your teacher, obtain the solutions you have chosen to use, to test for the presence of ions in the cleaning product.

3. Using different eyedroppers, place three drops of one of the testing solutions in the first well with the cleaning product. Record your observations.

4. Repeat step 3 with the other testing solutions. Record your observations.

Analysis

(b) Answer the Question.

(c) Give reasons why the qualitative analysis of consumer products is necessary for maintaining product standards.

(d) Write a net ionic equation for each reaction that occurred.

Evaluation

(e) Evaluate your selection of testing solutions. Make any suggestions for improvements.

Figure 1

Determining the Presence of Ions in a Solution

Being able to identify the ions that are present in a solution is a valuable skill in chemistry. In this activity, you will follow a Procedure outlined in a flow chart to determine whether the chloride, $Cl^-_{(aq)}$, sulfate, $SO_4^{2-}_{(aq)}$, and ferrocyanide, $Fe(CN)_6^{4-}_{(aq)}$, ions are present in a solution. You will record whether or not a precipitate forms when these ions are mixed with solutions that contain silver, $Ag^+_{(aq)}$, barium, $Ba^{2+}_{(aq)}$, or zinc, $Zn^{2+}_{(aq)}$, ions.

Question

Which ions (chloride, sulfate, and/or ferrocyanide ions) does a solution contain?

Experimental Design

(a) Study the flow chart in **Figure 1**. Plan an Experimental Design to test for sulfate, chloride, and ferrocyanide ions based on the information provided in the flow chart.

Materials

eye protection
lab apron
protective gloves
silver nitrate solution, $AgNO_{3(aq)}$
barium nitrate solution, $Ba(NO_3)_{2(aq)}$
zinc nitrate solution, $Zn(NO_3)_{2(aq)}$
3 eyedroppers
centrifuge
solution containing unidentified ions
test-tube rack
centrifuge tubes

 Silver nitrate, barium nitrate, and zinc nitrate are toxic. Wear protective gloves, and avoid contact with the skin. Silver nitrate stains the skin. Silver nitrate, barium nitrate, and zinc nitrate are harmful if swallowed.

 Silver solutions are corrosive and must be kept away from the eyes and skin. Silver nitrate is harmful if swallowed. Wear eye protection, and do not rub your eyes.

 Dispose of all waste substances in a special container labelled "Heavy Metal Waste."

Procedure

(b) Write a Procedure for your Experimental Design. Specify the equipment you will use in each step of your Procedure. Include any necessary safety precautions. With your teacher's approval, carry out your Procedure.

Analysis

(c) Analyze your experimental results, and then answer the Question.

(d) Write the net ionic equation for each double displacement reaction that occurred.

Evaluation

(e) Suggest sources of error, and describe changes to your Procedure that would help to reduce these sources of error.

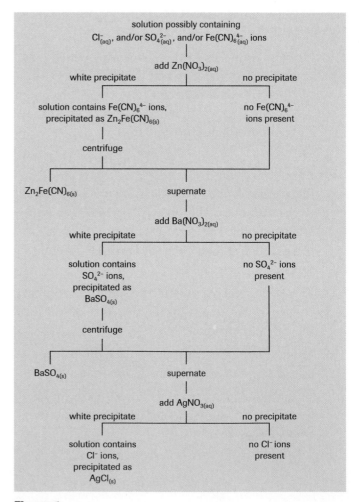

Figure 1

Key Understandings

1.1 Activity: Identifying a Mystery Powder

- Investigate different physical and chemical properties of matter that may be used to identify substances.
- Carry out qualitative analysis to identify a substance.

1.2 Building Scientific Knowledge

- Distinguish between observation and inference, using examples.
- Distinguish between empirical knowledge and theoretical knowledge, and their relationship to observation and inference.

1.3 Early Models of the Atom

- Gain an understanding of the tentative nature of scientific knowledge.
- Become familiar with Democritus', Dalton's, and Thomson's models of the atom.
- Conduct a "black box" activity, and relate it to the probing of an atom by a scientist.

1.4 The Electromagnetic Spectrum

- Distinguish between a continuous spectrum and a line spectrum.
- Explain the relationship between frequency and wavelength.

1.5 Activity: Identifying Gases Using Line Spectra

- Conduct a qualitative analysis investigation using a spectroscope.
- Describe an application of spectroscopy.
- Identify a gas sample by comparing its observed line spectrum with the line spectra of other gases.

1.6 The Bohr Model of the Hydrogen Atom

- Describe the Bohr model of the atom.
- Explain the difference between ground state and excited state.
- Relate observations from flame tests and absorption spectra to the concept of quanta of energy that was proposed by Bohr.

1.7 Activity: Flame Tests

- Describe how observations from flame tests can be used to identify a substance.
- Identify a metal using flame tests.

1.8 Case Study: Qualitative Analysis of Concrete

- Describe how qualitative analysis is used in the analysis of concrete.
- Describe how flame emission spectroscopy works.
- Describe the carbonation process in concrete.
- Describe an application of spectroscopy.

1.9 Tech Connect: Detecting Counterfeit Canadian Currency

- Describe the security features of Canadian currency.
- Describe some methods that are used to detect counterfeit Canadian currency.

1.10 Explore an Issue: MRI Clinics

- Describe how magnetic resonance imaging works and how it is used in a medical setting.
- Describe an application of spectroscopy.

1.11 The Formation of Ionic Compounds

- Demonstrate an understanding of the formation of ionic bonds between cations and anions, and relate the charge on an ion to the number of electrons lost or gained.
- Explain the difference between electrolytes and nonelectrolytes.
- Describe ionic bonding.
- Draw Lewis symbols for elements.

1.12 Covalent Bonding

- Use Lewis structures to explain covalent bonding in simple molecules.
- Describe the concept of electronegativity.
- Explain the difference between polar and nonpolar compounds.
- Draw Lewis structures of simple compounds.
- Describe intermolecular bonds.
- Use a molecular model kit to represent molecules.

1.13 Investigation: Classifying Solids Using Physical Properties

- Conduct an investigation to explore different physical and chemical properties of two types of solids.
- Classify an unidentified solid as ionic or molecular.

1.14 Chemical Reactions

- Define and classify synthesis, combustion, decomposition, single displacement, and double displacement reactions.
- Distinguish between a solute and a solvent.
- Describe and explain precipitation reactions.

1.15 Using Solubility Rules to Predict Precipitate Formation

- Predict the precipitate that will be formed in a chemical reaction by writing the equation for the double displacement reaction and the net ionic equation, and using the solubility rules.

- Understand how double displacement (precipitation) reactions are used in qualitative analysis.

1.16 Activity: Analyzing a Household Cleaning Product

- Select appropriate solutions to produce precipitates with suspected ions.
- Analyze a household chemical for the presence of specific ions.

1.17 Activity: Determining the Presence of Ions in a Solution

- Using a flow chart, write an experimental design and procedure to determine the presence of ions in a sample using precipitation reactions.
- Use scientific vocabulary to communicate ideas related to qualitative analysis.
- Conduct qualitative analysis investigations using a centrifuge.

Key Terms

1.2
observation
inference
empirical knowledge
theoretical knowledge
theory
model

1.3
atom
electron
nucleus
proton
neutron
isotope

1.4
electromagnetic energy
frequency
wavelength
nanometre
visible spectrum
continuous spectrum
line spectrum
spectroscope

1.6
quantized
excited state
ground state

1.8
alkali metal
flame emission
  spectroscopy
base
pH
indicator
acid

1.11
conductivity
electrolyte
nonelectrolyte
octet rule
ion
anion
cation
valence electrons
Lewis symbol
monatomic ion

polyatomic ion
ionic bond
ionic compound
ionic crystal
formula unit
dissociate
IUPAC

1.12
covalent bond
molecule
chemical entity
Lewis structure
lone pair
electronegativity
nonpolar covalent bond
polar covalent bond
polar molecule
nonpolar molecule
intermolecular bonds
dipole–dipole force
  (DDF)
London dispersion force
  (LDF)
van der Waals forces

1.14
chemical reaction
synthesis reaction
combustion reaction
decomposition reaction
enzyme
solution
solute
solvent
homogeneous mixture
aqueous solution
single displacement
  reaction
double displacement
  reaction
precipitate

1.15
solubility
net ionic equation
total ionic equation
spectator ion
centrifuge
supernate

Key Symbols and Equations

1.4
- λ
- nm

1.14
- synthesis reaction: $A + B \rightarrow AB$
- decomposition reaction: $AB \rightarrow A + B$
- single displacement reaction: $A + BC \rightarrow AC + B$
- double displacement reaction: $AB + CD \rightarrow AD + CB$

Problems You Can Solve

1.12
- Draw Lewis structures with single bonds.
- Draw Lewis structures with double bonds.
- Draw Lewis structures with triple bonds.
- Draw Lewis structures for molecules that consist of more than one type of atom.

1.15
- Predict the formation of a precipitate.
- Write total ionic equations and net ionic equations.

▶ *MAKE* a summary

In a table like **Table 1**, list the qualitative analysis techniques (such as a flame test) that were presented in this unit. Describe when each technique is used, and indicate whether it is based on a physical or chemical property of matter. In point form, outline the scientific theory on which each technique is based.

Table 1 Qualitative Analysis Techniques

Technique	Possible use	Physical or chemical property	Scientific theory

PERFORMANCE TASK

Identifying the Ions in a Mystery Solution

In many careers, the ability to identify matter on the basis of physical and chemical properties is very important. Astronomers try to determine the composition of distant stars, water quality analysts try to determine the contents of water samples, and forensic chemists conduct toxicology analyses of suspected poisons. In Unit 1, you have studied three qualitative analysis techniques: flame tests, spectroscopy (line spectra), and precipitation reactions. As well, you have probed matter for physical and chemical properties, such as conductivity, hardness, and solubility. Throughout the unit, you have acquired some theory to help you understand why matter behaves the way it does, producing the results you see in flame tests, spectroscopy, and precipitation reactions.

In this performance task, you will be given a mystery solution that may contain more than one type of ion. The ions that may be present are barium, $Ba^{2+}_{(aq)}$, silver, $Ag^{+}_{(aq)}$, and copper(I), $Cu^{+}_{(aq)}$. First you will confirm whether or not the solution contains any ions. Then you will try to identify the ions that are present using precipitation reactions with sodium chloride, $NaCl_{(aq)}$, sodium sulfate, $Na_2SO_{4(aq)}$, and sodium acetate, $NaC_2H_3O_{2(aq)}$, solutions. Finally, you will use flame tests to confirm the identity of any precipitates that form. From your observations, you will infer the contents of the solution.

Question

Which of the following ions are present in the mystery solution: barium, silver, and/or copper(I)?

Prediction

(a) Predict what you will observe if the mystery solution contains ions. Explain briefly.

(b) Based on the precipitation reactions you will use, predict what will happen if any barium, silver, or copper(I) ions are present in solution. Use **Figure 1** to help you formulate your Prediction. Write balanced chemical equations to support your Prediction.

(c) Predict the colours you will observe if precipitates that form are subjected to a flame test.

Experimental Design

(d) Design an experiment that will allow you to answer the Question. Your Experimental Design should consist of three parts. In the first part, you will investigate whether or not the solution contains ions. In the second part, you will determine which specific ions are present in the solution (barium, silver, and/or copper(I)) by following the Procedure outlined in **Figure 1**. In the third part, you will confirm the presence of these ions using flame tests. Only use experimental techniques that you have used in investigations and activities in this unit.

Figure 1

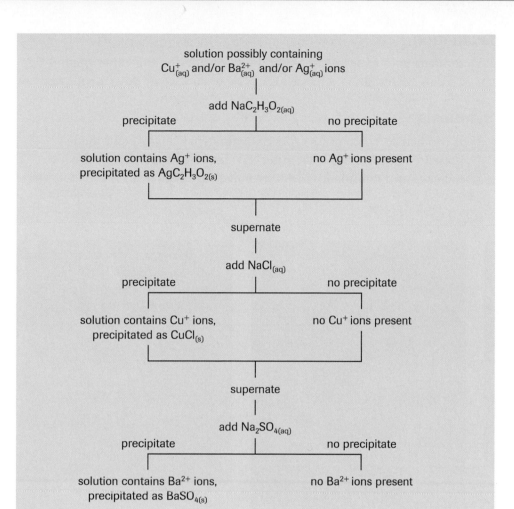

Materials

mystery solution
sodium acetate solution, $NaC_2H_3O_{2(aq)}$
sodium chloride solution, $NaCl_{(aq)}$
sodium sulfate solution, $Na_2SO_{4(aq)}$

(e) Complete the Materials list. Select equipment that is commonly available and familiar to you from previous experiments.

Procedure

(f) Write a step-by-step Procedure to carry out your Experimental Design. Include any necessary safety and disposal precautions.

(g) With your teacher's approval, carry out your Procedure. Record your observations at each step.

Analysis

(h) Analyze your observations, and then answer the Question.

Evaluation

(i) Evaluate your experiment by carefully considering the Experimental Design, Materials, and Procedure. How confident are you in your results?

Synthesis

(j) If you had access to other types of qualitative analysis equipment presented in this unit, what other tests could you perform in order to increase your confidence in your results? (See **Figure 2**.)

(a)

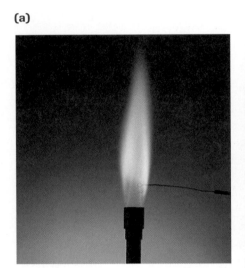

(b)

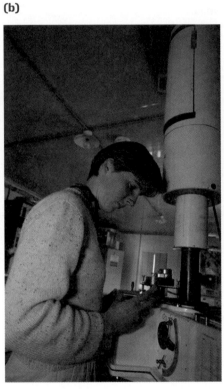

(c)

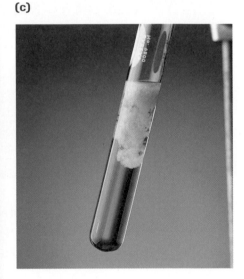

Figure 2
(a) Flame tests can be used to identify the composition of a substance. If a substance that contains copper ions is subjected to a flame test, the flame turns green.
(b) Spectrophotometry is a valuable tool for the analysis of matter. Many different types of spectrophotometric analysis are widely used in analytical chemistry.
(c) Ions can be removed from solution using a precipitation reaction. In the photo, solid iron(III) phosphate, $FePO_{4(s)}$, is formed when aqueous iron(III) nitrate, $Fe(NO_3)_{3(aq)}$, reacts with aqueous sodium phosphate, $Na_3PO_{4(aq)}$.

Understanding Concepts

1. Using examples, explain the difference between
   (a) inference and observation
   (b) empirical knowledge and theoretical knowledge
   (c) theory and model   (1.2)

2. The six terms in question 1 are interrelated. Create a concept map that illustrates the relationships among the six terms.   (1.2)

3. Many scientists have contributed to the development of the current model of the atom. Copy and complete **Table 1**, indicating the main contribution of each scientist.   (1.3)

Table 1   History of Atomic Theory

Scientist	Contribution to atomic theory
John Dalton	
J.J. Thomson	
Ernest Rutherford	
James Chadwick	
Niels Bohr	

4. Red light has a longer wavelength than blue light.
   (a) Which colour of light has a higher frequency?
   (b) Which colour of light has more energy? (1.4)

5. Explain why a rainbow is considered to be an example of a continuous spectrum.   (1.4)

6. Electrons can be found in either the ground state or the excited state.
   (a) Explain how an electron is promoted from the ground state to the excited state.
   (b) In which state does the electron possess more energy?
   (c) What happens when an electron returns to the ground state?   (1.6)

7. In your own words, explain what Bohr meant when he stated that the energy of an electron in an atom is quantized.   (1.6)

8. Describe how spectroscopy can be used to identify a gas.   (1.6)

9. "Matter emits light when it is subjected to a flame test." Explain this statement using Bohr's model of the atom.   (1.6)

10. Why may solutions that contain an ionic solute conduct electricity?   (1.11)

11. Draw a Lewis symbol for each of the following atoms and ions:
    (a) sodium ion
    (b) calcium ion
    (c) oxygen atom
    (d) phosphorus atom
    (e) neon atom
    (f) chloride ion   (1.11)

12. Identify which ions in question 11 are negatively charged and which ions are positively charged. Indicate how many electrons each atom has gained or lost when forming the ion.   (1.11)

13. Which of the following pairs of atoms would you expect to form ionic compounds? Give reasons for your answer.
    (a) sodium and fluorine
    (b) carbon and hydrogen
    (c) magnesium and chlorine   (1.11, 1.12)

14. Describe one similarity and one difference between a covalent bond and an ionic bond.
                                       (1.11, 1.12)

15. Draw a Lewis structure for the compound that consists of each of the following pairs of atoms:
    (a) hydrogen and nitrogen
    (b) oxygen and oxygen
    (c) hydrogen and oxygen
    (d) nitrogen and nitrogen   (1.12)

16. Why is water a polar molecule?   (1.12)

17. Why is carbon tetrachloride, CCl_4, a nonpolar molecule even though the C–Cl bond is polar?
                                           (1.12)

18. Why do nonpolar molecules composed of many atoms have higher melting points than nonpolar molecules composed of fewer atoms?   (1.12)

19. Classify each of the following reactions as a synthesis, decomposition, single displacement, or double displacement reaction:
   (a) $2\,Mg_{(s)} + O_{2(g)} \rightarrow 2\,MgO_{(s)}$
   (b) $Fe_{(s)} + CuSO_{4(aq)} \rightarrow FeSO_{4(aq)} + Cu_{(s)}$
   (c) $2\,NaHCO_{3(s)} \rightarrow Na_2CO_{3(s)} + H_2O_{(l)} + CO_{2(g)}$
   (d) $2\,NH_{3(g)} + H_2SO_{4(aq)} \rightarrow (NH_4)_2SO_{4(s)}$
   (e) $2\,KI_{(aq)} + Pb(NO_3)_{2(aq)} \rightarrow PbI_{2(s)} + 2\,KNO_{3(aq)}$
   (1.14)

20. Which reaction in question 19 could also be classified as a combustion reaction? (1.14)

21. Determine the products and write a balanced chemical equation to show the reaction (if any) that occurs when each pair of solutions is mixed. Use the solubility rules to help you predict precipitates.
   (a) sodium chloride and silver nitrate
   (b) copper(II) chloride and sodium nitrate
   (c) sodium sulfide and lead(II) nitrate
   (d) potassium hydroxide and ammonium chloride (1.15)

22. For any chemical reactions in question 21, write the total ionic equation and net ionic equation. (1.15)

23. Describe how each of the following procedures may be used in qualitative chemical analysis:
   (a) flame tests
   (b) spectroscopy
   (c) precipitation (1.15)

Applying Inquiry Skills

24. An unidentified substance appears on the surface of a city's water reservoir (**Figure 1**). What are some experimental techniques that could be used to help classify and identify the substance? (1.15)

25. Household cleaning products contain different chemicals. Obtain a cleaning product from your home, and list the ingredients. Using your knowledge of qualitative analysis techniques and the solubility rules, determine which ions could be used to test for the ions in the cleaning product. (1.16)

26. Molecular solids and ionic solids differ in many ways.
   (a) Given the information in **Table 2**, determine whether each solid is ionic or molecular.
   (b) What other tests could you conduct on solid A and solid B in order to support your classifications? Describe the expected results of these tests. (1.13)

Table 2   Molecular and Ionic Solids

Solid	Melting point (°C)	Boiling point (°C)	Conductivity in aqueous solution
A	776	1500	good
B	76	196	none

27. A forensic chemist is given samples of four unidentified solutions. The identities of these solutions could affect the outcome of a court case involving an electrocution. The chemist has reason to believe that the four solutions are sodium chloride, $NaCl_{(aq)}$, ethanol, $C_2H_5OH_{(aq)}$, hydrochloric acid, $HCl_{(aq)}$, and barium hydroxide (a base), $Ba(OH)_{2(aq)}$. The chemist designs an experiment to identify the chemicals. The chemist dissolves each sample in water and tests the solution with a conductivity apparatus and

Figure 1

litmus paper. (Acidic solutions turn blue litmus paper red. Basic solutions turn red litmus paper blue.) Complete the Analysis and Synthesis for the following lab report:

Question

What are the identities of the four substances, labelled 1 to 4 in **Table 3**?

Observations

Table 3   Evidence for Identifying Solutions

Solution	Electrical conductivity	Litmus
water	none	no change
1	high	no change
2	high	blue to red
3	none	no change
4	high	red to blue

Analysis

(a) Answer the Question.

Synthesis

(b) Why was the water that was used to prepare the solutions also tested?

(c) Which of the solutions (1, 2, 3, and/or 4) could have been involved in an electrocution?   (1.11, 1.12)

Making Connections

28. Some natural waters contain iron ions that affect the taste of the water and cause rust stains. Aeration converts any iron(II) ions into iron(III) ions. A basic solution (containing hydroxide ions) can then be added to produce a precipitate.
    (a) Write the net ionic equation for the reaction of aqueous iron(III) ions and aqueous hydroxide ions.
    (b) What separation method is most likely to be used during this water treatment process?   (1.15)

29. A heterogeneous mixture (a wet sample) of barium sulfate, $BaSO_{4(s)}$, is given to patients before an X-ray scan of the gastrointestinal tract. Barium sulfate is a white ionic compound with low solubility. Why is barium sulfate "sludge" safe to drink, even though barium ions are toxic?   (1.15)

30. Forensic scientists use many qualitative analysis techniques in order to identify a substance conclusively. Why would the use of just one qualitative analysis technique be open for questioning in a court of law?   (1.11)

31. The production and circulation of counterfeit money has increased dramatically over the last decade. Many governments, including the governments of Canada and the United States, have redesigned their currency to make it more difficult to counterfeit. Detection devices in stores now allow merchants to test for counterfeit money. Research the measures that are used to prevent the production of American counterfeit currency, and the measures that are used to detect it. Compare these measures with the measures that are used to prevent the counterfeiting of Canadian currency. Summarize your findings in paragraph form.   (1.9)

 www.science.nelson.com

32. Currently, in Ontario, there is a shortage of medical MRI technologists. The increasing demand for MRI by patients and doctors, together with the low number of qualified individuals who can perform MRI procedures, points to a healthy job market for medical MRI technologists. Research where MRI technologist programs are offered in Canada. Find out the courses taken in the program, and the duration of the program. Present your findings in paragraph form.   (1.10)

 www.science.nelson.com

Extension

33. Laundry detergents give better results when they are used with soft water rather than hard water. Today, many laundry detergents contain chemicals called zeolites, which soften the water. Research the difference between soft water and hard water. Also research how zeolites work to soften water. Write a paragraph to summarize your findings.

 www.science.nelson.com

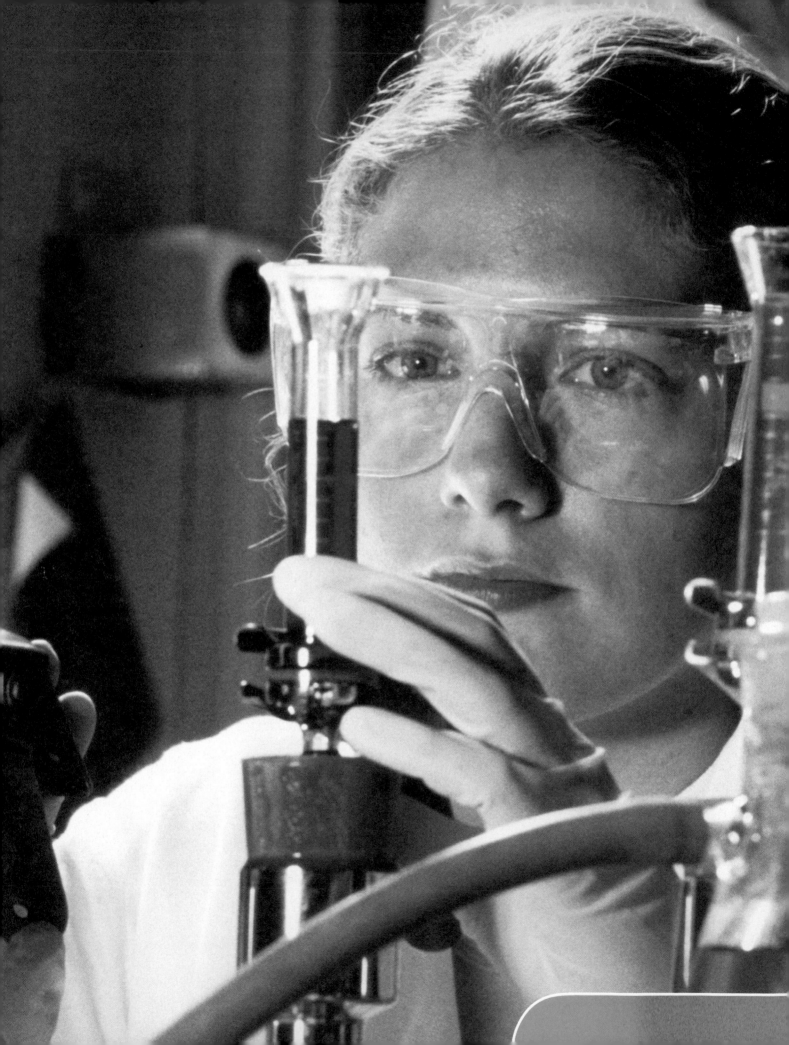

Quantities in Chemistry

Chemistry has greatly increased our understanding of the world around us. It is an exciting, ever-changing field of study that relies on the skills and creativity of many professionals.

Everything in the environment, whether natural or synthetic, is composed of chemicals. Large-scale chemical processes create many useful products, such as synthetic fibres, paints, drugs, adhesives, cosmetics, fertilizers, pesticides, and electronic components. Chemistry technicians specify the ingredients, mixing times, and temperatures of chemical manufacturing processes, and monitor reactions to ensure adequate product yields. Chemistry technicians also test samples of raw materials and finished products to ensure that they meet industry and government standards for purity and quality, including pollution standards.

One branch of chemistry, called analytical chemistry, is crucial in the pharmaceutical industry, where the molecular structures and chemical interactions of drugs must be well understood. Analytical chemists determine the composition of substances by identifying and describing the elements or compounds they are made of. Chemistry plays an important role in the criminal justice system, as well. Forensic chemists investigate homicides, thefts, fraud, arson, food poisoning, and environmental pollution, and produce scientific evidence that is used in court cases to determine if laws have been broken.

▶ Overall Expectations

In this unit, you will be able to

- demonstrate an understanding of the mole concept, as well as quantitative relationships in chemical reactions;
- use techniques of quantitative analysis in the preparation of standard solutions, and solve problems involving the analysis of quantities in chemical reactions, using both theoretical and experimentally measured quantities;
- explain the importance of quantitative chemical relationships in industry and in everyday life.

ARE YOU READY?

Knowledge and Understanding

1. Copy and complete **Table 1**.

Table 1 Chemical Elements and Compounds

IUPAC name	Chemical formula	Space-filling models	Element or compound
	O_2	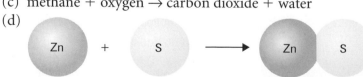	
ammonia			
			compound

2. Copy and complete **Table 2**.

Table 2 Ionic Compounds

Positive ion	Negative ion	Compound formula	IUPAC name
Na^+	Cl^-		
Ca^{2+}	Br^-		
Al^{3+}	S^{2-}		
Na^+	SO_4^{2-}		
NH_4^+	CO_3^{2-}		
K^+	ClO_3^-		
Cu^{2+}	PO_4^{3-}		

3. Classify each of the following reactions as a synthesis, decomposition, single displacement, double displacement, or combustion reaction:

(a)

(b) barium sulfide + potassium iodide ⟶ barium iodide + potassium sulfide

(c) methane + oxygen → carbon dioxide + water

(d)

4. A piece of magnesium is placed in a beaker that contains hydrochloric acid. Aqueous magnesium chloride and bubbles of hydrogen gas are produced. Write a word equation that represents this chemical change.

5. Balance each of the following chemical equations:
   (a) $Na_{(s)} + F_{2(g)} \rightarrow NaF_{(s)}$
   (b) $Al_{(s)} + S_{8(s)} \rightarrow Al_2S_{3(s)}$
   (c) $CO_{(g)} + H_{2(g)} \rightarrow C_7H_{14(l)} + H_2O_{(l)}$

6. Write a balanced symbol equation for each of the following word equations:
   (a) solid calcium + liquid water $\rightarrow$
   solid calcium hydroxide + hydrogen gas
   (b) aqueous lead(II) nitrate + aqueous potassium iodide $\rightarrow$
   solid lead(II) iodide + aqueous potassium nitrate

Inquiry and Communication

7. Match each WHMIS symbol in the first column of **Table 3** with the correct class and type of compound in the second column.

8. A student performs the following experiment:

Part 1: Empty flasks (121.08 g)   Part 2: Reactants A and B (219.17 g)   Part 3: Reaction   Part 4: Products in flask B

   (a) What evidence indicates that a chemical reaction occurs in Part 3?
   (b) Predict the final mass of flask B and its contents (Part 4).
   (c) Which law of chemistry is the student trying to test?
   (d) Suggest reasons why the total mass of the products (Part 4, flask B) could appear to be less than the total mass of the reactants (Part 2, flasks A and B).

Math Skills

9. Copy and complete **Table 4**.

10. Complete each of the following calculations:
    (a) Given $\dfrac{x}{y} = \dfrac{3}{2}$, find y when $x = 9$.
    (b) There are 436 boys and 657 girls in a school. Calculate the percentage of girls in the school.
    (c) $\dfrac{5.42 \times 10^8 \text{ g}}{7.34 \text{ mL}} = ?$

11. Graph the data in **Table 5** by plotting volume on the x-axis and mass on the y-axis. Draw a line of best fit. Remember to label the axes and include a title for your graph.

Table 3 Matching HMIS Symbols

Symbol	Class and type of compound
	Class B: Flammable and Combustible Materials
	Class C: Oxidizing Materials
	Class D: Toxic Materials Immediate and Severe
	Class F: Dangerously Reactive Materials

Table 4 Converting Measurements

Decimal notation	Scientific notation
0.010 m	
	4.01×10^2 mL
385.5 g	

Table 5 Data for Graph

Mass (g)	Volume (mL)
1.0	1.9
2.0	3.4
4.0	5.6
5.0	7.8
7.0	10.0
9.0	12.3

Chemistry has come a long way since the time when alchemists tried to turn common metals into gold (**Figure 1**). Early chemistry involved activities such as brewing, pottery, metallurgy (the production of metals), embalming, and the formulation of folk medicines and ointments. Modern chemistry is defined as the systematic study of matter and its changes. Chemistry is a systematic study because chemists use logical methods to solve problems. Chemists also use well-defined techniques, such as spectroscopy, to determine the structure of molecules and compounds and to measure the concentrations of substances in solutions.

Most of the matter on Earth is in the form of complex mixtures. Air, water, ink, gasoline, and cosmetics are all mixtures of elements and compounds. Separating a mixture into pure substances and then determining the identities and quantities of the pure substances in the mixture are some of the most important tasks of an analytical chemist.

In 1906, botanist Mikhail Tswett found that he could separate the different pigment molecules in autumn leaves by grinding up the leaves, adding a solvent, and then letting the mixture move through a tube full of powdered chalk (calcium carbonate). The various pigments separated into coloured bands. Tswett carefully removed the column of chalk from the tube and divided the chalk into individual sections, each with a different-coloured band. He used these sections to produce individual solutions of the different pigments. He called his new separation technique "chromatography" because it produced brightly coloured bands in the column. Although Twsett won

Figure 1
In medieval times, alchemists combined practical chemistry with magic to try to transform common metals into gold and to discover a tonic for longer life.

awards for his new separation technique, few scientists at the time considered it useful. Little did they know that chromatography would become one of the most powerful analytical tools in chemistry (**Figure 2**).

Also in the early 1900s, J.J. Thomson (best known for his discovery of the electron) invented the first mass spectrometer. A mass spectrometer is an instrument that can determine the mass of an element or a compound. One of the most useful instruments in modern chemistry, called the gas chromatograph–mass spectrometer (GC–MS), is a combination of a chromatograph and a mass spectrometer. A GC–MS can be used to separate, purify, identify, and quantify the components of a complex mixture of chemicals to a very high degree of accuracy and precision. It has been used to determine the types and quantities of pollutants in the environment, the concentrations of drugs in blood, and the chemical composition of moon rocks and meteorites. The GC–MS is just one of the many analytical instruments that are used in chemistry today.

Figure 2
Today, scientists use high pressure liquid chromatography (HPLC) to separate mixtures of molecules. The solution is forced through a column that contains microscopic beads. The components are absorbed by the beads to various extents, and the solution travels up the column at different rates, allowing the scientist to separate and identify the components of the mixture.

▶ TRY THIS activity — *The Mineral Content of Milk*

Milk is an important source of calcium and other minerals. Before milk reaches the stores, however, it needs to be homogenized and pasteurized so that it is safe for you to drink.

In homogenization, the butterfat particles are broken up to make the milk creamier. If milk was not homogenized, you would have to stir or shake it every time before drinking it. The milk is then pasteurized—a process named after French scientist Louis Pasteur. In 1865, Pasteur discovered that heating liquids to high temperatures kills bacteria.

In this activity, you will determine if processing affects the mineral content of milk, especially the amount of calcium.

Materials: evaporating dish, clay triangle support for dish, ring, crucible tongs, Bunsen burner, 10 mL whole milk, 10 mL skim milk

1. Determine the mass of the evaporating dish.

2. Add 10 mL of whole milk to the evaporating dish. Over a low flame, gently heat the milk until it is reduced to a black material.

 If foaming occurs, quickly and carefully remove the heat until the reaction subsides. Continue heating gently.

3. Heat the black residue with a roaring flame. Continue heating until only white ashes remain.

4. Cool the evaporating dish and contents for at least 5 min. Determine the mass of the evaporating dish and its contents.

5. Repeat steps 1 to 4 using skim milk.

(a) What difference, if any, is there between the mineral content (white ash) of skim milk and the mineral content of whole milk?

(b) What conclusions can you make, based on this activity?

In Unit 1, you learned that atoms contain protons and neutrons in the nucleus and that electrons surround the nucleus. Neutrons have approximately the same mass as protons, and protons and neutrons each have approximately 2000 times the mass of an electron. Thus, the mass of the electrons in an atom is relatively insignificant. The masses of the protons and neutrons (called *nucleons*) determine the mass of an atom.

Determining the Mass of Atoms

The masses of everyday objects—such as books, butter, and people—are determined by comparing them with a standard mass, called the kilogram. The mass of an object is obtained by placing it on one arm of an equal-arm balance, and placing enough kilograms on the other side to achieve a balance. Atoms and molecules are far too small for their masses to be determined by comparison with the kilogram. Instead, the mass of an atom is determined by comparison to a suitable *atomic mass standard*. In the past, chemists used the hydrogen atom as the standard because it is the smallest atom known, consisting of one proton and one electron. Thus, the hydrogen atom's mass is equal to the mass of one proton because, as already mentioned, the mass of an atom is essentially equal to the mass of its nucleons (protons and neutrons) only. A typical atom of sulfur has 16 protons and 16 neutrons in its nucleus, so it is 32 times heavier than a typical hydrogen atom. Thus, if the mass of a hydrogen atom is set at 1, the relative atomic mass of sulfur is 32.

For various technical and historical reasons, hydrogen is no longer used as the basis of the relative atomic mass scale for atoms. Instead, atomic masses are now based on the element carbon. Unfortunately, a sample of carbon consists of a mixture of isotopes containing 98.9% carbon-12 and 1.1% carbon-13. These percentages are known as the **isotopic abundance** of carbon.

The current system for determining the mass of all atoms is based on the use of an atom of the carbon-12 isotope as a standard. Carbon-12 atoms have been assigned a mass that is exactly equal to 12 (12.00). The **unified atomic mass unit (u)**, commonly called the atomic mass unit, is defined as the mass of one-twelfth of a carbon-12 atom. The other isotope of carbon, carbon-13, has an atomic mass of approximately 13 u. Since natural carbon consists of a mixture of these two isotopes, the relative atomic mass of carbon (the value given in the periodic table) is calculated as follows:

isotopic abundance the relative quantities of isotopes in a natural sample of an element, expressed as percentages

unified atomic mass unit (u) the mass of one-twelfth of a carbon-12 atom

$$m_C = \text{(percent abundance carbon-12} \times \text{atomic mass carbon-12)} +$$
$$\text{(percent abundance carbon-13} \times \text{atomic mass carbon-13)}$$
$$= (0.989 \times 12 \text{ u}) + (0.011 \times 13 \text{ u})$$
$$= 11.87 \text{ u} + 0.14 \text{ u}$$
$$m_C = 12.01 \text{ u}$$

Therefore, the atomic mass of carbon is 12.01 u. The masses of all other elements in the periodic table have been calculated in terms of the atomic mass unit. They are known as *relative atomic masses.*

Atomic Mass and Molecular Mass

The **atomic mass** of an element is the mass of one atom of the element, expressed in atomic mass units. The values of the atomic masses of all elements may be obtained from the periodic table. Thus, the atomic mass of hydrogen is 1.01 u, and the atomic mass of copper is 63.55 u. The **molecular mass** of a molecule is the mass of one molecule in atomic mass units. It is obtained by adding the atomic masses of all the atoms in the molecule. Thus, the molecular mass of water is calculated as follows:

$$m_{H_2O} = 2(m_H) + 1(m_O)$$
$$= 2(1.01 \text{ u}) + 1(16.00 \text{ u})$$
$$m_{H_2O} = 18.02 \text{ u}$$

One molecule of water has a mass of 18.02 u.

Since ionic compounds are represented by formula units, the **formula unit mass** of an ionic compound is obtained by adding the masses of all ions in the formula unit of the compound. Since electrons are considered to be massless in these calculations, we assume that the mass of an ion is equal to the mass of its corresponding neutral atom. The formula unit mass of calcium chloride, $CaCl_2$, is calculated as follows:

$$m_{CaCl_2} = 1(m_{Ca^{2+}}) + 2(m_{Cl^-})$$
$$= 1(40.08 \text{ u}) + 2(35.45 \text{ u})$$
$$m_{CaCl_2} = 110.98 \text{ u}$$

One formula unit of calcium chloride has a mass of 110.98 u.

atomic mass the mass of one atom of an element, expressed in atomic mass units, u

molecular mass the mass of one molecule, expressed in atomic mass units, u

formula unit mass the mass of one formula unit of an ionic compound, expressed in atomic mass units, u

---

▶ *Practice*

Understanding Concepts

1. Calculate the relative atomic mass of hydrogen if the relative abundance of hydrogen-1 is 99.985% and its mass is 1.01 u, and the relative abundance of hydrogen-2 is 0.015% and its mass is 2.01 u. (*Note:* There is a third isotope of hydrogen, called hydrogen-3, but its relative abundance is insignificant.)

2. Define the atomic mass unit.

3. Calculate the molecular mass or formula unit mass of each of the following compounds:
   (a) hydrogen bromide, $HBr_{(g)}$
   (b) glucose, $C_6H_{12}O_{6(s)}$
   (c) sodium hydrogen carbonate, $NaHCO_{3(s)}$
   (d) ammonium phosphate, $(NH_4)_3PO_{4(s)}$

Answers

1. 1.01 u

3. (a) 80.91 u
   (b) 180.18 u
   (c) 84.01 u
   (d) 149.12 u

Making an Omelette versus Making Water

Chemists, like bakers, chefs, and automakers, combine ingredients to produce useful products. In each case, it is important to know the quantities of materials used and produced. A chef who is preparing an omelette by mixing eggs, mushrooms, and peppers can be compared to a chemist who is preparing water by mixing hydrogen gas and oxygen gas (**Figure 1**). Both processes require raw materials and an input of energy.

A recipe book instructs a chef to make an omelette by mixing two eggs, three mushrooms, and one pepper, and heating the mixture until it is cooked. The following word equation describes the omelette-making process:

$$2 \text{ eggs} + 3 \text{ mushrooms} + 1 \text{ pepper} \xrightarrow{\text{heat}} 1 \text{ omelette}$$

Notice that the omelette contains the two eggs, three mushrooms, and pepper that are used to make it. The equation for making an omelette is balanced.

A chemist produces water by heating a mixture of hydrogen gas and oxygen gas. The following chemical equation describes this process:

$$2 \text{ H}_{2(g)} + \text{O}_{2(g)} \xrightarrow{\text{heat}} 2 \text{ H}_2\text{O}_{(l)}$$

The equation for water is balanced: all the atoms in the reactants exist in the products. For example, the two oxygen atoms in the oxygen molecule, $O_{2(g)}$, are found in the two water molecules, $2 \text{ H}_2\text{O}_{(l)}$. The four hydrogen atoms in the two hydrogen molecules, $2 \text{ H}_{2(g)}$, are also found in the two water molecules, $2 \text{ H}_2\text{O}_{(l)}$. Therefore, the total mass of the reactants ($2 \text{ H}_{2(g)}$ and $O_{2(g)}$) equals the total mass of the products ($2 \text{ H}_2\text{O}_{(l)}$).

One of the great advantages that chefs have over chemists is that chefs can see and count many of the ingredients they use in their recipes. Eggs, mushrooms, and peppers are all **macroscopic**: they can be seen with unaided eyes. Atoms and molecules, however, are submicroscopic. Even the largest molecules cannot be seen clearly with the most powerful microscopes. Nevertheless, chemists have developed methods that allow them to predict, with great accuracy, the numbers of entities (atoms, ions, or molecules) that take part in chemical reactions.

Grouping Entities

Since atoms and molecules are so small, chemists always work with extremely large numbers of chemical entities. Sometimes, chefs also work with large numbers of entities, especially when they are preparing food for a large number of people. A chef can use the basic recipe (equation) for making one omelette to describe the recipe for making two, 12, or any number of omelettes (**Table 1**).

The entities that are used to make 12 omelettes can be expressed as multiples of a dozen (**Table 2**). Notice that the coefficients in the three equations are the same whether you are dealing with individual entities or groups of entities, such as dozens.

(a)

(b)

Figure 1
(a) A chef combines ingredients to make products such as omelettes, cakes, and soups.
(b) A chemist combines reactants to make products such as medicines, plastics, and synthetic fibres.

macroscopic large enough to be seen with unaided eyes

Table 1 Analyzing the Coefficients of a Basic Recipe

Number of omelettes	Recipe
1	2 eggs + 3 mushrooms + 1 pepper → 1 omelette
2 (multiply by 2)	4 eggs + 6 mushrooms + 2 peppers → 2 omelettes
12 (multiply by 12)	24 eggs + 36 mushrooms + 12 peppers → 12 omelettes

Table 2 Grouping Entities: Dozens of Omelettes

Individual entities	2 eggs + 3 mushrooms + 1 pepper → 1 omelette
Multiples of 12	2 (12) eggs + 3 (12) mushrooms + 1 (12) peppers → 1 (12) omelettes
Multiples of a dozen	2 dozen eggs + 3 dozen mushrooms + 1 dozen peppers → 1 dozen omelettes

The equation for the formation of water can also be interpreted in terms of individual entities and groups of entities (**Table 3**). Again, notice that the coefficients are the same whether you are dealing with individual entities or groups of entities.

Table 3 Grouping Entities: Dozens of Water Molecules

Individual entities	$2\,H_{2(g)} + 1\,O_{2(g)} \rightarrow 2\,H_2O_{(l)}$
Multiples of 12	$2\,(12)\,H_{2(g)} + 1\,(12)\,O_{2(g)} \rightarrow 2\,(12)\,H_2O_{(l)}$
Multiples of a dozen	2 dozen $H_{2(g)} + 1$ dozen $O_{2(g)} \rightarrow 2$ dozen $H_2O_{(l)}$

Making one dozen omelettes is possible because you can isolate and count two dozen eggs or one dozen peppers. Molecules are so small, however, that it is impossible to isolate and count a dozen or two of them. What is a reasonable number of atoms or molecules for chemists to work with?

The Mole and Molar Mass

In 1811, a chemist named Amedeo Avogadro (**Figure 2**) realized that any convenient macroscopic quantity of matter must contain an enormous number of chemical entities (individual atoms, ions, formula units, or molecules). Using Avogadro's ideas, chemist Josef Loschmidt showed that approximately 602 000 000 000 000 000 000 000, or 6.02×10^{23}, is a convenient and measurable number of entities to deal with in chemistry. This quantity is called the mole. One **mole** of any entity is 6.02×10^{23} of the entity. Therefore, one mole of oxygen atoms is 6.02×10^{23} atoms of oxygen, one mole of sodium ions is 6.02×10^{23} sodium ions, and one mole of water molecules is 6.02×10^{23} molecules of water. The number 6.02×10^{23} is also called **Avogadro's constant (N_A)**, in honour of Amedeo Avogadro. The SI symbol for the mole is "mol." The word "amount" specifically refers to the number of moles of a chemical entity. **Figure 3** shows 1 mol of various elements and compounds.

Why is 6.02×10^{23} a more significant number than any other very large number of chemical entities? The results of many experiments show that 6.02×10^{23} atoms of any element (one mole of atoms) have a mass, in grams, that is equal to the numerical value of the element's atomic mass (in relative

Figure 2
Amedeo Avogadro (1776–1856)

mole 6.02×10^{23} entities

Avogadro's constant, N_A the number of entities in one mole; 6.02×10^{23}

Figure 3
These quantities of sugar, $C_{12}H_{22}O_{11(s)}$, table salt, $NaCl_{(s)}$, and carbon, $C_{(s)}$, each contain about one mole of entities. The mole is a convenient way to represent a specific quantity of a chemical.

atomic mass units, u). For example, one atom of carbon has a mass of 12.01 u, and one mole of carbon atoms has a mass of 12.01 g. One atom of iron has a mass of 55.85 u, and one mole of iron atoms has a mass of 55.85 g. Thus, the numerical value of an element's atomic mass (in u) is equal to the numerical value of the mass (in grams) of one mole (6.02×10^{23}) of the element's atoms. This similarity would not be true for any other number of atoms but Avogadro's number.

molar mass the mass, in grams, of one mole of a chemical entity; symbol M

The mass, in grams, of one mole of a chemical entity is called the **molar mass** (M), and it is measured in units of grams per mole, or g/mol. In other words, molar mass is equal to an entity's mass (in grams) divided by its amount (in moles), or molar mass $= \dfrac{\text{mass}}{\text{amount}}$ ($M = \dfrac{m}{n}$).

You may obtain the molar mass of an element directly from a periodic table. For example, one mole of hydrogen atoms (6.02×10^{23} H atoms) has a mass of 1.01 g. Thus,

$$M_{\text{H}} = 1.01 \text{ g/mol}$$

One mole of calcium atoms (6.02×10^{23} Ca atoms) has a mass of 40.08 g. Thus,

$$M_{\text{Ca}} = 40.08 \text{ g/mol}$$

Figure 4
One mole of water, 6.02×10^{23} molecules of water (approximately 18.02 g), is a specific and reasonable amount of water.

The Mole Is a Reasonable Quantity

The mole is a reasonable quantity in chemistry because a mole of any chemical entity can be observed and measured. One dozen (12) water molecules is an unreasonable quantity because it is too small a quantity to see or measure. One mole (6.02×10^{23}) of water molecules (**Figure 4**), however, is a large enough quantity to work with. The mole is not a reasonable quantity when dealing with macroscopic entities. For example, one mole of eggs (6.02×10^{23} eggs)contains enough eggs to cover the entire surface of Earth to a depth of approximately 60 km! Obviously, one mole of eggs is an unreasonably large number of eggs (**Figure 5**).

Moles and Chemical Equations

You can now express the equation for making water in terms of moles of entities (**Table 4**).

Table 4   Grouping Entities: Moles of Water Molecules

Individual entities	$2\,H_{2(g)} + 1\,O_{2(g)} \rightarrow 2\,H_2O_{(l)}$
Multiples of 6.02×10^{23}	$2\,(6.02 \times 10^{23})\,H_{2(g)} + 1\,(6.02 \times 10^{23})\,O_{2(g)} \rightarrow 2\,(6.02 \times 10^{23})\,H_2O_{(l)}$
Multiples of a mole	$2\,\text{mol}\,H_{2(g)} + 1\,\text{mol}\,O_{2(g)} \rightarrow 2\,\text{mol}\,H_2O_{(l)}$

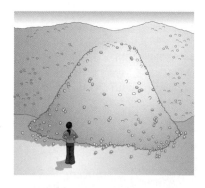

Figure 5
One mole of eggs would cover the entire surface of Earth to a depth of over 60 km.

The Molar Mass of Molecules and Ionic Compounds

The molar mass of a molecule, or of the formula unit of an ionic compound, is equal to the sum of the molar masses of the atoms in the molecule or ions in the ionic compound's formula unit. Therefore, the molar mass of water, M_{H_2O}, is equal to the sum of the molar mass of two hydrogen atoms and the molar mass of one oxygen atom. Thus,

$$M_{H_2O} = 2(M_H) + 1(M_O)$$
$$= 2(1.01\ g) + 1(16.00\ g)$$
$$M_{H_2O} = 18.02\ g$$

The molar mass of water is 18.02 g.

Similarly, the molar mass of sodium chloride, NaCl, is equal to the sum of the molar mass of one sodium ion and the molar mass of one chloride ion. Thus,

$$M_{NaCl} = M_{Na} + M_{Cl}$$
$$= 22.99\ g + 35.45\ g$$
$$M_{NaCl} = 58.44\ g$$

The molar mass of sodium chloride is 58.44 g.

▶ **Practice**

Understanding Concepts

4. Calculate the molar mass of each of the following molecules or formula units:
   (a) C_6H_6
   (b) $Ba(NO_3)_2$
   (c) K_3PO_4
   (d) NH_3

Answers

4. (a) 78.12 g
   (b) 261.35 g
   (c) 212.27 g
   (d) 17.04 g

When expressing the molar masses of elements, it is important to distinguish between elements that are composed of single atoms, and elements that are composed of molecules containing two or more atoms. For example, in nature, oxygen is found as $O_{2(g)}$ molecules. The molar mass of oxygen *atoms* (O) is 16.00 g/mol (from the periodic table). The molar mass of oxygen *molecules* (O_2) is 32.00 g/mol (two times the mass of an oxygen atom). When working with molecular elements, such as O_2, Cl_2, and S_8, one mole of the element usually means one mole of molecules. To avoid confusion, it is best to specify whether you are expressing moles of atoms or moles of molecules.

The Periodic Table and the Mass of Elements

In the periodic table (**Figure 6**), the symbol for each element has two numbers associated with it: the atomic number and the element's mass. The atomic number is usually written near the top of the symbol, and the element's mass is usually written near the bottom. The atomic number of an element is equal to the number of protons in the nucleus of an atom of the element. The number of protons determines the distinguishing characteristics of an

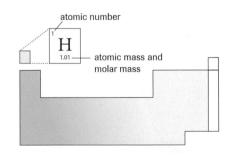

Figure 6
The atomic number and the element's mass are given in the periodic table. The mass value may be interpreted as atomic mass in atomic mass units or molar mass in grams. Thus, the atomic mass of hydrogen is 1.01 u, and the molar mass of hydrogen is 1.01 g.

element. The numerical value given as the element's mass in the periodic table may be interpreted as the element's atomic mass (the mass of one atom measured in atomic mass units) or its molar mass (the mass of one mole of atoms measured in grams).

As you will see in the following Sample Problem, you may calculate the molar mass of an entity (atom, ion, molecule, or formula unit) if you know the mass (in grams) of a given amount (in moles).

▶ SAMPLE problem 1

Calculating Molar Mass from Amount

Calculate the molar mass of magnesium, M_{Mg}, if 0.330 mol has a mass of 8.02 g.

From the definition of molar mass,

$$\text{molar mass} = \frac{\text{mass}}{\text{amount}}$$

$$M_{Mg} = \frac{8.02 \text{ g}}{0.330 \text{ mol}}$$

$$M_{Mg} = 24.3 \text{ g/mol}$$

The molar mass of magnesium is 24.3 g/mol.

Example

Calculate the molar mass of strontium if 1.65 mol has a mass of 144.57 g.

Solution

$$\text{molar mass} = \frac{\text{mass}}{\text{amount}}$$

$$M_{Sr} = \frac{144.57 \text{ g}}{1.65 \text{ mol}}$$

$$M_{Sr} = 87.6 \text{ g/mol}$$

The molar mass of strontium is 87.6 g/mol.

▶ Practice

Understanding Concepts

5. Calculate the molar mass of tin, if 1.010 mol has a mass of 119.88 g.

6. Calculate the molar mass of platinum, if a 49.74-g sample contains 0.2550 mol of platinum.

7. Calculate the molar mass of an element, if 2.220 mol has a mass of 26.66 g. Identify the element.

Answers

5. 118.7 g/mol

6. 195.1 g/mol

7. 12.01 g/mol; carbon

Calculating Number of Entities from Mass

Since chemists cannot count atoms or molecules one at a time, they must "count" them indirectly by measuring out a certain mass of substance and calculating the number of moles of chemical entities contained. Industries use a similar method to estimate the number of objects in a large collection of objects. The mass of a small number of objects is determined first, and this value is used to estimate the total number of objects in the whole collection. For example, knowing that 15 jellybeans (**Figure 7**) have a mass of 10.0 g allows a candy company to estimate the number of jellybeans in a box of jellybeans, as the following Sample Problem shows.

Figure 7
A balance is used to determine the mass of a small number of jellybeans. This mass is used to estimate the total number of jellybeans in a box.

> ▶ **SAMPLE** problem **2**
>
> ### Calculating the Number of Entities from Mass
>
> **How many jellybeans are in a box that contains 250 g of jellybeans, if 15 jellybeans have a mass of 10.0 g?**
>
> The following steps show how to solve this problem using the factor-label method:
>
> **Step 1: Identify Key Value and Conversion Factor Equation**
>
> mass of 1 box of jellybeans = 250 g (key value)
>
> mass of 15 jellybeans = 10.0 g (conversion factor equation)
>
> **Step 2: Identify Required Value**
> The required value is the number of jellybeans in the box.
>
> **Step 3: List Possible Conversion Factors**
> Using the conversion factor equation, the possible conversion factors are
>
> $$\frac{15 \text{ jellybeans}}{10.0 \text{ g}} \quad \text{or} \quad \frac{10.0 \text{ g}}{15 \text{ jellybeans}}$$
>
> **Step 4: Substitute Values into Solution Equation, and Solve**
> When substituting, remember that the units in the *denominator* of the conversion factor must be the same as the units of the key value.
>
> $$\text{required value} = \text{key value} \times \text{conversion factor}$$
>
> $$\text{number of jellybeans in box} = 250 \text{ g} \times \frac{15 \text{ jellybeans}}{10.0 \text{ g}}$$
>
> $$\text{number of jellybeans in box} = 375 \text{ jellybeans}$$
>
> There are 375 jellybeans in the box that contains 250 g of jellybeans.

Example 1

How many kernels of popcorn are in a bag that contains 454 g of unpopped popcorn, if one kernel has a mass of 0.110 g?

Solution

mass of 1 bag of popcorn = 454 g (key value)

mass of 1 popcorn kernel = 0.110 g (conversion factor equation)

$$\text{number of popcorn kernels} = 454 \text{ g} \times \frac{1 \text{ kernel}}{0.110 \text{ g}}$$

$$\text{number of popcorn kernels} = 4.13 \times 10^3 \text{ kernels}$$

There are 4.13×10^3 kernels of popcorn in a bag that contains 454 g of unpopped popcorn.

Example 2

How many dozen thumbtacks are in a box that contains 48.0 g of thumbtacks, if one dozen thumbtacks has a mass of 4.0 g?

Solution

mass of 1 box of thumbtacks = 48.0 g (key value)

mass of 1 dozen thumbtacks = 4.0 g (conversion factor equation)

$$\text{dozens of thumbtacks in box} = 48 \text{ g} \times \frac{1 \text{ dozen thumbtacks}}{4.0 \text{ g}}$$

$$\text{dozens of thumbtacks in box} = 12 \text{ dozen}$$

There are 12 dozen thumbtacks in a box that contains 48.0 g of thumbtacks.

▶ Practice

Understanding Concepts

Answers

8. 500 cotton swabs
9. 900 rubber bands
10. 31 dozen playing cards

8. How many cotton swabs are in a package that contains 240 g of cotton swabs, if one cotton swab has a mass of 0.480 g?

9. How many rubber bands are in a package that contains 729 g of rubber bands, if one rubber band has a mass of 0.81 g?

10. How many dozen playing cards are in a box that contains 558 g of playing cards, if one playing card has a mass of 1.50 g?

You can also use the factor-label method to determine the number or amount of chemical entities in a sample of pure matter, if you know the mass of the sample and the mass of a certain number of entities. Remember that *amount of chemical entities* specifically refers to the number of moles of chemical entities.

Symbols are used to represent amount in moles, mass, molar mass, number of entities, and Avogadro's constant. **Table 5** summarizes the quantity symbols and units that you will use in calculations throughout this unit.

Table 5   Quantity Symbols and Units

Symbol	Quantity	Unit
n	amount (in moles)	mol
m	mass	mg, g, kg
M	molar mass	g/mol
N	number of entities	atoms, ions, formula units, molecules
N_A	Avogadro's constant, 6.03×10^{23}	–

▶ *SAMPLE* problem *3*

Calculating the Number and Amount of Entities

(a) How many oxygen molecules, $O_{2(g)}$, are in a cylinder that contains 48.0 g of oxygen gas, if 6.02×10^{23} oxygen molecules have a mass of 32.00 g?

Step 1: Identify Key Value and Conversion Factor Equation

m_{O_2} = 48.0 g O_2 (key value)

6.02×10^{23} molecules O_2 = 32.00 g O_2 (conversion factor equation)

Step 2: Identify Required Value

The required value is the number of oxygen molecules, N_{O_2}, in the cylinder.

Step 3: List Possible Conversion Factors

The possible conversion factors are

$$\frac{6.02 \times 10^{23} \text{ molecules } O_2}{32.00 \text{ g } O_2} \text{ or } \frac{32.00 \text{ g } O_2}{6.02 \times 10^{23} \text{ molecules } O_2}$$

Step 4: Substitute Values into Solution Equation, and Solve

Remember that the units in the denominator of the conversion factor must be the same as the units of the key value.

required value = key value × conversion factor

$$N_{O_2} = 48.0 \text{ g } O_2 \times \frac{6.02 \times 10^{23} \text{ molecules } O_2}{32.00 \text{ g } O_2}$$

$$N_{O_2} = 9.03 \times 10^{23} \text{ molecules } O_2$$

There are 9.02×10^{23} molecules of oxygen in the cylinder.

(b) What amount of iodine molecules, $I_{2(s)}$, is in 1.231 kg of iodine, if 2 mol of iodine has a mass of 253.8 g?

Step 1: Identify Key Value and Conversion Factor Equation

m_{I_2} = 1.231 kg I_2 (key value)

2 mol I_2 = 253.8 g I_2 (conversion factor equation)

Notice that the unit of mass of the key value (kg) is different from the unit of mass in the conversion factor equation (g). In step 3, you will set up the solution equation so that like units cancel. To be able to cancel, you need to convert 1.231 kg to a value in grams, or 253.8 g to a value in kilograms. Let's

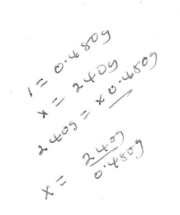

convert the key value, 1.231 kg, to a value in grams by multiplying it by the metric conversion factor $\dfrac{1000\ g}{1\ kg}$ (**Table 6**).

$$m_{I_2} = 1.231\ \cancel{kg}\,I_2 \times \frac{1000\ g\ I_2}{1\ \cancel{kg}\,I_2}$$

$$m_{I_2} = 1.231\ kg \times 10^3\ g\ I_2$$

Step 2: Identify Required Value
The required value is the amount of iodine, I_2, in moles.

Step 3: List Possible Conversion Factors
The possible conversion factors are

$$\frac{2\ mol\ I_2}{253.8\ g\ I_2} \quad or \quad \frac{253.8\ g\ I_2}{2\ mol\ I_2}$$

Step 4: Substitute Values into Solution Equation, and Solve
Remember that the units in the denominator of the conversion factor must be the same as the units of the key value. (Recall that the symbol n is used to represent amount.)

$$required\ value = key\ value \times conversion\ factor$$

$$n_{I_2} = 1.231 \times 10^3\ \cancel{g}\,I_2 \times \frac{2\ mol\ I_2}{253.8\ \cancel{g}\,I_2}$$

$$n_{I_2} = 9.70\ mol\ I_2$$

There is 9.70 mol of iodine in 1.231 kg of iodine.

Table 6 Metric Prefixes

Metric prefix	Symbol	Numerical value	Exponential value
mega	M	1 000 000	10^6
kilo	k	1000	10^3
hecto	h	100	10^2
deca	da	10	10^1
deci	d	0.1	10^{-1}
centi	c	0.01	10^{-2}
milli	m	0.001	10^{-3}
micro	μ	0.000 001	10^{-6}
nano	n	0.000 000 001	10^{-9}

Example
How many molecules of water are in a glass that contains 54.0 g of water, if 6.02×10^{23} molecules of water have a mass of 18.02 g? (Recall that the symbol N is used to represent number of entities.)

Solution

$m_{H_2O} = 54.0$ g (key term)

6.02×10^{23} molecules $H_2O = 18.02$ g H_2O (conversion factor equation)

$$N_{H_2O} = 54.0 \, \cancel{g \, H_2O} \times \frac{6.02 \times 10^{23} \text{ molecules } H_2O}{18.02 \, \cancel{g \, H_2O}}$$

$$N_{H_2O} = 1.80 \times 10^{24} \text{ molecules } H_2O$$

There are 1.80×10^{24} molecules of water in 54.0 g of water.

▶ **Practice**

Understanding Concepts

11. How many feathers are in a pillow that contains 200.4 g of feathers, if six feathers have a mass of 2.40 g?

12. How many pennies are in a piggy bank that contains 1.605 kg of pennies, if three pennies have a mass of 7.50 g?

13. How many atoms of sulfur are in a 48.0-g chunk of sulfur, if 6.02×10^{23} atoms of sulfur have a mass of 32.06 g?

14. How many moles of methane molecules, $CH_{4(g)}$, are in 52.0 g of methane, if one mole of methane has a mass of 16.05 g?

LEARNING TIP

Significant Digits
Be careful, when doing calculations, that you pay attention to significant digits. Refer to Appendix A5 for a review of significant digits.

Answers

11. 501 feathers

12. 642 pennies

13. 9.01×10^{23} atoms

14. 3.24 mol $CH_{4(g)}$

▶ **TRY THIS** activity — *Counting by Mass*

Chemists need to know the number of chemical entities to use in a chemical reaction. They use the mass of the substance they are working with to calculate the number of entities in the sample. Their counting technique is similar to the counting technique used by companies that need to count large numbers of small objects, such as paper clips, safety pins, and jellybeans. In this activity, you will calculate the number of paper clips in a box, using the mass of a small group of paper clips.

Materials: full box of paper clips (100), paper cup, electronic balance, calculator

1. Measure and record the mass of one dozen paper clips.

2. Tare the balance with a paper cup.

3. Pour most of the box of paper clips into the cup.

4. Measure and record the mass of the paper clips in the cup.

5. Calculate the number of paper clips in the box of paper clips using the masses you obtained in steps 1 and 4.

6. Count the paper clips in the cup. Record the number.

(a) Compare the number of paper clips you calculated in step 5 with the number of paper clips you counted in step 6. Account for any differences.

(b) If a chemist wanted to follow this procedure to determine the number of water molecules in a cup of water, which steps of the procedure would not be possible?

(c) What is the purpose of step 6?

(d) How does a chemist make up for the steps that are not possible?

(e) Using the same total number of paper clips, repeat the activity using one paper clip in step 1. Compare the results of the two calculations. Which procedure is better? Why? Which procedure is more like the work of a chemist? Explain.

Understanding Concepts

1. The following equation represents the production of ammonia, $NH_{3(g)}$, from nitrogen gas and hydrogen gas:

$$N_{2(g)} + 3\,H_{2(g)} \xrightarrow{\text{heat}} 2\,NH_{3(g)}$$

   (a) List the coefficients of nitrogen, hydrogen, and ammonia.
   (b) How many molecules of nitrogen are used to produce two molecules of ammonia?
   (c) How many dozen molecules of nitrogen are used to produce two dozen molecules of ammonia?
   (d) How many moles of nitrogen molecules are used to produce two moles of ammonia molecules?

2. (a) What are nucleons? Where are they found in an atom?
   (b) Why are electrons usually ignored when determining the mass of an atom?
   (c) Why is an atom of carbon approximately 12 times heavier than an atom of hydrogen?

3. In your own words, explain the meaning of the following statement: The natural abundance of magnesium-24 is 79%.

4. What is the symbol and value of Avogadro's constant?

5. Describe a mole of a substance in your own words.

6. What is the relationship between atomic mass and molar mass?

7. (a) What does the term "diatomic" mean?
   (b) Name eight elements that form diatomic molecules. (*Hint:* Think of elements with names that end in *-ine* and *-gen*.)
   (c) How many molecules are in one mole of any diatomic element?
   (d) How many atoms are in one mole of any diatomic element?

8. How many dimes are in a bag that contains 7.800 kg of dimes, if seven dimes have a mass of 12.30 g?

9. What amount, in moles, of hydrogen bromide molecules, $HBr_{(g)}$, are in a cylinder that contains 38.40 g of hydrogen bromide, if 2 mol of hydrogen bromide has a mass of 159.8 g?

10. (a) Define the term "molar mass." State its symbol and SI unit.
    (b) State the molar mass of carbon atoms by referring to the periodic table.
    (c) State the mass of 6.02×10^{23} atoms of zinc. How is this mass related to the molar mass of zinc?
    (d) If the molar mass of an element is 197 g/mol, what could the element be?

Applying Inquiry Skills

11. A piggy bank contains a large number of coins. There are equal numbers of pennies, nickels, dimes, and quarters.
    (a) Write a procedure you could use to estimate the total number of coins in the piggy bank.
    (b) How would you estimate the number of dimes in the piggy bank?

Making Connections

12. (a) How is the work of a chemist like the work of a chef? How is it like the work of an accountant?
    (b) How is the work of a chemist unlike the work of a chef? How is it unlike the work of an accountant?

13. (a) Scientists claim to have produced images of individual atoms using the scanning tunnelling microscope (STM). Research more about the STM. View some images taken with an STM and evaluate the claim.
    (b) If the claim is correct, will the STM be able to replace our present need to calculate quantities in chemistry? Explain.

 www.science.nelson.com

Calculations Involving the Mole *2.2*

When making chocolate chip cookies, it is more convenient to add a particular *mass* of chocolate chips to the mix than a particular *number* of chocolate chips (**Figure 1**). You have learned that the number of small objects in a collection can be calculated if you know the mass of one or a small number of the objects and the mass of the collection. For example, if you want to make 10 cookies with approximately 10 chocolate chips per cookie, you should add at least 100 chocolate chips to the mix. Instead of counting out 100 chocolate chips, you could calculate the mass that contains 100 chocolate chips. If the mass of one chocolate chip is 0.1 g, how many grams of chocolate chips must be added to the cookie mix?

Recall that the symbol *m* represents mass. In this example, the key value is 100 chocolate chips, the conversion factor equation is 1 chocolate chip = 0.1 g, and the required value is the mass of chocolate chips, $m_{\text{chocolate chips}}$.

$$m_{\text{chocolate chips}} = 100 \ \cancel{\text{chocolate chips}} \times \frac{0.1 \text{ g}}{1 \ \cancel{\text{chocolate chip}}}$$

$$m_{\text{chocolate chips}} = 10 \text{ g}$$

You would need to add 10 g of chocolate chips to the mix.

Figure 1
It is more convenient to determine the number of chocolate chips by mass than by counting.

Calculations Involving Atoms

When chemists need a certain amount of a chemical, they cannot count out the atoms, ions, molecules, or formula units. Instead, they calculate the mass of the chemical that contains the required amount, just as you would calculate the mass of chocolate chips you need to make cookies. Then, using a precise balance, they measure out the calculated mass of the chemical. The next three Sample Problems show the relationships among mass, molar mass, and amount when dealing with elements.

> ▶ **SAMPLE** problem 1
>
> ### Calculating Mass from Amount in Moles
>
> **Calculate the mass, in grams, of 2.00 mol of calcium atoms.**
>
> **Step 1: Identify Key Value and Conversion Factor Equation**
>
> n_{Ca} = 2.00 mol Ca (key value)
>
> Information for a conversion factor equation is not given in this problem. You can use the molar mass of the element, however, to produce the necessary equation. Look up the molar mass of calcium in the periodic table.
>
> M_{Ca} = 40.08 g/mol Ca
>
> 1 mol Ca = 40.08 g Ca (conversion factor equation)
>
> **Step 2: Identify Required Value**
>
> The required value is the mass of calcium atoms, m_{Ca}. (Remember that the symbol *m* is used to represent mass.)

Step 3: List Possible Conversion Factors

The possible conversion factors are

$$\frac{1 \text{ mol Ca}}{40.08 \text{ g Ca}} \quad \text{or} \quad \frac{40.08 \text{ g Ca}}{1 \text{ mol Ca}}$$

Step 4: Substitute Values into Solution Equation, and Solve

Remember that the units in the denominator of the conversion factor must be the same as the units of the key value.

$$\text{required value} = \text{key value} \times \text{conversion factor}$$

$$m_{Ca} = 2.00 \text{ mol Ca} \times \frac{40.08 \text{ g Ca}}{1 \text{ mol Ca}}$$

$$m_{Ca} = 80.2 \text{ g Ca}$$

The mass of 2.00 mol of calcium is 80.2 g.

Example

Calculate the mass of 0.200 mol of nitrogen atoms.

Solution

$n_N = 0.200 \text{ mol N}$

$M_N = 14.01 \text{ g/mol N}$

$1 \text{ mol N} = 14.01 \text{ g N}$

$$m_N = 0.200 \text{ mol N} \times \frac{14.01 \text{ g N}}{1 \text{ mol N}}$$

$$m_N = 2.80 \text{ g N}$$

The mass of 0.200 mol of nitrogen atoms is 2.80 g.

▶ **Practice**

Understanding Concepts

1. Calculate the mass of 1.60 mol of aluminum atoms.

2. Calculate the mass of 0.25 mol of sulfur atoms.

Answers

1. 43.2 g
2. 8.0 g

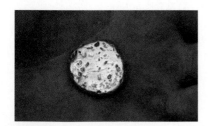

Figure 2
Gold is a precious metal that can be mixed with other metals, such as nickel and platinum, to create jewellery.

▶ **SAMPLE** problem 2

Calculating Amount in Moles from Mass

What amount of gold is in a 275.8-g nugget of pure gold (Figure 2)?

You can use the molar mass of an element to calculate the amount of atoms, in moles, in a sample of known mass.

Step 1: Identify Key Value and Conversion Factor Equation

$m_{Au} = 275.8 \text{ g Au}$ (key value)

$M_{Au} = 196.97 \text{ g/mol Au}$ (value from periodic table)

$1 \text{ mol Au} = 196.97 \text{ g Au}$ (conversion factor equation)

Step 2: Identify Required Value
The required value is the amount of gold, n_{Au}.

Step 3: List Possible Conversion Factors
The possible conversion factors are

$$\frac{1 \text{ mol Au}}{196.97 \text{ g Au}} \quad \text{or} \quad \frac{196.97 \text{ g Au}}{1 \text{ mol Au}}$$

Step 4: Substitute Values into Solution Equation, and Solve
required value = key value × conversion factor

$$n_{Au} = 275.8 \text{ g Au} \times \frac{1 \text{ mol Au}}{196.97 \text{ g Au}}$$

$$n_{Au} = 1.40 \text{ mol Au}$$

There is 1.40 mol of gold in a 275.8-g nugget of pure gold.

Example
What amount of helium is in a balloon that contains 1.60 g of helium gas (**Figure 3**)?

Solution

$m_{He} = 1.60 \text{ g He}$

$M_{He} = 4.00 \text{ g/mol He}$

$1 \text{ mol He} = 4.00 \text{ g He}$

$$n_{He} = 1.60 \text{ g He} \times \frac{1 \text{ mol He}}{4.00 \text{ g He}}$$

$$n_{He} = 0.400 \text{ mol He}$$

There is 0.400 mol of helium in a balloon that contains 1.60 g of helium gas.

Figure 3
Helium is less dense than air and will cause this balloon to float.

▶ Practice

Understanding Concepts
3. What amount of iron is in a 3.30-g iron nail?

4. What amount of silver is in a silver coin that contains 23.6 g of pure silver?

5. What amount of copper is in a bracelet that contains 7.65 g of pure copper?

Answers

3. 0.059 mol Fe

4. 0.219 mol Ag

5. 0.120 mol Cu

So far, you have learned to calculate

• the mass of an element when the amount is known;

• the amount (in moles) of an element when the mass is known.

You can also calculate

• the number of atoms in an element when the mass is known;

• the mass of an element when the number of atoms is known.

In Sample Problem 3, the element (gold) is the same as the element in Sample Problem 2, and it has the same mass, so you can compare the two solutions.

► **SAMPLE** problem 3

Calculating the Number of Atoms from Mass

How many atoms of gold are in a 275.8-g nugget of pure gold?

You can divide this problem into two parts to solve it. In the first part, the molar mass of the element is used to calculate the amount of gold in the nugget, as in Sample Problem 2. In the second part, the amount of gold is used to calculate the number of gold atoms in the nugget.

Part 1: Calculate Amount of Gold

Step 1: Identify Key Value and Conversion Factor Equation

m_{Au} = 275.8 g Au (key value)

M_{Au} = 196.97 g/mol Au

1 mol Au = 196.97 g Au (conversion factor equation)

Step 2: Identify Required Value

The required value for this part of the solution is the amount of gold atoms, n_{Au}.

Step 3: List Possible Conversion Factors

The possible conversion factors are

$$\frac{1 \text{ mol Au}}{196.97 \text{ g Au}} \quad \text{or} \quad \frac{196.97 \text{ g Au}}{1 \text{ mol Au}}$$

Step 4: Substitute Values into Solution Equation, and Solve

required value = key value × conversion factor

$$n_{Au} = 275.8 \text{ g Au} \times \frac{1 \text{ mol Au}}{196.97 \text{ g Au}}$$

$$n_{Au} = 1.40 \text{ mol Au}$$

Part 2: Calculate Number of Gold Atoms

Step 1: Identify Key Value and Conversion Factor Equation

The amount of gold that was calculated in Part 1 is the key value for Part 2.

n_{Au} = 1.40 mol Au (key value)

The conversion factor equation for this part is obtained from the fact that one mole of atoms contains Avogadro's number of atoms.

1 mol Au = 6.02×10^{23} atoms Au (conversion factor equation)

Step 2: Identify Required Value

The required value is the number of atoms of gold, N_{Au}. (Remember that N represents the number of entities.)

Step 3: List Possible Conversion Factors

The possible conversion factors are

$$\frac{1 \text{ mol Au}}{6.02 \times 10^{23} \text{ atoms Au}} \quad \text{or} \quad \frac{6.02 \times 10^{23} \text{ atoms Au}}{1 \text{ mol Au}}$$

Step 4: Substitute Values into Solution Equation, and Solve

$$n_{Au} = 1.40 \text{ mol Au} \times \frac{6.02 \times 10^{23} \text{ atoms Au}}{1 \text{ mol Au}}$$

$$n_{Au} = 8.43 \times 10^{23} \text{ atoms Au}$$

There are 8.43×10^{23} atoms of gold in a 275.8-g nugget of pure gold.

Parts 1 and 2 can be combined into a single-line calculation by substituting the key value given in the problem and the conversion factors from Parts 1 and 2. Notice that conversion factor 1 (from Part 1) converts the mass of gold into the amount of gold. Conversion factor 2 (from Part 2) converts the amount of gold into the number of gold atoms.

required value = key value × conversion factor 1 × conversion factor 2

$$N_{Au} = 275.8 \text{ g Au} \times \frac{1 \text{ mol Au}}{196.97 \text{ g Au}} \times \frac{6.02 \times 10^{23} \text{ atoms Au}}{1 \text{ mol Au}}$$

$$N_{Au} = 8.43 \times 10^{23} \text{ atoms Au}$$

There are 8.43×10^{23} atoms of gold in a 275.8-g nugget of pure gold.

Example

How many atoms of sulfur are in a 230.0-g sample of pure sulfur (**Figure 4**)?

Solution

Part 1: Calculate Amount of Sulfur

$m_S = 230.0$ g S

$M_S = 32.06$ g/mol S

1 mol S = 32.06 g S

$$n_S = 230.0 \text{ g S} \times \frac{1 \text{ mol S}}{32.06 \text{ g S}}$$

$$n_S = 7.17 \text{ mol S}$$

Part 2: Calculate Number of Sulfur Atoms

$$N_S = 7.17 \text{ mol S} \times \frac{6.02 \times 10^{23} \text{ atoms S}}{1 \text{ mol S}}$$

$$N_S = 4.32 \times 10^{24} \text{ atoms S}$$

There are 4.32×10^{24} atoms of sulfur in a 230.0-g sample of sulfur.

Alternative single-line solution:

$$N_S = 230.0 \text{ g S} \times \frac{1 \text{ mol S}}{32.06 \text{ g S}} \times \frac{6.02 \times 10^{23} \text{ atoms S}}{1 \text{ mol S}}$$

$$N_S = 4.32 \times 10^{24} \text{ atoms S}$$

There are 4.32×10^{24} atoms of sulfur in a 230.0-g sample of sulfur.

Figure 4
Sulfur is a component of black gunpowder, and it is used to improve the strength of natural rubber.

molecular element a molecule that contains two or more atoms of one type of element

compound a molecule that contains two or more atoms of different elements, or is a combination of oppositely charged ions

> ▶ **Practice**

Understanding Concepts

6. How many atoms of carbon are in a 3.30-g diamond (pure carbon)?

7. How many atoms of neon are in a neon sign that contains 6.80 g of neon?

8. How many atoms of mercury are in the bulb of a thermometer, if the mercury has a mass of 78.2 g?

The Molar Mass of Molecular Elements and Compounds

Molecular elements contain two or more atoms of one element only. Hydrogen, $H_{2(g)}$, fluorine, $F_{2(g)}$, ozone, $O_{3(g)}$, and sulfur, $S_{8(s)}$, are examples of molecular elements. **Compounds** are molecules that contain two or more atoms of different elements, or they are combinations of oppositely charged ions represented by formula units. Water, $H_2O_{(l)}$, carbon dioxide, $CO_{2(g)}$, and ammonia, $NH_{3(g)}$, are examples of molecular compounds. Sodium chloride, $NaCl_{(s)}$, is an example of an ionic compound.

The molar mass of a molecule is equal to the sum of the molar masses of all the atoms in the molecule. The molar mass of an ionic compound is equal to the sum of the molar masses of all the ions in the formula unit of the ionic compound. For example, the molar mass of hydrogen chloride, $HCl_{(g)}$, is equal to the sum of the molar mass of hydrogen and the molar mass of chlorine. The following Sample Problem shows how to calculate the molar mass of a compound.

> ▶ **SAMPLE** problem 4
>
> ### Calculating the Molar Mass of a Compound
>
> **Calculate the molar mass of water.**
>
> In 1 mol of water, there are 2 mol of hydrogen atoms and 1 mol of oxygen atoms. Add together the molar masses of these atoms to obtain the molar mass of the compound. Notice that you have to multiply the molar mass of hydrogen by 2 because there are two hydrogen atoms in every water molecule.
>
> $$M_{H_2O} = 2(M_H) + 1(M_O)$$
> $$= 2(1.01 \text{ g/mol}) + 1(16.00 \text{ g/mol})$$
> $$= 2.02 \text{ g/mol} + 16.00 \text{ g/mol}$$
> $$M_{H_2O} = 18.02 \text{ g/mol}$$
>
> The molar mass of water is 18.02 g/mol.
>
> *Example*
> Calculate the molar mass of sodium hydrogen carbonate (baking soda), $NaHCO_{3(s)}$.

Solution

$$M_{NaHCO_3} = 1(M_{Na}) + 1(M_H) + 1(M_C) + 3(M_O)$$
$$= 1(22.99 \text{ g/mol}) + 1(1.01 \text{ g/mol}) + 1(12.01 \text{ g/mol}) + 3(16.00 \text{ g/mol})$$
$$= 22.99 \text{ g/mol} + 1.01 \text{ g/mol} + 12.01 \text{ g/mol} + 48.00 \text{ g/mol}$$
$$M_{NaHCO_3} = 84.01 \text{ g/mol}$$

The molar mass of sodium hydrogen carbonate is 84.01 g/mol.

▶ **Practice**

Understanding Concepts

9. Calculate the molar mass of octane (a component of gasoline), $C_8H_{18(l)}$.
10. Calculate the molar mass of acetylsalicylic acid (Aspirin), $C_9H_8O_{4(s)}$.
11. Calculate the molar mass of calcium sulfate (gypsum), $CaSO_{4(s)}$.

Answers

9. 114.26 g/mol
10. 180.17 g/mol
11. 136.14 g/mol

Calculations Involving Molecules and Compounds

Earlier in this section, you learned how to perform various calculations involving elements. The following Sample Problems show how the same principles can be applied to molecules and ionic compounds.

▶ **SAMPLE** problem **5**

Calculating Mass from Amount in Moles

Sodium fluoride is added to toothpaste and tap water to help prevent tooth decay. Calculate the mass of 2.00 mol of sodium fluoride, $NaF_{(s)}$.

Step 1: Identify Key Value and Conversion Factor Equation

n_{NaF} = 2.00 mol NaF (key value)

Calculate the molar mass of the compound, M_{NaF}, to determine the conversion factor equation.

$$M_{NaF} = 1(22.99 \text{ g/mol}) + 1(19.00 \text{ g/mol})$$
$$M_{NaF} = 41.99 \text{ g/mol NaF}$$
1 mol NaF = 41.99 g NaF (conversion factor equation)

Step 2: Identify Required Value

The required value is the mass of sodium fluoride, m_{NaF}.

Step 3: List Possible Conversion Factors

The possible conversion factors are

$$\frac{1 \text{ mol NaF}}{41.99 \text{ g NaF}} \text{ or } \frac{41.99 \text{ g NaF}}{1 \text{ mol NaF}}$$

Step 4: Substitute Values into Solution Equation, and Solve

$$\text{required value} = \text{key value} \times \text{conversion factor}$$

$$m_{NaF} = 2.00 \cancel{\text{mol NaF}} \times \frac{41.99 \text{ g NaF}}{1 \cancel{\text{mol NaF}}}$$

$$m_{NaF} = 84.0 \text{ g NaF}$$

The mass of 2.00 mol of sodium fluoride is 84.0 g.

Example

Calculate the mass of 11.7 mol of ozone, $O_{3(g)}$.

Solution

$$n_{O_3} = 11.7 \text{ mol } O_3$$
$$M_{O_3} = 3(16.00 \text{ g/mol})$$
$$M_{O_3} = 48.00 \text{ g/mol } O_3$$
$$1 \text{ mol } O_3 = 48.00 \text{ g } O_3$$

$$m_{O_3} = 11.7 \cancel{\text{mol } O_3} \times \frac{48.00 \text{ g } O_3}{1 \cancel{\text{mol } O_3}}$$

$$m_{O_3} = 562 \text{ g } O_3$$

The mass of 11.7 mol of ozone is 562 g.

▶ **Practice**

Understanding Concepts

Answers

12. 15.3 g

13. 435 g

12. Calculate the mass of 0.900 mol of ammonia, $NH_{3(g)}$.

13. Calculate the mass of 3.60 mol of freon-12, $CCl_2F_{2(g)}$.

▶ **SAMPLE** problem 6

Calculating Amount in Moles from Mass

Iron(III) oxide, $Fe_2O_{3(s)}$, is more commonly known as rust (Figure 5). What amount of iron(III) oxide is in a 77.2-g sample?

Figure 5
Cars have steel frames and bodies that rust when exposed to water and oxygen.

Step 1: Identify Key Value and Conversion Factor Equation

$m_{Fe_2O_3} = 77.2$ g Fe_2O_3 (key value)

$M_{Fe_2O_3} = 2(55.85$ g/mol$) + 3(16.00$ g/mol$)$

$M_{Fe_2O_3} = 159.7$ g/mol Fe_2O_3

1 mol $Fe_2O_3 = 159.7$ g Fe_2O_3 (conversion factor equation)

Step 2: Identify Required Value

The required value is the amount of iron(III) oxide, $n_{Fe_2O_3}$.

Step 3: List Possible Conversion Factors

The possible conversion factors are

$$\frac{1 \text{ mol } Fe_2O_3}{159.7 \text{ g } Fe_2O_3} \quad \text{or} \quad \frac{159.7 \text{ g } Fe_2O_3}{1 \text{ mol } Fe_2O_3}$$

Step 4: Substitute Values into Solution Equation, and Solve

required value = key value $\times$ conversion factor

$$n_{Fe_2O_3} = 77.2 \text{ g } Fe_2O_3 \times \frac{1 \text{ mol } Fe_2O_3}{159.7 \text{ g } Fe_2O_3}$$

$$n_{Fe_2O_3} = 0.483 \text{ mol } Fe_2O_3$$

There is 0.483 mol of iron(III) oxide in a 77.2-g sample.

Example

What amount of propane, $C_3H_{8(g)}$, is in a tank that is filled with 5.00 kg of propane?

Solution

$$m_{C_3H_8} = 5.00 \text{ kg } C_3H_8 \times \frac{1000 \text{ g } C_3H_8}{1 \text{ kg } C_3H_8}$$

$$m_{C_3H_8} = 5.00 \times 10^3 \text{ g } C_3H_8$$

$$M_{C_3H_8} = 3(12.01 \text{ g/mol}) + 8(1.01 \text{ g/mol})$$

$$M_{C_3H_8} = 44.11 \text{ g/mol } C_3H_8$$

1 mol $C_3H_8 = 44.11$ g C_3H_8

$$n_{C_3H_8} = 5.00 \times 10^3 \text{ g } C_3H_8 \times \frac{1 \text{ mol } C_3H_8}{44.11 \text{ g } C_3H_8}$$

$$n_{C_3H_8} = 113 \text{ mol } C_3H_8$$

There is 113 mol of propane in a tank with 5.00 kg of propane in it.

Alternative single-line solution:

$$n_{C_3H_8} = 5.00 \text{ kg } C_3H_8 \times \frac{1000 \text{ g } C_3H_8}{1 \text{ kg } C_3H_8} \times \frac{1 \text{ mol } C_3H_8}{44.11 \text{ g } C_3H_8}$$

$$n_{C_3H_8} = 113 \text{ mol } C_3H_8$$

There is 113 mol of propane in a tank with 5.00 kg of propane in it.

▶ *Practice*

Understanding Concepts

14. What amount of magnesium hydroxide, $Mg(OH)_{2(s)}$, is in 204.0 g of magnesium hydroxide?

15. What amount of sucrose (table sugar), $C_{12}H_{22}O_{11(s)}$, is in a bag that contains 1.00 kg of sucrose?

You can also calculate the number of entities (molecules or formula units) in a sample of known mass using the factor-label method. Notice that single-line solutions are used in the following Sample Problem and Example.

▶ *SAMPLE* problem *7*

Calculating the Number of Entities from Mass

How many formula units of the ionic compound iron(III) oxide, $Fe_2O_{3(s)}$, are in 77.2 g of iron(III) oxide?

Step 1: Identify Key Value and Conversion Factor Equation

$$m_{Fe_2O_3} = 77.2 \text{ g } Fe_2O_3 \text{ (key value)}$$
$$M_{Fe_2O_3} = 2(55.85 \text{ g/mol}) + 3(16.00 \text{ g/mol})$$
$$M_{Fe_2O_3} = 159.7 \text{ g/mol } Fe_2O_3$$

The following two conversion factor equations will be used:

- conversion factor equation 1:
  1 mol Fe_2O_3 = 159.7 g/mol Fe_2O_3 (from $M_{Fe_2O_3}$)

- conversion factor equation 2:
  1 mol Fe_2O_3 = 6.02×10^{23} formula units Fe_2O_3 (from mole definition)

Step 2: Identify Required Value

The required value is the number of iron(III) oxide formula units, $N_{Fe_2O_3}$.

Step 3: List Possible Conversion Factors

The possible conversion factors from conversion factor equation 1 are

$$\frac{1 \text{ mol } Fe_2O_3}{159.7 \text{ g } Fe_2O_3} \text{ or } \frac{159.7 \text{ g } Fe_2O_3}{1 \text{ mol } Fe_2O_3}$$

The possible conversion factors from conversion factor equation 2 are

$$\frac{1 \text{ mol } Fe_2O_3}{6.02 \times 10^{23} \text{ formula units } Fe_2O_3} \text{ or } \frac{6.02 \times 10^{23} \text{ formula units } Fe_2O_3}{1 \text{ mol } Fe_2O_3}$$

Step 4: Substitute Values into Solution Equation, and Solve

Remember to select conversion factors such that the units in the denominator of conversion factor 1 match the units of the key value, and the units in the denominator of conversion factor 2 match the units in the numerator of conversion factor 1.

required value = key value $\times$ conversion factor 1 $\times$ conversion factor 2

$$N_{Fe_2O_3} = 77.2 \text{ g Fe}_2O_3 \times \frac{1 \text{ mol Fe}_2O_3}{159.7 \text{ g Fe}_2O_3} \times \frac{6.02 \times 10^{23} \text{ formula units Fe}_2O_3}{1 \text{ mol Fe}_2O_3}$$

$$N_{Fe_2O_3} = 2.91 \times 10^{23} \text{ formula units Fe}_2O_3$$

There are 2.91×10^{23} formula units of iron(III) oxide in a 77.2-g sample.

Example

Calcium hydroxide, $Ca(OH)_{2(s)}$, is an ionic compound that is used to make slaked lime, a substance that is used as a top-coat on the walls of many homes. How many formula units of calcium hydroxide are in a sample of slaked lime that contains 250.0 g of calcium hydroxide?

Solution

$$m_{Ca(OH)_2} = 250.0 \text{ g Ca(OH)}_2$$
$$M_{Ca(OH)_2} = 1(40.08 \text{ g/mol}) + 2(16.00 \text{ g/mol}) + 2(1.01 \text{ g/mol})$$
$$M_{Ca(OH)_2} = 74.1 \text{ g/mol Ca(OH)}_2$$

$$N_{Ca(OH)_2} = 250.0 \text{ g Ca(OH)}_2 \times \frac{1 \text{ mol Ca(OH)}_2}{74.1 \text{ g Ca(OH)}_2} \times \frac{6.02 \times 10^{23} \text{ formula units Ca(OH)}_2}{1 \text{ mol Ca(OH)}_2}$$

$$N_{Ca(OH)_2} = 2.03 \times 10^{24} \text{ formula units Ca(OH)}_2$$

There are 2.03×10^{24} formula units of calcium hydroxide in a sample that contains 250.0 g of calcium hydroxide.

▶ Practice

Understanding Concepts

16. How many molecules of water are in a bottle that contains 250.0 g of water?

17. How many formula units of cobalt(III) dichromate, $Co_2(Cr_2O_7)_{3(s)}$, are in a 3.30-kg sample?

Answers

16. 8.35×10^{24} molecules

17. 2.59×10^{24} formula units

Calculating the Number of Atoms from the Mass of Molecules

Calculating the number of atoms in a sample of molecules can be tricky. If the sample contains a molecular element, then the atoms are all the same. If the sample contains a compound, however, then there are different atoms to consider. The following Sample Problem involves a compound, and the Example involves a molecular element. Note the differences in the two calculations.

Calculating the Number of Atoms

Sand is composed of silicon dioxide, $SiO_{2(s)}$. How many atoms of oxygen are in a bag of pure sand, which contains 1.00 kg of silicon dioxide?

Step 1: Identify Key Value and Conversion Factor Equation

$m_{SiO_2} = 1.00$ kg SiO_2 (key value)

$M_{SiO_2} = 1(28.09$ g/mol$) + 2(16.00$ g/mol$)$

$M_{SiO_2} = 60.09$ g/mol SiO_2

The following conversion factor equations will be used in the solution:

1 kg $= 1000$ g (metric equality)

1 mol $SiO_2 = 60.09$ g SiO_2 (from M_{SiO_2})

1 mol $SiO_2 = 6.02 \times 10^{23}$ units SiO_2 (from mole definition)

2 atoms O $= 1$ unit SiO_2 (since there are two oxygen atoms per unit of silicon dioxide)

Step 2: Identify Required Value

The required value is the number of oxygen atoms, N_O.

Step 3: List Possible Conversion Factors

See the conversion factor equations in step 1 and the corresponding conversion factors in step 4.

Step 4: Substitute Values into Solution Equation, and Solve

There are four conversion factors in this solution.

$N_O =$ key value $\times$ conversion factor 1 $\times$ conversion factor 2 $\times$ conversion factor 3 $\times$ conversion factor 4

$$= 1.00 \text{ kg } SiO_2 \times \frac{1000 \text{ g } SiO_2}{1 \text{ kg } SiO_2} \times \frac{1 \text{ mol } SiO_2}{60.09 \text{ g } SiO_2} \times \frac{6.02 \times 10^{23} \text{ units } SiO_2}{1 \text{ mol } SiO_2} \times \frac{2 \text{ atoms O}}{1 \text{ unit } SiO_2}$$

$N_O = 2.00 \times 10^{25}$ atoms O

There are 2.00×10^{25} atoms of oxygen in a bag that contains 1.00 kg of pure sand.

Example

One form of solid sulfur is composed of $S_{8(s)}$ molecules. How many atoms of sulfur are in an 18.0-g chunk of solid sulfur?

Solution

$m_{S_8} = 18.0$ g S_8

$M_{S_8} = 8(32.06$ g/mol$)$

$M_{S_8} = 256.5$ g/mol S_8

1 mol $S_8 = 256.5$ g S_8

$$N_S = 18.0 \text{ g } S_8 \times \frac{1 \text{ mol } S_8}{256.5 \text{ g } S_8} \times \frac{6.02 \times 10^{23} \text{ molecules } S_8}{1 \text{ mol } S_8} \times \frac{8 \text{ atoms S}}{1 \text{ molecule } S_8}$$

$N_S = 3.38 \times 10^{23}$ atoms S

There are 3.38×10^{23} atoms of sulfur in an 18.0-g chunk of solid sulfur.

▶ **Practice**

Understanding Concepts

18. How many atoms of fluorine are in 4.4 g of fluorine gas?

19. How many atoms of nitrogen are in 1.26 kg of nitrogen gas?

20. How many atoms of hydrogen are in 29.5 g of ethene, $C_2H_{4(g)}$?

21. How many atoms of oxygen are in 0.170 mg of strontium hydroxide, $Sr(OH)_{2(s)}$?

Answers

18. 1.4×10^{23} F atoms

19. 5.41×10^{25} N atoms

20. 2.53×10^{24} H atoms

21. 1.68×10^{18} O atoms

SUMMARY   *Steps in the Factor-Label Method*

Step 1: Identify Key Value and Conversion Factor Equation
Step 2: Identify Required Value
Step 3: List Possible Conversion Factors
Step 4: Substitute Values into Solution Equation, and Solve

Use the following solution equation:

required value = key value × conversion factor × conversion factor

Note: Use as many conversion factors as required.

▶ **TRY THIS** activity   *Counting Atoms, Molecules, and Other Entities*

Use the following materials to measure and/or calculate the quantity described in each step below. Write an explanation of each calculation.

Materials: balance, graduated cylinder, beaker, disposable cups, copper pennies, iron nails, granulated sugar, table salt, chalk, water

1. Determine the mass of a drop of water by measuring the mass of 50 drops of water.

2. Place a single drop of water on the lab bench, and record the time the drop takes to evaporate completely. Calculate how many molecules of water evaporate per second.

3. Calculate the number of copper atoms in a penny. Use the number of copper atoms to calculate the monetary value of each atom of copper in the penny. Assume that the penny contains pure copper only.

4. Measure half a mole of sucrose molecules, $C_{12}H_{22}O_{11(s)}$, into a graduated cylinder.

5. Measure the quantity of sugar that contains two moles of carbon atoms into a graduated cylinder. Record the reading on the graduated cylinder.

6. Measure the mass of a piece of chalk. Use the piece of chalk to write your full name on the chalkboard. Measure the mass of the chalk again. Calculate the number of atoms that you used to write your name. (Assume that chalk is made entirely of calcium carbonate.)

7. Dissolve 3.00 g of table salt (assume $NaCl_{(s)}$) in 200 mL of water. Calculate the number of sodium ions in the salt solution.

8. Calculate the number of iron atoms in an iron nail.

9. Calculate the number of years in one mole of seconds.

Understanding Concepts

1. Magnesium hydroxide, $Mg(OH)_{2(s)}$, is a key ingredient in some antacid tablets. What is the molar mass of magnesium hydroxide?

2. Ozone, $O_{3(g)}$, is a molecule that protects you from dangerous solar radiation. What is the molar mass of ozone in the upper atmosphere?

3. If the molar mass of a substance is 67.2 g/mol, what is the mass of 8.0 mol of the substance?

4. Calculate the mass of a mole of sucrose, $C_{12}H_{22}O_{11(s)}$.

5. (a) How many atoms of aluminum are in 0.1 mol of aluminum?
   (b) How many formula units of magnesium chloride, $MgCl_{2(s)}$, are in 3.5 mol of magnesium chloride?

6. Calculate the amount of entities in each of the following samples of pure substances:
   (a) 5.00 kg of table sugar (sucrose), $C_{12}H_{22}O_{11(s)}$
   (b) 250 g of naphthalene moth balls, $C_{10}H_{8(s)}$
   (c) 35.0 g of propane, $C_3H_{8(g)}$, in a camp stove cylinder
   (d) 275 mg of acetylsalicylic acid (Aspirin), $C_9H_8O_{4(s)}$, in a headache relief tablet
   (e) 240 g of 2-propanol (rubbing alcohol), $C_3H_8O_{(l)}$

7. Calculate the mass, in grams, of each of the following substances:
   (a) 2.67 mol of ammonia in a window-cleaning solution
   (b) 0.965 mol of sodium hydroxide, $NaOH_{(s)}$, in a drain-cleaning solution
   (c) 19.7 mol of water vapour produced by a Bunsen burner
   (d) 3.85 mol of potassium permanganate (a fungicide), $KMnO_{4(s)}$
   (e) 0.47 mol of ammonium sulfate (a fertilizer), $(NH_4)_2SO_{4(s)}$

8. Calculate the number of molecules in each of the following samples:
   (a) 2.5 mol of solid carbon dioxide in dry ice
   (b) 2.5 g of ammonia gas in household cleaning solutions
   (c) 2.5 g of hydrogen chloride in hydrochloric acid

9. Calculate the mass, in grams, of 0.10 mol of each of the following substances:
   (a) carbon dioxide
   (b) glucose, $C_6H_{12}O_{6(s)}$
   (c) oxygen gas

10. Calculate the number of oxygen molecules in 2.7 mol of oxygen gas.

11. A daily vitamin tablet contains 90 mg of vitamin C. The chemical name for vitamin C is ascorbic acid, $C_6H_8O_{6(s)}$. If you take one vitamin tablet each day, how many molecules of vitamin C are you taking?

12. A recipe for a sweet and sour sauce calls for the following ingredients:
    450 g water
    100 g sugar, $C_{12}H_{22}O_{11(s)}$
    30 g vinegar (containing 2.4 g acetic acid, $HC_2H_3O_{2(aq)}$)
    2 g salt, $NaCl_{(s)}$
    Convert the recipe into amounts (in moles). For vinegar, use 2.4 g of acetic acid.

Applying Inquiry Skills

13. Silver ions in a solution of silver salts can be recovered by immersing copper metal in the solution (**Figure 6**). Crystals of pure silver are deposited on the copper metal. Design an experiment to determine the number of moles of silver atoms that form. Describe the procedure, materials, and safety procedures. Explain the required calculations.

Making Connections

14. The Academy of Science is giving a prestigious award to the most significant scientific concept. Write a brief paper, nominating the mole for this award. Cite the role and importance of the mole in the application of chemical reactions in society, industry, and the environment.

Figure 6
When copper metal is placed in a solution of silver ions, a single displacement reaction occurs. Copper ions go into solution, and silver crystals are formed.

Scientists estimate that new chemical compounds are being discovered at a rate of two every minute, which is about one million new compounds every year! Many compounds, such as the recently invented cancer drug *temozolomide*, are tested for possible medical benefits. Before a compound can be approved for use as a drug in Canada, however, the Food and Drug Act requires the drug company to conduct extensive tests to ensure that the drug is safe and effective. As part of the approval process, the drug company must submit the molecular formula of the compound to the appropriate department of the Ministry of Health. How is the molecular formula of a new compound determined? Analytical laboratories have sophisticated instruments that can determine almost any characteristic of an element or compound. All of these instruments operate on the basis of fundamental chemical principles, some of which were discovered hundreds of years ago.

In 1799, Joseph Proust (**Figure 1**) proposed the **law of constant composition**. This law states that *a compound contains elements in certain fixed proportions (ratios) and in no other combinations, regardless of how the compound is prepared or where it is found in nature.* Therefore, one molecule of methane (natural gas), $CH_{4(g)}$, always contains one atom of carbon and four atoms of hydrogen. One molecule of water always contains two atoms of hydrogen and one atom of oxygen. Adding or removing atoms from a compound changes the compound to a completely different substance. For example, adding an additional oxygen atom to a water molecule changes the compound to hydrogen peroxide, $H_2O_{2(l)}$. Whereas water is safe to consume, hydrogen peroxide is not. Hydrogen peroxide is a powerful bleaching agent and a potent disinfectant.

Since chemists cannot see individual molecules, they cannot directly determine the types and numbers of atoms in molecules. In the nineteenth century, chemists used several complex and time-consuming laboratory procedures to determine the molecular formula of a new compound. Today, most of the procedures are automated and efficiently performed by computerized equipment. Two key instruments that are used to determine the formula of a compound are the mass spectrometer and the combustion analyzer.

The Mass Spectrometer

A **mass spectrometer** (**Figure 2**) is used to measure the molar mass of a compound. A small sample of the compound is vaporized (turned into gas) and bombarded by a beam of electrons. The electrons cause the molecules to become electrically charged (ionized) and possibly broken up into a number of smaller fragments (**Figure 3**, on the next page). The charged fragments are accelerated by an electric field and deflected by a magnetic field. The amount of deflection depends on the mass and charge of the fragments. The molar mass of the original molecule can be determined from the molar mass of the

Figure 1
Joseph Proust (1754–1826)

law of constant composition A compound contains elements in certain fixed proportions (ratios), regardless of how the compound is prepared or where it is found in nature.

mass spectrometer laboratory instrument that is used to measure the molar mass of a compound

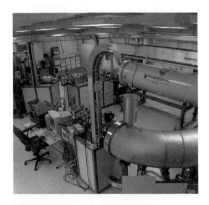

Figure 2
A mass spectrometer is used to identify the drugs that an athlete may have used before a race.

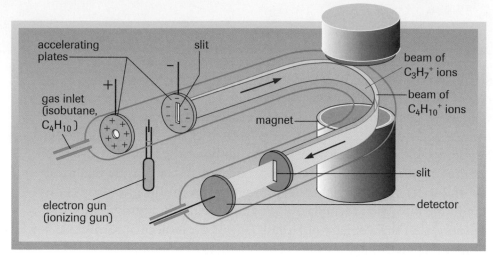

Figure 3
A mass spectrometer is used to determine the masses of ionized fragments by measuring the amount of deflection in the path of the fragments as they pass through a magnetic field.

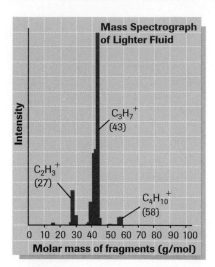

Figure 4
A mass spectrograph of lighter fluid is used to determine the molar mass of the chemical in the fluid. Expert analysis of the mass spectrograph provides the basis for a molecular model.

combustion analyzer laboratory instrument that is used to determine the percentages of carbon, hydrogen, oxygen, and nitrogen in a compound

largest fragment. **Figure 4** shows a mass spectrograph (a printout from a mass spectrometer) of butane (lighter fluid), $C_4H_{10(l)}$.

The Combustion Analyzer

A **combustion analyzer** is used to determine the percentages of carbon, hydrogen, oxygen, and possibly nitrogen in a compound. An accurately known mass of a pure compound is burned in a stream of pure oxygen gas at a temperature of approximately 980°C. Oxygen is used because it is required for the combustion of carbon-based compounds. All the carbon in the sample is incorporated into carbon dioxide gas. All the hydrogen is incorporated into water vapour (steam). For example, if the sample contains methane, the following combustion reaction occurs:

$$CH_{4(g)} + 2\,O_{2(g)} \rightarrow CO_{2(g)} + 2\,H_2O_{(g)}$$

Notice that

- every hydrogen atom (H) in the original sample ends up in an $H_2O_{(g)}$ molecule;
- every carbon atom (C) in the original sample ends up in a $CO_{2(g)}$ molecule;
- all the atoms, and therefore all the mass of the original sample, are accounted for in the products.

As the combustion products (carbon dioxide and water) pass through the system (**Figure 5**), the water vapour is absorbed by a water trap. The water trap is a tube that is packed with a compound, such as calcium chloride, $CaCl_{2(s)}$, that absorbs water. To determine the mass of water that is produced in the combustion reaction, the mass of the water trap before it has absorbed water vapour is subtracted from the mass of the water trap after it has absorbed water vapour.

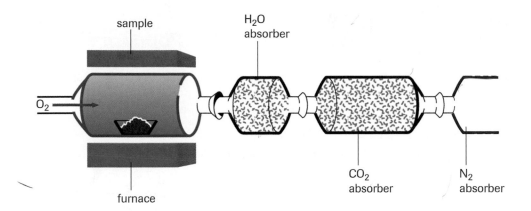

Figure 5
When a substance is burned in a combustion analyzer, oxides are produced. The oxides are captured by absorbers in chemical traps. The initial and final masses of each trap indicate the masses of the oxides produced. These masses are then used to calculate the percentage composition of the substance burned.

The remainder of the combustion gas now contains only carbon dioxide gas. The carbon dioxide travels through a carbon dioxide trap, which absorbs all the gas. The carbon dioxide trap is a tube that is packed with a compound that reacts completely with carbon dioxide, such as solid sodium hydroxide, $NaOH_{(s)}$. To determine the mass of carbon dioxide produced in the combustion reaction, the mass of the carbon dioxide trap before it has absorbed carbon dioxide is subtracted from the mass of the trap after it has absorbed carbon dioxide.

The masses of hydrogen and carbon in the sample compound are calculated as follows:

$$m_H = m_{H_2O} \times \frac{1 \text{ mol } H_2O}{18.02 \text{ g } H_2O} \times \frac{2.02 \text{ g } H}{1 \text{ mol } H_2O}$$

$$m_C = m_{CO_2} \times \frac{1 \text{ mol } CO_2}{44.01 \text{ g } CO_2} \times \frac{12.01 \text{ g } C}{1 \text{ mol } CO_2}$$

The masses of hydrogen and carbon are then used to calculate the percentage, by mass, of each element in the sample compound:

$$\% \text{ H} = \frac{m_H}{m_{sample}} \times 100\%$$

$$\% \text{ C} = \frac{m_C}{m_{sample}} \times 100\%$$

If the compound is a hydrocarbon (containing only hydrogen and carbon), the sum of % H and % C will be 100%. If the compound also contains oxygen, the sum of % H and % C will be less than 100%, with the difference equal to the percentage of oxygen in the compound. The percentage of oxygen in the compound is determined by subtraction because the oxygen atoms in water and carbon dioxide do not all come from the sample compound. Some oxygen atoms come from the oxygen gas that is used in the combustion process. The percentage, by mass, of each element in a compound is called the **percentage composition** of the compound.

percentage composition the percentage, by mass, of each element in a compound

Modern combustion analyzers are equipped with computers that perform these calculations automatically. **Table 1** illustrates a typical printout from a combustion analyzer for a compound that contains carbon, hydrogen, and oxygen. Ten samples of the compound were run through the combustion analyzer. The average percentage composition was calculated for each element in the compound. Note that the total of the average percentage composition values equals 100%.

Table 1 Combustion Analyzer Data

Sample	% C	% H	% O
1	38.70	9.71	51.60
2	38.69	9.71	51.58
3	38.71	9.71	51.58
4	38.71	9.73	51.59
5	38.73	9.72	51.58
6	38.72	9.71	51.57
7	38.71	9.71	51.58
8	38.71	9.70	51.58
9	38.71	9.70	51.56
10	38.71	9.70	51.58
Average % composition	**38.71**	**9.71**	**51.58**

Determining the Empirical Formula

Percentage composition values can be used to calculate the empirical formula of a compound. The **empirical formula** tells you the lowest ratio of atoms in a compound, but it does not necessarily tell you the exact number of each type of atom. The following Sample Problem and Example show how an empirical formula can be calculated using percentage composition data.

empirical formula a formula that gives the lowest ratio of atoms in a compound

---

▶ **SAMPLE** problem **1**

Determining the Empirical Formula of a Compound

(a) Use the average percentage composition values given in Table 1 to determine the empirical formula of the compound.

The empirical formula of a compound is based on its percentage composition. From **Table 1**, the percentage composition of the compound is 38.71% carbon, 9.71% hydrogen, and 51.58% oxygen.

Step 1: List Given Values

C = 38.71%

H = 9.71%

O = 51.58%

Step 2: Calculate Mass (m) of Each Element in 100-g Sample

$$m_C = \frac{38.71}{100} \times 100 \text{ g}$$

$$m_C = 38.71 \text{ g}$$

$$m_H = \frac{9.71}{100} \times 100 \text{ g}$$

$$m_H = 9.71 \text{ g}$$

$$m_O = \frac{51.58}{100} \times 100 \text{ g}$$

$$m_O = 51.58 \text{ g}$$

Step 3: Convert Mass (m) into Amount (n)

$$n_C = 38.71 \text{ g C} \times \frac{1 \text{ mol C}}{12.01 \text{ g C}}$$

$$n_C = 3.22 \text{ mol C}$$

$$n_H = 9.71 \text{ g H} \times \frac{1 \text{ mol H}}{1.01 \text{ g H}}$$

$$n_H = 9.61 \text{ mol H}$$

$$n_O = 51.58 \text{ g O} \times \frac{1 \text{ mol O}}{16.00 \text{ g O}}$$

$$n_O = 3.22 \text{ mol O}$$

Step 4: State Amount Ratio

$$n_C : n_H : n_O = 3.22 : 9.61 : 3.22$$

Step 5: Calculate Lowest Whole-Number Amount Ratio

Divide all the amounts by the smallest amount. Here the smallest amount is 3.22 mol O.

$$n_C : n_H : n_O = \frac{3.22}{3.22} : \frac{9.61}{3.22} : \frac{3.22}{3.22}$$

$$n_C : n_H : n_O = 1.00 : 2.98 : 1.00$$

Notice that the value for hydrogen is 2.98. Round 2.98 to the closest whole number, which is 3.

$$n_C : n_H : n_O = 1 : 3 : 1$$

Therefore, the empirical formula of the compound is CH_3O.

(b) The percentage composition of a compound is 69.9% iron and 30.1% oxygen. What is the empirical formula of the compound?

Step 1: List Given Values

Fe = 69.9%

O = 30.1%

Step 2: Calculate Mass (m) of Each Element in 100-g Sample

$$m_{Fe} = \frac{69.9}{100} \times 100 \text{ g}$$

$$m_{Fe} = 69.9 \text{ g}$$

$$m_O = \frac{30.1}{100} \times 100 \text{ g}$$

$$m_O = 30.1 \text{ g}$$

Step 3: Convert Mass (m) into Amount (n)

$$n_{Fe} = 69.9 \text{ g Fe} \times \frac{1 \text{ mol Fe}}{55.85 \text{ g Fe}}$$

$$n_{Fe} = 1.25 \text{ mol Fe}$$

$$n_O = 30.1 \text{ g O} \times \frac{1 \text{ mol O}}{16.00 \text{ g O}}$$

$$n_O = 1.88 \text{ mol O}$$

Step 4: State Amount Ratio

$$n_{Fe} : n_O = 1.25 : 1.88$$

Step 5: Calculate Lowest Whole-Number Amount Ratio

Divide all the amounts by the smallest amount. Here the smallest amount is 1.25 mol Fe.

$$n_{Fe} : n_O = \frac{1.25}{1.25} : \frac{1.88}{1.25}$$

$$n_{Fe} : n_O = 1.00 : 1.50$$

Notice that this ratio has one number that is not a whole number. To obtain a whole-number ratio, multiply both numbers by 2.

$$n_{Fe} : n_O = 2(1.00) : 2(1.50)$$
$$= 2.00 : 3.00$$
$$n_{Fe} : n_O = 2 : 3$$

Therefore, the empirical formula of the compound is Fe_2O_3.

Example

The percentage composition of a compound is 21.6% sodium, 33.3% chlorine, and 45.1% oxygen. What is the empirical formula of the compound?

Solution

$Na = 21.6\%$

$Cl = 33.3\%$

$O = 45.1\%$

$$m_{Na} = \frac{21.6}{100} \times 100 \text{ g}$$

$$m_{Na} = 21.6 \text{ g}$$

$$m_{Cl} = \frac{33.3}{100} \times 100 \text{ g}$$

$$m_{Cl} = 33.3 \text{ g}$$

$$m_O = \frac{45.1}{100} \times 100 \text{ g}$$

$$m_O = 45.1 \text{ g}$$

$$n_{Na} = 21.6 \text{ g Na} \times \frac{1 \text{ mol Na}}{22.99 \text{ g Na}}$$

$$n_{Na} = 0.940 \text{ mol Na}$$

$$n_{Cl} = 33.3 \text{ g Cl} \times \frac{1 \text{ mol Cl}}{35.45 \text{ g Cl}}$$

$$n_{Cl} = 0.939 \text{ mol Cl}$$

$$n_O = 45.1 \text{ g O} \times \frac{1 \text{ mol O}}{16.00 \text{ g O}}$$

$$n_O = 2.82 \text{ mol O}$$

$$n_{Na} : n_{Cl} : n_O = 0.940 : 0.939 : 2.82$$

$$= \frac{0.940}{0.939} : \frac{0.939}{0.939} : \frac{2.82}{0.939}$$

$$= 1.00 : 1.00 : 3:00$$

$$n_{Na} : n_{Cl} : n_O = 1 : 1 : 3$$

The empirical formula of this compound is $NaClO_3$.

Understanding Concepts

1. Calculate the empirical formula of a compound that, on analysis, is found to contain 2.2% hydrogen, 26.7% carbon, and 71.1% oxygen.

2. The percentage composition of a compound is 35.9% aluminum and 64.1% sulfur. What is the empirical formula of the compound?

SUMMARY	**Steps Used to Determine an Empirical Formula**

Step 1: List Given Values
Step 2: Calculate Mass (m) of Each Element in 100-g Sample
Step 3: Convert Mass (m) into Amount (n)
Step 4: State Amount Ratio
Step 5: Calculate Lowest Whole-Number Amount Ratio

Determining the Molecular Formula

molecular formula a formula that indicates the actual numbers of atoms in one molecule of a compound

The **molecular formula** of a compound tells you the exact number of atoms in one molecule of the compound. The molecular formula may be equal to the empirical formula, or it may be a multiple of this formula. The molecular formula of water, $H_2O_{(l)}$, cannot be reduced any further, so it is also the empirical formula. The molecular formula of hydrogen peroxide is $H_2O_{2(l)}$, which can be reduced to its empirical formula of HO. To determine the molecular formula of a compound, you need to know the empirical formula and the molar mass of the compound. As explained earlier, the molar mass can be obtained using a mass spectrometer.

▶ **SAMPLE** problem 2

Calculating the Molecular Formula of a Compound

(a) **The empirical formula of a compound is CH_3O, and its molar mass is 93.12 g/mol (as determined by a mass spectrometer). What is the molecular formula of the compound?**

Step 1: List Given Values

empirical formula of compound = CH_3O

$M_{compound}$ = 93.12 g/mol

Step 2: Determine Molar Mass of Empirical Formula

M_{CH_3O} = 1(12.01 g/mol) + 3(1.01 g/mol) + 1(16.00 g/mol)

M_{CH_3O} = 31.04 g/mol

Step 3: Determine Ratio of Molar Mass of Compound to Molar Mass of Empirical Formula

The $\dfrac{M_{compound}}{M_{empirical\ formula}}$ ratio gives the factor by which the molar mass of the compound is greater than the molar mass of the empirical formula.

$$\frac{M_{compound}}{M_{CH_3O}} = \frac{93.12\ \cancel{g/mol}}{31.04\ \cancel{g/mol}}$$

$$\frac{M_{compound}}{M_{CH_3O}} = 3$$

Step 4: Calculate Molecular Formula

Use the ratio determined in step 3 to calculate the molecular formula. The molar mass of the compound is three times greater than the molar mass of its empirical formula. Multiply the subscripts of the empirical formula by 3 to obtain the subscripts of the molecular formula.

$$molecular\ formula = 3(empirical\ formula)$$
$$= 3(CH_3O)$$
$$molecular\ formula = C_3H_9O_3$$

The molecular formula of the compound is $C_3H_9O_3$.

(b) The percentage composition of a compound, as determined by a combustion analyzer, is 40.03% carbon, 6.67% hydrogen, and 53.30% oxygen. Using a mass spectrometer, the molar mass of the compound is found to be 180.18 g/mol. What is the molecular formula of the compound?

Notice that the empirical formula is not given. You can use the percentage composition values, however, to determine the empirical formula of the compound. Then you can use the empirical formula and the molar mass of the compound to determine the molecular formula. The solution to this problem includes the calculations for determining the empirical formula of a compound (Part 1) and the calculations for determining the molecular formula (Part 2).

Part 1: Determine Empirical Formula

Step 1: List Given Values

C = 40.03%

H = 6.67%

O = 53.30%

$M_{compound}$ = 180.18 g/mol

Step 2: Calculate Mass (m) of Each Element in 100-g Sample

$$m_C = \frac{40.03}{100} \times 100\ g$$

$$m_C = 40.03\ g$$

$$m_H = \frac{6.67}{100} \times 100 \text{ g}$$

$$m_H = 6.67 \text{ g}$$

$$m_O = \frac{53.30}{100} \times 100 \text{ g}$$

$$m_O = 53.30 \text{ g}$$

Step 3: Convert Mass (m) into Amount (n)

$$n_C = 40.03 \text{ g C} \times \frac{1 \text{ mol C}}{12.01 \text{ g C}}$$

$$n_C = 3.33 \text{ mol C}$$

$$n_H = 6.67 \text{ g H} \times \frac{1 \text{ mol H}}{1.01 \text{ g H}}$$

$$n_H = 6.60 \text{ mol H}$$

$$n_O = 53.30 \text{ g O} \times \frac{1 \text{ mol O}}{16.00 \text{ g O}}$$

$$n_O = 3.33 \text{ mol O}$$

Step 4: State Amount Ratio

$$n_C : n_H : n_O = 3.33 : 6.60 : 3.33$$

Step 5: Calculate Lowest Whole-Number Amount Ratio

To obtain the lowest whole-number ratio of elements, divide all the amounts by the smallest amount. Here the smallest amount is 3.33 mol O.

$$n_C : n_H : n_O = \frac{3.33}{3.33} : \frac{6.60}{3.33} : \frac{3.33}{3.33}$$

$$n_C : n_H : n_O = 1.00 : 1.98 : 1.00$$

Notice that the value for hydrogen is 1.98. Round 1.98 to the closest whole number, which is 2.

$$n_C : n_H : n_O = 1 : 2 : 1$$

Therefore, the empirical formula of the compound is CH_2O.

Part 2: Determine Molecular Formula

Step 6: Determine Molar Mass of Empirical Formula

$$M_{CH_2O} = 1(12.01 \text{ g/mol}) + 2(1.01 \text{ g/mol}) + 1(16.00 \text{ g/mol})$$
$$M_{CH_2O} = 30.03 \text{ g/mol}$$

Step 7: Determine Ratio of Molar Mass of Compound to Molar Mass of Empirical Formula

$$\frac{M_{compound}}{M_{CH_2O}} = \frac{180.18 \text{ g/mol}}{30.03 \text{ g/mol}}$$

$$\frac{M_{compound}}{M_{CH_2O}} = 6$$

Step 8: Calculate Molecular Formula

The molar mass of the compound is six times greater than the molar mass of the empirical formula. Therefore, multiply the subscripts of the empirical formula by 6 to obtain the subscripts of the molecular formula.

$$\text{molecular formula} = 6(\text{empirical formula})$$
$$= 6(CH_2O)$$
$$\text{molecular formula} = C_6H_{12}O_6$$

The molecular formula of the compound is $C_6H_{12}O_6$.

Example

A combustion analyzer determines the percentage composition of a compound to be 32.0% carbon, 6.70% hydrogen, 42.6% oxygen, and 18.7% nitrogen. Using a mass spectrometer, the molar mass of the compound is found to be 75.08 g/mol. What is the molecular formula of the compound?

Solution

$C = 32.0\%$

$H = 6.70\%$

$O = 42.6\%$

$N = 18.7\%$

$M_{compound} = 75.08$ g/mol

$$m_C = \frac{32.0}{100} \times 100 \text{ g} \qquad\qquad m_O = \frac{42.6}{100} \times 100 \text{ g}$$

$$m_C = 32.0 \text{ g} \qquad\qquad\qquad m_O = 42.6 \text{ g}$$

$$m_H = \frac{6.70}{100} \times 100 \text{ g} \qquad\qquad m_N = \frac{18.7}{100} \times 100 \text{ g}$$

$$m_H = 6.70 \text{ g} \qquad\qquad\qquad m_N = 18.7 \text{ g}$$

$$n_C = 32.0 \text{ g C} \times \frac{1 \text{ mol C}}{12.01 \text{ g C}} \qquad\qquad n_O = 42.6 \text{ g O} \times \frac{1 \text{ mol O}}{16.00 \text{ g O}}$$

$$n_C = 2.66 \text{ mol C} \qquad\qquad\qquad n_O = 2.66 \text{ mol O}$$

$$n_H = 6.70 \text{ g H} \times \frac{1 \text{ mol H}}{1.01 \text{ g H}} \qquad\qquad n_N = 18.7 \text{ g N} \times \frac{1 \text{ mol N}}{14.01 \text{ g N}}$$

$$n_H = 6.63 \text{ mol H} \qquad\qquad\qquad n_N = 1.33 \text{ mol N}$$

$$n_C : n_H : n_O : n_N = 2.66 : 6.63 : 2.66 : 1.33$$

$$= \frac{2.66}{1.33} : \frac{6.63}{1.33} : \frac{2.66}{1.33} : \frac{1.33}{1.33}$$

$$= 2.00 : 4.98 : 2.00 : 1.00$$

$$n_C : n_H : n_O : n_N = 2 : 5 : 2 : 1$$

The empirical formula of the compound is $C_2H_5O_2N$.

$$M_{C_2H_5O_2N} = 2(12.01 \text{ g/mol}) + 5(1.01 \text{ g/mol}) + 2(16.00 \text{ g/mol}) + 1(14.01 \text{ g/mol})$$

$$M_{C_2H_5O_2N} = 75.00 \text{ g/mol } C_2H_5O_2N$$

$$\frac{M_{compound}}{M_{C_2H_5O_2N}} = \frac{75.00 \text{ g/mol}}{75.08 \text{ g/mol}}$$

$$\frac{M_{compound}}{M_{C_2H_5O_2N}} = 1$$

$$\text{molecular formula} = 1(\text{empirical formula})$$
$$= 1(C_2H_5O_2N)$$
$$\text{molecular formula} = C_2H_5O_2N$$

The molecular formula of the compound is $C_2H_5O_2N$.

Understanding Concepts

Answers

3. $C_8H_{12}O_2$
4. $C_{45}H_{85}O_5$
5. $K_2Cr_2O_7$
6. $C_{10}H_{14}N_2$

3. A combustion analyzer determines the percentage composition of a compound to be 68.54% carbon, 8.63% hydrogen, and 22.83% oxygen. A mass spectrometer determines its molar mass to be 140.20 g/mol. What is the molecular formula of the compound?

4. A fat that is used to make soap contains 76.5% carbon, 12.2% hydrogen, and 11.3% oxygen by mass. Determine the molecular formula of the fat if its molar mass is 706.3 g/mol.

5. A substance contains 26.65% potassium, 35.33% chromium, and 38.02% oxygen. Its molar mass is 294.20 g/mol. Determine the molecular formula of the compound.

6. The percentage composition of nicotine is 74.0% carbon, 8.7% hydrogen, and 17.3% nitrogen. Its molar mass is 162.26 g/mol. What is the molecular formula of nicotine?

Calculating Percentage Composition by Mass

When the chemical formula of a compound is known, its percentage composition may be calculated using the atomic masses of its elements and the molecular mass (or formula unit mass) of the compound.

▶ **SAMPLE problem 3**

Calculating the Percentage Composition of a Compound

Calculate the percentage composition of carbon dioxide, $CO_{2(g)}$.

Step 1: Calculate Total Mass of Each Element in Compound

$$m_C = 12.01 \text{ u}$$
$$m_O = 2(16.00 \text{ u})$$
$$m_O = 32.00 \text{ u}$$

Step 2: Calculate Molecular Mass (or Formula Unit Mass) of Compound

$$m_{CO_2} = m_C + m_O$$
$$= 12.01 \text{ u} + 32.00 \text{ u}$$
$$m_{CO_2} = 44.01 \text{ u}$$

Step 3: Calculate Percentage Composition by Mass of Compound

$$\% \text{ C} = \frac{m_C}{m_{CO_2}} \times 100\% \qquad\qquad \% \text{ O} = \frac{m_O}{m_{CO_2}} \times 100\%$$

$$= \frac{12.01 \text{ u}}{44.01 \text{ u}} \times 100\% \qquad\qquad = \frac{32.00 \text{ u}}{44.01 \text{ u}} \times 100\%$$

$$\% \text{ C} = 27.29\% \qquad\qquad\qquad \% \text{ O} = 72.71\%$$

The percentage composition by mass of carbon dioxide is 27.29% carbon and 72.71% oxygen.

Example

Calculate the percentage composition by mass of potassium sulfate, $K_2SO_{4(s)}$.

Solution

$$m_K = 2(39.10 \text{ u}) \qquad m_S = 32.06 \text{ u} \qquad m_O = 4(16.00 \text{ u})$$
$$m_K = 78.20 \text{ u} \qquad\qquad\qquad\qquad\quad m_O = 64.00 \text{ u}$$

$$m_{K_2SO_4} = 2(m_K) + (m_S) + 4(m_O)$$
$$= 2(39.10 \text{ u}) + (32.06 \text{ u}) + 4(16.00 \text{ u})$$
$$= 78.20 \text{ u} + 32.06 \text{ u} + 64.00 \text{ u}$$
$$m_{K_2SO_4} = 174.26 \text{ u}$$

$$\% \text{ K} = \frac{78.20 \text{ u}}{174.26 \text{ u}} \times 100\% \qquad\qquad \% \text{ S} = \frac{32.06 \text{ u}}{174.26 \text{ u}} \times 100\%$$

$$\% \text{ K} = 44.87\% \qquad\qquad\qquad\qquad \% \text{ S} = 18.40\%$$

$$\% \text{ O} = \frac{64.00 \text{ u}}{174.26 \text{ u}} \times 100\%$$

$$\% \text{ O} = 36.73\%$$

The percentage composition by mass of potassium sulfate is 44.87% potassium, 18.40% sulfur, and 36.73% oxygen.

▶ **Practice**

7. Calculate the percentage composition by mass of each of the following compounds:
  (a) $C_6H_8O_{6(s)}$
  (b) $Al_2O_{3(s)}$
  (c) $Zn(NO_3)_{2(s)}$

Answers

7. (a) 40.91% C, 4.59% H, 54.50% O
   (b) 52.92% Al, 47.08% O
   (c) 33.82% Zn, 14.94% N, 51.18% O

Steps Used to Determine a Molecular Formula

Part 1: Determine Empirical Formula
Step 1: List Given Values
Step 2: Calculate Mass (m) of Each Element in 100-g Sample
Step 3: Convert Mass (m) into Amount (n)
Step 4: State Amount Ratio
Step 5: Calculate Lowest Whole-Number Amount Ratio

Part 2: Determine Molecular Formula
Step 6: Determine Molar Mass of Empirical Formula
Step 7: Determine Ratio of Molar Mass of Compound to Molar Mass of Empirical Formula
Step 8: Calculate Molecular Formula

Steps Used to Calculate Percentage Composition by Mass

Step 1: Calculate Total Mass of Each Element in Compound
Step 2: Calculate Molecular Mass (or Formula Unit Mass) of Compound
Step 3: Calculate Percentage Composition by Mass of Compound

▶ Section 2.3 Questions

Understanding Concepts

1. State the law of constant composition, in your own words.

2. (a) What is a mass spectrometer used for? Briefly describe how it works.
   (b) What is a combustion analyser used for? Briefly describe how it works.

3. (a) What information do you need to determine the empirical formula of a compound?
   (b) What information, in addition to the empirical formula, do you need to determine the molecular formula of a compound?

4. Write the empirical formula for each of the following molecular formulas:
   (a) $C_2H_4O_2$
   (b) NH_3
   (c) C_6H_6

5. What is the empirical formula of a compound that contains 26.6% potassium, 35.4% chromium, and 38.1% oxygen?

6. The percentage composition values of two antibiotics are given. Determine the empirical formula of each antibiotic.

   (a) chloromycetin: 40.87% carbon, 3.72% hydrogen, 8.67% nitrogen, 24.77% oxygen. The rest is chlorine.
   (b) sulfanilamide: 41.86% carbon, 4.65% hydrogen, 16.28% nitrogen, 18.60% oxygen. The rest is sulfur.

7. Calculate the percentage composition by mass of each of the following compounds:
   (a) $H_2O_{(l)}$
   (b) $Ca(OH)_{2(s)}$

Applying Inquiry Skills

8. The analysis of a compound shows that it contains 21.9% sodium, 45.7% carbon, 1.9% hydrogen, and 30.5% oxygen. A mass spectrometer determines its molar mass to be 210 g/mol. What is the molecular formula of the compound?

9. The analysis of a newly discovered compound produced the following data:
   • combustion analyzer: 49.38% carbon, 3.55% hydrogen, 9.40% oxygen, 37.67% sulfur
   • mass spectrometer: molar mass = 170.2 g/mol
   Determine the molecular formula of the compound.

Inquiry Skills

○ Questioning ○ Planning ● Analyzing
○ Hypothesizing ● Conducting ● Evaluating
● Predicting ● Recording ● Communicating

Percentage Composition by Mass of Magnesium Oxide

Magnesium is a silvery metal that burns with such a bright flame that it was once used in flashbulbs for photography. Magnesium oxide, a white powder, is a product of the combustion reaction. In this investigation, you will use the combustion of magnesium to test the law of constant composition. You will determine the composition, by mass, of magnesium oxide and calculate its percentage composition. Then you will compare your results and other students' results with the predicted values.

Question

What is the percentage composition, by mass, of magnesium oxide?

Prediction

(a) Predict the percentage composition of magnesium oxide, based on the law of constant composition.

Materials

eye protection
lab apron
electronic balance
7–8 cm magnesium ribbon
steel wool
porcelain crucible and lid
Bunsen burner
retort stand
ring stand and clamp
clay triangle
crucible tongs
glass stirring rod
distilled water
10-mL graduated cylinder

Procedure

1. Using a balance, determine the mass of a clean and dry porcelain crucible and lid.

2. Polish the magnesium ribbon with the steel wool. Then fold the ribbon to fit into the bottom of the crucible.

3. Determine the combined mass of the crucible, lid, and magnesium ribbon.

4. Place the crucible securely on the clay triangle. Set the lid slightly off-centre on the crucible, to allow air to enter but to prevent the magnesium oxide from escaping.

5. Place the Bunsen burner under the crucible, light the burner, and begin to heat the crucible with a gentle flame (**Figure 1**).

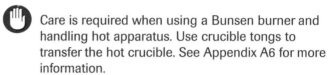 Care is required when using a Bunsen burner and handling hot apparatus. Use crucible tongs to transfer the hot crucible. See Appendix A6 for more information.

6. Gradually increase the flame intensity until all the magnesium turns into a white powder.

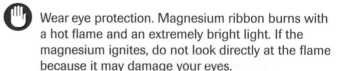 Wear eye protection. Magnesium ribbon burns with a hot flame and an extremely bright light. If the magnesium ignites, do not look directly at the flame because it may damage your eyes.

Figure 1

7. Cut the flow of gas to the burner. Allow the crucible, lid, and contents to cool.

8. Using the stirring rod, crush the contents of the crucible into a fine powder. Using the graduated cylinder, carefully add approximately 10 mL of distilled water to the powder. Use some of the water to rinse any powder on the stirring rod into the crucible.

9. With the lid slightly ajar, heat the crucible and contents gently for 3 min and strongly for an additional 7 min. Turn off the Bunsen burner.

10. Allow the crucible, lid, and contents to cool. Remove the crucible from the stand using crucible tongs (**Figure 2**).

11. Using the balance, determine the combined mass of the cooled crucible, lid, and contents.

12. Dispose of your materials as directed by your teacher.

Analysis

(b) What evidence did you obtain to indicate that a chemical reaction took place?

(c) Why were the contents of the crucible crushed with a stirring rod in step 8 and reheated in step 9?

(d) Calculate the mass of oxygen that reacted with magnesium.

(e) Answer the Question.

Evaluation

(f) If some of the magnesium oxide had escaped from the crucible, would your calculation of the percentage composition of magnesium be higher than your Prediction or lower than your Prediction? Explain.

(g) If the magnesium had reacted with some other component in the air, would your calculation of the percentage composition of magnesium be higher than your Prediction or lower than your Prediction? Explain.

(h) You polished the magnesium ribbon to remove any white film on its surface before beginning the experiment. Explain why you needed to remove the white film.

(i) Describe some sources of experimental error. Suggest modifications you could make to the Procedure to ensure that all the magnesium would completely react with oxygen.

(j) Evaluate your Prediction. Based on the evidence obtained from several groups, is the law of constant composition valid?

Synthesis

(k) Gemologists use analytical skills to distinguish between different gems, and weigh and appraise gems. Gemologists rely on the analysis of percentage composition to determine the value of a particular gem. They may work as designers, salespeople, or appraisers. They may also work with mining companies, analyzing the gems extracted from the ground. Research the educational requirements to become a gemologist, as well as some of the instruments that gemologists use when working with gems.

 www.science.nelson.com

Figure 2

When a driver is asked to breathe into a Breathalyzer, the instrument analyzes the breath sample and determines the quantity of alcohol in the driver's blood (**Figure 1**). A lifeguard at a swimming pool takes a sample of pool water to analyze the quantity of acid and determine whether any chemicals need to be added to the pool (**Figure 2**). Soil samples are sent to a laboratory to be analyzed for the quantity of toxins. Blood and urine are routinely analyzed in medical labs to determine the quantities of sugars, ions, gases, and fats they contain.

High levels of cholesterol in the blood have been linked to heart disease and stroke. The quantity of cholesterol in the blood increases after a meal because many foods contain cholesterol. As the cells of the body use cholesterol, its quantity in the blood decreases to a minimum level. To obtain an accurate measurement of the quantity of cholesterol in the blood, a patient is asked to fast for 12 to 14 h before a blood sample is drawn. The patient is also asked to stop taking medications that could affect the cholesterol level.

In many laboratory procedures, such as the cholesterol test, the *quantity* (usually concentration) of a chemical entity is measured. This type of analysis is called **quantitative analysis**.

Figure 1
A Breathalyzer provides an on-the-spot quantitative analysis of a person's blood-alcohol content by measuring the concentration of alcohol in a sample of exhaled air.

Concentrations of Solutions

In many cases, quantitative analysis involves the measurement of the concentration of a solute in a solution. The **concentration** of a solution is the ratio of the quantity of solute to the quantity of solution. In general, the concentration, *c*, may be expressed by the ratio

$$\text{concentration} = \frac{\text{quantity of solute}}{\text{quantity of solution}}$$

A solution is dilute if it has a relatively small quantity of solute per unit volume of solution (**Figure 3(a)**). A solution is concentrated if it has a relatively large quantity of solute per unit volume of solution (**Figure 3(b)**).

Figure 2
Quantitative analysis requires careful and precise work. It also requires perseverance and openness to unexpected results.

quantitative analysis
measurement of the quantity of a chemical entity

concentration a ratio of the quantity of solute to the quantity of solution; symbol *c*

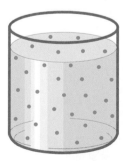

(a) dilute solution

(b) concentrated solution

Figure 3

An intravenous (IV) solution is a solution that is injected directly into a patient's veins. Knowing the concentration of an IV solution is critical in a hospital setting. The label on the container for a typical IV solution shows the concentration of the solute in the solution (**Figure 4**). The doctor or nurse is careful to administer the solution with the correct solute concentration to a patient. Giving an IV solution with the wrong concentration can sometimes be fatal. If the concentration of an IV solution is too low, for example, red blood cells swell and burst, causing a condition called hemolytic crisis. In hemolytic crisis, the cells of the body are deprived of nutrients and oxygen. Severe cases of hemolytic crisis can be fatal if not treated quickly.

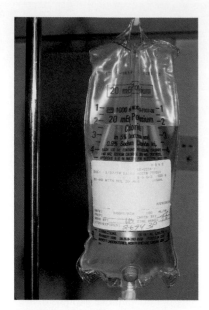

Figure 4
The label on this IV solution indicates that it contains 0.9% sodium chloride.

Percentage Concentration

The concentrations of solutions are sometimes expressed as percentages, especially on the labels of consumer products (**Figure 5**). For example, a vinegar label indicating 5% V/V acetic acid means that there are 5 mL of pure acetic acid dissolved in every 100 mL of vinegar solution. The V/V (volume by volume) notation indicates that two liquids (water and acetic acid) are mixed to make the solution. The first V refers to the volume of solute dissolved in the solution. The second V refers to the final volume of the solution.

When the solute is a solid, the percentage concentration is followed by W/V (weight by volume). Historically, the W in W/V stood for weight. Today, the W refers to the mass of the solid solute that is used to make the solution. The V refers to the total volume of solution made.

Simple equations are used to calculate percentage concentrations. For volume by volume (V/V) concentrations,

$$c_{solution} = \frac{v_{solute}}{v_{solution}} \times 100\%$$

where $c_{solution}$ is the concentration of the solution
v_{solute} is the volume of solute in the solution
$v_{solution}$ is the volume of the solution

For weight by volume (W/V) concentrations,

$$c_{solution} = \frac{m_{solute}}{v_{solution}} \times 100\%$$

where $c_{solution}$ is the concentration of the solution
m_{solute} is the mass of solute in the solution
$v_{solution}$ is the volume of the solution

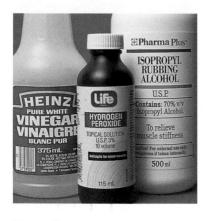

Figure 5
The concentrations of different consumer products depend on the product and sometimes the brand name. Concentrations are usually expressed as percentages.

▶ *SAMPLE* problem *1*

Calculating V/V Percentage Concentration

A photographic "stop bath" contains 160 mL of pure acetic acid, $HC_2H_3O_{2(l)}$, in 600 mL of solution. What is the V/V percentage concentration of acetic acid in the stop bath?

Step 1: List Given Values

$$v_{HC_2H_3O_2} = 160 \text{ mL}$$
$$v_{solution} = 600 \text{ mL}$$
$$c_{solution} = ?$$

Step 2: Write Percentage Concentration Equation, Substitute Values, and Solve

Here you need to use the V/V percentage concentration equation:

$$c_{solution} = \frac{v_{HC_2H_3O_2}}{v_{solution}} \times 100\%$$

$$= \frac{160 \text{ mL}}{600 \text{ mL}} \times 100\%$$

$$c_{solution} = 27\% \text{ V/V}$$

The concentration of the stop bath is 27% V/V.

Example

A salt solution is formed by mixing 2.80 g of sodium chloride, $NaCl_{(s)}$, in enough water to make exactly 250 mL of solution. What is the W/V percentage concentration of sodium chloride in the salt solution?

Solution

$$m_{NaCl} = 2.80 \text{ g}$$
$$v_{solution} = 250 \text{ mL}$$
$$c_{solution} = ?$$

$$c_{solution} = \frac{m_{solute}}{v_{solution}} \times 100\%$$

$$= \frac{2.80 \text{ g}}{250 \text{ mL}} \times 100\%$$

$$c_{solution} = 1.1\% \text{ W/V}$$

The concentration of the salt solution is 1.1% W/V.

molar concentration the concentration of a solution expressed as moles of solute per unit volume of solution (mol/L)

> ▶ **Practice**

Understanding Concepts

1. Gasohol, a solution of ethanol and gasoline, is considered to be a cleaner fuel than gasoline alone. A typical gasohol mixture, which is available across Canada, contains 4.1 L of ethanol in a 55-L tank of fuel. Calculate the V/V percentage concentration of ethanol in this mixture.

2. Solder flux, which is available at hardware and craft stores, contains 16 g of zinc chloride in 500 mL of solution. The solvent is hydrochloric acid. What is the W/V percentage concentration of zinc chloride in the solution?

3. An IV solution is prepared by dissolving 27.5 g of glucose, $C_6H_{12}O_{6(s)}$, in water to make 550 mL of solution. What is the W/V percentage concentration of glucose in the IV solution?

Molar Concentration

In laboratory chemistry, the concentration of a solution is more commonly expressed as the **molar concentration** (c). The molar concentration is the amount of solute, in moles, that is dissolved in one litre of solution.

$$\text{molar concentration} = \frac{\text{amount of solute (mol)}}{\text{volume of solution (L)}}$$

$$c = \frac{n}{v}$$

where c is the molar concentration in mol/L

  n is the amount of solute in mol

  v is the volume of solution in L

Molar concentration is sometimes indicated by the use of square brackets. For example, if the molar concentration of an ammonia solution is 0.5 mol/L, it may be written as $[NH_{3(aq)}] = 0.5$ mol/L.

> ▶ **SAMPLE** problem 2
>
> ### Calculating Molar Concentration of a Solution, Given the Amount of Solute
>
> **(a) A sodium hydroxide solution contains 0.186 mol of sodium hydroxide in 0.250 L of solution. Calculate the molar concentration of the sodium hydroxide solution.**
>
> **Step 1: List Given Values**
>
> $n_{NaOH} = 0.186$ mol
>
> $v_{NaOH} = 0.250$ L
>
> $c_{NaOH} = ?$

Step 2: Write Molar Concentration Equation, Substitute Values, and Solve

$$c_{NaOH} = \frac{n_{NaOH}}{v_{NaOH}}$$

$$= \frac{0.186 \text{ mol NaOH}}{0.250 \text{ L}}$$

$$c_{NaOH} = 0.744 \text{ mol/L}$$

The molar concentration of the sodium hydroxide solution is 0.744 mol/L.

(b) A solution is prepared by dissolving 1.68 g of copper(II) sulfate, $CuSO_{4(s)}$, in enough water to make 150 mL of solution. Calculate the molar concentration of the copper(II) sulfate solution. The molar mass of copper(II) sulfate is 159.6 g/mol.

In section 2.2, you learned that chemists measure the amount of a chemical (number of moles) by mass. Here you are given the mass of solute that is used to prepare a solution, and you need to calculate the concentration of the solution.

Step 1: List Given Values

$$m_{CuSO_4} = 1.68 \text{ g}$$
$$v_{CuSO_4} = 150 \text{ mL}$$
$$c_{CuSO_4} = ?$$

Since molar concentration is calculated per litre of solution, convert the given volume of solution from millilitres to litres.

$$v_{CuSO_4} = 150 \text{ mL} \times \frac{1 \text{ L}}{1000 \text{ mL}}$$

$$v_{CuSO_4} = 0.150 \text{ L}$$

Step 2: Convert Mass of Compound to Amount of Compound
In this problem, you are given the mass of the solute and the volume of the solution. To calculate the molar concentration of the solution, you must first calculate the amount of copper(II) sulfate in 1.68 g of the salt.

$$n_{CuSO_4} = 1.68 \text{ g } CuSO_4 \times \frac{1 \text{ mol } CuSO_4}{159.6 \text{ g } CuSO_4}$$

$$n_{CuSO_4} = 0.0105 \text{ mol } CuSO_4$$

Step 3: Write Molar Concentration Equation, Substitute Values, and Solve

$$c_{CuSO_4} = \frac{n_{CuSO_4}}{v_{CuSO_4}}$$

$$= \frac{0.0105 \text{ mol } CuSO_4}{0.150 \text{ L}}$$

$$c_{CuSO_4} = 0.0700 \text{ mol/L}$$

The molar concentration of the copper(II) sulfate solution is 0.0700 mol/L.

Example

Sodium carbonate, $Na_2CO_{3(s)}$, is a water softener that is a significant part of washing-machine detergent. A student dissolves 5.00 g of solid sodium carbonate in enough water to make 250 mL of solution. What is the molar concentration of the sodium carbonate solution?

Solution

$$m_{Na_2CO_3} = 5.00 \text{ g}$$

$$v_{Na_2CO_3} = 250 \text{ mL} \times \frac{1 \text{ L}}{1000 \text{ mL}}$$

$$v_{Na_2CO_3} = 0.250 \text{ L}$$

$$M_{Na_2CO_3} = 105.99 \text{ (calculated from periodic table values)}$$

$$c_{Na_2CO_3} = ?$$

$$n_{Na_2CO_3} = 5.00 \text{ g Na}_2\text{CO}_3 \times \frac{1 \text{ mol Na}_2CO_3}{105.99 \text{ g Na}_2\text{CO}_3}$$

$$n_{Na_2CO_3} = 0.0472 \text{ mol Na}_2CO_3$$

$$c_{Na_2CO_3} = \frac{n_{Na_2CO_3}}{v_{Na_2CO_3}}$$

$$= \frac{0.0472 \text{ mol Na}_2CO_3}{0.250 \text{ L}}$$

$$c_{Na_2CO_3} = 0.189 \text{ mol/L}$$

The molar concentration of the sodium carbonate solution is 0.189 mol/L.

> ### ▶ Practice

Understanding Concepts

Answers

4. 0.705 mol/L

5. 1.34 mol/L

6. 0.140 mol/L

7. 0.200 mol/L

4. Household bleach is an aqueous solution that contains 5.25 g of sodium hypochlorite, $NaOCl_{(aq)}$, per 100.0 mL of solution. What is the molar concentration of sodium hypochlorite in bleach?

5. A brine solution that is used in pickling contains 235 g of pure sodium chloride, $NaCl_{(s)}$, dissolved in 3.00 L of solution. What is the molar concentration of the sodium chloride?

6. A stock solution of hydrochloric acid, $HCl_{(aq)}$, is made by dissolving 7.66 g of hydrogen chloride, $HCl_{(g)}$, in enough distilled water to produce 1.50 L of solution. Calculate the concentration of aqueous hydrochloric acid in this stock solution.

7. Potassium dichromate solution, $K_2Cr_2O_{7(aq)}$, is used to analyze the alcohol content of wine. A lab technician dissolves 102.9 g of solid potassium dichromate in enough distilled water to produce 1.75 L of aqueous potassium dichromate. Determine the concentration of aqueous potassium dichromate in this solution.

You can also use the molar concentration equation to calculate the amount (number of moles) of solute in a solution, when you know the concentration and volume.

▶ *SAMPLE* problem *3*

Calculating Amount of Solute in a Solution

A sample of laboratory ammonia solution, $NH_{3(aq)}$, has a molar concentration of 14.8 mol/L (Figure 6). What amount of ammonia is present in a 1.50-L bottle? (Remember that *amount* means *number of moles.*)

Step 1: List Given Values

$$v_{NH_3} = 1.50 \text{ L}$$
$$c_{NH_3} = 14.8 \text{ mol/L}$$
$$n_{NH_3} = ?$$

Step 2: Write Molar Concentration Equation

$$c_{NH_3} = \frac{n_{NH_3}}{v_{NH_3}}$$

Step 3: Isolate Unknown Value on Left Side of Equation

To calculate the amount of ammonia, isolate the n_{NH_3} term on the left side of the equation.

$$n_{NH_3} = c_{NH_3} v_{NH_3}$$

Step 4: Substitute Values into Equation, and Solve

$$n_{NH_3} = c_{NH_3} v_{NH_3}$$
$$= \frac{14.8 \text{ mol } NH_3}{\cancel{L}} \times 1.50 \cancel{L}$$
$$n_{NH_3} = 22.2 \text{ mol } NH_3$$

There is 22.2 mol of ammonia in the bottle.

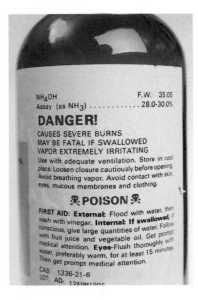

Figure 6
Aqueous ammonia is purchased for science laboratories as a concentrated solution.

Example

The concentration of sodium chloride, $NaCl_{(aq)}$, in typical blood serum is approximately 0.140 mol/L. What amount of sodium chloride is in 0.120 mL of blood serum?

Solution

$$v_{NaCl} = 0.120 \cancel{mL} \times \frac{1 \text{ L}}{1000 \cancel{mL}}$$
$$v_{NaCl} = 1.20 \times 10^{-4} \text{ L}$$
$$c_{NaCl} = 0.140 \text{ mol/L}$$
$$n_{NaCl} = ?$$

$$c_{NaCl} = \frac{n_{NaCl}}{v_{NaCl}}$$
$$n_{NaCl} = c_{NaCl} v_{NaCl}$$
$$= \frac{0.140 \text{ mol } NaCl}{\cancel{L}} \times 1.20 \times 10^{-4} \cancel{L}$$
$$n_{NaCl} = 1.68 \times 10^{-5} \text{ mol } NaCl$$

There is 1.68×10^{-5} mol of sodium chloride in the blood serum.

Answers

8. 0.0285 mol
9. 18.6 mol
10. 4.75 × 10^{-5} mol
11. 6.30 × 10^{-4} mol

> ▶ **Practice**

Understanding Concepts

8. What amount of silver nitrate, $AgNO_{3(aq)}$, is in 50.0 mL of a 0.570-mol/L solution?

9. The molar concentration of commercial hydrochloric acid, $HCl_{(aq)}$, is 12.4 mol/L. What amount of hydrogen chloride is in 1.50 L of commercial hydrochloric acid?

10. The hemoglobin in red blood cells carries oxygen to body tissues. What amount of hemoglobin is in 25.0 mL of a hemoglobin solution with a concentration of 1.90×10^{-3} mol/L?

11. A lab technician adds 35.8 mL of a 0.0176-mol/L solution of potassium hydroxide, $KOH_{(aq)}$, to a reaction mixture. What amount of potassium hydroxide does the technician add to the mixture?

The molar concentration equation may be used to calculate the volume of a solution when the concentration and amount of solute are known.

> ▶ **SAMPLE problem 4**
>
> **Calculating the Volume of a Solution**

What volume of a 5.0-mol/L glucose solution, $C_6H_{12}O_{6(aq)}$, contains 2.5 mol of glucose?

Step 1: List Given Values

$$c_{C_6H_{12}O_6} = 5.0 \text{ mol/L}$$
$$n_{C_6H_{12}O_6} = 2.5 \text{ mol}$$
$$v_{C_6H_{12}O_6} = ?$$

Step 2: Write Molar Concentration Equation

$$c_{C_6H_{12}O_6} = \frac{n_{C_6H_{12}O_6}}{v_{C_6H_{12}O_6}}$$

Step 3: Isolate Unknown Value on Left Side of Equation

To calculate the volume of solution, isolate the $v_{C_6H_{12}O_6}$ term on the left side of the equation.

$$v_{C_6H_{12}O_6} = \frac{n_{C_6H_{12}O_6}}{c_{C_6H_{12}O_6}}$$

Step 4: Substitute Values into Equation, and Solve

$$v_{C_6H_{12}O_6} = \frac{n_{C_6H_{12}O_6}}{c_{C_6H_{12}O_6}}$$

$$= \frac{2.5 \text{ mol}}{5.0 \text{ mol/L}}$$

$$v_{C_6H_{12}O_6} = 0.50 \text{ L}$$

Therefore, 0.50 L of 5.0-mol/L glucose solution contains 2.5 mol of glucose.

Example
What volume of a 0.120-mol/L copper(II) sulfate solution, $CuSO_{4(aq)}$, contains 0.150 mol of copper(II) sulfate?

Solution

$$c_{CuSO_4} = 0.120 \text{ mol/L}$$

$$n_{CuSO_4} = 0.150 \text{ mol}$$

$$v_{CuSO_4} = ?$$

$$c_{CuSO_4} = \frac{n_{CuSO_4}}{v_{CuSO_4}}$$

$$v_{CuSO_4} = \frac{n_{CuSO_4}}{c_{CuSO_4}}$$

$$= \frac{0.150 \text{ mol}}{0.120 \text{ mol/L}}$$

$$v_{CuSO_4} = 1.25 \text{ L}$$

Therefore, 1.25 L of copper(II) sulfate solution contains 0.150 mol copper(II) sulfate.

▶ **Practice**

Understanding Concepts

12. Seawater contains approximately 0.055 mol/L of magnesium chloride, $MgCl_{2(aq)}$. What volume of seawater contains 4.1 mol of magnesium chloride?

13. What volume of 7.6-mol/L hydrochloric acid, $HCl_{(aq)}$, must be poured into a flask to obtain 0.050 mol of hydrochloric acid?

14. A lab technician must add 1.25 mol of iron chloride, $FeCl_{3(aq)}$, to a flask using a 6.00-mol/L iron chloride stock solution. How many millilitres of the stock solution must the technician pour into the flask?

15. How many litres of a 0.0020-mol/L sodium dichromate solution, $Na_2Cr_2O_{7(aq)}$, contains 5.0 mol of sodium dichromate?

Answers

12. 75 L

13. 6.6 mL

14. 208 mL

15. 2500 L

Parts per Million

In environmental studies, very dilute solutions often need to be analyzed. For example, dioxin can be detected in lake water at concentrations as low as 3×10^{-6} mol/L. (Dioxin is a toxic pollutant that is produced by the incineration of plastics containing chlorine, and also by processes in the pulp and paper industry.) The human body is so sensitive to dioxin that, even at such low concentrations, it poses a health risk. Such low molar concentration values are cumbersome to deal with. Therefore, for very dilute solutions, chemists use **parts per million** (ppm), parts per billion (ppb), or parts per trillion (ppt). In general, one part per million of chlorine in a swimming pool corresponds to 1 g of chlorine in 1000 L of pool water, which is equivalent to 1 mg of chlorine per litre of water.

parts per million concentration unit that is used for very low concentrations; one part solute per million parts of solution; unit ppm

Concentrations in parts per million (ppm) can be expressed using a variety of units. When solving a problem, choose the units that match the information given in the problem. For aqueous solutions,

$$1 \text{ ppm} = 1 \text{ g}/10^6 \text{ mL}$$
$$= 1 \text{ g}/1000 \text{ L}$$
$$1 \text{ ppm} = 1 \text{ mg/L}$$

▶ **SAMPLE problem 5**

Calculating Concentration in Parts per Million

In a chemical analysis, 2.2 mg of oxygen was measured in 250 mL of pond water. What is the concentration of oxygen, in parts per million?

Step 1: List Given Values

$m_{O_2} = 2.2$ mg

$v_{H_2O} = 250$ mL

$c_{O_2} = ?$

Step 2: Write Concentration Equation

Since you are given a mass and a volume, use the W/V concentration equation:

$$c_{O_2} = \frac{m_{O_2}}{v_{H_2O}}$$

Step 3: Make Necessary Metric Conversions

Since 1 ppm = 1 mg/L, convert v_{H_2O} from millilitres to litres. The mass of oxygen is given in milligrams, so it does not have to be converted.

$$v_{H_2O} = 250 \text{ mL H}_2\text{O} \times \frac{1 \text{ L H}_2\text{O}}{1000 \text{ mL H}_2\text{O}}$$

$$v_{H_2O} = 0.25 \text{ L H}_2\text{O}$$

Step 4: Substitute Values into Equation, and Solve

$$c_{O_2} = \frac{m_{O_2}}{v_{H_2O}}$$

$$= \frac{2.2 \text{ mg O}_2}{0.25 \text{ L}}$$

$$= 8.8 \text{ mg/L}$$

$$c_{O_2} = 8.8 \text{ ppm}$$

The concentration of oxygen is 8.8 ppm.

Example

The maximum permitted mass of lead in 1.0 L of public drinking water is 5.0×10^{-5} g. What is this concentration, in parts per million?

Solution

$m_{Pb} = 5.0 \times 10^{-5}$ g

$v_{H_2O} = 1.0$ L

$c_{Pb} = ?$

$1 \text{ ppm} = 1 \text{ mg/L}$

$$m_{Pb} = 5.0 \times 10^{-5} \text{ g Pb} \times \frac{1000 \text{ mg Pb}}{1 \text{ g Pb}}$$

$$m_{Pb} = 0.050 \text{ mg Pb}$$

$$c_{Pb} = \frac{m_{Pb}}{v_{H_2O}}$$

$$= \frac{0.05 \text{ mg}}{1.0 \text{ L}}$$

$$= 0.05 \text{ mg/L}$$

$$c_{Pb} = 0.05 \text{ ppm}$$

The maximum permitted concentration of lead in public drinking water is 0.05 ppm.

▶ *Practice*

Understanding Concepts

16. Formaldehyde, $CH_2O_{(g)}$, is an indoor air pollutant that comes from synthetic materials and cigarette smoke. A dangerous level of formaldehyde is 3.2 mg in a 500.0-L sample of air. Express this concentration of formaldehyde in parts per million.

17. Copper is an element that is required in very small concentrations in the bodies of all animals. What is the concentration of copper, in parts per million, if 1.0 L of drinking water contains 3.0×10^{-5} g of copper?

18. Dissolved oxygen in natural water is an important measure of the health of the ecosystem. In a chemical analysis of the water from a pond, 350 mL of water is found to contain 1.8 mg of dissolved carbon dioxide. What is the concentration of dissolved carbon dioxide, in parts per million?

Answers

16. 6.4×10^{-3} ppm

17. 0.030 ppm

18. 5.1 ppm

SUMMARY   *Concentration of a Solution Equations*

Type	Equation	Units
percentage V/V	$c = \dfrac{v_{solute}}{v_{solution}} \times 100\%$	% V/V
percentage W/V	$c = \dfrac{m_{solute}}{v_{solution}} \times 100\%$	% W/V
very low (number)	$c = \dfrac{m_{solute}}{v_{solution}} \times 100\%$	mg/L = ppm µg/L = ppb ng/L = ppt
molar	$c = \dfrac{n_{solute}}{v_{solution}} \times 100\%$	mol/L

Diluting Aqueous Solutions

Experiments sometimes require several concentrations of the same solution. If you begin with a solution of known concentration (called a stock solution), you can prepare a solution of lower concentration by **dilution**. Diluting a stock solution is a faster and more accurate way to make lower concentrations of a solution than making different concentrations of the solution from scratch.

Calculating the concentration of a dilute solution is straightforward if you know the concentration of the original stock solution and the volume of solvent added. Adding more solvent does not change the quantity of solute, only its concentration. For example, if water is added to 6% hydrogen peroxide until the total volume is doubled, the concentration becomes one-half the original value, or 3% hydrogen peroxide (**Figure 7**).

dilution the process of decreasing the concentration of a solution by adding more solvent

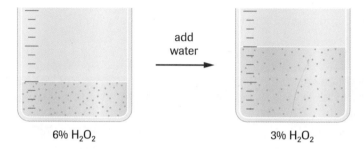

Figure 7
If water is added to 6% V/V hydrogen peroxide until the total volume is doubled, the concentration becomes one-half of the original concentration, or 3% V/V hydrogen peroxide.

You can calculate this concentration by using the following dilution equation:

$$c_i v_i = c_f v_f$$

where c_i is the concentration of the initial solution
v_i is the volume of the initial solution
c_f is the concentration of the final solution
v_f is the volume of the final solution

This equation allows you to solve for any one variable if the other three variables are known.

▶ **SAMPLE** problem 6

Calculations of Diluted Solutions

(a) Calculate the final concentration of a hydrogen peroxide solution if water is added to 100 mL of 6% V/V hydrogen peroxide until the total volume is 200 mL.

Step 1: List Given Values

c_i = 6% V/V

v_i = 100 mL

v_f = 200 mL

c_f = ?

Step 2: Write Dilution Equation

$$c_i v_i = c_f v_f$$

Step 3: Isolate Unknown Value on Left Side of Equation

Since you are asked to determine the concentration of the final solution, isolate c_f on the left side of the equation:

$$c_f = \frac{c_i v_i}{v_f}$$

Step 4: Substitute Values into Equation, and Solve

$$c_f = \frac{c_i v_i}{v_f}$$

$$= \frac{(6\%)(100 \text{ mL})}{200 \text{ mL}}$$

$$c_f = 3\%$$

The final concentration of the diluted solution is 3% V/V hydrogen peroxide.

(b) You are asked to dilute a 2.0-mol/L copper(II) sulfate solution to make 250 mL of 0.50-mol/L copper(II) sulfate solution. What volume of 2.0-mol/L copper(II) sulfate should you initially use? Describe the dilution procedure.

Step 1: List Given Values

$c_i = 2.0 \text{ mol/L}$

$c_f = 0.50 \text{ mol/L}$

$v_f = 250 \text{ mL}$

$v_i = ?$

Step 2: Write Dilution Equation

$$c_i v_i = c_f v_f$$

Step 3: Isolate Unknown Value on Left Side of Equation

Since you are asked to calculate the initial volume of the solution, isolate v_i on the left side of the equation:

$$v_i = \frac{c_f v_f}{c_i}$$

Step 4: Substitute Values into Equation, and Solve

$$v_i = \frac{c_f v_f}{c_i}$$

$$= \frac{(0.50 \text{ mol/L})(250 \text{ mL})}{2.0 \text{ mol/L}}$$

$$v_i = 62 \text{ mL}$$

Therefore, 62 mL of 2.0-mol/L copper(II) sulfate solution is needed to make 250 mL of 0.50-mol/L copper(II) sulfate solution.

Figure 8(a)

Figure 8(b)

Answers

19. 12% W/V glucose

20. 3.13 mL

21. 6.76 mL

22. 0.100 mol/L

LEARNING TIP

Diluting Solutions
When diluting aqueous solutions, other than acids, begin with the volume of stock solution calculated. Then add water to reach the final volume required.

Procedure: Since you are asked to make 250 mL of the final solution, you require a 250-mL volumetric flask. The flask has a graduation mark at the 250-mL level.

1. Place 62 mL of the 2.0-mol/L copper(II) sulfate solution into a 250-mL volumetric flask (**Figure 8(a)**).

2. Add distilled water up to the 250-mL mark on the flask (**Figure 8(b)**).

   The volume of the final solution is 250 mL, and its concentration is 0.5-mol/L copper(II) sulfate.

3. Place a stopper on the flask, and invert several times to thoroughly mix the contents.

Example
What volume of 10.0-mol/L sodium hydroxide solution, $NaOH_{(aq)}$, is needed to prepare 250 mL of 0.375-mol/L sodium hydroxide solution?

Solution

$c_i = 10.0$ mol/L

$c_f = 0.375$ mol/L

$v_f = 250$ mL

$v_i = ?$

$$c_i v_i = c_f v_f$$

$$v_i = \frac{c_f v_f}{c_i}$$

$$= \frac{(0.375 \text{ mol/L})(250 \text{ mL})}{10.0 \text{ mol/L}}$$

$$v_i = 9.38 \text{ mL}$$

Therefore, 9.38 mL of 10.0-mol/L sodium hydroxide solution is needed to make 250 mL of 0.375-mol/L sodium hydroxide solution.

> **Practice**

Understanding Concepts

19. Calculate the final concentration of a glucose solution if 240 mL of 15% W/V glucose is diluted with water to 300.0 mL.

20. A laboratory technician needs to make 500.0 mL of a 0.100-mol/L sulfuric acid solution. What volume of 16.0-mol/L sulfuric acid does the technician need to use?

21. How many millilitres of a 14.8-mol/L ammonia solution, $NH_{3(aq)}$, do you need in order to make 100.0 mL of 1.00-mol/L ammonia solution?

22. Calculate the final concentration of a 0.400 mol/L barium chloride solution, $BaCl_{2(aq)}$, when 125 mL of the solution is diluted with distilled water to a final volume of 500.0 mL.

► **Section 2.5 Questions**

Understanding Concepts

1. What mass of pure hydrogen peroxide, $H_2O_{2(l)}$, is needed to make 500 bottles that each contain 250 mL of 6% W/V hydrogen peroxide?

2. The maximum acceptable concentration of fluoride ions in municipal water supplies is 1.5 ppm. What is the maximum mass of fluoride ions you would get from a 250-mL glass of water?

3. How many grams of sucrose, $C_{12}H_{22}O_{11(s)}$, are in 50.0 mL of a 0.50-mol/L solution of sucrose in water?

4. What is the molar concentration of phosphoric acid in a solution that contains 40.2 g of hydrogen phosphate, $H_3PO_{4(s)}$, in 250 mL of solution?

5. (a) A lab technician dilutes 45.5 mL of a 1.50-mol/L sodium sulfate solution, $Na_2SO_{4(aq)}$, to a final volume of 200.0 mL. What is the concentration of the diluted solution?
   (b) Another lab technician dilutes 50.0 mL of a 3.50-mol/L nitric acid solution, $HNO_{3(aq)}$, to a 2.50-mol/L nitric acid solution. What is the final volume?

Applying Inquiry Skills

6. A lab technician uses a 0.25-mol/L sodium carbonate stock solution, $Na_2CO_{3(aq)}$, to prepare 250 mL of a 0.010-mol/L solution.

(a) Calculate the volume of 0.25-mol/L sodium carbonate stock solution needed to prepare the diluted solution.
(b) Describe the procedure that the technician follows to prepare the solution. Include safety precautions.

Making Connections

7. Toxicity of substances for animals is usually expressed by a quantity called LD_{50}. Research the use of this quantity. What does LD_{50} mean? What is the LD_{50}, in ppm, of a substance considered to be "highly toxic" and a substance considered to be "slightly toxic" for mice on the Hodge and Sterner scale? Assume that the substance is taken by mouth.

 www.science.nelson.com

8. Why is it important for nurses, doctors, and pharmacists to establish common systems for communicating the concentration of solutions? Conduct research to find out what systems they use. Create a wall chart that can be used to inform medical workers of the various units that are used for each system.

 www.science.nelson.com

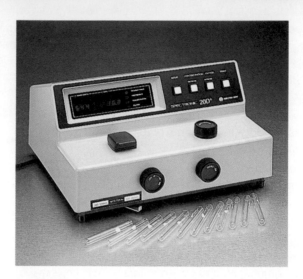

Figure 1
A Spectronic 20 spectrophotometer

Several laboratory methods allow chemists to determine the concentration of an aqueous solution. One method, called gravimetric analysis, involves precipitating the solutes out of solution, determining the mass of each solute, converting the mass to an amount in moles, and calculating the concentration of the original solution. This method can be quite complicated and time consuming, especially when the solute concentration is very low. Instead, a spectrophotometer can be used to determine the concentration of the solution. ⬇️▮

A spectrophotometer measures the amount of light energy that is absorbed or transmitted by dissolved solutes as a beam of light passes through the solution (**Figure 1**). Inside the spectrophotometer, a prism disperses the white light into its various wavelengths (colours), and a movable slit allows the individual wavelengths of light to pass through a sample of the solution (**Figure 2**). The solution is kept in a

special colourless test tube, called a cuvette. The cuvette is placed in a chamber that positions the solution directly in the path of the beam of light. The transmitted light strikes a photoelectric cell that converts the light energy into electric current. The amount of current is measured by a galvanometer, and the value is displayed on a meter as percent transmittance or percent absorbance.

A spectrophotometer does not measure concentration directly. It measures the percentage of light of a particular wavelength that a solution absorbs (percent absorbance) or the percentage of light that a solution transmits (percent transmittance). Different solutes absorb and transmit light of different wavelengths.

Either percent absorbance or percent transmittance can be used to determine the solute concentration of a solution. The percent absorbance (or percent transmittance) of various samples of a solution at different known concentrations can be measured with the spectrophotometer. These values can be graphed to produce a standard curve, which is used to determine the concentration of a sample of unknown concentration (**Figure 3**). In section 2.7, you will have an opportunity to measure the concentration of a solution using a spectrophotometer.

A solution must be able to absorb some visible light for a spectrophotometer to be useful, and only coloured solutes absorb

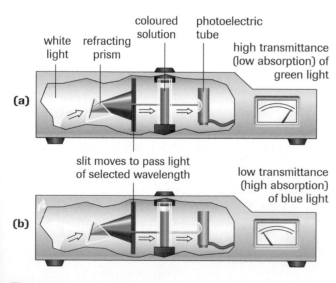

Figure 2
How a spectrophotometer works

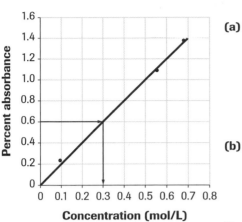

Figure 3

The concentration of a solution can be determined using a percent absorbance curve. If the percent absorbance of a solution of unknown concentration is 0.6, for example, the concentration of the solution is 0.3 mol/L.

(a)

percent absorbance low

(b)

percent absorbance high

Figure 4

Tech Connect

visible light. Therefore, a spectrophotometer cannot be used to determine the concentration of a colourless solution. Higher concentrations of coloured solutes absorb more (and transmit less) visible light than lower concentrations (**Figure 4**). If a solute does not produce an intensely coloured solution, it can sometimes be converted to brightly coloured complex ions that absorb more light. A typical example is the pale blue copper ion, $Cu^{2+}_{(aq)}$, which is converted to the intensely coloured $Cu(NH_3)_{4(aq)}^{2+}$ ion by the addition of concentrated aqueous ammonia, $NH_{3(aq)}$.

▶ **Tech Connect 2.6** *Questions*

Understanding Concepts

1. (a) What is gravimetric analysis?
   (b) When is gravimetric analysis not useful for determining the concentration of a solution?

2. (a) Describe how a spectrophotometer works.
   (b) How are the measurements that are produced by a spectrophotometer used to determine the concentration of a solution?

3. (a) Why can a spectrophotometer measure the concentrations of coloured solutions only?
   (b) Why are the cuvettes that are used in a spectrophotometer colourless?

GO   www.science.nelson.com

Making Connections

4. Research answers to the following questions about the use of spectrophotometers in forensics laboratories:
   (a) What types of investigations are conducted in forensics labs?
   (b) Why do some forensic laboratories use infrared spectrophotometers?
   (c) What types of evidence can spectrophotometers provide in the field of forensics?

GO   www.science.nelson.com

Determining the Concentration of a Solution

As you learned in section 2.6, chemists can use a spectrophotometer to determine the concentration of a solution. A spectrophotometer measures the amount of light energy that is absorbed or transmitted by dissolved solutes as light passes through the solution. In this activity, you will use a spectrophotometer to determine the concentration of a copper(II) sulfate solution, $CuSO_{4(aq)}$.

Question

What is the molar concentration of a copper(II) sulfate solution?

Materials

eye protection
lab apron
spectrophotometer
eight 50-mL beakers
burette stand
two 50-mL burettes
10-mL graduated cylinder
glass stirring rod
marking pens
10 test tubes or cuvettes (to fit the spectrophotometer)
100 mL 0.50-mol/L copper(II) sulfate solution, $CuSO_{4(aq)}$
5 mL copper(II) sulfate solution of unknown concentration

Procedure

 Copper(II) sulfate is toxic. Avoid contact with skin.

1. Copy **Table 1**, allowing enough space to record your observations for ten samples.

Table 1   Observation Table

Test tube	$[Cu^{2+}_{(aq)}]$	Absorbance

(Assume that $[Cu^{2+}_{(aq)}] = [CuSO_{4(aq)}]$.)

2. Fill a clean burette with 50 mL of distilled water.

3. Fill another clean burette with 50 mL of 0.50-mol/L copper(II) sulfate solution.

4. Label eight clean, dry 50-mL beakers with the numbers 1 to 8.

5. Use the eight labelled beakers to prepare increasingly dilute copper(II) sulfate solutions, according to **Table 2**.

Table 2   Concentrations of Dilute Copper(II) Sulfate Solutions

Beaker	$[CuSO_{4(aq)}]$ (mol/L)
1	0.50
2	0.25
3	0.14
4	0.07
5	0.04
6	0.02
7	0.002
8	0.0002

6. Label eight test tubes with the numbers 1 to 8. Label a ninth test tube "distilled water." (This test tube will act as a "blank.") Label a tenth test tube "unknown."

7. Pour 5.0 mL of each copper(II) sulfate solution that you prepared in step 5 into its corresponding test tube. Pour 5.0 mL of distilled water into test tube 9.

8. Use the operating instructions for your spectrophotometer to set the desired wavelength and zero the instrument. Ask your teacher for assistance if necessary. Keep the spectrophotometer well-covered at all times to prevent stray light from entering the instrument.

9. Place test tube 1 in the well of the spectrophotometer, and read the absorbance for the solution. Repeat for test tubes 2 through 9. Record the absorbance values in your observation table (**Table 1**).

10. Pour 5.0 mL of copper(II) sulfate of unknown concentration into test tube 10. Measure the absorbance for the solution, and record this value in your observation table.

11. Dispose of all solutions according to your teacher's instructions. Clean up your work area.

Analysis

(a) Plot a graph of absorbance versus concentration of $Cu^{2+}_{(aq)}$ by plotting absorbance on the vertical axis. Draw a line of best fit through the points on the graph. If more points are needed to produce a smooth line, prepare additional concentrations of $Cu^{2+}_{(aq)}$ ions, measure the absorbance values, and add these values to your graph.

(b) What does your graph tell you about the relationship between absorbance and concentration?

(c) Answer the Question.

Evaluation

(d) Why did you zero the spectrophotometer initially? Explain why you used a "blank."

(e) Describe possible sources of experimental error in this activity.

(f) Suggest improvements to the Procedure that would help to reduce error.

Figure 1
Kelly Guest

On July 14, 2002, Canadian triathlete Kelly Guest (**Figure 1**) finished his event at a World Cup meet in Edmonton and proceeded to the drug-testing centre to provide a mandatory urine sample. Two weeks later, while training for the Commonwealth Games in Manchester, England, Guest was informed that the Edmonton drug-testing centre had found trace concentrations of the banned muscle-building steroid *nandrolone* in his urine. Guest denied taking any banned substance, but he was immediately removed from the Canadian team and sent home just before the opening ceremonies.

A few of Canada's star athletes, including sprinter Ben Johnson, snowboarder Ross Rebagliati, rower Silken Laumann, and weightlifter Jacques Demers, have tested positive for banned substances at international games, such as the Olympics. Some of these athletes, such as Jacques Demers, have had their titles revoked. Repeat offenders, such as Ben Johnson, have been permanently banned from competition. Ross Rebagliati, who was stripped of his gold medal at the 1998 Nagano Olympics for testing positive for marijuana, had his medal reinstated when the Olympic Arbitration Board accepted the possibility that he had inhaled secondhand marijuana smoke at a party in Canada.

The Canadian Centre for Ethics in Sport (CCES) regulates doping control and provides programs to educate athletes about drug-free sport. The CCES defines doping in sport as "the use by an athlete of a substance or method banned by the International Olympic Committee, or prohibited by an International Sport Governing Body." Substances that are banned at the Olympics include anabolic (body-building) steroids, hormones that increase stamina and endurance (such as growth hormone), stimulants (such as caffeine and cocaine), diuretics (which help an athlete lose weight quickly), narcotics (such as morphine), and recreational drugs (such as marijuana).

For a urine drug test, an athlete provides two 75-mL to 100-mL urine samples, an A-sample and a B-sample. The A-sample is tested first. The B-sample is used to confirm the results for the A-sample, if requested by the athlete. The urine tests are extremely sensitive for detecting narcotics and steroids. For example, rower Silken Laumann and three teammates were stripped of a gold medal at the 1995 Pan-American Games after Laumann tested positive for a banned substance. This substance is found in an over-the-counter cold medicine called Benadryl.

Determining whether or not an athlete has been exposed to a banned substance is no easy task for the laboratory technicians. In addition to experimental error, there are many ways in which false positives can occur.

False positives are test results that mistakenly indicate the presence of banned substances in blood. There are many situations that may produce false positive test results:

- Athletes sometimes take nutritional supplements that they believe are free of banned substances. In some cases, chemicals in the supplements are converted into banned compounds within the body.

- Some farm animals are routinely injected with steroids or vitamins, which can increase the concentrations of banned substances in the bodies of athletes who eat meat.

- Women who take oral contraceptives may be found to have illegal concentrations of testosterone.

- Poppy seeds on breads, bagels, and pastries contain trace amounts of morphine that can lead to a positive test for banned opiates.

Blood tests are the newest form of drug testing. Blood tests for banned compounds were introduced at the 2000 Summer Olympic Games in Sydney, primarily to measure the concentration of substances such as erythropoietin (EPO), a drug that boosts muscle endurance by increasing the concentration of oxygen in blood.

▶ *Understanding the Issue*

Understanding Concepts

1. Why was Kelly Guest banned from competing at the 2002 Commonwealth Games in Manchester, England?

2. List five types of substances that are banned by the International Olympic Committee.

3. (a) Why are two samples of urine provided for urine testing?
   (b) Describe three ways in which drug tests can produce false positives.

4. Why has blood-testing been introduced?

▶ *Take a Stand*

Disciplining Athletes

Statement: Competitive athletes should be disciplined based on the results of blood and urine tests for banned substances.

In your group, research the issue. Search for information in newspapers, periodicals, and CD-ROMs, and on the Internet.

 www.science.nelson.com

Decision-Making Skills

- ● Define the Issue
- ● Defend a Decision
- ● Analyze the Issue
- ○ Identify Alternatives
- ● Research
- ○ Evaluate

Identify individuals, organizations, and government agencies that have addressed the issue. Identify the perspectives of opposing positions, and arrange the different perspectives in a suitable graphic organizer. As a concerned Canadian who financially supports Olympic athletes, write a position paper that summarizes your opinion.

Whether you are making omelettes in a kitchen or soap in a factory, you need to know the quantities of ingredients required to produce a certain quantity of product. For example, a manufacturing company needs to know how much raw material to buy to make the quantities of products ordered by its customers. By applying the mole concept (section 2.1) to balanced chemical equations, you can make quantitative predictions such as these.

To predict the masses of reactants and products, you always begin with a balanced equation for the reaction. A balanced equation tells you the ratio of the amounts of reactants and products that take part in a chemical reaction.

Determining Ratios from Balanced Chemical Equations

Consider the reaction between nitrogen gas and hydrogen gas to produce ammonia, $NH_{3(g)}$. A balanced equation for the production of ammonia is given below:

$$N_{2(g)} + 3\,H_{2(g)} \rightarrow 2\,NH_{3(g)}$$

The coefficients in the balanced equation may be analyzed in terms of numbers of entities or amounts (number of moles) of entities. The equation indicates that two molecules of ammonia are produced for every molecule of nitrogen that reacts with three molecules of hydrogen according to the molecule ratio

$$1 \text{ molecule } N_{2(g)} : 3 \text{ molecules } H_{2(g)} : 2 \text{ molecules } NH_{3(g)}$$

Considering amounts of reactants and products, the equation tells you that two moles of ammonia molecules are produced for every mole of nitrogen molecules that react with three moles of hydrogen molecules. The ratio of moles in a balanced chemical equation is called the **mole ratio**. The mole ratio for the reaction between nitrogen and hydrogen is

mole ratio the ratio of the amount, in moles, of reactants and products in a chemical reaction

$$1 \text{ mol } N_{2(g)} : 3 \text{ mol } H_{2(g)} : 2 \text{ mol } NH_{3(g)}$$

or simply 1 : 3 : 2. To summarize,

	$N_{2(g)}$	+	$3\,H_{2(g)}$	→	$2\,NH_{3(g)}$
molecule ratio	1 molecule	:	3 molecules	:	2 molecules
mole ratio	1 mol	:	3 mol	:	2 mol

Calculations Involving Mole Ratios

The procedure for calculating quantities of reactants and products in chemical reactions is called **stoichiometry**. In stoichiometry, you use mole ratios in balanced equations to calculate the quantities of reactants used or products formed.

stoichiometry mathematical procedures for calculating the quantities of reactants and products involved in chemical reactions

▶ *SAMPLE* problem *1*

Calculations Involving Mass of Reactants

Propane, $C_3H_{8(g)}$, is a gas that is commonly used in barbecues. Calculate the mass of oxygen that is needed to burn 15 g of propane.

Step 1: Write Unbalanced Equation

Remember that in the complete combustion of a hydrocarbon, carbon dioxide gas and water vapour are the only products.

$$C_3H_{8(g)} + O_{2(g)} \rightarrow CO_{2(g)} + H_2O_{(g)}$$

Step 2: Balance Equation, List Given Values and Molar Masses

To organize the information, list the given values and calculated molar masses below their corresponding entities:

Balanced equation	$C_3H_{8(g)}$	$+ \; 5\,O_{2(g)}$	$\rightarrow \; 3\,CO_{2(g)}$	$+ \; 4\,H_2O_{(g)}$
Given mass	15 g			
Molar mass	44.11 g/mol	32.00 g/mol		

Step 3: Convert Mass of Given Substance to Amount of Given Substance

In this problem, the given substance is propane. Using the molar mass of propane, convert the mass of propane (given) to an amount (in moles).

$$n_{C_3H_8} = 15 \text{ g } C_3H_8 \times \frac{1 \text{ mol } C_3H_8}{44.11 \text{ g } C_3H_8}$$

$$n_{C_3H_8} = 0.34 \text{ mol } C_3H_8$$

Step 4: Convert Amount of Given Substance to Amount of Required Substance

To perform the conversion, first determine which mole ratio you need to use.

The mole ratio can be expressed as $\dfrac{1 \text{ mol } C_3H_8}{5 \text{ mol } O_2}$ or $\dfrac{5 \text{ mol } O_2}{1 \text{ mol } C_3H_8}$. Since you are

converting amount of propane to amount of oxygen, use the form $\dfrac{5 \text{ mol } O_2}{1 \text{ mol } C_3H_8}$.

$$n_{O_2} = 0.34 \text{ mol } C_3H_8 \times \frac{5 \text{ mol } O_2}{1 \text{ mol } C_3H_8}$$

$$n_{O_2} = 1.7 \text{ mol } O_2$$

Step 5: Convert Amount of Required Substance to Required Value

In this case, the required value is the mass of oxygen. To find the mass of oxygen, you must multiply the amount of oxygen by the molar mass of oxygen.

$$m_{O_2} = 1.7 \text{ mol } O_2 \times \frac{32.00 \text{ g } O_2}{1 \text{ mol } O_2}$$

$$m_{O_2} = 54 \text{ g } O_2$$

Therefore, 54 g of oxygen is needed to burn 15 g of propane.

The calculations in steps 3, 4, and 5 can be combined as follows:

$$m_{O_2} = 15 \text{ g } C_3H_8 \times \frac{1 \text{ mol } C_3H_8}{44.11 \text{ g } C_3H_8} \times \frac{5 \text{ mol } O_2}{1 \text{ mol } C_3H_8} \times \frac{32.00 \text{ g } O_2}{1 \text{ mol } O_2}$$

$$m_{O_2} = 54 \text{ g } O_2$$

The stoichiometric calculation can be extended to determine the number of entities (N) that have reacted or are produced. The next Sample Problem involves this type of calculation.

▶ **SAMPLE** problem 2

Calculations Involving Numbers of Entities and Mass of Reactants

How many molecules of oxygen are produced from the decomposition of 12 g of water into its elements?

Step 1: Write Unbalanced Equation

$$H_2O_{(l)} \rightarrow H_{2(g)} + O_{2(g)}$$

Step 2: Balance Equation, List Given Values and Molar Masses

Balanced equation	$2 H_2O_{(l)}$	$\rightarrow 2 H_{2(g)}$	$+ O_{2(g)}$
Given mass	12 g		
Molar mass	18.02 g/mol		32.00 g/mol

Step 3: Convert Mass of Given Substance to Amount of Given Substance
Use the molar mass of the measured substance.

$$n_{H_2O} = 12 \text{ g } H_2O \times \frac{1 \text{ mol } H_2O}{18.02 \text{ g } H_2O}$$

$$n_{H_2O} = 0.67 \text{ mol } H_2O$$

Step 4: Convert Amount of Given Substance to Amount of Required Substance
First determine which mole ratio to use. The mole ratio for oxygen and water in the balanced equation is 2 mol H_2O : 1 mol O_2. This ratio can be expressed as $\frac{2 \text{ mol } H_2O}{1 \text{ mol } O_2}$ or $\frac{1 \text{ mol } O_2}{2 \text{ mol } H_2O}$. Use the form $\frac{1 \text{ mol } O_2}{2 \text{ mol } H_2O}$ to convert the amount of water into the amount of oxygen.

$$n_{O_2} = 0.67 \text{ mol } H_2O \times \frac{1 \text{ mol } O_2}{2 \text{ mol } H_2O}$$

$$n_{O_2} = 0.34 \text{ mol } O_2$$

Step 5: Convert Amount of Required Substance to Required Value

In this case, the required value is the number of molecules of oxygen. Use the equation 1 mol O_2 = 6.02×10^{23} molecules O_2 to form a suitable conversion factor.

$$N_{O_2} = 0.34 \text{ mol } O_2 \times \frac{6.02 \times 10^{23} \text{ molecules } O_2}{1 \text{ mol } O_2}$$

$$N_{O_2} = 2.0 \times 10^{23} \text{ molecules } O_2$$

When 10.0 g of water decomposes into its elements, 2.0×10^{23} molecules of oxygen are formed.

The combined calculation is given below:

$$N_{O_2} = 12 \text{ g } H_2O \times \frac{1 \text{ mol } H_2O}{18.02 \text{ g } H_2O} \times \frac{1 \text{ mol } O_2}{2 \text{ mol } H_2O} \times \frac{6.02 \times 10^{23} \text{ molecules } O_2}{1 \text{ mol } O_2}$$

$$N_{O_2} = 2.0 \times 10^{23} \text{ molecules } O_2$$

Example

Magnesium metal reacts with hydrochloric acid to produce aqueous magnesium chloride and hydrogen gas. How many hydrogen molecules are produced when 60.0 g of magnesium reacts with excess hydrochloric acid?

Solution

Balanced equation	$Mg_{(s)}$	+ 2 $HCl_{(aq)}$	→	$MgCl_{2(aq)}$	+	$H_{2(g)}$
Given mass	60.0 g					
Molar mass	24.31 g/mol					2.02 g/mol

$$n_{Mg} = 60.0 \text{ g } Mg \times \frac{1 \text{ mol } Mg}{24.31 \text{ g } Mg}$$

$$n_{Mg} = 2.47 \text{ mol } Mg$$

$$n_{H_2} = 2.47 \text{ mol } Mg \times \frac{1 \text{ mol } H_2}{1 \text{ mol } Mg}$$

$$n_{H_2} = 2.47 \text{ mol } H_2$$

$$N_{H_2} = 2.47 \text{ mol } H_2 \times \frac{6.02 \times 10^{23} \text{ molecules } H_2}{1 \text{ mol } H_2}$$

$$N_{H_2} = 1.49 \times 10^{24} \text{ molecules } H_2$$

When 60.0 g of magnesium reacts with hydrochloric acid, 1.49×10^{24} hydrogen molecules are produced.

The combined calculation is given below:

$$N_{H_2} = 60.0 \text{ g } Mg \times \frac{1 \text{ mol } Mg}{24.31 \text{ g } Mg} \times \frac{1 \text{ mol } H_2}{1 \text{ mol } Mg} \times \frac{6.02 \times 10^{23} \text{ molecules } H_2}{1 \text{ mol } H_2}$$

$$N_{H_2} = 1.49 \times 10^{24} \text{ molecules } H_2$$

Figure 1 summarizes the steps in a stoichiometric calculation.

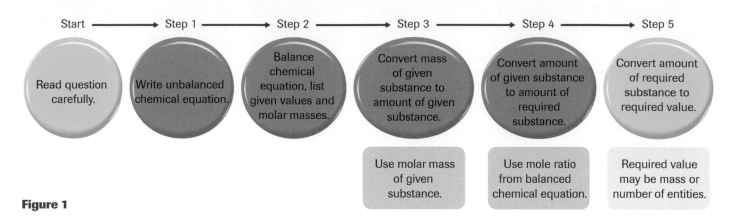

Start → Step 1 → Step 2 → Step 3 → Step 4 → Step 5

| Read question carefully. | Write unbalanced chemical equation. | Balance chemical equation, list given values and molar masses. | Convert mass of given substance to amount of given substance. | Convert amount of given substance to amount of required substance. | Convert amount of required substance to required value. |

Use molar mass of given substance.

Use mole ratio from balanced chemical equation.

Required value may be mass or number of entities.

Figure 1

▶ *Practice*

Understanding Concepts

1. Bauxite ore contains aluminum oxide, $Al_2O_{3(s)}$, which is decomposed using electricity to produce aluminum metal (**Figure 2**) and oxygen. What mass of aluminum metal can be produced from 125 g of aluminum oxide?

2. Potassium metal, $K_{(s)}$, reacts with hydrochloric acid to produce aqueous potassium chloride and hydrogen gas. How many grams of potassium are required to produce 5.00 g of hydrogen gas?

3. Potassium chlorate, $KClO_{3(s)}$, decomposes when heated to form solid potassium chloride and oxygen gas. How many formula units of potassium chlorate must decompose to produce 0.96 g of oxygen?

Figure 2
An aluminum refinery

▶ *Section 2.9 Questions*

Understanding Concepts

1. What is a mole ratio?

2. In a balanced equation, the total number of atoms in the reactants equals the total number of atoms in the products. Does the equality of atoms mean that the total amount (in moles) before and after a chemical reaction is the same? Give an example to illustrate your answer.

3. Write a balanced equation for the reaction between nitrogen dioxide gas and water to produce nitric acid and nitrogen monoxide gas. State the mole ratios of the reactants and the products.

4. What does the term "stoichiometry" mean?

5. How many molecules of hydrogen are produced from the decomposition of 12.0 g of water into its elements?

There are many everyday situations that require you to add enough reactants to ensure a complete reaction. When you wash dishes, for example, you must add enough dishwashing detergent to the water to react with all the grease on the dishes. Assume that one molecule of detergent reacts with one molecule of grease. If there are more grease molecules on the dishes than detergent molecules in the water, the grease molecules are in excess and the detergent molecules are limiting (**Figure 1**). Therefore, not all the dishes will be cleaned. The excess reactant is commonly called the **excess reagent**, and it is not completely consumed in the reaction. The other reactant is called the **limiting reagent**. The limiting reagent is completely consumed in the reaction, so it limits the amount of product. If there are more detergent molecules in the water than grease molecules on the dishes, there will be excess detergent molecules left in the water. The grease molecules will be limiting (**Figure 2**).

If there are equal numbers of grease molecules and detergent molecules (**Figure 3**), there are no excess molecules and no limiting molecules.

In wine making, the alcohol content of the wine can be controlled by making sugar the limiting reagent. The sugar is a reactant in the fermentation reaction. When the sugar is used up, the reaction stops. In baking, the baking soda is often the limiting reagent in the production of carbon dioxide bubbles. The baking soda controls the texture, resulting in a chewy cookie or a spongy cake. The quantity of the limiting reagent must be carefully calculated and measured, because it determines the quantity of all the products that are formed in a reaction.

excess reagent the reactant that is present in more than the required amount for a complete reaction to occur

limiting reagent the reactant that is completely consumed in a chemical reaction

Figure 1
There are not enough detergent molecules to react with all the grease molecules. The grease molecules are in excess, and the detergent molecules are limiting.

Figure 2
There are more detergent molecules than grease molecules. The detergent molecules are in excess, and the grease molecules are limiting.

Figure 3
There are equal numbers of detergent molecules and grease molecules. There are no excess or limiting molecules.

(a)

(b)

Figure 4
(a) A dissolved substance can be precipitated out of solution.
(b) A precipitate is filtered and dried, and its mass is measured to determine the amount of substance that was dissolved in the original solution.

Analysis by Precipitation

If you wish to analyze a substance that is dissolved in a solution, you can add a reactant to combine with the dissolved substance and form a precipitate (**Figure 4(a)**). The solid precipitate is filtered and dried, and its mass is measured. The mass of the precipitate is used to calculate the amount of substance that was dissolved in the original solution (**Figure 4(b)**). For this method to work, you must add enough reactant to the solution to precipitate *all* the dissolved substance. If you don't add enough reactant, some of the dissolved substance will remain in solution and will be unaccounted for in the calculation.

Limiting and Excess Reagents

As mentioned in section 2.9, stoichiometric calculations are always based on a balanced chemical equation. The coefficients in the balanced equation determine the ratio of moles in which reactants become products. Consider the balanced equation for the formation of water from its elements, hydrogen and oxygen:

$$2\,H_{2(g)} + O_{2(g)} \rightarrow 2\,H_2O_{(l)}$$

Using the molar masses of each entity, you can describe the amounts (in moles) and the masses (in grams) that are involved in this reaction.

Balanced equation	$2\,H_{2(g)}$	$+$ $O_{2(g)}$	$\rightarrow$ $2\,H_2O_{(l)}$
Moles	2 mol	1 mol	2 mol
Molar mass	2.02 g/mol	32.00 g/mol	18.02 g/mol
Mass	4.04 g	32.00 g	36.04 g

Therefore, if two moles of hydrogen gas react with exactly one mole of oxygen gas, two moles of water are formed. Using the masses of entities instead of moles, the equation tells you that 4.04 g of hydrogen reacts with 32.00 g of oxygen to produce 36.04 g of water. The quantitative changes that occur in the reaction can be determined by analyzing the quantities before, during, and after the reaction.

Balanced equation	$2\,H_{2(g)}$	$+$ $O_{2(g)}$	$\rightarrow$ $2\,H_2O_{(l)}$
Before reaction	2 mol (4.04 g)	1 mol (32.00 g)	0 mol (0.0 g)
Reaction according to balanced equation	2 mol (4.04 g)	$+$ 1 mol (32.00 g)	$\rightarrow$ 2 mol (36.04 g)
After reaction	0 mol (0.0 g)	0 mol (0.0 g)	2 mol (36.04 g)

In this example, you begin with 2 mol $H_{2(g)}$ and 1 mol $O_{2(g)}$. Notice that all the oxygen molecules and all the hydrogen molecules react to form water. No hydrogen or oxygen molecules are left over. The reaction is complete. What will happen if 3 mol $H_{2(g)}$ are mixed with 1 mol $O_{2(g)}$ and allowed to react? There are more hydrogen molecules than are required for a complete reaction. The quantities of entities before, during, and after the reaction are analyzed on the next page.

Balanced equation	2 H$_{2(g)}$	+	O$_{2(g)}$	→	2 H$_2$O$_{(l)}$
Before reaction	3 mol (6.06 g)		1 mol (32.00 g)		0 mol (0.0 g)
Reaction according to balanced equation	2 mol (4.04 g)	+	1 mol (32.00 g)	→	2 mol (36.04 g)
After reaction	1 mol (2.02 g)		0 mol (0.0 g)		2 mol (36.04 g)

As before, one mole of oxygen molecules reacts with two moles of hydrogen molecules to produce two moles of water. Here, however, all the oxygen is used up, but one mole of hydrogen molecules is left unreacted. Therefore, hydrogen is the excess reagent because there is more of it than is required for a complete reaction. Oxygen is a limiting reagent because it is completely consumed in the reaction, and it runs out before all the hydrogen is used up. Oxygen limits, or determines, the amount of product (water) that can be formed.

The following Sample Problem and Example show how to identify the limiting and excess reagents in a chemical reaction and how to calculate the quantities of product formed, based on the amount of limiting reagent available.

> ▶ **SAMPLE** problem

> ### Mass of Product Formed with a Limiting Reagent

Table salt, NaCl$_{(s)}$, can be formed by the reaction of sodium metal and chlorine gas:

> **2 Na$_{(s)}$ + Cl$_{2(g)}$ → 2 NaCl$_{(s)}$**

A reaction mixture contains 45.98 g of sodium and 142.0 g of chlorine. Calculate the mass of sodium chloride that is produced.

Step 1: List Given Values

m_{Na} = 45.98 g

m_{Cl_2} = 142.0 g

Step 2: Complete Reaction Chart
Use this chart to show the amounts before, during, and after the reaction.

Balanced equation	2 Na$_{(s)}$	+	Cl$_{2(g)}$	→	2 NaCl$_{(s)}$
Before reaction	45.98 g		142.0 g		0 g
Reaction according to balanced equation	2 mol (45.98 g)	+	1 mol (70.9 g)	→	2 mol
After reaction	0 g		71.1 g		? g

Step 3: Identify Limiting and Excess Reagents
According to the reaction chart, sodium is used up in the reaction and is therefore the limiting reagent. Some chlorine is left over after the reaction and is therefore the excess reagent. The mass of available sodium determines the mass of sodium chloride formed.

Step 4: Calculate Amount of Limiting Reagent

Use the mass and molar mass of the limiting reagent.

$$n_{Na} = 45.98 \text{ g Na} \times \frac{1 \text{ mol Na}}{22.99 \text{ g Na}}$$

$$n_{Na} = 2.000 \text{ mol Na}$$

Step 5: Calculate Amount of Product

Use the amount of limiting reagent from step 4 and the appropriate form of the mole ratio from the balanced equation.

$$n_{NaCl} = 2.000 \text{ mol Na} \times \frac{2 \text{ mol NaCl}}{2 \text{ mol Na}}$$

$$n_{NaCl} = 2.000 \text{ mol NaCl}$$

Step 6: Calculate Mass of Product

Use the appropriate form of the molar mass ratio of sodium chloride.

$$m_{NaCl} = 2.000 \text{ mol NaCl} \times \frac{58.44 \text{ g NaCl}}{1 \text{ mol NaCl}}$$

$$m_{NaCl} = 116.9 \text{ g NaCl}$$

When 45.98 g of solid sodium reacts with 142.0 g of chlorine gas, 116.9 g of solid sodium chloride is produced.

Example

Aluminum and oxygen react to form aluminum oxide, as shown in the following balanced chemical equation:

$$4 \text{ Al}_{(s)} + 3 \text{ O}_{2(g)} \rightarrow 2 \text{ Al}_2O_{3(s)}$$

A reaction mixture contains 134.9 g of aluminum and 96.0 g of oxygen. Calculate the mass of aluminum oxide that is produced.

Solution

$$m_{Al} = 134.9 \text{ g}$$
$$m_{O_2} = 96.0 \text{ g}$$

Balanced equation	$4 \text{ Al}_{(s)}$	$+ \; 3 \text{ O}_{2(g)}$	$\rightarrow 2 \text{ Al}_2O_{3(s)}$
Before reaction	134.9 g	96.0 g	0 g
Reaction according to balanced equation	4 mol (107.9 g)	+ 3 mol (96.00 g)	→ 2 mol
After reaction	26.98 g	0.0 g	? g

According to the reaction chart, oxygen is the limiting reagent and aluminum is the excess reagent.

$$n_{O_2} = 96.0 \text{ g O}_2 \times \frac{1 \text{ mol O}_2}{32.00 \text{ g O}_2}$$

$$n_{O_2} = 3.00 \text{ mol O}_2$$

$$n_{Al_2O_3} = 3.00 \ \text{mol} \ O_2 \times \frac{2 \ \text{mol} \ Al_2O_3}{3 \ \text{mol} \ O_2}$$

$$n_{Al_2O_3} = 2.00 \ \text{mol} \ Al_2O_3$$

$$m_{Al_2O_3} = 2.00 \ \text{mol} \ Al_2O_3 \times \frac{101.96 \ \text{g} \ Al_2O_3}{1 \ \text{mol} \ Al_2O_3}$$

$$m_{Al_2O_3} = 204 \ \text{g} \ Al_2O_3$$

When 134.9 g of aluminum reacts with 96.0 g of oxygen, 204 g of aluminum oxide is produced.

▶ **Practice**

Understanding Concepts

1. (a) Identify the limiting reagent and excess reagent when 10.0 g of hydrogen gas and 32.00 g of oxygen gas react to form water as the only product.
   (b) Determine the mass of water that is obtained from the reaction.

2. Determine the mass of carbon monoxide that is produced when 32.1 g of methane, $CH_{4(g)}$, undergoes incomplete combustion with 160.0 g of oxygen gas. (Assume that carbon monoxide, $CO_{(g)}$, and water vapour, $H_2O_{(g)}$, are the only products.)

3. Sulfur dioxide gas and oxygen gas react to produce gaseous sulfur trioxide.
   (a) Identify the limiting reagent and excess reagent when 192.18 g of sulfur dioxide reacts with 32.00 g of oxygen to produce sulfur trioxide as the only product.
   (b) What amount of sulfur trioxide is produced?
   (c) What mass of sulfur trioxide is produced?

4. Phosphorus, $P_{4(s)}$, reacts with chlorine gas to produce solid phosphorus pentachloride, $PCl_{5(s)}$, as the only product. Determine the mass of phosphorus pentachloride that is produced from a reaction between 123.88 g of phosphorus and 1.00 kg of chlorine.

Answers

1. (b) 36.04 g
2. 56.0 g
3. (b) 2.00 mol
   (c) 160.1 g
4. 832.9 g

▶ **Section 2.10 Questions**

Understanding Concepts

1. Distinguish between the excess reagent and the limiting reagent in a chemical reaction.

2. Is there always a limiting reagent in a chemical reaction? Explain.

3. Imagine that you are trying to determine the amount of substance A by allowing it to react with a solution of substance B. Which substance must be the limiting reagent? Which substance must be in excess? Explain.

4. (a) How many grams of oxygen are needed to react completely with 6.4 g of methane to produce carbon monoxide and water? The balanced chemical equation is

$$2 \ CH_{4(g)} + 3 \ O_{2(g)} \rightarrow 2 \ CO_{(g)} + 4 \ H_2O_{(g)}$$

   (b) How many grams of carbon monoxide are formed?

5. How many grams of aluminum sulfate, $Al_2(SO_4)_{3(aq)}$, are formed when 6.71 g of aluminum reacts with excess sulfuric acid, $H_2SO_{4(aq)}$? The balanced chemical equation is

$$2 \ Al_{(s)} + 3 \ H_2SO_{4(aq)} \rightarrow Al_2(SO_4)_{3(aq)} + 3 \ H_{2(g)}$$

6. Determine the mass of copper that is obtained from a reaction between 286.2 g of copper(I) oxide, $Cu_2O_{(s)}$, and 286.2 g of copper(I) sulfide, $Cu_2S_{(s)}$. The balanced chemical equation is

$$2 \ Cu_2O_{(s)} + Cu_2S_{(s)} \rightarrow 6 \ Cu_{(s)} + SO_{2(g)}$$

2.11 Investigation

The Limiting Reagent in a Chemical Reaction

Inquiry Skills
○ Questioning   ● Planning   ● Analyzing
○ Hypothesizing   ● Conducting   ● Evaluating
● Predicting   ● Recording   ● Communicating

When conducting experiments, it is necessary to determine which substance is the limiting reagent and which substance is the excess reagent. In this investigation, you will predict the mass of the precipitate that forms in a double displacement reaction, based on the quantities of reactants used and the balanced chemical equation. Then you will design and conduct an experiment to test your prediction and determine which reagent is in excess.

Strontium chloride, $SrCl_{2(aq)}$, reacts with copper(II) sulfate, $CuSO_{4(aq)}$, to produce strontium sulfate, $SrSO_{4(s)}$, as a precipitate. The other product is copper(II) chloride, $CuCl_{2(aq)}$, which remains in solution. The balanced chemical equation is

$$SrCl_{2(aq)} + CuSO_{4(aq)} \rightarrow SrSO_{4(s)} + CuCl_{2(aq)}$$

Question

What is the mass of the precipitate that is produced by the reaction of 2.00 g of strontium chloride with excess copper(II) sulfate in 75 mL of water?

 Strontium chloride is moderately toxic. Copper(II) sulfate is a strong irritant and is toxic if ingested. Lab aprons and eye protection must be worn.

Prediction

(a) Determine which reactant is the limiting reagent, and predict the mass of the precipitate that will form.

Experimental Design

(b) Design an experiment in which you will determine the limiting reagent in the reaction of strontium chloride with copper(II) sulfate from the mass of the precipitate, strontium sulfate.

Materials

(c) List any necessary chemicals and equipment.

Procedure

(d) Write a step-by-step Procedure, including safety precautions. Also include disposal instructions, which you can obtain from your teacher. Have your Procedure approved by your teacher before carrying it out.

Observations

(e) Design a table that you can use to record both qualitative and quantitative observations.

Analysis

(f) Analyze your Observations.

(g) Answer the Question.

Evaluation

(h) Discuss sources of experimental error. Suggest changes to the Procedure that may help to reduce these sources of error.

(i) Evaluate your Prediction based on your Observations and on an analysis of the Experimental Design and your Procedure.

(j) Evaluate the stoichiometric method, as used to predict the masses of reactants and products.

The quantities of reactants and products you calculate on the basis of the mole ratios in a balanced equation are called theoretical quantities. Theoretical quantities are the quantities that should be used or produced in a chemical reaction. They are not necessarily the quantities that are used or produced when you mix the reactants and collect the products.

The quantity of product produced in a chemical reaction is called the **yield**. When the reaction is carried out in a laboratory, the quantity of product that is obtained and measured at the end of the reaction is called the **actual yield**. The quantity of product (the yield) can also be calculated using the balanced equation. The calculated quantity is called the **theoretical yield**.

Theoretically, each and every entity that is used as a reactant in a chemical reaction should be accounted for in the products. Often, however, the actual yield in a chemical reaction turns out to be less than the theoretical yield. There are several reasons why not all the reactants end up in the products. The most common reason is related to experimental procedures. For example, loss of product may occur when transferring solutions or filtering precipitates, or from splattering during heating. This kind of loss can be reduced by improving technical skills, improving the equipment used, or reducing the number of steps in the experimental design.

Low yield can also be due to impurities in the reagents used. Chemicals come in a wide variety of grades, or purities (**Figure 1**). Some low-purity or technical grades may only be 80% to 90% pure. If impurities are not accounted for in the amount of reactant used, the actual yield will be less than the theoretical yield. Impurities may also result from other processes. For example, a metal, such as magnesium, readily reacts with air to form a layer of magnesium oxide on the surface. This impurity is included in the mass of the reactant but does not proceed to form the products collected, thus causing the actual yield to differ from the theoretical yield.

Another cause of low yield is the occurrence of side reactions that form products other than the desired products. As already mentioned, magnesium ribbon reacts with the oxygen in air to form magnesium oxide. Air contains other gases, however, such as nitrogen. A side reaction may occur, in which some of the magnesium reacts with nitrogen to form magnesium nitride. If the product that is collected is presumed to be only magnesium oxide but also contains magnesium nitride, the actual yield will be different from the theoretical yield. To correct this problem, the reaction may be carried out in pure oxygen instead of air.

Low yield may also occur because most reactions are reversible. In reversible reactions, products may react with one another to form the reactants. Environmental conditions determine the extent to which reactions proceed in one direction or the other. When products accumulate in closed containers, they have a tendency to react with each other to regenerate the original reactant molecules. As a result, the amount of product that is

yield the quantity of product produced in a chemical reaction

actual yield the quantity of product that is actually produced in a chemical reaction

theoretical yield the quantity of product calculated from a balanced equation

---

LEARNING *TIP*

Different Yields?
Actual yield is sometimes referred to as experimental yield.

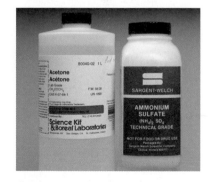

Figure 1
Chemicals come in a wide variety of grades, or purities. The purity of a chemical can significantly affect experimental results in a chemical reaction.

collected may be less than the amount predicted by simple stoichiometry calculations. To minimize these losses, the conditions for the reaction may need to be changed to allow the forward reaction to go to completion. For example, the reaction may be carried out in an open container.

To compare the actual yield and the theoretical yield, the percentage yield is calculated. The **percentage yield** is obtained by dividing the actual yield by the theoretical yield:

$$\text{percentage yield} = \frac{\text{actual yield}}{\text{theoretical yield}} \times 100\%$$

If, for example, the theoretical yield in a certain reaction is 10 kg and the actual yield is 9 kg, the percentage yield can be calculated as follows:

$$\text{percentage yield} = \frac{9\,\cancel{kg}}{10\,\cancel{kg}} \times 100\%$$

$$\text{percentage yield} = 90\%$$

The percentage yield is 90%.

percentage yield the ratio, expressed as a percentage, of the actual or experimental quantity of product obtained (actual yield) to the maximum possible quantity of product (theoretical yield) derived from a stoichiometric calculation

---

▶ *SAMPLE problem*

Calculating the Percentage Yield

Iron is produced from its ore, hematite, $Fe_2O_{3(s)}$, by heating hematite with carbon monoxide in a blast furnace. The IUPAC name for hematite is iron(III) oxide. If 635 kg of iron is obtained from 1150 kg of hematite, what is the percentage yield of iron? The equation for the reaction is

$$Fe_2O_{3(s)} + 3\,CO_{(g)} \rightarrow 2\,Fe_{(s)} + 3\,CO_{2(g)}$$

To calculate the percentage yield, you need to know the actual yield and the theoretical yield. In this problem, you are given the actual yield produced and the mass of hematite used. Therefore, you need to calculate the theoretical yield using the balanced equation provided. Use the steps you learned in section 2.9 (Stoichiometry) to calculate the mass of iron that should be produced. Hematite is the given substance.

Step 1: Write Unbalanced Equation
Since a balanced equation is provided, proceed to step 2.

Step 2: Balance Equation, List Given Values and Molar Masses

Balanced equation	$Fe_2O_{3(s)}$	$+ 3\,CO_{(g)}$	$\rightarrow$	$2\,Fe_{(s)}$	$+ 3\,CO_{2(g)}$
Given mass (kg)	1150			635	
Given mass (g)	1.150×10^6			6.35×10^5	
Molar mass (g/mol)	159.70			55.85	

Step 3: Convert Mass of Given Substance to Amount of Given Substance

Use the molar mass of the given substance:

$$n_{Fe_2O_3} = 1.150 \times 10^6 \text{ g Fe}_2\text{O}_3 \times \frac{1 \text{ mol Fe}_2\text{O}_3}{159.7 \text{ g Fe}_2\text{O}_3}$$

$$n_{Fe_2O_3} = 7201 \text{ mol Fe}_2\text{O}_3$$

Step 4: Convert Amount of Given Substance to Amount of Required Substance

Use a mole ratio that allows you to cancel mol Fe_2O_3. From the balanced equation, the mole ratio of hematite and iron is 1 mol Fe_2O_3 : 2 mol Fe.

$$n_{Fe} = 7201 \text{ mol Fe}_2\text{O}_3 \times \frac{2 \text{ mol Fe}}{1 \text{ mol Fe}_2\text{O}_3}$$

$$n_{Fe} = 1.440 \times 10^4 \text{ mol Fe}$$

Step 5: Convert Amount of Required Substance to Mass of Required Substance

Use the molar mass of iron to calculate the mass of iron.

$$m_{Fe} = 1.440 \times 10^4 \text{ mol Fe} \times \frac{55.85 \text{ g Fe}}{1 \text{ mol Fe}}$$

$$m_{Fe} = 8.04 \times 10^5 \text{ g Fe}$$

Convert grams to kilograms.

$$m_{Fe} = 8.04 \times 10^5 \text{ g Fe} \times \frac{1 \text{ kg Fe}}{1000 \text{ g Fe}}$$

$$= 8.04 \times 10^2 \text{ kg Fe}$$

$$m_{Fe} = 804 \text{ kg Fe}$$

The theoretical yield of iron is 804 kg.

Step 6: Calculate Percentage Yield

$$\text{percentage yield} = \frac{\text{actual yield}}{\text{theoretical yield}} \times 100\%$$

$$= \frac{635 \text{ kg}}{804 \text{ kg}} \times 100\%$$

$$\text{percentage yield} = 79\%$$

The percentage yield of iron is 79%.

Example

The most common ore of arsenic, $FeSAs_{(s)}$, can be heated to produce arsenic, $As_{(s)}$:

$$FeSAs_{(s)} \rightarrow FeS_{(s)} + As_{(s)}$$

When 250 kg of this ore was processed industrially, 95.3 kg of arsenic was obtained. Calculate the percentage yield of arsenic.

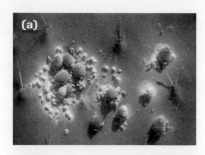

Figure 2
(a) Cochineal insects live in prickly pear cacti in the high desert plains of the Peruvian Andes.
(b) (c) A vivid red dye is made of the dried and crushed bodies of female cochineal insects.

Answers

1. 74.9%

2. (b) 81.0%

3. 82.5%

Solution

Balanced equation	$FeSAs_{(s)}$	→ $FeS_{(s)}$	+ $As_{(s)}$
Given mass (kg)	250		95.3
Given mass (g)	2.50×10^5		9.53×10^5
Molar mass (g/mol)	162.83		74.92

$$n_{FeSAs} = 2.50 \times 10^5 \text{ g FeSAs} \times \frac{1 \text{ mol FeSAs}}{162.83 \text{ g FeSAs}}$$

$$n_{FeSAs} = 1535 \text{ mol FeSAs}$$

The mole ratio from the balanced equation is 1 mol FeSAs : 1 mol As.

$$n_{As} = 1535 \text{ mol FeSAs} \times \frac{1 \text{ mol As}}{1 \text{ mol FeSAs}}$$

$$n_{As} = 1535 \text{ mol As}$$

$$m_{As} = 1535 \text{ mol As} \times \frac{74.92 \text{ g As}}{1 \text{ mol As}}$$

$$= 1.150 \times 10^5 \text{ g As}$$

$$= 1.150 \times 10^5 \text{ g As} \times \frac{1 \text{ kg As}}{1000 \text{ g As}}$$

$$m_{As} = 115.0 \text{ kg}$$

$$\text{percentage yield} = \frac{\text{actual yield}}{\text{theoretical yield}} \times 100\%$$

$$= \frac{95.3 \text{ kg}}{115.0 \text{ kg}} \times 100\%$$

$$\text{percentage yield} = 82.9\%$$

The percentage yield of arsenic was 82.9%.

▶ Practice

Understanding Concepts

1. Methyl salicylate, $C_8H_8O_{3(l)}$, is the chemical that is responsible for wintergreen flavouring. It can be prepared by heating salicylic acid, $C_7H_6O_{3(s)}$, with methanol, $CH_3OH_{(l)}$:

$$C_7H_6O_{3(s)} + CH_3OH_{(l)} \rightarrow C_8H_8O_{3(l)} + H_2O_{(l)}$$

If 2.00 g of salicylic acid reacts with excess methanol, and the yield of wintergreen is 1.65 g, what is the percentage yield?

2. Aluminum metal reacts with liquid bromine to produce solid aluminum bromide as the only product.
(a) Write a balanced chemical equation for this reaction.
(b) When 53.7 g of bromine reacts with excess aluminum, 48.4 g of aluminum bromide is produced. Calculate the percentage yield for this reaction.

3. Zinc reacts with hydrochloric acid, as shown in the following balanced chemical equation:

$$Zn_{(s)} + 2 HCl_{(aq)} \rightarrow ZnCl_{2(aq)} + H_{2(g)}$$

Calculate the percentage yield if 1.541 g of zinc chloride, $ZnCl_{2(aq)}$, is produced when 0.999 g of hydrochloric acid reacts with excess zinc.

4. A student produces acetylsalicylic acid (Aspirin), $C_9H_8O_{4(s)}$, and acetic acid from the reaction of salicylic acid, $C_7H_6O_{3(s)}$, with acetic anhydride, $C_4H_6O_{3(aq)}$. The balanced chemical equation is

$$C_7H_6O_{3(s)} + C_4H_6O_{3(aq)} \rightarrow C_9H_8O_{4(s)} + HC_2H_3O_{2(aq)}$$

(a) What is the theoretical yield of Aspirin, if 213.0 g of salicylic acid reacts with excess acetic anhydride?

(b) What is the percentage yield of Aspirin, if 189.3 g of Aspirin is produced?

Answers

4. (a) 277.8 g
   (b) 68.14%

▶ *Section 2.12* Questions

Understanding Concepts

1. Distinguish between the terms "actual yield" and "theoretical yield."

2. Can the actual yield ever be greater than the theoretical yield? Explain.

3. In an experiment, 5.00 g of silver nitrate, $AgNO_{3(s)}$, is added to a solution that contains an excess of sodium bromide, $NaBr_{(s)}$, and 5.03 g of silver bromide, $AgBr_{(s)}$, is produced.
   (a) Write a balanced equation for the reaction.
   (b) What is the theoretical yield of silver bromide?
   (c) What is the actual yield of silver bromide in the experiment?
   (d) What is the percentage yield in the experiment?

4. In an experiment, 16.1 g of iron sulfide, $FeS_{(s)}$, is added to excess oxygen, and 14.1 g of iron(III) oxide, $Fe_2O_{3(s)}$, is produced. The balanced equation for the reaction is

$$4\ FeS_{(s)} + 7\ O_{2(g)} \rightarrow 2\ Fe_2O_{3(s)} + 4\ SO_{2(g)}$$

Calculate the percentage yield of iron(III) oxide in the experiment.

Applying Inquiry Skills

5. In an experiment to recover a precipitate that was formed in a chemical reaction, a chemistry student followed the procedure below:
   1. The mass of a reactant was determined using weighing paper on an electronic balance.
   2. The reactant was transferred to a large beaker.
   3. A second reactant, an aqueous solution used in excess, was measured using a graduated cylinder and added to the beaker.

   4. The mixture was stirred, placed in an evaporating dish, and heated to dryness on a laboratory hot plate.
   5. The precipitate was transferred from the evaporating dish to the weighing paper, and the mass was determined.

   Suggest ways in which this procedure could be modified to improve the percentage yield.

Making Connections

6. When you consume a beverage or candy that was artificially coloured with red food dye, you may be ingesting a chemical that was produced by a tiny red insect from Peru. Several synthetic red dyes have been found to be carcinogenic (cancer-causing), but a vivid red dye called carmine has been approved for use in foods, drugs, and cosmetics. Carmine is made from cochineal insects (**Figure 2**). Two Canadian chemists developed the process that is used to extract the red dye from the insects. The process has been improved to increase the purity and the yield of the product. Research answers to the following questions, and summarize your findings in a one-page report.
   (a) What is the typical percentage yield of carmine in the extraction process? How is carmine extracted?
   (b) What effect has the industrial production of carmine had on the people of Peru?

 www.science.nelson.com

2.13 Investigation

The Percentage Yield of a Chemical Reaction

Aluminum metal reacts with aqueous copper(II) chloride dihydrate to produce aqueous aluminum chloride, copper metal, and water.

Inquiry Skills

○ Questioning   ○ Planning   ● Analyzing
○ Hypothesizing ● Conducting ● Evaluating
● Predicting   ● Recording   ● Communicating

Question

What mass of copper is formed when excess aluminum reacts with a given mass of copper(II) chloride dihydrate?

Prediction

(a) Use the following balanced chemical equation to calculate the theoretical yield of copper:

$$3 \, CuCl_2 \cdot 2H_2O_{(aq)} + 2 \, Al_{(s)} \rightarrow 3 \, Cu_{(s)} + 2 \, AlCl_{3(aq)} + 6 \, H_2O_{(l)}$$

Based on your calculation, predict the actual yield of copper.

Materials

eye protection
lab apron
8-cm by 8-cm piece of aluminum foil
2.00 g copper(II) chloride dihydrate, $CuCl_2 \cdot 2H_2O_{(aq)}$

two 150-mL beakers	50-mL graduated cylinder
stirring rod	ruler
forceps	hot plate
ring stand	iron ring
wire gauze	watch glass
crucible tongs	electronic balance

Procedure

1. Measure and record the mass of an empty beaker.

2. Measure out 2.00 g of the copper salt, and dissolve it in 50 mL of water in a second beaker.

 Copper chloride dihydrate is toxic and must not be ingested.

3. Fold the aluminum foil lengthwise twice to make a 2-cm by 8-cm strip. Coil the strip loosely to fit into the copper chloride solution in the beaker. Ensure that the strip is entirely immersed.

4. Heat the beaker gently on the hot plate until the blue colour in the solution has disappeared completely (approximately 5 min). Continue to heat gently for an additional 5 min. Allow the beaker and its contents to cool.

 Care must be taken when handling hot equipment. Eye protection and lab aprons must be worn.

5. Use the forceps to shake loose all the copper that formed on the aluminum foil. Carefully transfer the copper to the beaker from step 1. Rinse the copper with water.

6. Pour off as much of the rinse water as possible. Spread the copper on the bottom of the beaker.

7. Cover the beaker containing the wet copper with a watch glass. Gently heat the beaker to drive off the water. Reduce the heat if the copper begins to turn black.

8. When the copper is dry, determine the mass of the copper.

Analysis

(b) Identify the limiting reagent and the excess reagent in this reaction. What visible evidence is there to confirm your identification?

(c) Answer the Question.

(d) Determine the percentage yield of copper in this experiment.

Evaluation

(e) If the percentage yield is less than 100%, identify some sources of experimental error. If the percentage yield is greater than 100%, suggest specific factors that may account for this high yield.

(f) What steps did you take to ensure that the reaction went to completion?

(g) If you wanted to use the other reactant as the limiting reagent, what changes in the Procedure would you need to make? What visible evidence would you look for to ensure that the reaction had gone to completion?

In the late nineteenth century, rapid population growth in Europe and North America began to outstrip the supply of fresh food. Scientists knew that if they added nitrogen-based fertilizers (such as sodium nitrate, $NaNO_{3(s)}$, or ammonium nitrate, $NH_4NO_{3(s)}$) to the soil, crop yields would increase and a worldwide food shortage could be prevented. The world supply of fertilizers could not keep up with the growing demand for food, however, because large amounts of sodium nitrate were also being used to produce explosives such as gunpowder and nitroglycerine. Additional sources of ammonia or nitrate had to be found to avoid a global catastrophe.

In 1909, a leading German chemical company, Badische Anilin und Soda Fabrik (BASF), started to investigate the possibility of producing ammonia, $NH_{3(g)}$, from atmospheric nitrogen, $N_{2(g)}$. Little did they know that one year earlier, Fritz Haber, a professor at a technical college in Karlsruhe, Germany, had discovered a method for doing just that (**Figure 1**).

Haber realized, after much experimentation, that nitrogen gas and hydrogen gas react to form ammonia as the only product. Optimum conditions for the reaction included a closed container, a suitable catalyst (such as iron oxide, $Fe_2O_{3(s)}$), a temperature of 500°C, and a pressure of 40 MPa.

$$N_{2(g)} + 3\,H_{2(g)} \xrightarrow{Fe_2O_{3(s)}} 2\,NH_{3(g)}$$

Haber's method for producing ammonia is now called the Haber process. BASF bought the rights to the Haber process. With the help of Carl Bosch, BASF's chief chemical engineer, BASF built a giant industrial plant that was capable of producing 10 000 t of ammonia per year. Today, ammonia is in sixth position in a ranking of chemicals produced worldwide, with over 80 billion kilograms produced each year.

The Temperature–Pressure Puzzle

The reaction of nitrogen and hydrogen at low temperatures is so slow that the process becomes uneconomical. Adding heat increases the rate of the reaction (**Figure 2**), which is important in any industrial process. In this reaction, however, the higher the temperature, the lower the percentage yield of ammonia is. The relationship between percentage yield and temperature is shown in **Figure 3**, on the next page.

Haber had to balance the rate of the reaction (increased by higher temperatures) against the decrease in percentage yield of ammonia at higher temperatures. He discovered that using an iron oxide catalyst eliminates the need for excessively high temperatures. Without the catalyst, the production of

Figure 1
Fritz Haber discovered a method for converting atmospheric nitrogen into ammonia at a technical college in Karlsruhe, Germany. He was awarded the Nobel Prize in chemistry in 1918 for discovering the process that now bears his name.

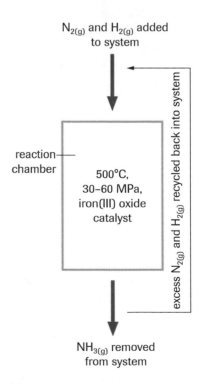

Figure 2
Outline of the Haber process

significant amounts of ammonia is too slow to be economical. Haber also discovered that the pressure in which the reaction is allowed to occur affects the percentage yield of ammonia. In general, higher pressures increase the percentage yield (**Figure 4**).

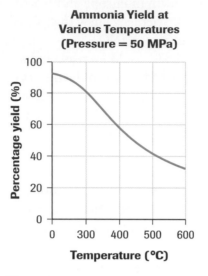

Ammonia Yield at Various Temperatures (Pressure = 50 MPa)

Ammonia Yield at Various Pressures (Temperature = 500°C)

Figure 3
The percentage yield of ammonia decreases with increasing temperature.

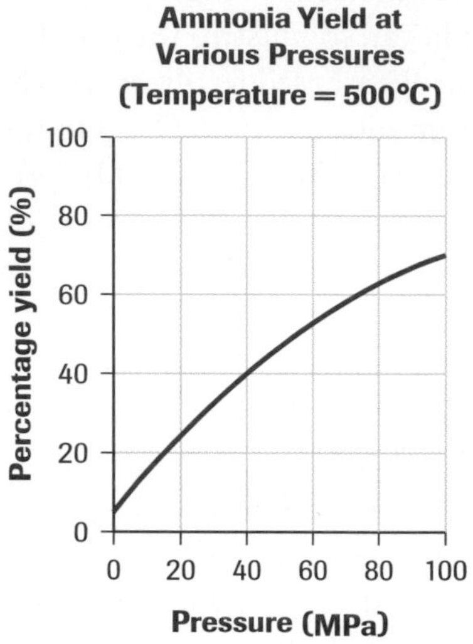

Figure 4
The percentage yield of ammonia increases with increasing pressure.

Haber and his students carried out the reaction under various conditions of temperature and pressure. They found that a satisfactory percentage yield of ammonia could be obtained at a temperature of 500°C and a pressure of 40 MPa. After a suitable length of time under these conditions, the yield of ammonia is about 40%.

Today, the Haber process is used to produce ammonia from its elements in over 335 active synthetic ammonia plants worldwide. Much of the ammonia that is produced is used in agriculture (**Figure 5**). As a fertilizer, the ammonia dissolves in moisture that is present in the soil. If the soil is slightly acidic, the ammonia is converted to nitrate ions by soil bacteria. Nitrate ions are absorbed by the roots of plants and used in the synthesis of proteins, chlorophyll, and nucleic acids. Without a source of nitrogen, plants do not grow but produce yellow leaves and die prematurely.

Figure 5
Ammonia fertilizer can be added directly to the soil.

Case Study 2.14 *Questions*

Understanding Concepts

1. Suggest four factors that could affect the production of ammonia in the Haber process.

2. Why is a low temperature, which gives a higher percentage yield of ammonia, not used in the Haber process?

3. What role does iron oxide play in the Haber process?

4. Create a concept map of the Haber process, including raw materials, their sources, and at least two end uses of the product of this process.

Making Connections

5. Ammonia can be oxidized to nitric acid, the raw material that is used to manufacture explosives.
   (a) Conduct library and/or Internet research to determine the most common types of explosives that are produced with nitric acid.
   (b) Draw structural formulas for the three most common nitrogen-based explosives. What are the specific uses of each explosive?
   (c) What is gun cotton? What are its uses? How is it made?

 www.science.nelson.com

6. Imagine that you have been hired as an efficiency consultant by a plant that produces ammonia using the Haber process.
   (a) What advice would you give the company regarding environmental conditions that would maximize the percentage yield of ammonia?
   (b) In what ways might the ideal conditions suggested in your answer to (a) be less than ideal for the company?

   (c) What additional advice could you give the company to help reduce the costs associated with your answer to (a)?

7. The Haber process requires nitrogen and hydrogen as reactants.
   (a) Suggest reasonable sources for each of these elements.
   (b) Conduct library and/or Internet research to learn how modern ammonia production facilities obtain pure hydrogen and nitrogen for the Haber process.

 www.science.nelson.com

8. Terrestrial (land) plants require nitrogen for many cellular functions, including growth and reproduction. Approximately 78% of the air in the atmosphere is composed of nitrogen. To be used by growing plants and other organisms, elemental nitrogen, $N_{2(g)}$, must first be converted into another form (such as ammonia) in a natural process called nitrogen fixation. The Haber process is a synthetic form of nitrogen fixation.
   (a) How do bacteria fix nitrogen naturally?
   (b) Currently, which of the two processes, synthetic or natural, fixes the most nitrogen?
   (c) What problems have arisen from the dramatic increase in nitrogen fixation in the last century? Pick one of these problems, and suggest some remedies.

 www.science.nelson.com

Key Understandings

2.1 Amounts in Chemistry: Mass, Moles, and Molar Mass
- Define the terms "mole" and "Avogadro's constant (N_A)," and discuss how they relate to each other.
- Interpret the coefficients in a chemical equation as individual entities or groups of entities.

2.2 Calculations Involving the Mole
- State the difference between molecular elements and compounds.
- Use the periodic table to calculate the relative atomic mass and the molar mass (M) of an entity.
- Count a large number of entities using the mass of a small group of the same entities.

2.3 Determining Chemical Formulas
- State the law of constant composition.
- Explain the function of a mass spectrometer and a combustion analyzer.
- Define the percentage composition of a compound.
- State the difference between the empirical formula and the molecular formula of a compound.

2.4 Investigation: Percentage Composition by Mass of Magnesium Oxide
- Collect experimental evidence to determine the composition, by mass, of magnesium oxide, and calculate the percentage composition.
- Use results to test the validity of the law of constant composition.
- Research information about the career of a gemologist.

2.5 Quantitative Analysis: Concentrations of Solutions
- State the difference between % V/V and % W/V concentrations of a solution.
- Discuss when to use ppm, ppb, and ppt to describe concentration.
- Define the term "molar concentration."

2.6 Tech Connect: The Spectrophotometer
- Explain how a spectrophotometer works and how results can be used to determine the concentration of a sample.

- Research information about the career of a chemical laboratory technician.

2.7 Activity: Determining the Concentration of a Solution
- Collect experimental evidence to determine the molar concentration of a copper(II) sulfate solution.

2.8 Explore an Issue: Drug Testing
- List the types of substances that are banned by the International Olympic Committee, and understand why they are banned.
- Describe ways in which drug tests can produce false positives.

2.9 The Mole and Chemical Equations: Stoichiometry
- Determine mole ratios from balanced chemical equations.
- Solve stoichiometric problems.

2.10 Limiting and Excess Reagents
- Discuss the roles of the limiting reagent and excess reagent in a chemical reaction.
- Describe how the process of analysis by precipitation works.
- Identify the limiting reagent and excess reagent in a chemical reaction, and perform calculations based on this information.

2.11 Investigation: The Limiting Reagent in a Chemical Reaction
- Design and conduct an experiment to determine the excess reagent.
- Organize qualitative and quantitative observations.
- Evaluate the stoichiometric method, as used in the prediction of masses of reactants and products.

2.12 Percentage Yield
- Distinguish between the terms "theoretical yield" and "actual yield."
- Understand the advantages of obtaining maximum percentage yield, particularly as it relates to industry.

2.13 Investigation: The Percentage Yield of a Chemical Reaction
- Collect experimental evidence to determine and evaluate the actual yield, theoretical yield, and percentage yield of copper(II) chloride dihydrate and aluminum metal.

- Identify sources of experimental error.

2.14 Case Study: The Haber Process
- Understand the Haber process, as well as the factors that affect the Haber process.

Key Terms

2.1
isotopic abundance
unified atomic mass
  unit (u)
atomic mass
molecular mass
formula unit mass
macroscopic
mole
Avogadro's constant (N_A)
molar mass

2.2
molecular element
compound

2.3
law of constant
  composition
mass spectrometer
combustion analyzer
percentage composition
empirical formula
molecular formula

2.5
quantitative analysis
concentration
molar concentration
parts per million
dilution

2.9
mole ratio
stoichiometry

2.10
excess reagent
limiting reagent

2.12
yield
actual yield
theoretical yield
percentage yield

Problems You Can Solve

2.1
- Use mass and molar mass to calculate the number of entities.
- Use the periodic table to calculate atomic mass and molecular mass.

2.2
- Use mass, amount in moles, molar mass, and number of entities to calculate an unknown value.

2.3
- Calculate the empirical formula of a compound.
- Calculate the molecular formula of a compound.
- Calculate the percentage composition by mass of a compound.

2.5
- Use the percentage concentration formula to calculate % V/V and % W/V of a solution.
- Use the molar concentration formula to solve for the unknown value.
- Use the dilution concentration formula to solve for the unknown value.

2.9
- Use a balanced chemical equation to calculate the masses of the reactants.
- Use a balanced chemical equation to calculate the numbers of entities and the masses of the reactants.

2.10
- Use a balanced equation to determine the limiting reagent in a reaction and calculate the mass of a product formed.

2.12
- Use actual yield and theoretical yield to calculate percentage yield.

> ▶ **MAKE** a summary

In this unit, you learned how to calculate a number of different quantities related to compounds and chemical reactions. To summarize your learning, construct two master flow charts on separate sheets of paper. In one flow chart, summarize the calculations related to individual compounds (sections 2.1 to 2.5). In the other flow chart, summarize the calculations based on a balanced chemical equation (sections 2.9 to 2.12).

PERFORMANCE TASK

Assessment

Your completed task will be
assessed according to the
following criteria:

Process

- Ask questions.
- Make and evaluate
  predictions.
- Design appropriate
  experimental procedures.
- Choose and safely use
  laboratory materials.
- Analyze results using
  qualitative and quantitative
  methods.
- Evaluate experimental
  procedures, and suggest
  improvements.

Product

- Demonstrate an
  understanding of the
  concepts presented in this
  unit.
- Use appropriate terms,
  measurements, and SI
  symbols and units correctly.
- Prepare a formal lab report
  to communicate the
  questions, predictions,
  materials, experimental
  designs, procedures, results,
  analyses, and evaluations of
  your experiments, as well as
  your recommendation of the
  best procedure.
- Carry out procedures,
  measurements, and
  calculations with accuracy
  and precision.

Production Control Technician

Sodium acetate, $NaC_2H_3O_{2(s)}$, is used in many industrial processes. For example, it is used with acetic acid, $HC_2H_3O_{2(aq)}$, as a buffer to control the pH of food during various stages of processing. It is used for manufacturing plastics, tanning leather, treating wastewater, and producing petroleum, and in medicines and medical procedures.

Imagine that you are a newly hired senior production control technician in a company that produces large quantities of sodium acetate. Your company produces sodium acetate from the reaction of a solution of sodium hydrogen carbonate (baking soda), $NaHCO_{3(aq)}$, with acetic acid according to the following equation:

$$NaHCO_{3(aq)} + HC_2H_3O_{2(aq)} \rightarrow NaC_2H_3O_{2(aq)} + CO_{2(g)} + H_2O_{(l)}$$

In the industrial process, large quantities of vinegar and baking soda are mixed in a large steel vat and allowed to react. As the reaction proceeds, the carbon dioxide gas bubbles out of the reaction mixture, leaving behind an aqueous solution of sodium acetate. When the reaction is complete, the sodium acetate solution is pumped into a large desiccating (drying) vessel. The solution is heated to allow the water to evaporate, leaving behind solid sodium acetate. The solid sodium acetate is removed from the desiccating vessel and packaged for transport to factories around the world.

One of your duties as the senior production control technician is to conduct small-scale controlled experiments, from time to time, to determine whether the percentage yield of sodium acetate in the production process is as high as it could be. Since you are new to the company, you do not have the results of past experiments to use for comparison. You do, however, have a laboratory logbook that contains the procedure your predecessor used for small-scale testing of the production process the company is presently using to make sodium acetate.

In this performance task, you will conduct a preliminary percentage yield test by carrying out the Procedure in the logbook and determining the percentage yield of sodium acetate. You will evaluate the Procedure by identifying sources of experimental error and environmental conditions that may be limiting the percentage yield. You will then predict how possible changes to the Materials and/or Procedure may affect percentage yield. Based on your Evaluation and Predictions, you will design and perform one or two additional experiments that may increase the percentage yield. Although you must use the same raw materials (sodium hydrogen carbonate and acetic acid) in all your experiments, you may vary the quantities, concentrations, and sources of these materials. You may also change the reaction time, heating method, heating rate, mixing method, and any other technique or environmental condition. Finally, you will write a comprehensive lab report to submit to your supervisor, the vice-president of production.

Laboratory Logbook Procedure for Testing the Percentage Yield of Sodium Acetate

Use the following Procedure for your preliminary percentage yield test.

Materials

eye protection
lab apron
balance
scoopula
stirring rod
hot plate
100-mL graduated cylinder
500-mL beaker
watch glass large enough to fit over the mouth of a 500-mL beaker
sodium hydrogen carbonate, $NaHCO_{3(s)}$
vinegar (5% W/V $HC_2H_3O_{2(aq)}$)
distilled water

Procedure

1. Measure and record the mass of a clean, dry 500-mL beaker.

2. Add approximately 30 mL of distilled water to the pre-massed beaker. Dissolve 4.2 g of sodium hydrogen carbonate in the water. Stir the solution until all of the solid is dissolved.

3. Place 100 mL of vinegar in a clean, dry graduated cylinder.

4. Very slowly add all the vinegar to the sodium hydrogen carbonate solution. You will observe the formation of carbon dioxide bubbles when the acetic acid is added to the sodium hydrogen carbonate solution. When you have added all the acetic acid, stir for 2 min.

5. Place another 100 mL of vinegar into the graduated cylinder you used in step 4. Very slowly add it to the solution in the 500-mL beaker, in 10-mL increments. Continue adding vinegar until further additions no longer cause the mixture to bubble.

6. When bubbling has completely stopped, place the beaker on a hot plate. Heat the solution to a gentle boil. Make sure that the solution does not boil over. Reduce the heat if boiling becomes violent. You may set a suitable watch glass on the mouth of the beaker to prevent loss of solution from splattering.

7. When all the liquid part of the solution has evaporated, remove the beaker from the hot plate using protective gloves. Place the beaker on a countertop, and cool its contents to room temperature.

8. Measure and record the mass of the beaker and its contents (sodium acetate).

9. Rinse the beaker and its contents with lots of tap water in a sink. All the waste can go down the drain. Return the laboratory equipment, and wash your hands with soap and water.

Analysis

(a) Determine the limiting reagent in the vinegar and baking soda reaction you carried out. Then calculate the percentage yield of sodium acetate.

Evaluation

(b) Identify sources of experimental error and environmental conditions that may have limited the percentage yield of sodium acetate.

Maximizing Percentage Yield

(c) Predict how possible changes to materials, environmental conditions, and/or the procedure may affect the percentage yield.

(d) Design and perform one or two additional experiments, on the basis of your predictions, that may increase the percentage yield. Your experimental designs must include all relevant safety precautions, and they must be approved by your teacher before you try them.

(e) Determine the percentage yield of sodium acetate in each experiment. Evaluate your experimental designs and procedures by comparing the percentage yields in all the experiments.

Writing a Lab Report

(f) Write a comprehensive lab report that includes questions, predictions, materials, experimental designs, procedures, results, analyses, and evaluations of all of the experiments you carried out. You may follow the general lab report guidelines in Appendix A4 to complete your lab report. Your lab report must also contain a final section called "Recommendation," in which you recommend the best procedure for producing sodium acetate and provide reasons for your recommendation.

Understanding Concepts

1. The following equation represents the "rusting" of aluminum:

$$4\ Al_{(s)} + 3\ O_{2(g)} \rightarrow 2\ Al_2O_{3(s)}$$

   (a) List the coefficients in the equation.
   (b) Describe the reaction in words.　　(2.1)

2. (a) How many doughnuts are in one dozen doughnuts? How many doughnuts are in one mole of doughnuts?
   (b) Calculate the mass, in grams, of one mole of doughnuts, if one doughnut has a mass of 70 g.
   (c) Is one mole of doughnuts a reasonable number of doughnuts? Explain.　　(2.1)

3. (a) How many atoms of mercury are in one mole of mercury atoms?
   (b) What is the mass, in grams, of one mole of mercury atoms?
   (c) Is one mole of mercury atoms a reasonable number of mercury atoms? Explain.　　(2.1)

4. Why do we use the value of Avogadro's constant, 6.02×10^{23}, when working with atoms or molecules?　　(2.1)

5. The term "carat" is used to indicate the mass of a diamond. This term comes from the name of the carob bean. In the past, gem dealers used carob beans to balance their scales because all carob beans have approximately the same mass. A one-carat diamond (pure carbon) has a mass of 0.2 g. The term "karat" is used to indicate the purity of gold. Pure gold is 24 karats. The ring in **Figure 1** consists of a 0.50-carat diamond and 6.50 g of 18-karat gold.
   (a) How many moles of carbon are in the diamond in **Figure 1**? How many atoms?
   (b) How many moles of gold are in the ring in **Figure 1**? How many atoms?　　(2.2)

6. (a) Calculate the molar mass of 1,4-benzenedicarboxylic acid, $C_8H_6O_{4(aq)}$, a raw material that is used to make Dacron. Dacron is a synthetic fibre that is found in many types of clothing.

Figure 1

   (b) A patient is prescribed 1.5×10^{-3} mol of acetaminophen (Tylenol), $C_8H_9NO_{2(s)}$. How many grams of Tylenol should the patient take?
   (c) How many moles of butane, $C_4H_{10(l)}$, are in a lighter (**Figure 2**), if the butane has a mass of 0.95 g?
   (d) How many atoms of carbon are in a vitamin C tablet (ascorbic acid), $C_6H_8O_{6(s)}$, that contains 0.5 g of ascorbic acid?　　(2.2)

7. Distinguish between molecular elements and compounds. Provide an example of each.　　(2.2)

8. (a) What information does a mass spectrometer provide for determining the molecular formula of a compound?
   (b) How are the carbon dioxide and water traps in a combustion analyzer used to measure the masses of carbon and hydrogen in a sample of a hydrocarbon?　　(2.3)

Figure 2

9. A compound is found to contain 38.72% carbon, 9.72% hydrogen, and 51.56% oxygen.
   (a) Calculate the empirical formula of the compound.
   (b) State two possible molecular formulas for the compound.
   (c) What additional information do you need to determine the molecular formula of the compound? (2.3)

10. Calculate the molecular formulas of organic compounds A and B, given the information in **Table 1**. (2.3)

11. Why is the molar concentration of an aqueous solution measured as moles of solute per litre of solution, instead of moles of solute per litre of water? (2.5)

12. Distinguish between a dilute solution and a concentrated solution. (2.5)

13. A sample of drinking water has a nitrate concentration of 2.3 ppm (a level that is considered safe for drinking). Calculate the mass, in grams, of nitrate ions in a 250-mL glass of this water. (2.5)

14. Calculate the molar concentration of each of the following aqueous solutions:
   (a) 12.0 g of sodium hydroxide dissolved in water to make 2.5 L of solution
   (b) 2.28 g of potassium hydrogen tartrate, $KC_4H_5O_{6(s)}$, dissolved in water to form 100.0 mL of solution
   (c) 0.08 g ethanol, $C_2H_6O_{(l)}$, in 100 mL of blood (the legal limit of blood alcohol concentration in Canada when driving a car) (2.5)

15. A sodium hydroxide solution was prepared by transferring 0.40 g of sodium hydroxide, $NaOH_{(s)}$, to a 100-mL volumetric flask and filling the flask with water to the 100-mL mark.
   (a) Calculate the molar concentration and the weight by volume (W/V) concentration of the sodium hydroxide solution.
   (b) A 10-mL sample of the sodium hydroxide solution was transferred to a 50-mL volumetric flask, and the solution was diluted to the 50-mL mark. Calculate the molar concentration and weight by volume concentration of the final sodium hydroxide solution. (2.5)

16. Balance each of the following unbalanced equations:
   (a) $Fe + H_2O \rightarrow Fe_3O_4 + H_2$
   (b) $H_2SO_4 + NaOH \rightarrow H_2O + Na_2SO_4$
   (c) $Cu + O_2 \rightarrow Cu_2O$
   (d) $Fe_2(SO_4)_3 + KSCN \rightarrow K_3Fe(SCN)_6 + K_2SO_4$ (2.9)

17. (a) Write a balanced chemical equation to represent the complete combustion of ethanol, $C_2H_6O_{(l)}$.
   (b) Nitroglycerine (a drug used for heart conditions), $C_3H_5N_3O_9$, is produced by mixing glycerine, $C_3H_8O_3$, and nitric acid, HNO_3. Water is also a product. Write a balanced chemical equation to represent this reaction. (2.9)

18. The conversion of iron ore, $Fe_2O_{3(s)}$, into iron occurs in several steps. The first step involves the partial combustion of coal, $C_{(s)}$, to give carbon monoxide:

$$2\ C_{(s)} + O_{2(g)} \rightarrow 2\ CO_{(g)}$$

In a number of additional steps, carbon monoxide acts on iron ore, with the following overall result:

$$Fe_2O_{3(s)} + 3\ CO_{(g)} \rightarrow 2\ Fe_{(s)} + 3\ CO_{2(g)}$$

Table 1 Percent Composition and Molar Mass for Compounds A and B

| Compound | Percentage composition (%) | | | | Molar mass (g/mol) |
	Carbon	Hydrogen	Oxygen	Nitrogen	
A	64.6	10.8	24.6	none	260.0
B	38.67	16.22	none	45.11	31.06

(a) In a small-scale laboratory test of the conversion process, a 300-g sample of iron ore is converted into iron. How many moles of iron are produced? How many grams of iron are produced?

(b) The actual yield of iron in this test is 178 g. What is the percentage yield? (2.12)

19. Aluminum oxide (a polishing powder), $Al_2O_{3(s)}$, is made by combining 5.00 g of aluminum with oxygen. Calculate how much oxygen is needed, in moles and in grams. (2.10)

20. The thermite reaction (**Figure 3**) has been used to weld railroad rails, to make certain bombs, and to ignite solid-fuel rocket motors. The balanced chemical equation for this reaction is

$$Fe_2O_{3(s)} + 2\,Al_{(s)} \rightarrow 2\,Fe_{(l)} + Al_2O_{3(s)}$$

(a) What is the maximum mass of aluminum oxide that can be produced with 135.0 g of aluminum?

(b) How much aluminum oxide is produced if the yield is 87%? (2.10, 2.12)

Figure 3
The thermite reaction is used to weld railroad rails.

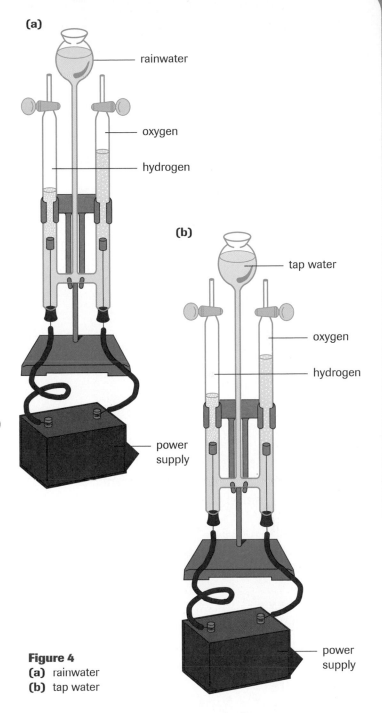

Figure 4
(a) rainwater
(b) tap water

Applying Inquiry Skills

21. To test the law of constant composition, a student uses a Hoffman apparatus to decompose a sample of tap water and a sample of rainwater into hydrogen and oxygen, under the same environmental conditions (**Figure 4**).

(A Hoffman apparatus uses electricity to decompose water into its elements.) The volumes of hydrogen gas and oxygen gas that are formed in the reactions are measured directly on the calibrated gas collection tubes. Complete the Analysis and Evaluation in the following lab report.

Question

Does the law of constant composition hold for water molecules?

Prediction

All water molecules have the same composition.

Observations

Table 1

Sample	rainwater	tap water
Volume of $H_{2(g)}$ produced (mL)	23.72	8.39
Volume of $O_{2(g)}$ produced (mL)	11.80	4.18

Analysis

(a) Calculate the hydrogen-to-oxygen ratio for rainwater and tap water.

(b) Answer the Question.

Evaluation

(c) What assumptions, if any, must be made in order to answer the Question?

(d) Evaluate the Prediction.   (2.3)

22. Calcium is a silvery white metal (**Figure 5(a)**) that burns readily in air to produce calcium oxide, $CaO_{(s)}$ (**Figure 5(b)**), as the only product.

(a) Describe an experimental design that may be used to determine the percentage composition by mass of calcium oxide.

(b) Octane, $C_8H_{18(l)}$, is a liquid that burns readily in air to produce carbon dioxide and water. Can you use the same experimental design to determine the percentage composition by mass of octane? Explain.

(2.13)

23. (a) Describe the steps in a procedure for preparing 100 mL of a 0.15-mol/L copper(II) nitrate solution, including safety precautions.

(b) Describe the steps in a procedure for diluting the solution you prepared in (a) to form 1.0 L of a copper(II) nitrate solution with a concentration of 0.03 mol/L.   (2.5)

24. The percent absorbance of several dilutions of the heart drug atropine was measured on a spectrophotometer. The data are listed in **Table 3**.

Table 3   Spectrophotometer Data for Atropine

Absorbance (%)	[Atropine] (μg/L)
0	0
0.10	0.62
0.14	1.2
0.23	1.9
0.37	3.1
0.64	5.5
0.76	6.3
1.0	8.0

(a)

(b)

Figure 5
(a) Calcium is a silvery white metal.
(b) Calcium oxide is the only product that is formed when calcium burns in air.

(a) Prepare a standard curve by graphing the data in **Table 3**, with percent absorbance plotted on the vertical axis and concentration plotted on the horizontal axis.

(b) Determine the concentration of an atropine solution whose percent absorbance is 0.50%.

(c) Can the standard curve you sketched in (a) be used to determine the concentration of a solution of digitalis, a different heart medication? Explain. (2.7)

25. When lead(II) nitrate solution reacts with sodium sulfate solution, a black precipitate of lead(II) sulfate is formed (**Figure 6**). The balanced chemical equation for the double displacement reaction is

$$Pb(NO_3)_{2(aq)} + Na_2SO_{4(aq)} \rightarrow PbSO_{4(s)} + 2\,NaNO_{3(aq)}$$

A student adds 10.0 mL of the sodium sulfate solution to 4.60 mL of the lead(II) nitrate solution. The student filters and dries the precipitate and measures its mass. Can the student use this mass to calculate the concentration of the lead(II) nitrate solution? Explain. (2.7)

Making Connections

26. A Breathalyzer is a portable breath alcohol testing device that is used by police officers to determine a driver's blood alcohol concentration (BAC). Many drivers who are charged with drunk driving challenge the results of the Breathalyzer test in court. Conduct library and/or Internet research to answer the following questions about Breathalyzers and Breathalyzer testing:

(a) Briefly explain how a Breathalyzer works.

(b) What is the legal BAC limit for drivers in Ontario?

(c) Distinguish between a portable Breathalyzer and a stationary Breathalyzer.

(d) What reasons do drivers use to challenge the results of Breathalyzer tests in court?

(e) Write a one-page paper, stating your position on the reliability of Breathalyzer tests. (2.5)

 www.science.nelson.com

Figure 6
When lead(III) nitrate solution and sodium sulfate solution are mixed, a precipitate of lead(II) sulfate is formed. Sodium nitrate remains dissolved in water.

Organic Chemistry

In a supermarket or a pharmacy, the term "organic" describes products that are grown entirely through natural biological processes. The farmers use no synthetic materials, apply no synthetic fertilizers or pesticides to "organic" fruits and vegetables, and feed no antibiotics to their "organic" chickens and cows.

As the expanding "organic foods" market indicates, some consumers believe that molecules made by a living plant or animal are different from molecules made in a laboratory. In fact, this belief was common in the early 1800s. At the time, most chemists believed that compounds produced by living systems could not be duplicated by any laboratory procedure. Chemists coined the term "organic" to distinguish between compounds obtained from living organisms and compounds obtained from mineral sources.

This belief was proved wrong in the mid-1800s when urea, until then only made by living organisms, was produced using common chemicals and lab equipment. Today, organic chemistry is defined as the study of compounds in which carbon is the principal element.

In this unit, you will examine the characteristic structures and properties of families of organic molecules. You will start with simple molecules of hydrogen and carbon, and end with complex molecules that make up plastics and foods.

▶ Overall Expectations

In this unit, you will be able to

- demonstrate an understanding of the names and properties of organic compounds and some of their reactions;
- carry out various laboratory tests and reactions involving organic compounds;
- describe the importance of organic compounds in consumer products, technological devices, and biochemical applications;
- explain some of the issues related to the environmental and social impact of organic compounds.

ARE YOU READY?

Knowledge and Understanding

1. Look at **Figure 1**. For each test, name the gas.

(a) When a glowing splint is placed in this gas, the splint bursts into flame.

(b) When this gas is bubbled into limewater, the limewater solution turns milky. When a flaming splint is held in this gas, the flame is extinguished.

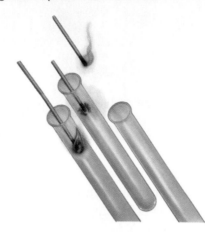

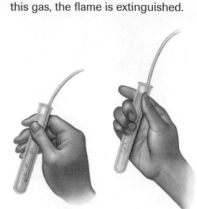

(c) When a flaming splint is held at the mouth of a test tube of this gas, a small explosion produces a popping sound.

(d) When cobalt chloride test paper is placed in this gas, the test paper changes from blue to pink.

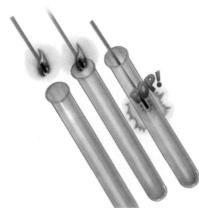

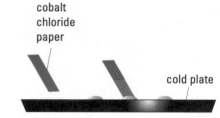

cobalt chloride paper

cold plate

Figure 1

2. Complete the following chemical equations for the neutralization reactions indicated:
   (a) $HCl_{(aq)} + NaOH_{(aq)} \rightarrow$
   (b) $H_2SO_{4(aq)} + 2\,KOH_{(aq)} \rightarrow$

3. Are the following pairs of atoms more likely to form ionic bonds or covalent bonds? Give reasons for your answers.
   (a) chlorine and chlorine
   (b) potassium and iodine
   (c) carbon and oxygen
   (d) magnesium and fluorine

4. Draw a Lewis structure for each of the following molecules:
   (a) O_2
   (b) CH_4
   (c) NH_3

5. Use the electronegativity table in section 1.12 to identify the more polar bond in each of the following pairs:
   (a) C–H and O–H
   (b) C–O and N–O
   (c) C–C and C–H
   (d) S–H and O–H
   (e) H–Cl and H–F

6. Balance each of the following equations.
   (a) $NH_{3(g)} + O_{2(g)} \rightarrow NO_{(g)} + H_2O_{(l)}$
   (b) $NO_{2(g)} + H_2O_{(l)} \rightarrow HNO_{3(g)} + NO_{(g)}$
   (c) $C_{12}H_{22}O_{11(s)} + O_{2(g)} \rightarrow CO_{2(g)} + H_2O_{(l)}$
   (d) $C_6H_{6(l)} + Cl_{2(g)} \rightarrow C_6H_3Cl_{3(l)} + HCl_{(g)}$

Technical and Safety Skills

7. Copy **Table 1** into your notebook. For each WHMIS symbol that is shown, identify the type of compound, the risks, and the precautions needed.

Table 1   WHMIS Symbols

Class and type of compound	WHMIS symbol	Risks	Precautions
Class A			
Class B			
Class C			
Class D			
Class E			

Figure 1
The properties of plastic make it suitable for many functions.

If you look around, you will likely find yourself surrounded by organic compounds, ranging from foods and medications to shoes and computers. Some of these molecules—table sugar, alcohol in wine or beer, natural gas for a stove, gasoline in a car—are relatively small, containing up to 10 or 20 carbon atoms as well as hydrogen and sometimes oxygen atoms. Other organic compounds are made up of long chains of carbon atoms, each chain connected to other chains through different types of intermolecular forces. Large organic molecules include starches and proteins, and synthetic polymers such as polyethylene and plastics (**Figure 1**).

The number of different organic compounds is limitless. Carbon atoms are able to form covalent bonds with up to four other carbon atoms. Therefore, they can continue to add more carbon atoms and form long straight chains, branched chains, and even ring structures. As well, they can bond with other atoms, to give an even greater variety of molecules with unique properties.

In this unit, you will learn that the structure of a molecule greatly affects its properties. You will begin your study of organic chemistry by looking at how compounds with similar groupings of atoms exhibit similar properties. For example, small compounds with only carbon and hydrogen atoms tend to be gases and liquids that are insoluble in water. Compounds that also contain oxygen atoms tend to be more soluble in water and have higher boiling points. Ethyne (acetylene) is an organic compound that is used in welding torches. It is a gas that is insoluble in water. Alcohol in wine and acetic acid in vinegar mix well with water. All three compounds consist of two carbon atoms bonded to hydrogen or oxygen atoms in different ways.

Understanding the relationships among the properties of compounds and the sizes and structures of molecules has many benefits. It allows you not only to predict the properties of organic molecules, but also to design and prepare compounds with the properties that you want in a product. Let's say that you wanted to produce a material that you could use to make a strong and flexible water bottle. You could take small molecules containing carbon and hydrogen and join them together in very long chains, to make the material strong and insoluble in water. Then you could join the chains to each other at intervals to form sheets of the material. Since some movement is possible between the chains, the material becomes flexible. The many synthetic materials that you use—such as Styrofoam, Teflon, and other plastics—are made in this way.

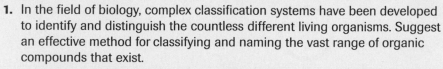

💡 **REFLECT** on your learning   ▼

1. In the field of biology, complex classification systems have been developed to identify and distinguish the countless different living organisms. Suggest an effective method for classifying and naming the vast range of organic compounds that exist.

2. From your knowledge of the bonds within molecules, and the forces of attraction between molecules, describe features in the molecular structure of a compound that would account for its solubility in water and its melting and boiling points.

3. What does "organic" mean? Give as many definitions as you can.

Organic compounds form the vast majority of all the chemical compounds that exist, and the number is continually increasing to meet consumer demand for new products. In this unit, you will examine some of the compounds used in artificial sweeteners, contact lens materials, nicotine patches, and the many different types of materials that make up a disposable diaper.

▶ **TRY THIS** activity   *Keeping Baby Dry with Polymers*

For many parents of a new baby, disposable diapers are a welcome convenience. They do not need laundering, are easy to put on without safety pins, and, most importantly, keep the baby comfortably dry. What types of materials contribute to this marvel of infant care? Dissect a diaper to find out.

The typical disposable diaper has many components. Most of these components are synthetic plastics, which are designed with the properties that are desirable for their functions.

- *polyethylene film*: cloth-like outer surface, which is impermeable to liquids

- *adhesives*: different types of glues to hold the various components together

- *polypropylene sheet*: soft and waterproof lining of the leg cuffs; also the porous inner surface sheet that is next to the baby's skin

- *polyurethane, rubber, and Lycra*: stretchy elastic for the leg cuffs and waistband

- *cellulose*: fluffy filling made from wood pulp

- *polymethylacrylate*: super-absorbent polymer crystals in the centre of the diaper, which hold water between the long polymer chains, turning the crystals into a gel. (Manufacturers claim that the crystals can absorb up to 400 times their own mass in water. Less urine can be absorbed than water, however, because urine contains sodium ions that reduce the retention of water by the polymer chains.)

Materials: eye protection; thin ("ultra") disposable diaper; sharp scissors; plastic grocery bag; 3 paper cups or beakers; water; table salt; sugar; calcium chloride, $CaCl_{2(s)}$

1. Cut a 2 cm by 2 cm sample of the first five components that are listed above (**Figure 2**). Examine the properties of each of these samples. (They are all polymers, either synthetic or natural.)

2. Cut down the centre of the inside surface. Remove the fluffy filling and put it in a plastic grocery bag. Pull the filling apart into smaller pieces, and tie a knot to close the bag. Shake the bag to dislodge the absorbent crystals within.

Figure 2

 Do not ingest the crystals or touch your face or eyes after coming in contact with them. The crystals cause irritation and dehydration.

3. Open and tilt the bag to collect the dislodged crystals in a corner of the bag. Cut a small hole in this corner, and empty the crystals into a paper cup or beaker.

4. Add about 100 mL of water to the crystals, and observe. After a few minutes, divide the material formed into three equal portions in three cups. To one portion, add about 2 mL of table salt. To the second portion, add 2 mL of sugar. To the third portion, add 2 mL of calcium chloride.

(a) Note any changes, and suggest an explanation for these changes.

5. Dispose of the diaper pieces in the garbage. Clean your work area, and wash your hands thoroughly.

(b) Design an experiment to determine the maximum volume of water that a diaper can hold without leaking. Describe the steps you would take and the measurements you would make. If possible, and with your teacher's approval, carry out the experiment.

Figure 1
Hydrocarbons are found as solids, liquids, and gases. All hydrocarbons burn to produce carbon dioxide, water, and large amounts of light and heat energy.

hydrocarbon an organic compound that contains only carbon and hydrogen atoms in its molecular structure

alkane a hydrocarbon that has only single bonds between carbon atoms; general formula C_nH_{2n+2}

alkene a hydrocarbon that contains at least one C=C double bond; general formula C_nH_{2n}

alkyne a hydrocarbon that contains at least one C≡C triple bond; general formula C_nH_{2n-2}

You will begin your study of organic compounds with hydrocarbons. **Hydrocarbons** are compounds that contain *only hydrogen and carbon atoms*. There is an enormous variety of hydrocarbons, from small molecules that are mainly gases to very large molecules that tend to be solids. Hydrocarbons include the natural gas that is piped to our homes and schools, propane in barbecue tanks, butane in cigarette lighters, and gasoline for cars. The asphalt that is used for our roofs and roads, candle wax, and tar are also hydrocarbons (**Figure 1**).

How can only two elements, carbon and hydrogen, combine to form so many different compounds? The answer is found in the ability of a carbon atom to form covalent bonds with up to four other carbon atoms. Most other elements cannot do this. In all hydrocarbon molecules, the carbon atoms are joined to each other with single, double, or triple bonds to form a *backbone*. The backbone may be in the form of a straight chain, a branched chain, or a ring structure (**Table 1**). Hydrogen atoms are attached to the carbon backbone so that each carbon atom forms a total of four bonds, as shown below. (You learned how to draw Lewis structures in section 1.12.)

Hydrocarbons are classified by the kinds of carbon–carbon bonds in their molecules. In **alkanes**, all carbon atoms are bonded to other atoms by *single bonds*. In **alkenes**, one or more of the carbon–carbon bonds are *double bonds*. In **alkynes**, one or more of the carbon–carbon bonds are *triple bonds*.

> ▶ **TRY THIS** activity

Hydrocarbons Are Made of ...

The main components of candle wax, natural gas, and cigarette lighter fuel are *organic* molecules. In other words, these products all contain carbon. You can "prove" it!

Materials: candle, cigarette lighter, Bunsen burner, metal spoon

1. Light a candle. Hold a metal spoon in the candle flame for a few seconds.
(a) What product is formed on the spoon?
2. Clean the spoon. Repeat step 1 using a yellow flame from a Bunsen burner (methane gas) instead of the candle flame.
3. Repeat, this time with a yellow flame from a butane cigarette lighter.

Table 1 Examples of Hydrocarbons

Hydrocarbon group	Example	Formula	Spacefill diagram	Bond diagram
Aliphatic				
alkane	ethane	C_2H_6	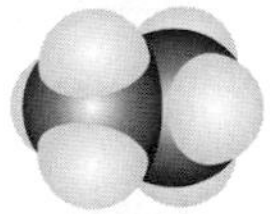	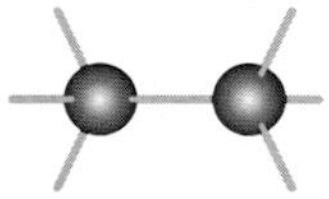
	cyclohexane	C_6H_{12}	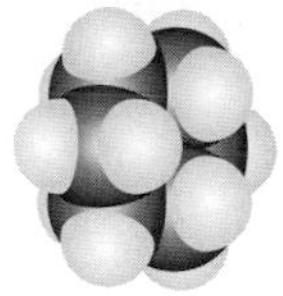	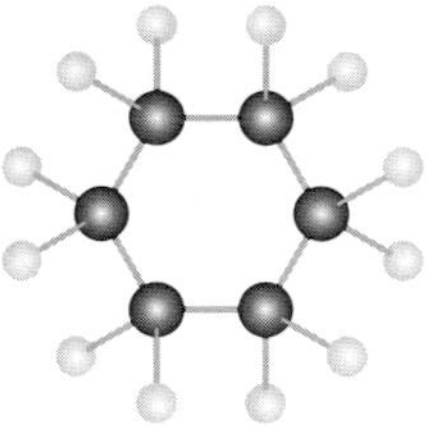
alkene	ethene	C_2H_4	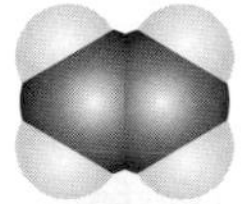	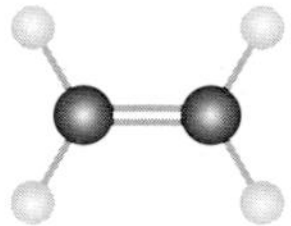
alkyne	ethyne	C_2H_2		
Aromatic				
	benzene	C_6H_6	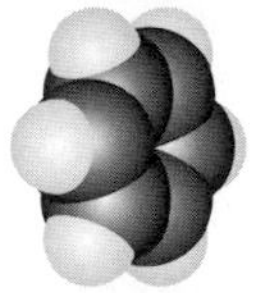	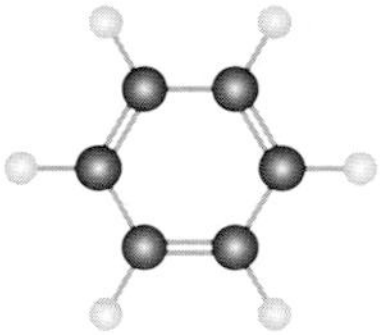

LEARNING TIP

Representing Molecules
Remember the three different ways to represent an organic molecule:

- **structural formula**, such as

$$H - {_1}C = {_2}C - {_3}C - {_4}C - H$$

with H atoms attached

- **condensed structural formula**, such as CH_2=$CHCH_2CH_3$
- **molecular formula**, such as C_4H_8

All three examples above represent the compound 1-butene.

Naming Hydrocarbons

As with inorganic compounds, there are so many organic compounds that scientists need a systematic way to classify and name them. As you will see, the IUPAC name of an organic compound tells you how the compound is "put together." In this textbook, you will learn the common names of some compounds, as well. The common name of a compound was often derived from the compound's properties, use, or original source.

The IUPAC names of hydrocarbons have two parts. The *first part* of each name (the prefix) indicates the number of carbon atoms in the longest carbon chain. The *second part* of each name (the suffix) indicates whether the hydrocarbon is an alkane, an alkene, or an alkyne. The names of alkanes all end in *-ane*, the names of alkenes all end in *-ene*, and the names of alkynes all end in *-yne*.

Alkanes

The prefixes for compounds with one to ten carbon atoms are shown in the first column of **Table 2**, on the next page. Take, for example, the name of methane, the main component of natural gas. The first part, *meth-*, tells you

The Smell of Methane

Methane is the main component of natural gas, which is used in homes and schools. Methane itself is odourless. The familiar smell of natural gas is due to trace amounts of a thiol, a smelly sulfur-containing compound that is added as a safety feature, to allow detection of any gas leaks.

LEARNING TIP

Prefixes
You will find it very useful to memorize the prefixes of hydrocarbon chains. You will use them repeatedly in this unit.

Butane Lighters

These lighters contain butane. Butane boils at –0.5°C, so it is a gas at room temperature and pressure. When stored at high pressure, as in these lighters, it becomes a liquid.

Table 2 Naming Alkanes

Prefix	IUPAC name	Formula
meth–	methane	$CH_{4(g)}$
eth–	ethane	$C_2H_{6(g)}$
prop–	propane	$C_3H_{8(g)}$
but–	butane	$C_4H_{10(g)}$
pent–	pentane	$C_5H_{12(l)}$
hex–	hexane	$C_6H_{14(l)}$
hept–	heptane	$C_7H_{16(l)}$
oct–	octane	$C_8H_{18(l)}$
non–	nonane	$C_9H_{20(l)}$
dec–	decane	$C_{10}H_{22(l)}$

that there is one carbon atom. The second part, *-ane*, tells you that the compound is an alkane. Since the single carbon atom needs four hydrogen atoms to fill its four bonds, the formula for methane is CH_4.

▶ **SAMPLE** problem *1*

Drawing Structural Formulas for Alkanes

Propane is a fuel that is commonly used in gas barbecues. Draw a structural formula for propane, and write its formula.

Step 1: Write Carbon Backbone of Appropriate Length
The first half of the name "propane" is *prop-*. This prefix tells you that propane has three C atoms.

$$C — C — C$$

Step 2: Add Single Lines to Each Carbon, to a Total of Four Lines
The second half of the name "propane" is *-ane*. This suffix tells you that propane is an alkane with only single bonds. Therefore, propane is a molecule with three C atoms joined by single bonds. Each C atom forms four bonds.

$$
\begin{array}{ccccc}
 & | & & | & & | \\
- & C & - & C & - & C & - \\
 & | & & | & & | \\
\end{array}
$$

Step 3: Fill Remaining Bonds with Hydrogen Atoms
Eight H atoms are needed to fill all the bonds. Therefore, the formula for propane is C_3H_8.

$$
\begin{array}{ccccccc}
 & H & & H & & H \\
 & | & & | & & | \\
H - & C & - & C & - & C & - H \qquad \text{propane, } C_3H_8 \\
 & | & & | & & | \\
 & H & & H & & H
\end{array}
$$

Example

Draw a structural formula and write the formula for each of the following alkanes:

(a) Butane, commonly used in lighters, contains a four-carbon chain joined by single bonds.

(b) Octane, found in gasoline, contains an eight-carbon chain joined by single bonds.

(c) Methane, the natural gas that is used in homes and school science labs, is a single carbon atom with four hydrogen atoms.

Solution

(a) butane, C_4H_{10}

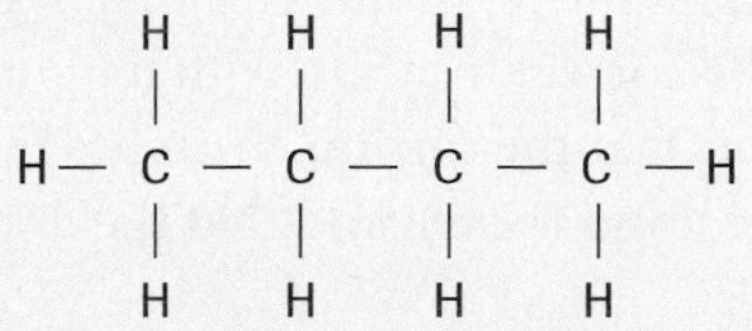

```
    H    H    H    H
    |    |    |    |
H — C — C — C — C —H
    |    |    |    |
    H    H    H    H
```

(b) octane, C_8H_{18}

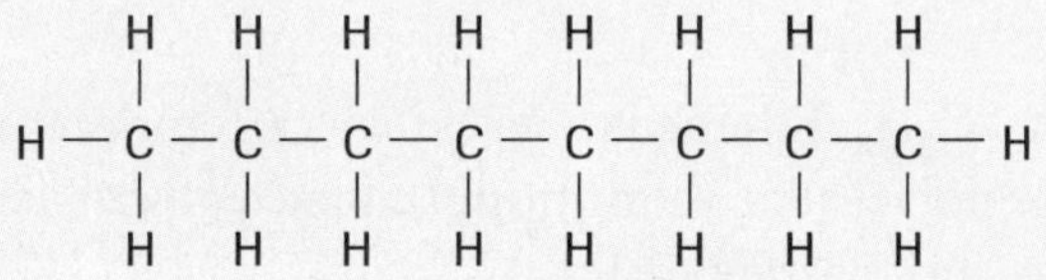

```
    H   H   H   H   H   H   H   H
    |   |   |   |   |   |   |   |
H — C — C — C — C — C — C — C — C — H
    |   |   |   |   |   |   |   |
    H   H   H   H   H   H   H   H
```

(c) methane, CH_4

```
     H
     |
H — C — H
     |
     H
```

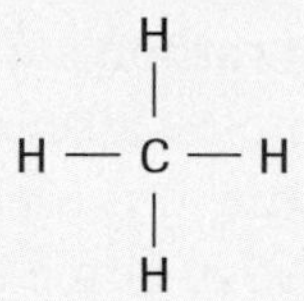

▶ **Practice**

Understanding Concepts

1. Draw a structural formula and write the molecular formula for each of the following alkanes:
   (a) ethane
   (b) pentane
   (c) hexane
   (d) decane

2. Write the IUPAC name and the molecular formula for each of the following alkanes:
   (a)

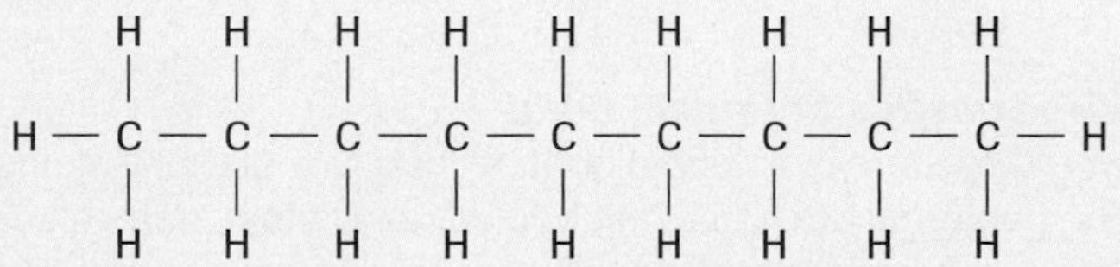

   ```
       H   H   H   H   H   H   H   H   H
       |   |   |   |   |   |   |   |   |
   H — C — C — C — C — C — C — C — C — C — H
       |   |   |   |   |   |   |   |   |
       H   H   H   H   H   H   H   H   H
   ```

   (b)

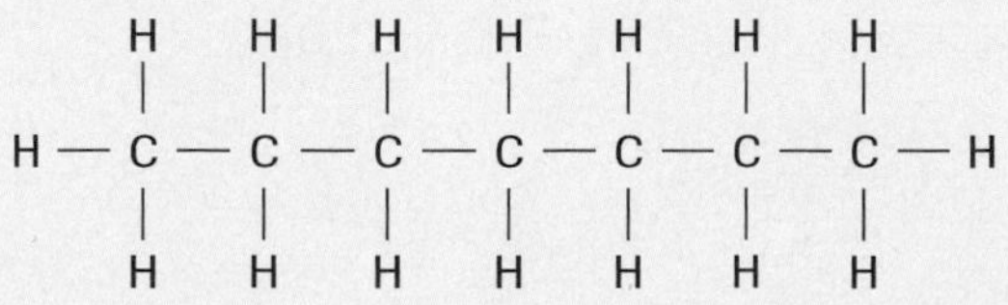

   ```
       H   H   H   H   H   H   H
       |   |   |   |   |   |   |
   H — C — C — C — C — C — C — C — H
       |   |   |   |   |   |   |
       H   H   H   H   H   H   H
   ```

LEARNING TIP

A Lot in Common
Some alkenes and alkynes have common names.

IUPAC name	Common name
ethene	ethylene
propene	propylene
ethyne	acetylene

Organic molecules are often represented as condensed structural formulas. In a condensed structural formula, the bonds are not shown. The hydrogen atoms are collected together and written after the carbon atom to which they are attached. Therefore, the condensed structural formulas for nonane and heptane (in Practice question 2 on the previous page) are $CH_3CH_2CH_2CH_2CH_2CH_2CH_2CH_2CH_3$ and $CH_3CH_2CH_2CH_2CH_2CH_2CH_3$, respectively. Sometimes structural formulas are even further condensed, to $CH_3(CH_2)_7CH_3$ and $CH_3(CH_2)_5CH_3$, for example.

Alkenes and Alkynes

The general rules for naming alkenes and alkynes are similar to the general rules for naming alkanes. You use *–ene* or *–yne*, however, instead of *–ane* at the end of the name. As well, if there is more than one possibility for the location of the double bond, you need to indicate the correct location with *a number*.

For example, the name *2-pentene* tells you that the carbon chain has five carbon atoms (*pent-*), that there is a double bond (*-ene*), and that the double bond starts at the second carbon atom (*2-*).

Similarly, the name *1-butyne* tells you that the carbon chain has four carbon atoms (*but-*), that there is a triple bond (*-yne*), and that the triple bond starts at the first carbon atom (*1-*).

The simplest alkene is ethene, C_2H_4. Ethene is a gas at room temperature and pressure. It is used to make polyethene (commonly called polyethylene), a plastic used in home insulation and plastic bottles.

The simplest alkyne is ethyne, C_2H_2. Ethyne (commonly called acetylene) is also a gas at room temperature and pressure. You may have seen or even used an oxy-acetylene torch in a welding shop.

▶ *SAMPLE* problem *2*

Drawing Structural Formulas for Alkenes and Alkynes

Draw a structural formula and write the molecular formula for 1-hexene.

Step 1: Write Carbon Backbone of Appropriate Length
Start with the name "hexene." The *hex-* tells you that there are six C atoms in a chain.

$$— C — C — C — C — C — C —$$

Step 2: Establish Location of Double Bond
The *-ene* tells you that there is a double bond linking two of the C atoms. The number at the beginning of the name tells you which C atom *starts* the double bond. (In the IUPAC system, the molecule is numbered from the end that gives the *lowest number* for the double bond.) Since the alkene is *1*-hexene, the double bond starts at the *first C atom* and ends at the second C atom.

$$\overset{1}{C} = \overset{2}{C} - \overset{3}{C} - \overset{4}{C} - \overset{5}{C} - \overset{6}{C} -$$

Step 3: Fill Remaining Bonds with Hydrogen Atoms

Note that each C atom forms four bonds.

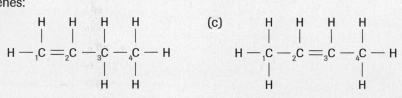

1–hexene, C_6H_{12}

There are 6 C atoms and 12 H atoms in total, so the molecular formula is C_6H_{12}.

Example 1

Write the IUPAC name and the molecular formula for each of the following alkenes:

(a)

$$H - {}_1C = {}_2C - {}_3C - {}_4C - H$$

(c)

$$H - {}_1C - {}_2C = {}_3C - {}_4C - H$$

(b)

$$H - {}_4C - {}_3C - {}_2C = {}_1C - H$$

> **LEARNING TIP**
>
> **Use the Lowest Number**
> The carbon backbone of alkenes and alkynes is not always numbered from the left. Check the location of the double or triple bond before deciding how to number the carbon atoms.

Solution

(a) 1-butene, C_4H_8

(b) 1-butene, C_4H_8

(c) 2-butene, C_4H_8

Example 2

Draw a structural formula and write the molecular formula for each of the following hydrocarbons:

(a) ethene

(b) ethyne

(c) 2-hexene

Solution

(a)

$$H - C = C - H$$

ethene, C_2H_4

(b) $H - C \equiv C - H$

ethyne, C_2H_2

(c)

$$H - C - C = C - C - C - C - H$$

2–hexene, C_6H_{12}

Understanding Concepts

3. Draw a structural formula and write the formula for each of the following hydrocarbons:
   (a) 3-hexene
   (b) 1-pentene
   (c) 2-butyne
   (d) 2-pentene
   (e) propyne
   (f) 3-octene

4. Write the IUPAC name and the molecular formula for each of the following hydrocarbons:

(a)

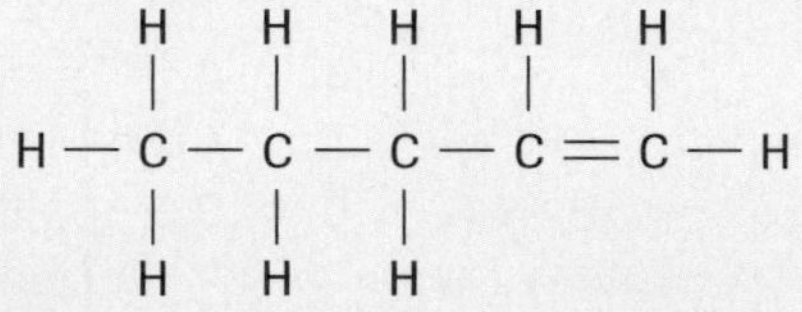

```
      H    H    H    H    H
      |    |    |    |    |
  H — C — C — C — C = C — H
      |    |    |
      H    H    H
```

(b)

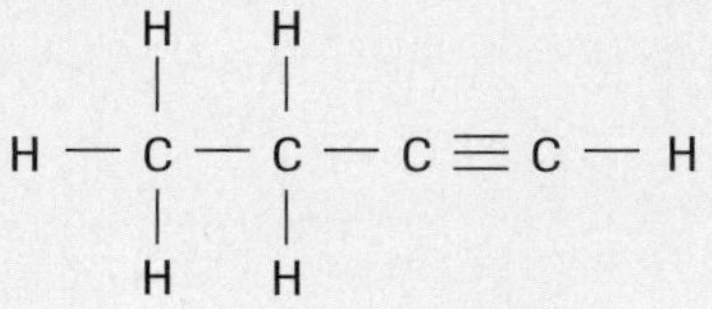

```
      H    H
      |    |
  H — C — C — C ≡ C — H
      |    |
      H    H
```

(c)

```
      H    H    H    H    H    H    H
      |    |    |    |    |    |    |
  H — C — C = C — C — C — C — C — H
      |         |    |    |    |
      H         H    H    H    H
```

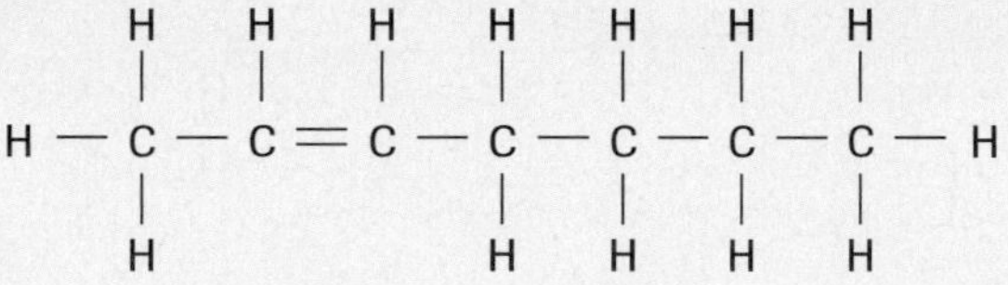

In Example 1 on the previous page, did you notice that alkanes (a) and (b) have the same name (1-butene) and molecular formula (C_4H_8)? In fact, they are the same compound. The structural formula in (b) may be obtained by simply flipping the structural formula in (a), left to right. Did you also notice that the structure of 1-butene is different from the structure of 2-butene, even though they have the same molecular formula, C_4H_8? These compounds are different because the different arrangement of atoms and bonds gives them different chemical and physical properties. Compounds that have the same formula but different structures and properties, such as 1-butene and 2-butene, are called **structural isomers**.

The carbon atoms in hydrocarbon molecules may link to form straight chains. As well, they may form branches at one or more points in the carbon chain. In some molecules, several carbon atoms form a ring structure. These molecules are called *cyclo*hydrocarbons. A special type of ring structure is found in the hydrocarbon *benzene*, an organic solvent (see **Table 1** on page 181). Compounds that contain the benzene ring structure are called *aromatic* hydrocarbons, because of their distinctive odours.

structural isomer a compound that has the same molecular formula as another compound but a different molecular structure

Physical Properties of Hydrocarbons

Recall, from section 1.12, that since hydrocarbons are nonpolar molecules, their physical properties are largely governed by the intermolecular bonds known as London dispersion forces. The more intermolecular bonds there are, the more energy that is required to separate the molecules and, thus, the higher the boiling point is. Therefore, large hydrocarbons are solids, while smaller hydrocarbons are liquids and gases.

Combustion Reactions of Hydrocarbons

All hydrocarbons will burn in air to produce large amounts of light and heat. This reaction with the oxygen in air makes hydrocarbons useful fuels. It is an example of a *combustion reaction*. (You learned about combustion reactions in section 1.14.) When enough oxygen is available for complete combustion, the only products that are formed are carbon dioxide gas and water vapour. These are the same two products that you exhale when your cells "burn" the molecules from the food you eat, using inhaled oxygen.

complete combustion

hydrocarbon + oxygen → carbon dioxide + water

You have probably seen the white fumes from car exhaust on a cold winter morning. These fumes are a result of the water vapour condensing in the cold air. Although you cannot see the carbon dioxide gas in the exhaust, the carbon dioxide levels in our atmosphere are increasing with the burning of fossil fuels. The increasing carbon dioxide levels may be contributing to global climate changes. You will examine these reactions later in the unit. The reaction of propane gas, commonly used in gas barbecues, is shown below.

$$C_3H_{8(g)} + 5\ O_{2(g)} \rightarrow 3\ CO_{2(g)} + 4\ H_2O_{(g)}$$

Many other organic compounds, such as alcohols, also undergo combustion reactions.

▶ **SAMPLE** problem **3**

Representing Combustion Reactions of Hydrocarbons

Write a balanced chemical equation to represent the complete combustion of 1-hexene. Use a condensed structural formula to represent 1-hexene.

Step 1: Write Word Equation for Reaction

1-hexene + oxygen → carbon dioxide + water

Step 2: Write Chemical Formulas

$CH_2\text{=}CHCH_2CH_2CH_2CH_3 + O_{2(g)} \rightarrow CO_{2(g)} + H_2O_{(g)}$

▶ **Practice**

Understanding Concepts

5. Write a balanced equation to represent the complete combustion of each hydrocarbon below. Represent each hydrocarbon using a condensed structural formula.
   (a) methane
   (b) ethane
   (c) propyne
   (d) 3-hexene

addition reaction a reaction of an alkene or alkyne in which a molecule, such as hydrogen or a halogen, is added to a double or triple bond

unsaturated containing at least one double or triple bond between carbon atoms

saturated containing only single C–C bonds

Figure 2
The reaction of cyclohexene and bromine water, $Br_{2(aq)}$, is rapid, forming a layer of colourless brominated cyclohexane.

Addition Reactions of Alkenes and Alkynes

Alkenes and alkynes are much more chemically reactive than alkanes, because their double and triple bonds are readily converted to single bonds. When one of the bonds in the double or triple bond breaks apart, the carbon atoms involved can bond with other atoms. The other atoms might be hydrogen atoms, halogens, or small groups such as –OH.

The equation below shows part of the double bond of ethene breaking, to allow the addition of two hydrogen atoms to two carbon atoms. The addition of H_2 to ethene, C_2H_4, produces ethane, C_2H_6.

$$
\begin{array}{cccccc}
& H & H & & & H & H \\
& | & | & & & | & | \\
H - & C & = C & - H + H - H \longrightarrow H - & C & - C & - H \quad \text{ethane, } C_2H_6 \\
& & & & & | & | \\
& & & & & H & H
\end{array}
$$

This type of reaction, in which entities (atoms or molecules) are added across a multiple bond, is called an **addition reaction**. Alkenes and alkynes can undergo addition reactions with hydrogen or with other substances such as hydrogen bromide, water, and bromine (**Figure 2**).

When C=C or C≡C bonds are present, the molecule contains less than the maximum number of hydrogen (or other) atoms. Alkenes and alkynes are said to be **unsaturated**. When a hydrocarbon molecule can no longer add more atoms, the molecule is referred to as **saturated**: that is, all the C–C bonds are single bonds and no more entities can be added to them. Therefore, *alkanes are saturated*, and *alkenes and alkynes are unsaturated*.

▶ **TRY THIS** activity   *Testing Fats and Oils*

Look around your kitchen for several fats and oils, such as lard, butter, margarine, corn oil, and canola oil. Read the label on each container to find out whether the fat or oil is saturated or unsaturated, or a combination of both. The terms "saturated" and "unsaturated" refer to the long carbon chains in the fatty acids that make up the molecules. You can use an addition reaction to test for saturated and unsaturated fatty acids. The presence of an unsaturated carbon chain turns potassium permanganate brown.

Materials: eye protection; lab apron; samples of various edible fats and oils; test tubes; potassium permanganate solution, $KMnO_{4(aq)}$

1. Place about 5 mL of each fat or oil in a separate test tube.
2. To each test tube, add a few drops of potassium permanganate solution.

 Potassium permanganate solution may stain skin and clothing. Wear a lab apron, and avoid contact with skin.

3. Place each test tube in a hot-water bath.
4. Watch for any colour change.

(a) Which samples contained only saturated carbon chains?
(b) Which samples contained unsaturated carbon chains?
(c) What properties, if any, are common to each group of fats or oils?

DID YOU KNOW ?

Margarine
Vegetable oils consist of molecules with long hydrocarbon chains that contain many double bonds. These oils are called "polyunsaturated." They are "hardened" by undergoing hydrogenation reactions to produce more saturated molecules, similar to those in animal fats such as lard.

▶ **SAMPLE** problem **4**

Representing Addition Reactions of Hydrocarbons

Draw a structural formula equation for each of the following addition reactions:

(a) 2-butene and hydrogen chloride, HCl

The two C atoms in the double bond allow the addition of an H atom and a C atom.

$$
\begin{array}{c}
\;\;\;\;\text{H}\;\;\;\;\text{H}\;\;\;\;\text{H}\;\;\;\;\text{H} \\
\;\;\;\;|\;\;\;\;\;|\;\;\;\;\;|\;\;\;\;\;| \\
\text{H} - \text{C} - \text{C} = \text{C} - \text{C} - \text{H} + \text{HCl} \longrightarrow \\
\;\;\;\;|\;\;\;\;\;|\;\;\;\;\;|\;\;\;\;\;| \\
\;\;\;\;\text{H}\;\;\;\;\text{H}\;\;\;\;\;\;\;\;\;\text{H}
\end{array}
\qquad
\begin{array}{c}
\text{H}\;\;\;\;\text{H}\;\;\;\;\text{H}\;\;\;\;\text{H} \\
|\;\;\;\;\;|\;\;\;\;\;|\;\;\;\;\;| \\
\text{H} - \text{C} - \text{C} - \text{C} - \text{C} - \text{H} \\
|\;\;\;\;\;|\;\;\;\;\;|\;\;\;\;\;| \\
\text{H}\;\;\;\;\text{H}\;\;\;\;\text{Cl}\;\;\;\;\text{H}
\end{array}
$$

(b) 2-butene and water, H₂O

Water, H_2O, may be rewritten as HOH. The two C atoms in the double bond allow the addition of an H atom to one and an –OH group to the other.

$$
\begin{array}{c}
\;\;\;\;\text{H}\;\;\;\;\text{H}\;\;\;\;\text{H}\;\;\;\;\text{H} \\
\;\;\;\;|\;\;\;\;\;|\;\;\;\;\;|\;\;\;\;\;| \\
\text{H} - \text{C} - \text{C} = \text{C} - \text{C} - \text{H} + \text{HOH} \longrightarrow \\
\;\;\;\;|\;\;\;\;\;|\;\;\;\;\;|\;\;\;\;\;| \\
\;\;\;\;\text{H}\;\;\;\;\text{H}\;\;\;\;\;\;\;\;\;\text{H}
\end{array}
\qquad
\begin{array}{c}
\text{H}\;\;\;\;\text{H}\;\;\;\;\text{H}\;\;\;\;\text{H} \\
|\;\;\;\;\;|\;\;\;\;\;|\;\;\;\;\;| \\
\text{H} - \text{C} - \text{C} - \text{C} - \text{C} - \text{H} \\
|\;\;\;\;\;|\;\;\;\;\;|\;\;\;\;\;| \\
\text{H}\;\;\;\;\text{H}\;\;\;\;\text{OH}\;\;\;\;\text{H}
\end{array}
$$

Example

Draw structural formulas to show the addition reaction between ethyne and hydrogen gas, $H_{2(g)}$, to form a saturated compound.

Solution

$$H - C \equiv C - H + 2\,H_{2(g)} \longrightarrow \begin{array}{ccc} & H & H \\ & | & | \\ H - & C - & C - H \\ & | & | \\ & H & H \end{array}$$

▶ **Practice**

Understanding Concepts

6. Draw structural formulas to show each of the following addition reactions:
   (a) ethene and hydrogen bromide, HBr
   (b) 3-hexene and water
   (c) 2-pentene and bromine, Br_2
   (d) 2-butyne and chlorine, Cl_2

7. Name three different alkenes that you could use for an addition reaction with hydrogen to produce heptane.

▶ *Section 3.1 Questions*

Understanding Concepts

1. Give three reasons why carbon atoms can form long chains. Draw a Lewis structure for a hydrocarbon to illustrate your answer.

2. Explain how you would recognize the structural formula of an
   (a) alkene
   (b) alkyne

3. Explain why alkenes and alkynes are generally more chemically reactive than alkanes.

4. Give an example of a saturated hydrocarbon and an unsaturated hydrocarbon. For each example, draw a structural formula and give its IUPAC name.

5. Explain, with a diagram, why the simplest alkene is ethene.

6. Draw structural formulas for 1-hexene, 2-hexene, and 3-hexene.

7. Why is no number used in the names "ethene" and "propene"?

8. Why is "3-pentene" an incorrect name? What is the correct name?

9. Write a balanced chemical equation to show the combustion of octane, one of the components of gasoline.

10. Draw structural formulas to show an addition reaction between propyne and hydrogen to form an alkane.

11. Draw structural formulas to show an addition reaction that will produce octane. Name the reactants in your reaction.

12. Throughout this unit, you will be asked to prepare an index card for each organic family, as a summary activity and a review resource. Obtain three index cards: one for alkanes, one for alkenes, and one for alkynes. On the front of each card,
    • write the name of the organic family;
    • give an example, including the IUPAC name, common name, and structural formula.

    On the back of each card,
    • list the characteristic properties, such as relative boiling points and solubility in polar and nonpolar solvents;
    • name the characteristic functional groups (such as double or triple bonds);
    • list the types of intermolecular forces.

▶ ***Section 3.1*** *Questions continued*

Making Connections

13. Research the common name and one use for each of the following hydrocarbons: ethene, propene, ethyne. Present your findings in a table.

 www.science.nelson.com

14. Research a use for each of the first 10 alkanes. Suggest why each alkane is appropriate for this use.

 www.science.nelson.com

15. The burning of fossil fuels, such as natural gas and gasoline, may be contributing to an environmental problem called global warming. Explain what this problem is and how it is related to the burning of fossil fuels.

16. Research and explain how the various parts of a butane lighter work (**Figure 3**). Describe necessary safety precautions for the use and storage of a butane lighter.

 www.science.nelson.com

Extension

17. Not all hydrocarbons form straight or branched chains. Research hydrocarbons with other shapes, and draw structural formulas for two examples of these molecules.

 www.science.nelson.com

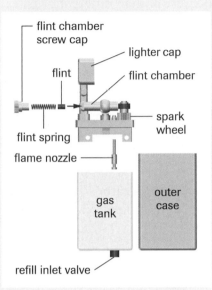

Figure 3
Workings of a butane lighter

Building Molecular Models

In this activity, you will build molecular models of hydrocarbons and write their IUPAC names and their formulas. As well, you will identify hydrocarbons that are isomers. As you learned in section 3.1, *isomers* are molecules that have the same formula but are put together differently. Because of their different structures, isomers often have very different properties.

Question

How many different hydrocarbons can you form using the same number of carbon atoms and the same number of hydrogen atoms?

Materials

molecular model kit containing 4 carbon pieces, 10 hydrogen pieces, and connecting pieces for single and multiple bonds

Procedure

(a) Create an observation table like **Table 1**.

Table 1 Sample Observation Table

Name	Formula	Structural formula

1. From a molecular model kit, obtain 4 carbon pieces and 10 hydrogen pieces.

2. Build a straight-chain hydrocarbon using all 14 pieces (**Figure 1**).

(b) Write the name, formula, and structural formula for your hydrocarbon in your observation table.

3. Put away two hydrogen pieces. Again, build and name as many different straight-chain structures as possible, using *all* of the remaining pieces. Repeat (b) for each isomer.

4. Repeat step 3 until you can no longer construct any hydrocarbon molecules.

5. Use some of the model pieces to build a model of ethene and one of hydrogen, H_2. Use the models to illustrate the addition reaction between

Figure 1
This molecule has five carbon atoms in a chain. Your model will have four carbon atoms.

hydrogen and ethene. Write a structural formula equation for the reaction.

6. Repeat step 5, first with propene and then with 2-butene instead of ethene.

Synthesis

(c) State how many isomers of a straight-chain hydrocarbon with four carbon atoms you could create, if the hydrocarbon is
  (i) an alkane
  (ii) an alkene with one double bond
  (iii) an alkyne with one triple bond

(d) Explain why each of the following names is incorrect:
  (i) 1-hexane
  (ii) 4-hexene
  (iii) methyne

(e) Your molecular model kit allows you to represent a double bond and a triple bond. How does this representation illustrate why alkenes and alkynes are more reactive than alkanes?

Extension

(f) Cyclohexane is an alkane with six carbon atoms linked in a ring structure. Each carbon atom is bonded to two other carbon atoms. Draw the structural formula for cyclohexane, and name an isomer of cyclohexane.

Fractional Distillation and Cracking   3.3

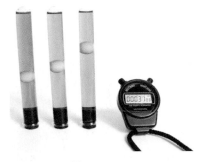

Figure 1
Be sure that you invert all your test tubes at the same time.

People who have the good fortune to "strike oil" have probably drilled deep into the ground and hit upon a complex mixture of hydrocarbon molecules, formed from prehistoric plants and animals. This mixture, referred to as **petroleum,** contains gases, liquids, and dissolved solids composed of many different hydrocarbon molecules, some of which may be up to 40 carbon atoms long.

petroleum a mixture of gases and liquids, composed of hydrocarbon molecules up to 40 carbon atoms long

As you learned earlier, small hydrocarbon molecules (such as methane, ethane, propane, and butane) exist as gases. Most larger hydrocarbon molecules are liquids, from light to heavy oils. The heaviest oils are asphalts and tars. The most valuable hydrocarbons in petroleum are the hydrocarbons with 5 to 12 carbons, because they are the components of gasoline.

How are the various hydrocarbons separated so that they can be sorted, and sold, by size? It so happens that molecules of different sizes have different boiling points. The smallest molecules have the lowest boiling points, which is why methane, ethane, and propane are all gases at room temperature. They have already boiled and evaporated at room temperature. The largest molecules have boiling points over 400°C. Therefore, asphalt can be heated to high temperatures to pave roads, without evaporating.

Why do hydrocarbons show this correlation between size and boiling point? Recall, from section 3.1, that the answer is found in the forces of attraction *between* neighbouring molecules, called intermolecular bonds. As you learned in section 1.12, nonpolar molecules like hydrocarbons are attracted to each other by relatively weak London dispersion forces. As the length of hydrocarbon molecules increases, the number of intermolecular forces between the molecules increases as well. Therefore, more heat is required to pull the molecules apart, meaning that higher temperatures are required to pull the molecules far enough apart to change into a gas.

Hydrocarbons are most useful to us when they are relatively pure. We do not want asphalt in our natural gas, and barbecues are designed to run on propane, not ethyne. How can petroleum be efficiently separated into its useful components?

Fractional Distillation

A method called **fractional distillation** is used to separate the many components of petroleum. Essentially, molecules of various sizes are separated into portions called *fractions*. Each fraction contains similar-sized molecules. The lighter fractions boil at lower temperatures, and the heavier fractions boil at higher temperatures.

In fractional distillation, the entire mixture of hydrocarbons is first heated to very high temperatures, high enough to evaporate nearly all of the hydrocarbons, small and large. Then the hot gases are allowed to rise in a tall fractionation tower (**Figure 2**). The upper parts of the tower are cooler than the lower parts. Each gas condenses at its own boiling point. As the hot gases travel up through the lower, warmer sections, the larger molecules condense. The smaller molecules with their low boiling points are still gases and ascend higher, to the top of the tower where the temperatures are lowest (**Figure 3**). As each fraction condenses, liquid forms on a tray and is collected. **Table 1** shows the various types of hydrocarbons, their boiling points, and their end uses.

fractional distillation the separation of components of petroleum by distillation, using differences in boiling points

Figure 2
Fractionation towers look like tall columns, with exterior stairs and platforms for maintenance.

Table 1 Uses of Hydrocarbons

Number of C atoms	Boiling point	End use
1–5	under 30°C	fuels for heat and cooking
5–6	30°C–90°C	camping fuel and dry-cleaning solvents
5–12	30°C–200°C	gasoline
12–16	175°C–275°C	kerosene and diesel fuel
15–18	250°C–375°C	furnace oil
16–22	over 400°C	heavy greases for lubricating
over 20	over 450°C	waxes, cosmetics, and polishes
over 26	over 500°C	asphalt and tar for roofs and roads

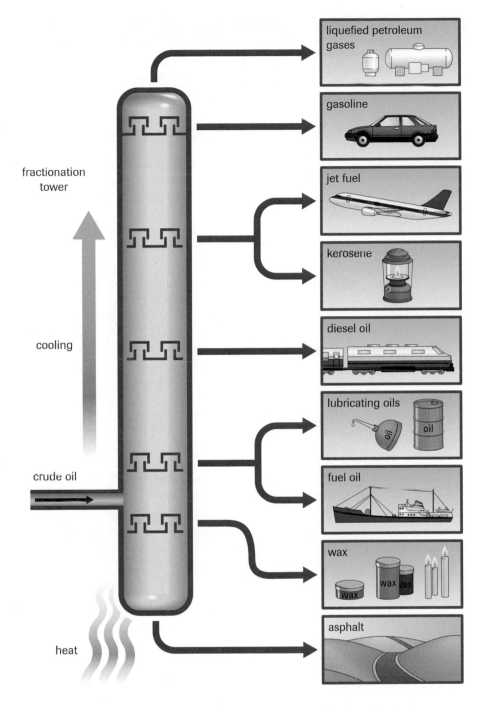

fractionation tower

cooling

crude oil

heat

liquefied petroleum gases

gasoline

jet fuel

kerosene

diesel oil

lubricating oils

fuel oil

wax

asphalt

Figure 3
Crude oil, the liquid component of petroleum, is heated and fed to the bottom of a fractionation tower. The gas mixture cools as it rises, allowing each component of the mixture to condense at its boiling point. The smallest hydrocarbons, with the lowest boiling points, condense at the top of the tower where the temperatures are lowest.

Figure 4
The higher the octane number of a gasoline, the more efficiently the gasoline burns to produce power and, thus, the less "knocking" in the engine. Highly branched alkanes have high octane numbers. For example, iso-octane is assigned an octane number of 100, while straight-chain heptane is assigned an octane number of 0.

cracking the process in which large straight-chain hydrocarbon molecules are converted into smaller branched-chain hydrocarbon molecules, usually by catalytic heating

Cracking

Of the many fractions collected, the most valuable and profitable is gasoline. Particularly in demand is *high-octane gasoline*, which contains highly branched alkanes (**Figure 4**). Fractionation of petroleum produces the less useful *straight-chain* hydrocarbons. A process called **cracking** is used to convert these straight-chain hydrocarbons into shorter branched-chain alkanes. In this process, the hydrocarbons are mixed with a catalyst and heated to temperatures of 400°C to 500°C. Cracking is also used to break apart larger hydrocarbon molecules, such as kerosene, into smaller molecules, such as ethene and propene (used in the production of plastics).

Understanding Concepts

1. What is the source of petroleum? What are some of its components?

2. (a) Describe and explain the relationship between the sizes of hydrocarbon molecules and their boiling points.
   (b) Describe how this relationship can be used to separate the components of petroleum.

3. A fraction of hydrocarbons with a boiling point of 10°C is collected in a fractional distillation process. Is this fraction a liquid or a gas at ordinary temperatures of about 20°C? Explain.

4. Why is cracking an important technological process?

5. The propane in a gas barbecue tank and the wax in a candle both consist of hydrocarbons. Give the approximate boiling point of each of these substances (refer to **Table 1**). Then give a theoretical explanation for the difference between their boiling points.

6. Name two substances you have used that have been made from a fraction of crude oil condensed in a fractionation tower, near
   (a) the top
   (b) the middle
   (c) the bottom

Applying Inquiry Skills

7. Design a laboratory set-up that would enable you to separate a mixture of two hydrocarbons with different boiling points. Explain the reasons for the apparatus you would need in your set-up. Draw and label a diagram of your set-up.

Making Connections

8. Research the production of crude oil in Canada. Write a short report to explain where crude oil is found and what its main uses are.

 www.science.nelson.com

9. Motor oils are available in different blends, which are appropriate for different seasons. Visit a gas station or hardware store to research the different brands and types of motor oils that are available for the different seasons in Canada. Present your information in a table. Include composition, brand, seasonal usage, and cost.

10. Gasoline and home heating oils are referred to as fossil fuels. Explain why this term applies to these fuels and why there is a concern about their use.

11. Petroleum production and processing not only provides Canadians with useful products, it also generates many jobs in the manufacture of these products. **Figure 5** shows a *petrochemical footprint*: an overview of the types and numbers of different jobs related to the use of ethene (ethylene). Use a telephone directory to list and count the number of businesses in your community that provide, directly or indirectly, products or services related to the petrochemical footprint shown. For each business that you list, describe the types of jobs involved. ✦▮

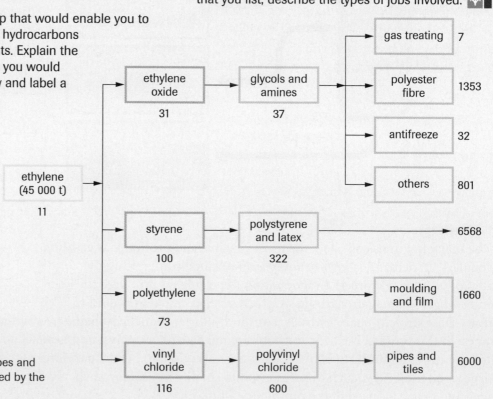

Figure 5
A petrochemical footprint: the types and numbers of jobs that are generated by the ethene (ethylene) industry

Separating a Mixture by Distillation

Organic chemists often need to purify a chemical substance before they can study its physical and chemical properties or before it can be used as a reactant or a product in a chemical reaction.

In this investigation, you will use boiling points to separate a mixture of hexane, 2-methyl-2-propanol (a branched alcohol), and paraffin wax. This technique is called separation by distillation. You will follow the progress of the distillation by recording the vapour temperature of the mixture at regular time intervals, and then plotting a graph of your observations.

Question

In what order are the components of a mixture recovered during the separation technique of distillation?

Prediction

(a) Predict an answer to the Question, and give reasons for your prediction.

(b) Look up the boiling points of hexane, 2-methyl-2-propanol, and paraffin wax (pentacosane) from a reference source. Predict the general shape of a graph with the temperature of the mixture plotted against the time.

Materials

eye protection
lab apron
round-bottom flask
fractionating column to fit flask
thermometer to fit fractionating column
condenser with rubber tubing
2 Erlenmeyer flasks (250 mL)
heating mantle and Variac
boiling chips
prepared mixture of hexane, C_6H_{14}, 2-methyl-2-propanol, $C_4H_9OH_{(l)}$, and paraffin wax, $C_{25}H_{52(s)}$
100-mL graduated cylinder
large beaker
tongs to handle flask
watch or clock with second hand

2 retort stands
2 clamps

Inquiry Skills

○ Questioning	○ Planning	● Analyzing
○ Hypothesizing	● Conducting	● Evaluating
● Predicting	● Recording	○ Communicating

 This investigation must be carried out in a well-ventilated area.

Procedure

1. Copy **Table 1** to record your observations.

Table 1 Temperature of the Vapour during Distillation

Time (min)	Temperature (°C)	Time (min)	Temperature (°C)
0.0		16.0	
0.5		16.5	
1.0		17.0	
1.5		17.5	
2.0		18.0	
continue to …		continue to …	
15.5		30.0	

2. Set up the distillation apparatus as illustrated in **Figure 1**, on the next page. Adjust the position of the thermometer so that the bulb sits just below the exit point of the vapours from the distillation column. Ensure that the water enters the condenser at the bottom and exits at the top. Before beginning the distillation, ensure that all the pieces of apparatus are well-positioned and secure.

3. Using the graduated cylinder, add 50 mL of the hexane/2-methyl-2-propanol/paraffin wax mixture to the round-bottom flask.

 Hexane and 2-methyl-2-propanol are highly flammable and must be kept away from open flames.

4. Add about six to eight boiling chips to the mixture in the flask. Attach the flask to the distillation column.

5. Place the heating mantle under the round-bottom flask. Plug the heating mantle into a Variac, not directly into an electrical outlet.

 The heating mantle must be plugged into a Variac, not directly into an electrical outlet.

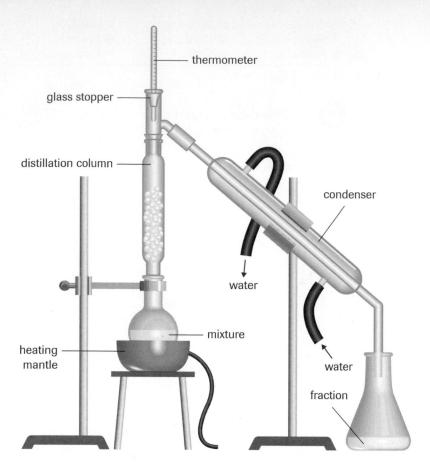

thermometer

glass stopper

distillation column

condenser

water

mixture

heating mantle

water

fraction

Figure 1
A typical fractional distillation apparatus used in a laboratory

10. Turn off the heat. Allow the apparatus to cool.

11. Carefully waft the contents of the two Erlenmeyer flasks, A and B, and record the odour of each.

12. Place a small drop of the contents of each Erlenmeyer flask on a clean surface of your lab bench. Observe and compare how quickly each sample evaporates at room temperature.

13. Using the tongs, carefully pour the remaining liquid contents of the distillation flask into a large beaker at room temperature. Try to pour the mixture so that it runs down the side of the beaker. Record your observations.

14. Dispose of your materials, and clean and return the equipment, according to your teacher's instructions.

Analysis

(c) Draw a graph of the vapour temperature (dependent variable) versus time (independent variable), for all the observations that you recorded in your table (**Table 1**).

(d) On your graph, draw an arrow to indicate the point at which the first fraction had completely boiled off. How does this point relate to the shape of the graph? Suggest an explanation for your answer.

(e) On your graph, indicate the section when the second fraction was being collected.

(f) From your graph, estimate the boiling points of the first two components of the mixture.

Evaluation

(g) Discuss whether your observations agree with your Prediction for this investigation. Give reasons for any discrepancies.

(h) Identify any possible sources of experimental error. Suggest improvements to the Procedure.

6. Turn on the Variac so that the heating mantle heats at medium heat. Adjust the setting so that the liquid mixture boils gently.

7. Begin recording the temperature when the heat is turned on. Record this temperature as zero time. Take and record temperature readings every 0.5 min until the distillation is complete.

8. After most of the first fraction has boiled off, and the column temperature is starting to rise noticeably, remove the Erlenmeyer flask and label it "A." Replace it with the other, empty Erlenmeyer flask, labelled "B." Note the time at which you replace the flask.

9. Continue the distillation until all of the next fraction has been collected and the temperature is starting to rise again.

With the huge number of organic compounds in existence, it would be very difficult for you to memorize the properties of each compound separately. Fortunately the compounds fall into **organic families**, which contain certain *combinations of atoms* in each molecule. These combinations are called **functional groups** (**Figure 1**).

Functional groups help to explain why molecules dissolve readily in some solvents and not in others, why they have high or low melting and boiling points, and how they react with other molecules. Understanding the effects of the functional groups allows chemists to predict the properties of organic molecules, which in turn, helps chemists design new compounds, ranging from high-tech fabrics to "designer drugs." Developing new products is one of the most important aspects of modern organic chemistry (**Figure 2**).

Most functional groups are made up of one or more of three main types. You will examine these types in this section.

Carbon–Carbon Double or Triple Bonds

One main type of functional group is a double or triple bond between carbon atoms (**Figure 1(a)**). When a carbon atom is single-bonded to another carbon atom, the bond is a strong covalent bond that is difficult to break. Therefore, saturated hydrocarbons are not reactive. Double or triple bonds between carbon atoms are more reactive, however, because the second and third bonds that are formed in a multiple bond are not as strong as the first bond and are more easily broken. For this reason, addition reactions (which you learned about in section 3.1) can take place at the multiple bonds.

Carbon Bonded to Oxygen or Nitrogen by a Single Bond

Another type of functional group is a carbon atom bonded to a more electronegative atom. Recall, from Unit 1, that the covalent bond between atoms with different electronegativites is *polar* (**Figure 1(b)**). Oxygen, nitrogen, and the halogens are more electronegative than carbon (**Table 1**, on the next page). Therefore, the shared electrons are held more closely to oxygen, nitrogen, or halogen atoms and farther away from the carbon atoms, leaving the oxygen, nitrogen, or halogen atoms with a partial negative charge and the carbon atoms with a partial positive charge. The presence of polar bonds in a molecule tends to increase the strength of the intermolecular forces: that is, the strength of the forces of attraction *between* molecules, which can be explained by the attraction between partial positive charges and partial negative charges. As a result of polar bonds, more force is required to separate the molecules from each other, resulting in higher melting and boiling points.

organic family a group of organic compounds with common structural features that impart characteristic physical and chemical properties

functional group a particular combination of atoms that contributes to the physical and chemical characteristics of a substance

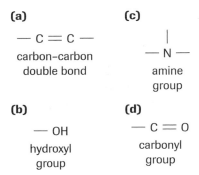

(a)

— C = C —
carbon–carbon double bond

(b)

— OH
hydroxyl group

(c)

|
— N —
|
amine group

(d)

— C = O
carbonyl group

Figure 1
Four common functional groups

Figure 2
The design and synthesis of new materials with specific properties, such as the plastic in this artificial ski run, is a key focus of the chemical industry.

hydrogen bond an intermolecular attraction between –OH and –NH groups in different molecules

Table 1 Electronegativities of Common Elements

Element	Electronegativity
H	2.1
C	2.5
N	3.0
O	3.5

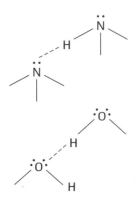

Figure 3
A hydrogen bond forms between an electronegative N, O, or F atom of one molecule and the more positively charged H atom of an adjacent molecule.

In addition, the presence of an –OH or –NH group enables an organic molecule to form **hydrogen bonds** with other such groups. The formation of hydrogen bonds further increases intermolecular attractions (**Figure 3**). It also enables organic molecules to mix readily with other polar molecules. There is a saying "like dissolves like." This saying refers to the observation that polar molecules dissolve readily in other polar molecules, and nonpolar molecules dissolve readily in other nonpolar molecules.

▶ **TRY THIS** activity **Bending Water**

Water from a tap flows straight down into your glass, doesn't it? Try this simple experiment using liquids of polar and nonpolar molecules, and see what happens.

Materials: eye protection; lab apron; 50-mL burette; clamp and stand for burette; 250-mL beaker; plastic object (such as a ruler or a comb); paper towel; 50 mL water (or water running from a tap); 50 mL hexane; 50 mL alcohol (such as ethanol or rubbing alcohol); 50 mL acetone (such as nail polish remover)

🔥 These liquids are highly flammable. Work in a fume hood or well-ventilated area, and keep away from an open flame.

1. Clamp the burette carefully onto the stand. Fill the burette with one of the four liquids. Place an empty beaker under the burette.
2. Charge the plastic object by rubbing it with the paper towel for a few seconds.
3. Allow a thin stream of the liquid to run from the burette into the beaker. At the same time, hold the charged plastic object close to the liquid stream. Observe any effect.
4. Repeat with each of the other liquids.
5. Dispose of the materials as directed by your teacher. Wash your hands thoroughly.

(a) Look up the structural formula for each liquid, and identify any functional groups present.

(b) Explain your observations in this activity with reference to the functional groups present and the polarity of the molecule.

Carbon Bonded to Oxygen by a Double Bond

The third type of functional group consists of a carbon atom double-bonded to an oxygen atom (**Figure 1(d)**). The double covalent bond between carbon and oxygen requires four electrons to be shared between the atoms. All four of these electrons are more strongly attracted to the oxygen atom, making the C=O bond strongly polar and accounting for the observations that compounds with C=O groups have higher boiling and melting points than other compounds, as well as increased solubility in polar solvents.

 *Functional Groups*

Double and triple bonds between carbon atoms:

–C=C– and
–C≡C–

Unlike single C–C bonds, double and triple bonds allow atoms to be added to the chain.

A carbon atom bonded to a more electronegative atom (oxygen, nitrogen, or a halogen):

C–O
C–N
C–Cl, C–Br,
C–F

Unequal sharing of electrons results in a polar bond, increasing intermolecular attraction and raising the boiling point and melting point of the compound.

A carbon atom double-bonded to an oxygen atom:

C=O

The resulting polar bond increases the boiling point and melting point of the compound.

▶ *Section 3.5* **Questions**

Understanding Concepts

1. Explain the meaning of the term "functional group."

2. Are double and triple bonds between carbon atoms more reactive or less reactive than single bonds? Explain.

3. Do polar organic molecules always have higher boiling points than nonpolar organic molecules? Explain.

4. Describe the three main functional groups in organic molecules.

5. Describe and explain the effect of the presence of an –OH group or an –NH group on
   (a) the melting point and boiling point of the molecule
   (b) the solubility of the molecule in polar solvents

6. Identify all the functional groups in each of the following structural diagrams. Predict the solubility of each substance in water.

   (a) CH₃ — O — H

   (b) CH₃ — CH ═ CHCH₃

   (c) CH₃CH ═ O

   (d) CH₃CH₂C ═ O
       |
       OH

7. Water is formed when an oxygen atom bonds with hydrogen atoms. Ammonia is formed when a nitrogen atom bonds with hydrogen atoms. Methane is formed when a carbon atom bonds with hydrogen atoms.
   (a) Write the chemical formula for each of the three compounds.
   (b) Predict, with reference to electronegativities and intermolecular forces, the solubility of each compound in the other two compounds.
   (c) Identify which of the three compounds are found in or produced by living organisms, and classify each compound as organic or inorganic. Justify your classifications. (*Hint:* Methane comes from cows.)

 www.science.nelson.com

Figure 1
Having cold drinks on hand costs more than just dollars.

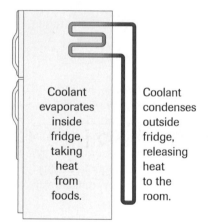

Coolant evaporates inside fridge, taking heat from foods.

Coolant condenses outside fridge, releasing heat to the room.

Figure 2
The coolant circulates through the coils, repeatedly evaporating and condensing.

When you take a nice cold drink from your fridge on a hot summer day (**Figure 1**), you may be increasing your risk of getting a sunburn outside. What is the connection between your fridge and getting a sunburn?

First consider how a refrigerator works. A fridge is cooled by a liquid, called a *coolant*, that repeatedly evaporates and condenses in the cooling coils. The coolant absorbs heat when it evaporates. This heat is extracted from the foods and drinks in the fridge, thus cooling them (**Figure 2**). Choosing a coolant that is safe to the consumer and to the environment has always been a concern.

In the late 1800s, refrigerators were cooled using ammonia, methyl chloride, or sulfur dioxide. These gases are toxic, however, and several fatal accidents occurred in the 1920s as a result of leaked coolant. In 1930, the DuPont company manufactured a chemical they named Freon. Freon, CF_2Cl_2, belongs to a group of organic compounds called chlorofluorocarbons (CFCs), which contain chlorine and fluorine atoms bonded to carbon. CFCs are members of a larger group of organic compounds called organic halides, many of which are toxic to some degree. Freon is chemically unreactive and, thus, was considered very safe. Its use spread to many other applications, such as aerosol sprays and paints.

Unfortunately, scientists discovered in the 1970s that large "ozone holes" had formed in the upper atmosphere, particularly over the polar regions (**Figure 3**). These ozone holes leave us unprotected from harmful UV radiation, which can cause painful sunburns. The source of the damage to the ozone layer was traced to the *halogen* atoms in the popular CFCs. As a result, newer coolants, containing *hydrogen* in addition to carbon and halogens (HCFCs and HFCs), have been promoted in recent years. In 2001, the Ontario government introduced legislation requiring all automobile air conditioners that need refilling to be adapted to use one of the new alternative coolants.

The need for coolants that do not contain halogens has resulted in the promotion of a substance called *Greenfreeze*, which is made up of hydrocarbons such as propane, butane, and cyclopentane. The major advantage of hydrocarbons is that they are harmless to the ozone layer, because they do not contain halogens. Since the mid-1990s, Greenfreeze refrigerators have become very popular throughout Europe. They are sold in Austria, Britain, Denmark, France, Germany, Italy, the Netherlands, and Switzerland. In fact, the Environment ministers of Britain, Denmark, and the Netherlands have all bought Greenfreeze refrigerators for their personal use.

There are also concerns regarding the use of Greenfreeze. As you know, hydrocarbons such as propane and butane are commonly used as fuels. They are highly flammable, which is not a good property for a chemical that circulates in a large electrical appliance. The manufacturers argue that the quantity of propane or butane in a Greenfreeze refrigerator is very small—barely enough to fill two cigarette lighters. They estimate that the risk of an explosion being caused by such a small quantity of fuel is minimal. In addition, ignition sources, such as electrical switches, are kept sealed within the insulation of the refrigerator. The German safety and standards institution has given approval to Greenfreeze refrigerators, and this approval is valid for all European Union countries.

North America, however, has been slow to switch over to Greenfreeze technology. North American consumers prefer larger refrigerators and additional features, such as automatic defrost, that require much larger quantities of coolant. North American manufacturers are also reluctant to completely abandon HFCs and HCFCs, because they have invested a great deal of time and money in this technology.

Ultimately it will be up to us to decide, through our purchases, which technology we will support. What are the short-term and long-term costs of coolant technology? What costs are we willing to pay?

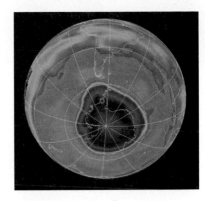

Figure 3
The ozone "hole" (blue) forms over the Antarctic every spring (September and October).

▶ *Understanding the Issue*

1. Explain how a coolant keeps foods cold in a refrigerator.

2. Name the different coolants that have been used in North America, from the late 1800s to the present. Describe any safety hazards that are associated with each coolant.

3. Write the names and chemical formulas of two compounds in Greenfreeze coolant. Explain how these compounds may be less harmful to the environment than other coolants are.

4. Describe two reasons why North American manufacturers are slower than European manufacturers to adopt Greenfreeze as a coolant for refrigerators.

5. Consumers can directly and indirectly influence the decisions of manufacturers regarding the types of products they make. Discuss this statement with a partner or in a small group. Together, compile a list of ways in which consumers can exert an influence.

 www.science.nelson.com

▶ *Role Play*

Choosing a Refrigerant

Decision-Making Skills

- Define the Issue
- Analyze the Issue
- Research
- Defend a Decision
- Identify Alternatives
- Evaluate

When refrigerator manufacturers are designing a new model of fridge, they need to consider the effects of every feature in their design. Imagine that a committee is set up to decide whether the next new model of fridge should use existing HFC or HCFC coolants, or the hydrocarbon Greenfreeze coolant.

Committee members include a union representative for the production-line workers, the local MP, an environmentalist, a reporter from a consumer product magazine, a physician, a representative from the Health and Safety Organization, an advertising executive, and shareholders in the company.

(a) Each coolant has a dollar cost. Cost can be measured in many other ways, however, such as socially, environmentally, and politically. Choose one other way of measuring cost. Collect and sort information to help you decide whether the cost of each of the three types of coolants is justified.

(b) Select a role for yourself from the list of committee members above. Consider how this committee member might feel about the choice of coolant. Keep in mind the cost that you have researched.

(c) Role play the meeting, with everyone taking a turn to put forward his or her position on whether or not the new model of fridge should use Greenfreeze coolant. Be sure to defend your position with arguments supported by facts.

(d) After the "meeting," discuss and summarize the most important points made. If possible, come to a consensus (unanimous decision) about the issue.

 www.science.nelson.com

alcohol an organic compound that is characterized by the presence of a hydroxyl functional group; general formula R–OH

ether an organic compound that has two alkyl groups (the same or different) attached to an oxygen atom; general formula R–O–R

alkyl group a substitution group or branch derived from a hydrocarbon; general formula R

Alcohols and **ethers** both contain a carbon atom bonded to an oxygen atom with a single bond. In fact, their structures are similar to the structure of water. If you think of water as being HOH, then an alcohol is simply water with one of the H atoms replaced by an alkyl group, R. An **alkyl group** is a hydrocarbon group derived from an alkane (**Table 1**). Thus, the general formula for an alcohol can be written ROH. Similarly, an ether is water with *both* H atoms replaced by alkyl groups. Thus, the general formula for an ether is ROR. The two R groups may be identical to each other or different.

HOH	ROH	ROR
water	alcohol	ether
	example: CH_3OH	example: CH_3-O-CH_3

Alcohols

You encounter many different alcohols in your daily life. The simplest alcohol is methanol. Methanol is essentially a methane molecule with one of the hydrogens replaced by an –OH group (**Figure 1**). It is highly toxic. Drinking even small quantities can lead to blindness and death. Ethanol, found in wines and beer, is less toxic but is a depressant and also a poison. The ethanol that is used in science laboratories has been purposely mixed with methanol, benzene, or other toxic materials to make it undrinkable. Another alcohol, 2-propanol, is commonly called "rubbing alcohol." Car antifreeze is also an alcohol—a very toxic alcohol with a sweet taste. The taste makes antifreeze appealing to animals, so it is important to clean up any spills to avoid poisoning accidents.

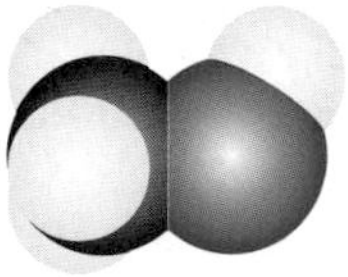

Table 1 Alkyl Groups

Alkyl group	Alkyl formula
methyl-	$-CH_3$
ethyl-	$-C_2H_5$
propyl-	$-C_3H_7$
butyl-	$-C_4H_9$
pentyl-	$-C_5H_{11}$
hexyl-	$-C_6H_{13}$
heptyl-	$-C_7H_{15}$
octyl-	$-C_8H_{17}$
nonyl-	$-C_9H_{19}$
decyl-	$-C_{10}H_{21}$

R – O – H

Figure 1
Molecular model for methanol, CH_3OH

> ▶ **TRY THIS** activity   *Yeast at Work*
>
> When the sugars in fruit juice ferment, with the help of yeast, they form a two-carbon alcohol called ethanol.
>
> **Materials:** 1 cup of fresh or canned apple juice, glass jar with lid, 10 mL (2 tsp) dry yeast
>
> 1. Pour the apple juice into the jar, and add the yeast.
> 2. Put the lid on the jar loosely. Keep the jar at room temperature for several days, to allow fermentation to take place.
> 3. Stir the contents of the jar each day. Note the presence of CO_2 bubbles and the odour of ethanol—the waste products of the yeast.

Table 2 shows the toxicity of several common alcohols *per kilogram* of human body mass. As you can see, all alcohols are toxic to varying degrees. In fact, any chemical, even table salt, can be toxic if taken in sufficient quantities.

Naming Alcohols

The functional group that is common to all alcohols is the –OH group, called the *hydroxyl group*. An alcohol consists of a hydrocarbon chain with a hydroxyl group attached somewhere on the chain in place of a hydrogen atom.

In the IUPAC system, each alcohol name consists of three parts. The first part of the name tells you the number of carbon atoms in the longest carbon chain. For example, *meth-* means one carbon, *eth-* means two carbons, *prop-* means three carbons, and so on. The middle part of the name, *-an-*, tells you that the hydrocarbon chain is saturated. The last part of the name, *-ol*, refers to the hydroxyl group. If the name of the alcohol begins with *a number, n,* then the hydroxyl group is attached on the *n*th carbon atom in the chain. The following Sample Problems show how to name alcohols and draw their structures.

Table 2 Alcohol Toxicity for Humans

Alcohol	LD$_{50}$ (g/kg) body mass
methanol	0.07
ethanol	13.7
1-propanol	1.87
2-propanol (rubbing alcohol)	5.8
glycerol (glycerine)	31.5
ethylene glycol (car antifreeze)	<1.45
propylene glycol (plumber's antifreeze)	30

The value for LD$_{50}$ stands for the estimated mass of the alcohol that, if given to test subjects, would kill 50% of them.

▶ **SAMPLE** problem

Naming and Drawing Alcohols

(a) Draw the structural formula and write the formula for 1-butanol.

Step 1: Draw Carbon Backbone

The prefix *but-* tells you that there are four C atoms with no multiple bonds.

— C — C — C — C —

Step 2: Add –OH Group in Appropriate Position

The *-anol* tells you that there is an –OH group. The number *1-* tells you that the –OH group is on carbon number 1. So you add –OH to the first C atom. (You can number from either end.)

Step 3: Complete Remaining Bonds with H Atoms

$$H — \underset{4}{C} — \underset{3}{C} — \underset{2}{C} — \underset{1}{C} — OH$$

[or CH$_3$CH$_2$CH$_2$CH$_2$OH] 1-butanol, C$_4$H$_9$OH

The structural formula can also be written as a condensed structural formula, as shown in the square brackets.

(b) Draw the structural formula and write the formula for 2-pentanol.

Step 1: Draw Carbon Backbone

The prefix *pent-* indicates that there are five C atoms.

— C — C — C — C — C —

Step 2: Add –OH Group in Appropriate Position

The name *2-pentanol* tells you that the –OH group is attached to the *second* C atom.

$$\underset{1}{C} — \underset{2}{C} — \underset{3}{C} — \underset{4}{C} — \underset{5}{C}$$
with OH on carbon 2

DID YOU KNOW ?

Toxic Alcohol

The alcohol in wine and beer is formed as a waste product of a fungus called yeast. The alcohol content in wine is limited to about 13% (26 proof) because, once the concentration of ethanol reaches this level, the yeast cannot survive and fermentation stops.

DID YOU KNOW ?

Grape Expectations

The cosmetics industry has developed a line of "anti-aging" creams. The active ingredient in these creams is listed as polyphenols (alcohols extracted from grape seeds and skins, and also from olives). Manufacturers claim that polyphenols, described as antioxidants, reverse signs of aging in the skin and make the skin look younger.

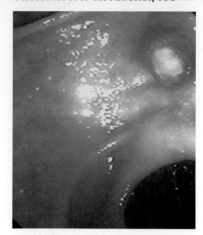
Step 3: Complete Remaining Bonds with H Atoms

Each C atom must have four occupied bonds.

$$H - {}_1C - {}_2C - {}_3C - {}_4C - {}_5C - H$$

(with H and OH substituents shown) [or $CH_3CH(OH)CH_2CH_2CH_3$] 2-pentanol, $C_5H_{11}OH$

The condensed structural formula shows the attached –OH group in brackets, *immediately after* the C atom to which it is attached.

(c) Write the name of the alcohol that has the condensed structural formula $CH_3CH_2CH(OH)CH_3$.

Step 1: Count Carbon Atoms to Establish Prefix

There are four C atoms, so the prefix is *but-*.

Step 2: Look for Functional Group to Establish Suffix

You are told (and can see from the –OH) that the compound is an alcohol, so the suffix is *-anol*. The compound is therefore a butanol.

Step 3: If Necessary, Add Number

There are several possible butanol isomers. You need to add a number to clarify which isomer the compound is. Number the carbon chain from right to left, to give the lowest number to the C atom to which the –OH group is attached.

$$\underset{4\quad\;3\quad\;2\qquad\;1}{CH_3CH_2CH(OH)CH_3}$$

Therefore, the –OH group is on the second C atom. The name of this alcohol is 2-butanol.

Example

Draw the condensed structural formula and write the formula for

(a) ethanol

(b) 3-hexanol

Solution

(a) CH_3CH_2OH, C_2H_5OH

(b) $CH_3CH_2CH(OH)CH_2CH_2CH_3$, $C_6H_{13}OH$

▸ Practice

Understanding Concepts

1. Draw a structural formula and write the formula for
   (a) 1-propanol
   (b) 2-propanol
   (c) 1-pentanol
   (d) 3-octanol

2. Name each of the following alcohols:
   (a) $CH_3CH_2CH(OH)CH_2CH_3$
   (b) $CH_3CH_2CH(OH)CH_2CH_2CH_2CH_3$

Properties of Alcohols

The presence of the polar hydroxyl group allows hydrogen bonds to form among molecules. These hydrogen bonds give alcohols boiling points that are much higher than the boiling points of comparable alkanes. For example, ethanol boils at 78°C, compared with ethane, which boils at −89°C. The longer the alcohol molecule, the higher the boiling point is (**Table 3**).

Alcohols are much more soluble in water than alkanes are. Methanol, ethanol, and rubbing alcohol all mix readily with water. The long hydrocarbon portion of larger alcohol molecules allows them to mix with nonpolar substances as well, making these alcohols useful solvents for both water-based and oil-based substances.

Like hydrocarbons, alcohols will burn to produce only carbon dioxide and water. You may have used an alcohol lamp or an ethanol-blend gasoline. You must be very careful to keep all alcohols away from open flames.

$$2\ CH_3CH_2CH_2OH_{(g)} + 9\ O_{2(g)} \rightarrow 8\ H_2O_{(g)} + 6\ CO_{2(g)}$$

1-propanol   oxygen   water   carbon dioxide

Table 3   Boiling Points of Alcohols

Name	Formula	Boiling point (°C)
methanol	CH_3OH	65
ethanol	C_2H_5OH	78
1-propanol	C_3H_7OH	97
1-butanol	C_4H_9OH	117

▶ TRY THIS activity   *Burning Paper*

Here's a real magic trick to try. Be sure to follow all the safety rules.

Materials: eye protection, lab apron, large drinking glass or beaker, 50 mL rubbing alcohol, 50 mL water, table salt, tongs, piece of paper (or $10 bill!), lighter or match

 Alcohol is highly flammable. Dispose of the alcohol mixtures from every beaker in the room before using an open flame.

1. In a large drinking glass, mix together 50 mL of rubbing alcohol, 50 mL of water, and a pinch of table salt.
2. Using tongs, dip a piece of paper about the size of a $10 bill into the solution until it is well soaked. Take out the paper with the tongs.
3. Flush the alcohol mixture down the sink, followed by a lot of water.
4. With the tongs, hold the soaked paper over the sink. Ignite the paper using a lighter or a match.

(a) Explain what happens to the paper.

Ethers

The functional group in ethers is an oxygen atom bonded to two carbon atoms. The general formula is ROR. An example of an ether is ethoxyethane, more commonly called diethyl ether or simply ether. Ethoxyethane, $CH_3CH_2OCH_2CH_3$, is a volatile and highly flammable liquid that was once a common anesthetic.

Naming Ethers

Ethers are named by adding *oxy* to the prefix of the *smaller* hydrocarbon group and joining the new prefix to the alkane name of the *larger*

DID YOU KNOW ?

No Laughing Matter

Many different compounds have been tried as anesthetics over the years. Dinitrogen monoxide, N_2O, was first tested in 1844 when a Boston dentist demonstrated a "painless" tooth extraction. Unhappily for the patient, the tooth was pulled before the "laughing gas" had taken effect, and the public was not impressed. Later diethyl ether was used with more reliable results.

Table 4 Boiling Points of Similar Compounds

Compound	Structure	Boiling point (°C)
ethane	$CH_3–CH_{3(g)}$	–89
methoxymethane (dimethyl ether)	$CH_3–O–CH_{3(g)}$	–23
ethanol	$CH_3–CH_2–O–H_{(l)}$	78.5
water	$H–O–H_{(l)}$	100

hydrocarbon group. Hence the IUPAC name for $CH_3OCH_2CH_3$ is *methoxy*ethane (not *ethoxy*methane). Ethers are often given common names derived from the two alkyl groups, followed by the term *ether*. Methoxyethane would thus be methyl ethyl ether.

Properties of Ethers

Since ethers do not contain any hydroxyl groups, they cannot form hydrogen bonds. The polar C–O bonds and molecular shape of ether molecules, however, do make them more polar than hydrocarbons. The boiling points of ethers are slightly higher than the boiling points of hydrocarbons, but lower than the boiling points of alcohols (**Table 4**). (Note that H_2O, with two O–H bonds per molecule, has the highest boiling point in this series.) Like alcohols, ethers are good solvents for organic reactions because they mix readily with both polar and nonpolar substances.

▶ **Section 3.7 Questions**

Understanding Concepts

1. Briefly explain why methanol has a higher boiling point than methane.

2. Draw a structural formula for
   (a) 1-propanol
   (b) 2-heptanol

3. Explain why "4-hexanol" is an incorrect name. What is the correct name for this alcohol?

4. Arrange the following compounds in order of increasing boiling point: butane, 1-butanol, octane, 1-octanol. Give reasons for your order.

5. Only a few of the simple alcohols are used in combustion reactions. Alcohol-gasoline mixtures, known as gasohol, make use of the flammability of simple alcohols. Write a balanced chemical equation, using condensed structural formulas, for the complete combustion of each of the following alcohols:
   (a) ethanol (in gasohol)
   (b) 2-propanol (isopropyl alcohol or rubbing alcohol)

6. Consider the two compounds $CH_3CH_2CH_2CH_2CH_2OH$ and $CH_3CH_2OCH_2CH_2CH_3$. Name the compound that
   (a) will evaporate at a lower temperature
   (b) has higher solubility in a nonpolar solvent
   Give reasons for your choices.

7. Prepare one index card for alcohols and another index card for ethers. On the front of each card,
   • write the name and general formula of the organic family;
   • give an example, including the IUPAC name, common name, and structural formula.
   On the back of each card,
   • list the characteristic properties, such as relative boiling points and solubility in polar and nonpolar solvents;
   • name the characteristic functional groups;
   • list the types of intermolecular forces.

8. Glycerol is more viscous than water, and it can lower the freezing point of water. When added to biological samples, it helps to keep the tissues from freezing, thereby reducing damage. From your knowledge of the molecular structure of glycerol, suggest reasons to account for these properties of glycerol.

Making Connections

9. Ethylene glycol is an alcohol that is commonly available in hardware stores. Research ethylene glycol and report on
   (a) its IUPAC name
   (b) its common usage
   (c) its physical properties and how they are related to its molecular structure

 www.science.nelson.com

Properties of Alcohols

Part 1: Trends in Properties of Alcohols

You might expect to see a trend in properties within a chemical family, such as alcohols. For example, is there a link between molecular size and physical properties in the first four primary alcohols? In this investigation, you will first use your knowledge of intermolecular forces of the hydrocarbon components and the hydroxyl functional group to predict trends. You will then test your predictions experimentally.

Question

What trends, if any, do three alcohols show in the following physical properties: melting point, boiling point, solubility in nonpolar and polar solvents, and acidity?

Prediction

(a) Predict possible trends shown by ethanol, 1-propanol, and 1-butanol in the properties listed in the Question.

Hypothesis

(b) Give theoretical reasons for your predictions.

Materials

eye protection
lab apron
blue and pink litmus paper
3 test tubes
2 mL ethanol
2 mL 1-propanol
2 mL 1-butanol
test-tube rack
test-tube holder
10-mL graduated cylinder
3 mL mineral oil (hydrocarbons)
3 mL water

Procedure

1. Copy **Table 1** into your notebook to record your observations.

2. Label each test tube with the name of one of the alcohols you are testing.

Inquiry Skills

○ Questioning   ○ Planning   ● Analyzing
● Hypothesizing   ● Conducting   ○ Evaluating
● Predicting   ● Recording   ○ Communicating

Table 1   Properties of Three Alcohols

Property	Ethanol	1-Propanol	1-Butanol
structural formula			
melting point			
boiling point			
solubility in mineral oil			
solubility in water			
colour with litmus			

3. Place 1 mL of each alcohol in the appropriate test tube. To each alcohol, add 1 mL of mineral oil. Observe and record any evidence of each liquid's ability to mix with mineral oil.

 All three alcohols are highly flammable. Do not use them near an open flame.

4. Follow your teacher's instructions for disposing of the contents of the test tubes and for cleaning the test tubes.

5. Repeat step 3, using water instead of mineral oil.

6. Before disposing of the mixtures in the test tubes, add a small piece of blue and red litmus paper to each mixture. Record the colour of the litmus paper.

7. Dispose of the mixtures as directed by your teacher.

(c) Draw a structural formula for each alcohol in your observation table (**Table 1**).

(d) Using a print or electronic reference source, find the melting point and boiling point of each alcohol you tested. Record the information in your observation table.

 www.science.nelson.com

Analysis

(e) Use your completed observation table to summarize any trends in the properties of the alcohols you tested.

(f) Do your observations agree with your predictions? Discuss, with reference to your theoretical knowledge of intermolecular forces and functional groups.

Synthesis

(g) Use reference sources to look up the boiling points of several alcohols with straight carbon chains and several alcohols with branched carbon chains. Compare these boiling points, and explain any trends.

 www.science.nelson.com

(h) Many alcohols contain not only stretches of straight chains, but also many branched chains. The shapes of these alcohol molecules may be more spherical than linear. Explain the boiling points of alcohols in relation to their molecular shapes.

Part 2: Alcohol and Alkane Combustion

Many of the fuels we use, such as propane in homes and octane in gasoline, consist of hydrocarbons. There is an increased interest in replacing these fossil fuels with renewable organic compounds, such as methanol or ethanol. Blended "gasohol" mixtures are already available at many gas stations. In this part of the investigation, you will test and compare the products formed by the combustion of an alcohol and the combustion of a hydrocarbon. Then you will write balanced chemical equations for these two reactions.

Question

What products are formed when ethanol and hexane undergo combustion?

Prediction

(i) Predict the products of the complete combustion of ethanol and the complete combustion of hexane.

Materials

eye protection
lab apron
5 mL ethanol
5 mL hexane
50-mL beaker
250-mL beaker
watch glass
ice cube
aluminum foil
wick (for example, cotton string or rolled paper towel strip)
match
30 mL limewater
cobalt chloride paper
tongs

Procedure

(j) Copy **Table 2** into your notebook, allowing enough space to record your observations.

Table 2 Identification of Products of Combustion

Property	Ethanol	Hexane
structural formula		
organic family		
cobalt chloride test		
limewater test		

8. Place 5 mL of ethanol in the 50-mL beaker, and 15 mL of limewater in the 250-mL beaker.

 Ethanol and hexane are highly flammable. Dispense small quantities from larger containers in the fume hood.

9. Set up the two beakers, aluminum foil, and wick as shown in **Figure 1**. Ignite the wick with a match.

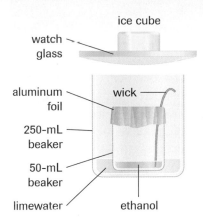

watch glass

ice cube

aluminum foil

wick

250-mL beaker

50-mL beaker

limewater

ethanol

Figure 1
Alcohol burner apparatus

10. Place a clean, cool watch glass slightly ajar on the larger beaker. Do not cover the beaker completely. Place an ice cube on the watch glass.

11. After 30 s, use cobalt chloride paper to test any droplets that have formed on the underside of the watch glass. Record your observations in your observation table (**Table 2**).

12. Replace the watch glass. Allow the flame to burn for another 5 min.

13. Blow out the flame, and use tongs to remove the smaller beaker. Replace the watch glass on the larger beaker, and allow the contents to cool.

14. Gently swirl the contents of the covered larger beaker for about 30 s, to allow the limewater to mix with any gases produced. Record your observations.

15. Repeat steps 8 to 14 using hexane instead of ethanol.

16. Dispose of the contents of the beakers as directed by your teacher.

(k) Complete your observation table.

Analysis

(l) Using your observations, identify the products of the combustion of ethanol and the combustion of hexane. Write a balanced chemical equation for each reaction.

(m) Why should the larger beaker not be covered completely with the watch glass?

Synthesis

(n) Why does the wick itself not burn when ignited? Relate your explanation to the Try This Activity: Burning Paper in section 3.7.

(o) Compare the products formed from the combustion of hydrocarbons and the combustion of alcohols. Suggest reasons why alcohols are sometimes preferred as fuels.

One group of organic compounds, called *pheromones*, is used by many insects as chemical signals, or sex attracters. Pheromones are specific to a species and have a powerful effect. Foresters use pheromones to attract and trap insect pests, such as the gypsy moth (**Figure 1**). Many pheromones belong to a family of organic compounds called **ketones**.

Closely related to the ketones are the **aldehydes**. The smaller aldehydes have strong, unpleasant odours. Methanal (commonly called formaldehyde) is the simplest aldehyde. It is used as an antiseptic and a disinfectant. The next simplest aldehyde is ethanal (acetaldehyde), which is a colourless liquid used as a preservative and in the synthesis of resins and dyes. Aldehydes of higher molecular mass have pleasant, flowery odours. They are often found in the essential oils of plants. The essential oils of plants are used for their fragrance in perfumes and aromatherapy products. For example, the oil of bitter almond is benzaldehyde, which is an aldehyde of benzene (**Figure 2**).

The functional group in ketones and aldehydes is the carbonyl group. The **carbonyl group** consists of *a carbon atom bonded to an oxygen atom with a double bond*. In ketones, the carbonyl group occurs somewhere in the *interior* of a carbon chain. In aldehydes, the carbonyl group occurs at the *end* of a carbon chain.

Figure 1
Ants, bees, and moths (such as the gypsy moths shown here) produce and detect minute quantities of ketones to communicate the presence of food or water, or the availability of a mate.

ketone an organic compound in which the carbon atom of a carbonyl group is bonded to two carbon atoms

aldehyde an organic compound that is characterized by a terminal carbonyl functional group; that is, the carbon atom of a carbonyl group is bonded to at least one hydrogen atom

carbonyl group a functional group that contains a carbon atom joined to an oxygen atom with a double covalent bond; general formula C=O

General formulas:

$$\underset{\text{ketone}}{R - \overset{\displaystyle \overset{O}{\|}}{C} - R} \qquad \underset{\text{aldehyde}}{R - \overset{\displaystyle \overset{O}{\|}}{C} - H}$$

Structural formulas:

$$\underset{\text{carbonyl group}}{- \overset{\displaystyle \overset{O}{\|}}{C} -} \qquad \underset{\text{butanone, a ketone}}{CH_3\overset{\displaystyle \overset{O}{\|}}{C}CH_2CH_3} \qquad \underset{\text{butanal, an aldehyde}}{CH_3CH_2CH_2\overset{\displaystyle \overset{O}{\|}}{C}H}$$

Naming Aldehydes and Ketones

The IUPAC names for aldehydes are formed by taking the parent alkane name, dropping the final *–e*, and adding *–al*. The simplest aldehyde has one carbon atom. Thus, the parent alkane is methane, and the aldehyde is methanal. Methanal is commonly known as formaldehyde. The aldehyde with two carbon atoms is ethanal, more commonly known as acetaldehyde. Because the carbonyl group in an aldehyde is, by definition, on the terminal carbon of the chain, a number is not necessary to give the position of the carbonyl group.

Ketones are named by replacing the *–e* ending of the parent alkane name with *–one*. The simplest ketone has three carbon atoms and is therefore named propanone, commonly called acetone. The position of the carbonyl group must be indicated by a number, unless there is only one possible position in the molecule.

$$\underset{\substack{\text{methanal}\\\text{(formaldehyde)}}}{H-\overset{\displaystyle O}{\overset{\|}{C}}-H} \qquad \underset{\substack{\text{ethanal}\\\text{(acetaldehyde)}}}{CH_3-\overset{\displaystyle O}{\overset{\|}{C}}-H} \qquad \underset{\substack{\text{propanone}\\\text{(acetone)}}}{CH_3-\overset{\displaystyle O}{\overset{\|}{C}}-CH_3} \qquad \underset{\substack{\text{butanone}}}{CH_3-\overset{\displaystyle O}{\overset{\|}{C}}-CH_2-CH_3}$$

(a)

▶ TRY THIS activity   *Where's the Cup?*

Find out what nail polish remover can do, besides dissolving nail polish.

Materials: Styrofoam cup, glass or ceramic container or plate, nail polish remover (containing acetone)

Hold a Styrofoam cup over a larger glass container and pour a little nail polish remover (about 5 mL) into the Styrofoam cup.

(a) Describe what happens.

(b) Is acetone polar or nonpolar? Explain.

(c) Is Styrofoam polar? Explain.

(b)

Properties of Aldehydes and Ketones

Aldehydes and ketones have lower boiling points than alcohols of similar sizes, and they are less soluble in water than alcohols are. These properties are to be expected because aldehydes and ketones do not contain –OH hydroxyl groups and, thus, do not form hydrogen bonds. The carbonyl group, C=O, is a strongly polar group, however, due to the large difference in electronegativity between carbon and oxygen. Therefore, aldehydes and ketones are more soluble in water than hydrocarbons are. The ability of these compounds to mix with both polar and nonpolar substances makes them good solvents.

Figure 2
Many essential oils contain aldehydes, which contribute to their pleasant fragrances **(a)**. Benzaldehyde, for example, is used to scent soaps and flavour candies. It is commonly called "oil of bitter almond." It is formed by grinding almonds **(b)** or apricot pits, and boiling them in water. Hydrogen cyanide, a poisonous gas, is also produced in this process.

▶ Section 3.9 Questions

Understanding Concepts

1. Arrange the following compounds in order of increasing boiling points. Give reasons for your order.

$$\underset{A}{CH_3CH_2\overset{\displaystyle O}{\overset{\|}{C}}H} \qquad \underset{B}{CH_3CH_2CH_3} \qquad \underset{C}{CH_3CH_2CH_2OH}$$

2. Predict the relative solubility of the following compounds in water, by listing the compounds in order of increasing solubility. Give reasons for your order.

$$\underset{A}{\underset{\displaystyle O}{\underset{\|}{CH_3CCH_2CH_3}}} \qquad \underset{B}{CH_3CH_2CH_2CH_2OH} \qquad \underset{C}{CH_3CH_2CH_2CH_3}$$

DID YOU *KNOW* ⁇

Steroids

Many steroids are ketones. Some examples are testoster*one* and progester*one* (male and female sex hormones) and anti-inflammatory agents such as cortis*one*. Oral contraceptives include two synthetic steroids. Some athletes use anabolic steroids to enhance their muscle development and physical performance, but such use may cause permanent damage.

testosterone

progesterone

cortisone

3. (a) From a molecular model kit, obtain two carbons, six hydrogens, and one oxygen. Build two different structures, each using all nine atoms. Draw the structural formula for each structure, and identify the organic family to which it belongs.

   (b) Repeat (a) using one additional carbon atom, so that your structure has a total of ten atoms.

   (c) Identify the functional group in each molecule in (a) and (b).

4. Prepare one index card for aldehydes and another index card for ketones. On the front of each card,
   • write the name and general formula of the organic family;
   • give an example, including the IUPAC name, common name, and structural formula.

On the back of each card,
   • list the characteristic properties, such as relative boiling points and solubility in polar and nonpolar solvents;
   • name the characteristic functional groups;
   • list the types of intermolecular forces.

Making Connections

5. Many organic compounds have been in everyday use for many years, so they are often known by common (nonsystematic) names. Make a list of common names of organic compounds that are found in solvents, cleaners, and other household items. Research the IUPAC names of five of these compounds, and identify the functional groups that are present in each compound. Discuss the useful properties that these functional groups may give to each compound.

> GO   www.science.nelson.com

6. The smell of formaldehyde was once common in high-school hallways, because formaldehyde was used to preserve biological specimens. This use has largely been discontinued.

   (a) What is the IUPAC name and structure of formaldehyde?

   (b) Why was its use as a preservative stopped? What substances are being used in its place?

> GO   www.science.nelson.com

7. Pheromones are powerful attractants in the insect world. Do they have a similar effect on humans? Research this topic, and write a report on how pheromones affect humans.

> GO   www.science.nelson.com

Many organic solvents, such as alcohols, aldehydes, and ketones, are flammable or combustible liquids. **Flammable liquids** readily ignite and burn at normal working temperatures. **Combustible liquids** ignite and burn at higher temperatures. Both flammable and combustible liquids are commonly used in homes and workplaces, as paint thinners, polish removers, furniture polishes, fuel, and organic solvents. Therefore, it is important to understand their hazards and to use proper safety procedures when handling, storing, and disposing of them (**Figure 1**).

When flammable and combustible liquids ignite, it is because they have evaporated and mixed with air. Gasoline, for example, is a *flammable* liquid with a **flashpoint** of –40°C. In other words, at –40°C, enough liquid gasoline will vaporize and mix with air to form a mixture that will burn. Phenol is an alcohol whose molecular structure contains a benzene ring. Phenol is a *combustible* liquid and has a much higher flashpoint of 79°C.

phenol

Hazards

Organic solvents generally flow easily. A minor spill can spread quickly over a large surface area, vaporizing into a dangerous invisible mixture. The spilled liquid can soak into fabrics, carpets, cardboard, newspapers, and wooden floors and furniture. Even when apparently dry, the materials containing the spilled liquid can continue to give off hazardous vapours that can ignite and burn. If the spilled liquid catches fire while it is flowing, the flames can travel quickly into other work areas or rooms.

Spray cans, such as those used for spot removers and furniture polishes, are particularly hazardous. The fine mist of flammable or combustible droplets travels a large distance and immediately mixes with air. The droplets will ignite if there is an open flame or an electrical spark in the vicinity.

Vapours that are formed by stored solvents can collect dangerously over time, particularly in low areas, because these vapours are denser than air and sink to the bottom of any room or building. Just as water will spread over an entire basement floor, a layer of burnable vapour will spread and form an invisible trail leading back to the storage container. If a spark occurs or a match is lit, the vapour—which might be hundreds of metres away from the stored solvent—will catch fire. The fire will travel very quickly back to the stored liquid, where it can cause an explosion or a major fire.

In some cases, no ignition source is needed to start a fire. Some organic solvents react slowly with air, giving off heat. When the temperature gets high

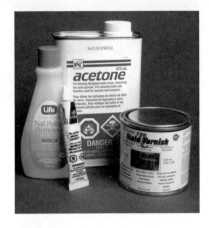

Figure 1
Propanone (acetone) is an effective organic solvent that is found in nail polish removers, plastic cements, resins, and varnishes. Like many ketones, acetone is both volatile and flammable. It should be used only in well-ventilated areas.

flammable liquid a liquid that will readily ignite and burn at room temperature

combustible liquid a liquid that will ignite and burn at temperatures higher than normal working temperatures

flashpoint the lowest temperature at which a flammable or combustible liquid will vaporize sufficiently to form a burnable mixture with air

spontaneous combustion igniting and burning by itself, without an outside ignition source

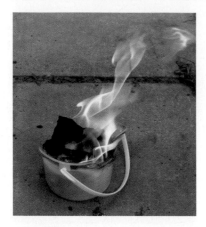

Figure 2
Rags soaked in linseed oil, which is commonly used to clean paint brushes, may spontaneously combust if stored in a confined area with access to air. The slow reaction with oxygen produces heat. If the heat cannot escape, it may warm the oil to its flashpoint.

enough, they will ignite all by themselves. This process is called **spontaneous combustion**. For example, rags soaked in motor oil or linseed oil may spontaneously combust if they are stuffed in a container, such as a plastic pail (**Figure 2**). The rags should be hung on a clothesline, to allow any heat that is produced to dissipate safely.

Organic solvents pose other dangers, besides being a fire hazard. They can cause health problems (such as damage to the skin or eyes), allergic reactions if inhaled or ingested, and overall toxicity. For example, inhalation of the common organic solvent 2-propanol (rubbing alcohol) may cause headaches, nausea, dizziness, drowsiness, poor coordination, and general confusion.

Safe Practice

When you work with organic solvents, always select a location with a good ventilation system, or work in a fume hood. Never inhale organic solvents.

Store organic solvents in appropriate cabinets, away from other highly reactive substances (**Figure 3**) and away from areas where there might be an open flame or electrical spark. Keep organic solvents away from sunlight and other heat sources to avoid a buildup of vapour, which will increase the pressure in the container.

Store the minimum quantities of solvents, and transfer only what you need to your work area. Return any unused solvent immediately to the proper storage container. Do not store solvents in basements or other below-ground locations. Solvent vapours, being more dense than air, cannot flow upward and are thus difficult to remove.

Dispose of waste organic solvents through hazardous waste collection and disposal companies or sites. Organic solvents should never be poured down the sink because they can be harmful to the environment. There are laws to ensure that proper procedures are followed regarding their disposal.

Consult the Material Safety Data Sheets (MSDS) for the chemicals that you use. (See Appendix B2.) These sheets have specific sections on fire and explosion hazards, the types of fire extinguishers to use, procedures for cleaning up spills, and first aid instructions.

DID YOU KNOW?

Not Opposites!
The terms "flammable" and "inflammable" have the same meaning: easily set on fire.

(a)

Class B: Flammable and Combustible Materials
Store in designated areas.
Work in well-ventilated areas.
Avoid heating.
Avoid sparks and flames.
Ensure that electrical sources are safe.

(b)

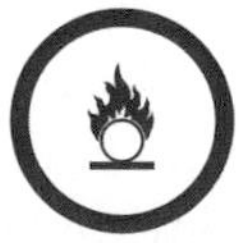

Class C: Oxidizing Materials
Store away from combustibles.
Wear body, hand, face, and eye protection.
Store in container that will not rust or oxidize.

Figure 3
WHMIS information for flammable and combustible materials **(a)** and oxidizing materials **(b)**

Safe Use of Organic Solvents

Always follow these basic safety procedures when using organic solvents:

- Carefully read and follow the recommendations on the MSDS for every chemical substance that you use. Be aware of the flammability and combustibility of solvents.

- Use organic solvents in a well-ventilated location or in a fume hood, away from ignition sources such as electrical sparks, open flames, and hot surfaces.

- Do not store organic solvents in direct sunlight or near heat sources.

- Do not use or store organic solvents in basements.

- Return unused solvents immediately to the appropriate storage containers.

- Dispose of waste solvents according to environmental restrictions. Never pour them down the sink.

CAREER CONNECTION

Firefighters must have a good background in organic chemistry, because they must be able to identify potential fire hazards. As well, they have to operate various chemical appliances to extinguish fires or to disperse or neutralize dangerous substances, such as organic solvents and petroleum. Firefighters must also be physically fit and able to work under pressure, often at great heights or in confined spaces.

(i) Research the training that is needed to become a firefighter. Locate the training facility that is closest to your home.

(ii) Research the typical starting salary for firefighters. Compare this figure with the national average for all careers.

 www.science.nelson.com

▶ Section 3.10 Questions

Understanding Concepts

1. Describe the difference between flammable liquids and combustible liquids.

2. Take a walk through your home, and list all the organic solvents you can find, their uses, and their storage locations. Assess the safety of the handling and storage of these solvents in your home.

3. What is meant when an organic solvent is said to have a flashpoint of $-9°C$?

Applying Inquiry Skills

4. Suppose that you are preparing to carry out an experiment with ethanol in the school laboratory. List all the safety precautions that you should follow.

Making Connections

5. Talk to the members of the Health and Safety committee at your school, and find out what rules are in place regarding food and clothing in laboratories, ventilation, and emergency response teams.

6. House-painters may use organic solvents to thin paint or clean brushes. Create an information brochure for painters, alerting them to the hazards of organic substances and the precautions they should take.

Because carboxylic acids have distinctive odours, police dogs can follow the characteristic blend of carboxylic acids in a person's sweat. Similarly, they can find illegal drug laboratories by following the odour of acetic acid. Acetic acid is formed when morphine, collected from opium poppies, is treated to produce heroin. Dogs' noses have been estimated to be from 300 to 10 000 times as sensitive as humans' noses.

carboxylic acid an organic compound that is characterized by the presence of a carboxyl group; general formula R–COOH

carboxyl group a functional group that consists of a hydroxyl group attached to the carbon atom of a carbonyl group; general formula –COOH

$$CH_3 - \overset{\overset{\displaystyle OH}{|}}{C} - \overset{\overset{\displaystyle OH}{|}}{\underset{\underset{\displaystyle H}{|}}{C}} = O$$

lactic acid

Figure 1
Lactic acid is produced in muscles when the oxygen supply cannot keep up with demand during strenuous exercise.

When wine is opened and left in contact with air for a period of time, it will likely turn into vinegar. Vinegar is a weak solution of a **carboxylic acid**, commonly called *acetic acid*. Grocery stores sell wine vinegar for cooking or making salad dressings. The chemical reaction in this souring process involves the *oxidation* of ethanol. (You will learn more about oxidation reactions in Unit 5.) Carboxylic acids are generally found in citrus fruits, crab apples, rhubarb, and other foods with a sour, tangy taste.

Sour milk and yogurt contain lactic acid (**Figure 1**), which is produced by bacteria. If you have ever felt your muscles ache after prolonged exertion, you have experienced the effect of lactic acid in your muscles. The gamey taste of meat from animals that were killed after a long hunt is due to the high concentration of lactic acid in their muscles.

Naming Carboxylic Acids

The functional group of carboxylic acids is the **carboxyl group**, written as –COOH. This functional group is a combination of two other functional groups that you are already familiar with: the *carbonyl* (–C=O) group in aldehydes and ketones, and the *hydroxyl* (–OH) group in alcohols.

The general structural formula for a carboxylic acid is

$$R \text{ or } H - \overset{\overset{\displaystyle O}{\|}}{C} - OH$$

The IUPAC name for a carboxylic acid has two parts. The first part tells you the number of carbon atoms in the longest carbon chain. The second part ends in *-anoic acid*. So the acid with *one carbon* is called *methanoic acid*. Similarly, the acid with *two carbons* is called *ethanoic acid*, and the acid with *three carbons* is called *propanoic acid*.

Methanoic acid, HCOOH, is commonly called formic acid. Its name comes from the Latin word *formica*, which means "ant." Not surprisingly, the first source of this acid was ants (**Figure 2**). Methanoic acid is used to remove hair from hides and to recycle rubber. Ethanoic acid is commonly called acetic acid, which is the acid found in vinegar.

Some other common acids contain two or more carboxyl groups. (The structures for several of these acids are shown at the top of the next page. For simplicity, only their common non-IUPAC names will be used in this textbook.) For example, spinach and rhubarb contain oxalic acid, which is used in rust removers and brass cleaners. You may have used tartaric acid in baking. It reacts with baking soda to produce bubbles of carbon dioxide, making a cake rise. Citric acid is responsible for the sour taste of citrus fruits. Vitamin C, or ascorbic acid, is found in many fruits and vegetables. Acetylsalicylic acid (ASA) is the active ingredient in Aspirin. You may have experienced its sour taste when swallowing a tablet.

oxalic acid   tartaric acid   citric acid   ascorbic acid (vitamin C)   acetylsalicylic acid (ASA)

Properties of Carboxylic Acids

Like all other acids, the solutions of carboxylic acids characteristically turn blue litmus red. They also react with bases in neutralization reactions. (See sections 3.13 and 3.15.)

The carboxyl groups of carboxylic acids make them polar molecules, capable of hydrogen-bonding with each other and with water molecules. The smaller acids are soluble in water. The longer carbon chains in the larger acids make them relatively insoluble. The polar carboxyl groups also account for the high melting points of acids, compared with the melting points of similar hydrocarbons. It also appears that the number of carboxyl groups per molecule affects the melting point of a carboxylic acid (**Table 1**).

Figure 2
Most ants and ant larvae are edible and are, in fact, considered quite delicious. They have a vinegary taste because they contain methanoic acid, HCOOH, commonly called formic acid. In some countries, large ants are squeezed directly over a salad to add the tangy ant juice as a dressing.

Table 1   Melting Points of Some Carboxylic Acids

Number of COOH groups	Carboxylic acid	Melting point (°C)
1	methanoic acid	8
1	ethanoic acid	17
2	oxalic acid	189

▶ TRY THIS activity   *Making a Bath Bomb*

Have you ever tried one of those scented bath bombs that fizz around in the bathtub (**Figure 3**)? The bubbles are formed by the reaction of a carboxylic acid and a base, right in your bath water. Here is a recipe to try.

Materials: eye protection, 40 mL citric acid (powder), 125 mL sodium hydrogen carbonate (baking soda), 40 mL cornstarch, 60 mL vegetable oil (for example, olive oil, coconut oil, or corn oil), fragrance (optional), food colouring (optional), large bowl, small cup, metal fork, 250-mL graduated cylinder or measuring cup, waxed paper

1. In the large bowl, mix together the dry ingredients (citric acid, cornstarch, and sodium hydrogen carbonate).
2. In the small cup, mix 60 mL of vegetable oil and a few drops of the fragrance and food colouring (optional).
3. Pour 40 to 60 mL of the oil mixture into the dry ingredients. Mix well with the fork, until the mixture is the consistency of dry pastry dough.
4. Form the mixture into a ball or another shape of your choice. Set your bath bomb on a sheet of waxed paper, and allow it to dry completely.
5. Place your bath bomb in warm water, and enjoy!

(a) Write a word equation to explain the reaction that you observe.

Figure 3
The "fizz" produced by a bath bomb comes from the neutralization reaction as the bath bomb dissolves.

▶ *Section 3.11* *Questions*

Understanding Concepts

1. (a) Draw a structural formula for the carboxyl group in carboxylic acids.
   (b) Explain the effect of the components of this functional group on the properties of carboxylic acids.

2. Why are carboxylic acids polar molecules?

3. Ethanoic acid is the two-carbon acid in vinegar. Oxalic acid is the two-carbon acid in rhubarb leaves. (See **Table 1**.)
   (a) Draw the structures of these two acids.
   (b) Explain why ethanoic acid is a liquid and oxalic acid is a solid at room temperature.

4. The labels have fallen off two bottles. Bottle A contains a gas and bottle B contains a liquid. The labels indicate that both compounds have the same number of carbon atoms. One compound is an alkane and one is a carboxylic acid. Suggest the identity of the compound in each bottle, and give reasons for your answer.

5. Prepare an index card for carboxylic acids. On the front of the card,
   • write the name and general formula of the organic family;
   • give an example, including the IUPAC name, common name, and structural formula.

   On the back of the card,
   • list the characteristic properties, such as boiling point and solubility in polar and nonpolar solvents;
   • name the characteristic functional group;
   • name the type of intermolecular force.

6. Some cosmetic facial creams contain ingredients that manufacturers call "alpha hydroxy acids," which are designed to remove wrinkles. Alpha hydroxy acids are carboxylic acids that also contain a hydroxyl group attached to the carbon atom adjacent to the carboxyl group. The alpha hydroxy acids in cosmetics may include glycolic acid, lactic acid, malic acid, and citric acid. Research and draw structural diagrams for these compounds.

 www.science.nelson.com

Applying Inquiry Skills

7. Suppose that you are given two unidentified colourless liquids. You are told that one liquid is a ketone and the other liquid is a carboxylic acid. Describe how you would distinguish between the two liquids. Give reasons for your strategy.

Properties of Carboxylic Acids

A carboxylic acid is identified by the presence of a carboxyl group. The physical properties and reactivity of carboxylic acids are related to the combination of their polar functional group and their nonpolar hydrocarbon components.

In this investigation, you will predict and then test some of the properties of two carboxylic acids: ethanoic (acetic) acid and octadecanoic (stearic) acid (**Figure 1**). The properties you will investigate are melting point, boiling point, solubility, acidity, and reaction with a base. You will obtain the melting point and boiling point of each acid from reference resources. You will determine the solubility of each acid in polar and nonpolar solvents by mixing the acid with water and with oil. Finally, you will observe and compare the reaction, if any, of each acid with a basic solution of sodium hydrogen carbonate, $NaHCO_{3(aq)}$.

Figure 1
Stearic acid is used to harden soaps, particularly those made with vegetable oils, that otherwise tend to be very soft.

Question

What are the similarities and differences in melting point, boiling point, solubility, and reaction with a base of ethanoic acid and octadecanoic acid?

Inquiry Skills
○ Questioning ○ Planning ● Analyzing
○ Hypothesizing ● Conducting ● Evaluating
● Predicting ● Recording ○ Communicating

Prediction

(a) Use your knowledge of the structure and functional groups of carboxylic acids to predict an answer to the Question. (For some properties, such as melting point and boiling point, you will only be able to give relative answers.)

Materials

eye protection
lab apron
ethanoic acid (concentrated acetic acid), $CH_3COOH_{(l)}$
dilute ethanoic acid (5% acetic acid: vinegar), $CH_3COOH_{(aq)}$
octadecanoic acid (stearic acid), $CH_3(CH_2)_{16}COOH_{(s)}$
water
vegetable oil
pH meter or universal indicator
sodium hydrogen carbonate, $NaHCO_{3(aq)}$ (saturated aqueous solution)
2 test tubes
test-tube holder
test-tube rack
pipette and bulb
eye dropper
10-mL graduated cylinder

Procedure

(b) Copy **Table 1** into your notebook, leaving enough space for your answers.

Table 1 Properties of Carboxylic Acids

Property	Acetic acid	Stearic acid
structural formula		
molar mass		
melting point		
boiling point		
solubility in water		
solubility in vegetable oil		
reaction with base		

(c) Determine the structural formulas and calculate the molar masses of acetic acid and stearic acid. Then complete the first and second rows of your table (**Table 1**).

(d) Using print or electronic resources, find the melting point and boiling point of acetic acid and stearic acid. Then complete the third and fourth rows of your table (**Table 1**).

1. Add 5 mL of water to one test tube and 5 mL of oil to another test tube. In the fume hood, use a pipette to add one drop of ethanoic acid (concentrated acetic acid) to each test tube (**Figure 2**). Shake each test tube very carefully to mix the contents.

Concentrated acetic acid is corrosive. Avoid contact with skin and eyes. This acid is also volatile, so be careful to avoid inhalation. Wear eye protection and a lab apron.

(e) Make and record observations on how well acetic acid dissolves in water and oil.

2. While still in the fume hood, add a drop of pH indicator to each test tube in step 1, or use a pH meter to measure the pH. Record the results.

3. Follow your teacher's instructions to dispose of the contents of each test tube, and clean the test tubes.

4. Repeat steps 1, 2, and 3, using a small amount of solid stearic acid (enough to cover the tip of a toothpick). These steps do not need to be performed in the fume hood.

5. Place about 2 mL of saturated sodium hydrogen carbonate solution, $NaHCO_{3(aq)}$, in each of two test tubes. Add 2 mL of dilute ethanoic acid (vinegar) to one test tube. Add a small amount of solid stearic acid to the other test tube. Shake the test tubes gently to mix, and observe for formation of bubbles. Record your observations.

Analysis

(f) Compare the melting points and boiling points of acetic acid and stearic acid. Account for the differences in these properties in terms of the molecular structure and intermolecular forces of each acid.

Figure 2
Carry out all work with concentrated acetic acid in the fume hood.

(g) Compare the solubilities of acetic acid and stearic acid in water and in oil. Account for any difference in their solubilities in terms of molecular structure and intermolecular forces.

(h) Do acetic acid and stearic acid appear to be organic acids in this investigation? Explain why or why not, with reference to the substances and experimental conditions.

Evaluation

(i) Did the Procedure enable you to collect appropriate evidence? Explain. Make suggestions for any necessary improvements.

(j) Compare the answers obtained in your Analysis to your Prediction. Which chemical properties did your theoretical model of carboxylic acids allow you to predict? Give reasons for your answer.

Esters occur naturally in many plants (**Figure 1**), and they are responsible for many of the odours of fruits and flowers. Synthetic esters are often added as flavourings to processed foods and as scents to cosmetics and perfumes. **Table 1** shows the main esters that are used to create certain artificial flavours.

Table 1 The Odours of Esters

Odour	Name	Molecular formula
apple	methyl butanoate	$CH_3CH_2CH_2COOCH_3$
apricot	pentyl butanoate	$CH_3CH_2CH_2COOCH_2CH_2CH_2CH_3$
banana	3-methylbutyl ethanoate	CH_3 | $CH_3COOCH_2CH_2CHCH_3$
cherry	ethyl benzoate	$C_6H_5COOC_2H_5$
orange	octyl ethanoate	$CH_3COOCH_2CH_2CH_2CH_2CH_2CH_2CH_3$
pineapple	ethyl butanoate	$CH_3CH_2CH_2COOCH_2CH_3$
red grape	ethyl heptanoate	$CH_3CH_2CH_2CH_2CH_2CH_2COOCH_2CH_3$
rum	ethyl methanoate	$HCOOCH_2CH_3$

An **ester** is an organic compond that contains a carbonyl group bonded to an oxygen atom. It is formed when a carboxylic acid reacts with an alcohol. A reaction in which an ester is formed is called **esterification**. (Esterification is an example of a type of reaction called *condensation*, in which a larger molecule is produced, with the release of a small molecule, such as water.)

The general reaction between a carboxylic acid and an alcohol is represented below. An acid catalyst, such as sulfuric acid, and heat are generally required. The other product that is formed, besides the ester, is water.

$$\underset{\text{acid}}{RC\overset{O}{\overset{\|}{-}}O\boxed{H}} + \boxed{OH}\underset{\text{alcohol}}{-R'} \xrightarrow[\text{heat}]{\text{conc. } H_2SO_4} \underset{\text{ester}}{RC\overset{O}{\overset{\|}{-}}O-R'} + \underset{\text{water}}{H_2O}$$

As you can see, the general formula for an ester is a combination of an organic acid and an alcohol.

Naming Esters

The name of an ester has two parts. The first part comes from the alcohol, and the second part comes from the carboxylic acid. The ending of the acid name is changed from *–anoic acid* to *–anoate*. For example, in the reaction of ethanol and butanoic acid, the ester that is formed is ethyl butanoate (an ester with a pineapple odour). Note that the acid is written as the *first part* of the formula for an ester, but as the *second part* of its name.

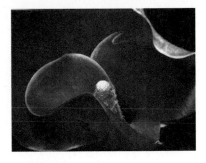

Figure 1
The rich scent of the lily is at least partially due to the esters that are produced in the flowers.

ester an organic compound that is characterized by the presence of a carbonyl group bonded to an oxygen atom

esterification a condensation reaction in which a carboxylic acid and an alcohol combine to produce an ester and water

$$CH_3CH_2CH_2C-O\boxed{H + OH}-CH_2CH_3 \longrightarrow CH_3CH_2CH_2C-O-CH_2CH_3 + H_2O$$

butanoic acid	ethanol	ethyl butanoate	water
(acid)	(alcohol)	(ester)	

Reactions and Properties of Esters

The esterfication reaction can be reversed: that is, the ester can be split into its acid and alcohol components. This type of reaction is called **hydrolysis**.

hydrolysis a reaction in which a bond is broken by the addition of the components of water, resulting in the formation of two or more products

$$RC-O-R' + H_2O \xrightarrow{\text{heat}} R-C-OH + HO-R'$$

carboxylic   alcohol
acid

Fats and oils are esters of long-chain carboxylic acids. When these esters are heated with a strong base, such as sodium hydroxide, NaOH, a hydrolysis reaction occurs. The sodium salts of the resulting acids are what we call soap.

$$\text{fats or oils (esters)} + NaOH \xrightarrow{\text{heat}} \text{soap} + \text{alcohol}$$

$$RC-O-R' + Na^+ + OH^- \longrightarrow RC-O^- + Na^+ + R'OH$$

ester   acid   alcohol

An ester still has a carbonyl group but, unlike a carboxylic acid, it has no hydroxyl group. Therefore, intermolecular forces are weaker in esters. The weaker intermolecular forces explain why esters are less soluble in water and have lower melting and boiling points than corresponding carboxylic acids. In addition, the acidity of carboxylic acids is due to the hydrogen atom on the hydroxyl group. Since esters have no hydroxyl group, they are not acidic.

It is the smaller esters that you can detect by scent, because they are gases at room temperature. The larger, heavier esters more commonly occur as waxy solids.

▶ *Section 3.13* **Questions**

Understanding Concepts

1. In what way is the functional group of an ester different from the functional group of a carboxylic acid? In what way is it similar?

2. Why are esters less soluble in water than carboxylic acids are?

3. How is an esterification reaction similar to the neutralization reaction between hydrochloric acid and sodium hydroxide?

4. What is a hydrolysis reaction? How is it similar to, or different from, an esterification reaction?

5. Prepare an index card for esters. On the front of the card,
   • write the name and general formula of the organic family;
   • give an example, including the IUPAC name, common name, and structural formula.

   On the back of the card,
   • list the characteristic properties, such as boiling point and solubility in polar and nonpolar solvents;
   • name the characteristic functional group;
   • name the type of intermolecular force.

Applying Inquiry Skills

6. Suppose that you are given two unlabelled liquids. You know that one liquid is a carboxylic acid and the other liquid is an ester. The two liquids are made up of molecules of similar size. Describe three simple tests you could perform to identify the organic family of each liquid.

Making Connections

7. Tannic acid was originally obtained from the wood and bark of certain trees. For centuries, it has been used to "tan" leather (**Figure 2**).
   (a) Give the chemical formula for tannic acid.
   (b) What effect does tannic acid have on animal hides? Explain your answer, referring to the chemical reactions that take place.

 www.science.nelson.com

Figure 2
Tanneries are notorious for the bad smells they produce. The bad smells are a result of the chemical reactions between the animal hides and the chemicals that are used to process the hides.

3.14 Activity

Synthesis of Esters

Many esters are found naturally in fruits, and they are responsible for some of the pleasant fruity odours. Synthetic esters are produced from condensation reactions between alcohols and carboxylic acids. They are used to add scents to many products (**Figure 1**). In this activity, you will synthesize several esters by combining various alcohols with a carboxylic acid. Then you will identify the odours of the esters you have synthesized.

Materials

eye protection
lab apron
500-mL beaker
water
hot plate
ethanol, $CH_3CH_2OH_{(aq)}$
2-propanol, $CH_3CH(OH)CH_{3(l)}$
1-pentanol, $CH_3CH_2CH_2CH_2CH_2OH_{(l)}$
concentrated acetic acid, $CH_3COOH_{(l)}$
10-mL graduated cylinder
2 mL concentrated sulfuric acid, $H_2SO_{4(aq)}$
3 small test tubes
test-tube rack
wax pencils
test-tube holder
10-mL graduated pipette and bulb
evaporating dish or petri dish

Figure 1
Many artificial flavours and scents are produced by synthetic esters.

Table 1 Contents of Test Tubes for Synthesizing Esters

Contents	Test tube 1	Test tube 2	Test tube 3
alcohol (1 mL)	ethanol	2-propanol	1-pentanol
acid (1 mL)	conc. acetic acid	conc. acetic acid	conc. acetic acid
catalyst (0.5 mL)	conc. $H_2SO_{4(aq)}$	conc. $H_2SO_{4(aq)}$	conc. $H_2SO_{4(aq)}$

Procedure

1. Prepare a boiling-water bath by half-filling a 500-mL beaker with water and heating the beaker carefully on a hot plate until the water comes to a gentle boil.

2. Number three test tubes in a test-tube rack. Use the graduated cylinder to measure 1 mL of a different alcohol into each test tube, as indicated in **Table 1**.

 All three alcohols are highly flammable. Do not use them near an open flame.

3. In a fume hood, use a pipette to add 1 mL of concentrated acetic acid and 0.5 mL of concentrated sulfuric acid to each test tube, as indicated in **Table 1**. Gently shake each test tube to mix the contents.

 Concentrated acetic acid and concentrated sulfuric acid are corrosive. Avoid contact with skin and eyes. Concentrated acetic acid is also volatile, so be careful to avoid inhalation. Wear eye protection and a lab apron.

4. Carefully return the test tubes to your lab bench. Place all three test tubes in the hot-water bath. Be careful not to point the test tubes at anybody. After about 5 min of heating, remove the test tubes from the heat and put them back in the rack.

5. Pour the contents of the first test tube into an evaporating dish or a glass petri dish that is half-filled with cold water. Identify the odour of the ester, as instructed by your teacher (**Figure 2**). Repeat for each ester.

6. Follow your teacher's instructions for handing in your esters or for disposing of them.

Analysis

(a) Copy and complete **Table 2**.

(b) Draw structural formula equations to represent each of the three esterification reactions in this activity. Write the IUPAC name of each reactant and product below the molecular formula.

(c) What was the function of the concentrated sulfuric acid in these reactions?

(d) What evidence did you find to show that the esters produced in this activity are soluble or insoluble in aqueous solution? Explain the evidence in terms of the molecular structure of the esters.

Figure 2
Keep your head well back from any substance you are smelling.

Table 2   Summary of Condensation Reactions

	Reaction 1	Reaction 2	Reaction 3
IUPAC name of alcohol used			
Structural formula of alcohol used			
IUPAC name of carboxylic acid used			
Structural formula of carboxylic acid used			
IUPAC name of ester produced			
Structural formula of ester produced			
Odour of ester produced			

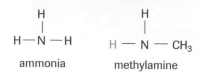

Figure 1
Methylamine is like ammonia with one hydrogen atom substituted with a methyl group.

amine an organic compound in which one or more hydrogen atoms are replaced by alkyl groups

amide an organic compound in which the carbon atom of a carbonyl group (C=O) is bonded to a nitrogen atom

Figure 2
Amines with low molar masses are partly responsible for the characteristic "fishy" smell of seafood. Lemon juice is often provided in restaurants to neutralize the bitter taste of these amines, which are weak bases.

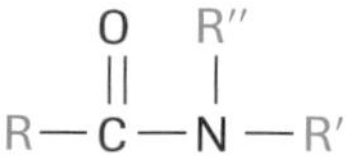

Figure 3
General formula of an amide: The R groups may be the same or different.

Living organisms produce many complex organic molecules, such as proteins and DNA, that contain nitrogen as well as carbon, hydrogen, and oxygen. When the organisms die and decompose, these nitrogenous molecules are broken down into simpler organic compounds called amines. An **amine** may be considered to be an ammonia molecule in which one or more hydrogen atoms are replaced by alkyl groups. An example of an amine is aminomethane, or methylamine (**Figure 1**).

Like many compounds of nitrogen, such as ammonia, NH_3, amines often have an unpleasant odour. For example, the smell of fish is due to a mixture of amines (**Figure 2**). The putrid odour of decomposing animal tissue is caused by amines appropriately called putrescine and cadaverine, which are produced by bacteria.

$$H_2N—CH_2—CH_2—CH_2—CH_2—NH_2$$

putrescine

Amines (or ammonia) can react with carboxylic acids to form nitrogenous compounds called **amides**. This reaction is similar to esterification, in which *alcohols* react with carboxylic acids to form *esters*.

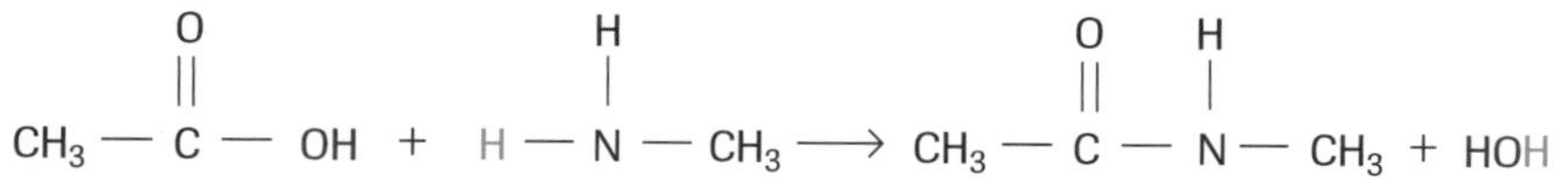

ethanoic acid   methylamine   methylethanamide   water

The functional group of amides consists of a carbonyl group (C=O) directly attached to a nitrogen atom (**Figure 3**). The amide group has major importance in biological systems, because it forms the backbone of all protein molecules.

Naming Amines

Amines may have one, two, or three alkyl groups attached to the nitrogen atom. These amines are called primary, secondary, and tertiary amines, respectively (**Figure 4**).

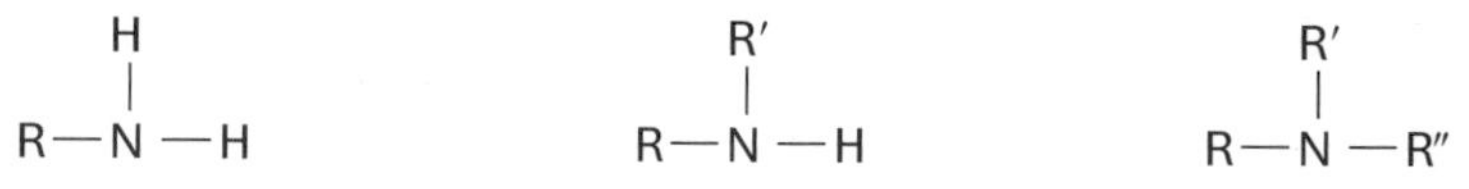

primary amine (1°)   secondary amine (2°)   tertiary amine (3°)

Figure 4
Notice how all the molecules have a similar structure in relation to the nitrogen atom.

The IUPAC names of amines have two parts. The first part is *amino-*, and the second part comes from the longest-chain hydrocarbon. For example, CH_3NH_2 has one carbon atom, so its name is *aminomethane*. Similarly, $CH_3CH_2NH_2$ is named aminoethane. When there are more than two carbon atoms in the chain, however, you have to specify the carbon to which the nitrogen is attached. Therefore, a three-carbon amine could be 1-aminopropane or 2-aminopropane.

Another system for naming amines includes the suffix *-amine*, preceded by the name of the alkyl group (which is the hydrocarbon prefix plus *-yl*). In other words, CH_3NH_2 can also be called *methylamine*. This system is often more convenient to use, particularly for naming secondary and tertiary amines. In this unit, you will be naming only primary amines.

Naming Amides

Naming amides is similar to naming esters. While esters end in *-oate*, amides end in *-amide*.

butanoic acid + methanol → methyl butanoate + water
(an ester)

butanoic acid + aminomethane → methyl butanamide + water
(an amide)

Properties of Amines and Amides

Amines have higher boiling points and melting points than similar-sized hydrocarbons, and smaller amines are readily soluble in water. These properties can be explained by the presence of two types of polar bonds in amines: the C–N bonds and any N–H bonds. Both types of bonds are polar because nitrogen is more electronegative than either carbon or hydrogen. The polar bonds increase intermolecular forces of attraction. Therefore, relatively high temperatures are required to melt or to vaporize amines.

Where N–H bonds are present, hydrogen bonding also occurs with water molecules. This hydrogen bonding accounts for the high solubility of amines in water. Since N–H bonds are less polar than O–H bonds, amines boil at lower temperatures than similar-sized alcohols do (**Table 1**).

Amides are generally insoluble in water. Smaller amides are slightly soluble in water, however, because of the N–H bonds, which allow hydrogen bonding.

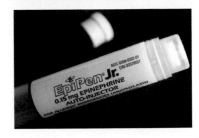

Some hormones, such as epinephrine, are amines. People with known allergic reactions carry "Epipens": syringes that contain measured doses of epinephrine, which controls the extent of an allergic reaction when injected immediately.

Table 1 Boiling Points of Corresponding Hydrocarbons, Amines, and Alcohols

Hydrocarbon	b.p. (°C)	Amine	b.p. (°C)	Alcohol	b.p. (°C)
CH_3CH_3	−89	CH_3NH_2	−6	CH_3OH	65
$C_2H_5CH_3$	−42	$C_2H_5NH_2$	16	C_2H_5OH	78
$C_3H_7CH_3$	−0.5	$C_3H_7NH_2$	48	C_3H_7OH	97
$C_4H_9CH_3$	36	$C_4H_9NH_2$	78	C_4H_9OH	117

Understanding Concepts

1. Why is the formation of an amide from a carboxylic acid and an amine a condensation reaction?

2. Why do amines generally have lower boiling points than corresponding alcohols?

3. Classify each of the following compounds as an amine or an amide:
   (a) $CH_3CH_2CH_2NH_2$
   (b) 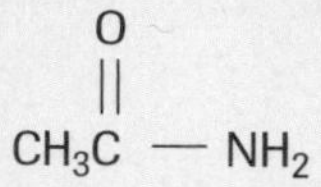
   $$CH_3C \overset{\overset{\textstyle O}{\|}}{} - NH_2$$

4. Arrange each pair of compounds in order of increasing solubility in nonpolar solvents. Give reasons for your order.
   (a) a corresponding amine and alcohol
   (b) a corresponding hydrocarbon and amine

5. What features in the molecular structure of a compound affect its solubility and its melting and boiling points? Illustrate your answer with examples of hydrocarbons, alcohols, carboxylic acids, amines, and amides. Include the structural formulas for your examples.

6. Prepare one index card for amines and another index card for amides. On the front of each card,
   • write the name and general formula of the organic family;
   • give an example, including the IUPAC name, common name, and structural formula.

   On the back of each card,
   • list the characteristic properties, such as relative boiling points and solubility in polar and nonpolar solvents;
   • name the characteristic functional groups;
   • list the types of intermolecular forces.

7. Protein molecules in all living organisms are long molecules made up of small units called amino acids. From what you have learned in this section,
   (a) deduce the structural features of amino acids
   (b) predict some physical and chemical properties of amino acids

Making Connections

8. In some cuisines, fish recipes include lemon garnish or a vinegar sauce, such as sweet-and-sour sauce. Suggest a reason why these common culinary techniques might be used. Support your answer with chemical information.

Many artificial sweeteners are synthetic compounds that often bear little chemical resemblance to sugar. Sugar molecules are **carbohydrates**: compounds that contain carbon, hydrogen, and oxygen atoms. These atoms are arranged as a carbon backbone with hydroxyl (−OH) groups and carbonyl (−C=O) groups attached to the carbon atoms. Some sugars, such as glucose, are aldehydes. Other sugars, such as fructose, are ketones.

carbohydrate a compound of carbon, hydrogen, and oxygen; general formula $C_x(H_2O)_y$

amino acid a compound in which an amino group and a carboxyl group are attached to the same carbon atom

glucose

fructose

amino acid

sodium cyclamate

Several artificial sweeteners have been widely used in North America in the last few decades. Supporters of sugar substitutes list benefits such as reduction of tooth decay and better food choices for diabetics, not to mention the fight against obesity. The use of sodium cyclamate was banned in 1970 in the United States because of a link to increased cases of liver cancer.

The compound aspartame, $C_{14}H_{18}N_2O_5$, was discovered in 1965, quite accidentally, by a chemist named James Schlatter. Schlatter made an ester using methanol and two amino acids. **Amino acids** are the building blocks of proteins. They contain both an amino group (−NH$_2$) and a carboxyl group (−COOH) attached to the same carbon atom. Amino acids link together to form amides.

When Schlatter tasted the ester he had made, he found it to be very sweet. Since the starting materials are amino acids that are identical to those found in foods such as meat and eggs, aspartame appears to be a safe alternative to sugars and other sugar substitutes.

As with most food substitutes, however, there is some controversy surrounding the use of aspartame. Enzymes in the digestive tract break aspartame down into its three components: two amino acids and methanol. The two amino acids are generally safe for most people. One of them, however, phenylalanine, is a danger to people with phenylketonuria (PKU, a rare genetic disease that is diagnosed in infancy). These people cannot break down phenylalanine, and the accumulation of phenylalanine can lead to brain damage.

A more widespread concern with aspartame is the release of methanol as one of the breakdown products. As you may recall from your study of

aspartame

alcohols, methanol is much more toxic than ethanol. When methanol breaks down in the body, methanal (formaldehyde) is formed. Methanal acts on proteins, changing their structure and causing drastic changes in their important functions in our cells.

The methanal is further changed to methanoic (formic) acid. This acid disrupts the function of mitochondria, the part of a living cell that produces energy.

When these breakdown products are formed in the cells of the retina and the optic nerve of the eye, they cause irreparable damage. Blindness is one of the first detectable symptoms of methanol poisoning.

Manufacturers of aspartame claim that the amounts of methanol produced from consuming normal amounts of the sweetener are so small that any harmful effects are negligible. Indeed, methanol is naturally produced in many fruits and vegetables, and it is also present in low concentrations (up to 300 mg/L) in wine and beer.

Is aspartame dangerous? If so, why is it so readily available? Aspartame has been approved for use in more than 90 countries.

▶ Understanding the Issue

1. Name some artificial sweeteners that have been developed in the past 100 years.

2. (a) What are the advantages of having artificially sweetened foods available?

   (b) What are some drawbacks?

3. (a) Draw the structure of aspartame, and calculate its molar mass.

   (b) Calculate the percentage of methanol in aspartame, by mass. Assume that one molecule of aspartame contains one molecule of methanol.

   (c) One can of a diet soft drink contains approximately 200 mg of aspartame. Calculate the mass of methanol produced from the ingestion of one can of a diet soft drink. Assume that one molecule of aspartame produces one molecule of methanol.

   (d) The toxicity of a substance is usually measured by the LD_{50} rating (the dosage that would kill 50% of the test subjects), which is expressed in g/kg of body mass. The LD_{50} for methanol is 0.07 g/kg. Calculate the lethal dose for a 70-kg person.

   (e) How many cans of diet pop would a person have to drink to ingest this lethal dose?

▶ Take a Stand

Health Benefit or Health Hazard?

Decision-Making Skills

○ Define the Issue   ● Analyze the Issue   ● Research
● Defend a Decision   ○ Identify Alternatives   ○ Evaluate

Aspartame, sold as NutraSweet, is probably the most widely used sugar substitute in North America. More than 200 studies have been done on the safety of its use. Some studies claim that it produces side effects ranging from headaches to brain tumours. Other studies claim that there is no evidence of any harmful effects.

(a) Research some of these studies, or studies about the use of another food substitute or additive (such as an artificial flavour or food colour). Write brief summaries of the arguments for and against the use of your chosen food substitute or additive.

(b) In small groups, discuss the factors that you would consider important to the validity of any research study on the safety of the use of a food substitute or additive.

(c) Analyze the risks and benefits of the use of your chosen food substitute or additive. Writing for a general interest magazine, prepare a report on the results of your analysis. In your report, make a recommendation on its use, with supporting arguments.

 www.science.nelson.com

Classifying Plastics

In recent times, we have been throwing away massive quantities of plastics. To reduce the amount of waste going to landfill sites, communities are starting to implement recycling programs. Because different types of plastics are made of different components, effective recycling requires that the different types of plastics be identified and separated. To aid this identification, the Society of the Plastics Industry, Inc. (SPI) established a resin identification coding system (**Figure 1**) in 1988. **Table 1** gives the properties and end products of resins identified by their SPI codes.

The different compositions and structures of plastics allow us to tell them apart by their properties. In this activity, you will differentiate among several unidentified samples of plastics by their density (by placing them in liquids of different densities), flame colour (a green flame indicates chlorine), solubility in acetone, and resistance to heating. You will then use your results to identify the plastics and their SPI codes.

Question

What is the SPI resin code for each of the six unidentified plastic samples?

Materials

eye protection
lab apron
1 cm by 1 cm samples (unidentified) of the six categories of plastics, each cut into an identifiable shape
water
60 g 2-propanol, $CH_3CH(OH)CH_{3(l)}$
corn oil
50 mL acetone solution (such as nail polish remover), $CH_3CH(O)CH_{3(aq)}$
three 250-mL beakers
100-mL beaker
glass stirring rod
15 cm copper wire
cork or rubber stopper
tongs
paper towel
hot plate
Bunsen burner

Figure 1
The triangular symbols on plastic containers allow us to identify the kind of plastic from which they are made.

Procedure

Part 1: Testing for Density

1. Obtain one sample of each of the six plastic materials.

2. Place all six samples in a 250-mL beaker containing 100 mL of water, and stir with a stirring rod. Allow the samples to settle. Use tongs to separate the samples that float from the samples that sink. Dry each sample with a paper towel.

3. Prepare an alcohol solution by weighing out 60 g of 2-propanol (rubbing alcohol) in a 250-mL beaker and adding water to make a total of 100 g. Mix well.

 2-propanol is highly flammable, so it must be kept well away from open flames.

4. Take any samples that float in water, and place them in the alcohol solution. Stir, and then allow the samples to settle. Use tongs to separate the samples that float from the samples that sink. Dry each sample with a paper towel.

5. Take any samples that float in the alcohol solution, and place them in a 250-mL beaker containing 100 mL of corn oil. Stir, and then allow the samples to settle for a few minutes. Note any samples that sink.

Table 1 Codes on Everyday Plastics

SPI resin code	Structure	Density (g/cm³)	Properties	End products
1 **PETE** polyethylene terephthalate		1.38–1.39	transparent and strong; impermeable to gas and oils; softens at approximately 100°C	bottles for carbonated drinks; containers for peanut butter and salad dressings
2 **HDPE** high density polyethylene		0.95–0.97	naturally milky white in colour; strong and tough; readily moulded; resistant to chemicals; permeable to gas	containers for milk, water, and juice; grocery bags; toys; liquid detergent bottles
3 **PVC** vinyl (polyvinyl chloride)		1.16–1.35	transparent and tough; stable over a long time, not flammable; electrical insulator	construction pipes and siding; carpet backing and window-frames; wire and cable insulation; floor coverings; medical tubing
4 **LDPE** low density polyethylene		0.92–0.94	transparent, tough, and flexible; low melting point; electrical insulator	dry-cleaning bags; grocery bags; wire and cable insulation; flexible containers and lids
5 **PP** polypropylene		0.90–0.91	excellent chemical resistance; strong; low density; high melting point	ketchup bottles; yogurt and margarine containers; medicine bottles
6 **PS** polystyrene		1.05–1.07	transparent, hard, and brittle; poor barrier to oxygen and water vapour; low melting point; may be in rigid or foam form; softens in acetone	cases for compact discs; knives, forks, and spoons; cups; grocery-store meat trays; fast-food sandwich containers

Part 2: Testing for Flame Colour

6. Take the samples that sank in water, and test each one for flame colour in a fume hood. Use a 15-cm length of copper wire attached to a cork or rubber stopper. Holding the cork, heat the free end of the copper wire in a Bunsen burner flame until the wire glows. Touch the hot end of the copper wire to each sample so that a small amount melts and attaches to the wire. Heat the melted sample that is attached to the copper wire in a flame. Record the colour of the flame.

Part 3: Testing with Acetone

7. Ensure that all open flames are extinguished. Obtain a fresh sample of any material that did not burn with a green flame in step 6. Use tongs to test each sample for softness. Then place each of these fresh samples in a 100-mL beaker that contains 50 mL of nail polish remover (acetone solution). Watch the sample for a few minutes, and note any colour change. Remove the sample with tongs, and test it for increased softness.

Acetone is highly flammable and must be kept well away from open flames.

Part 4: Testing for Resistance to Melting

8. Heat a 250-mL beaker that is half-filled with water on a hot plate until the water comes to a rolling boil. Place any sample that remained unchanged in step 7 into the boiling water. Keep the water at a boil for a few minutes. Note any change in the shape and softness of the sample.

9. Dispose of the waste materials as directed by your teacher.

Observations

(a) Copy **Table 2**, on the next page, and complete it with your own observations.

Analysis

(b) From your observations in **Table 2** and the flow chart in **Figure 2**, identify and give possible SPI codes for each of the six samples you tested.

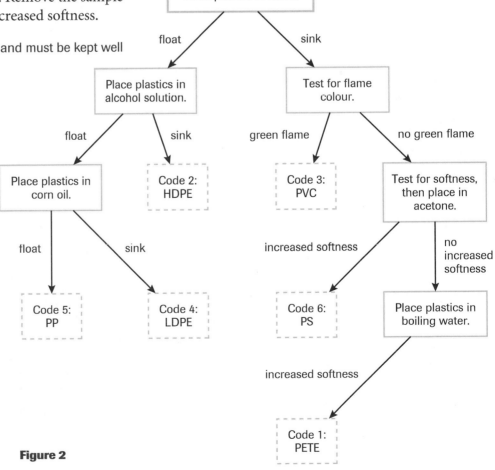

Figure 2

Organic Chemistry **235**

Table 2 Summary of Observations and Possible SPI Codes

Sample tested	Density	Flame colour	Acetone	Melting point	Possible SPI code
1					
2					
3					
4					
5					
6					

Evaluation

(c) Obtain the actual SPI resin codes for each sample from your teacher, and evaluate the reliability of your results. Suggest any changes to the Procedure that would improve the reliability of your results.

Synthesis

(d) Research the recycling operations in your school or local community. Find out what types of materials are collected, the amounts of each type, the participation rate in the program, problems that may be encountered in running a recycling program, and the destination of the materials collected. You may wish to use newspaper articles or Internet resources, visit a local recycling plant (**Figure 3**), or gather data about the recycling program at your school. Prepare and present a report on your findings.

 www.science.nelson.com

(e) Identify some issues related to the growing use of plastics, such as the consumption of fossil fuels and waste disposal. Suggest alternative materials that could be used instead of synthetic polymers such as polyester fabrics, plastic cutlery, Styrofoam cups, and disposable diapers.

(f) Working in small groups, design and make informative and attractive posters to increase awareness and participation in the recycling program at your school. Obtain permission to put up your posters around the school.

(g) Recycling only works if the substance, such as plastic, is made into a product and used again. Research at least three products that are made with recycled plastics. Create a table that includes the source of the "raw" material, the treatment it receives, and the cost per kilogram to make the new end product.

 www.science.nelson.com

(h) Select a career that you may be interested in pursuing in the plastic recycling industry. Find out the qualifications and training that are needed. Prepare a résumé (fictional) to apply for a position in your selected career. Include a cover letter with your résumé.

Figure 3
Plastics are sorted at a recycling facility.

Plastics belong to a group of substances called polymers. **Polymers** are large molecules that are made by linking together many smaller molecules, much like paper clips in a long chain. (In fact, the word *polymer* comes from two Greek words meaning "many parts.") Different types of small molecules form links in different ways. The types of small units and linkages can be selected to produce materials with desired properties, such as strength, flexibility, transparency, and chemical stability. As consumer needs change, new polymers are designed and manufactured.

Plastics are synthetic polymers, but there are many natural polymers as well. Our own cells manufacture complex carbohydrates (polymers of sugars), proteins (polymers of amino acids), and DNA (polymers of small units of genetic material).

plastic a synthetic substance that can be moulded (often under heat and pressure) and retains the shape it is moulded into

polymer a molecule that consists of many repeating subunits

Addition Polymers

Addition polymers result from the *addition reactions* of small subunits that contain double or triple carbon–carbon bonds. The small subunits that make up a polymer are called **monomers**. The monomers in a polymer may all be identical, or two or more monomers may occur in a repeating pattern.

Polyethene (commonly called polyethylene) is used for insulating electrical wires and for making plastic containers. Its monomers are ethene (ethylene) molecules that undergo addition reactions with other ethene molecules. As a result, the double bond in each ethene monomer becomes a single bond. The polymerizaton reactions may continue until thousands of ethene molecules have joined the chain.

addition polymer a polymer that is formed when monomer units are linked through addition reactions; all atoms in the monomer are retained in the polymer

monomer a molecule that is linked with other similar molecules to form a polymer

ethene   ethene   ethene   polyethene (polyethylene)

Similarly, propene undergoes addition polymerization to produce polypropene, commonly called polypropylene (**Figure 1**, on the next page). You may have used a polypropylene rope or walked on a polypropylene carpet. The propene molecule can be thought of as an ethene molecule with a methyl, CH_3, group attached. Therefore, the polymer that is formed contains a long carbon chain with methyl groups attached to every other carbon atom in the chain.

propene monomers   polypropene (polypropylene)

Figure 1
When the double bonds in propene molecules undergo addition reactions with other propene molecules, the polymer that is formed is structurally strong. This polymer, when used in polypropene ropes, has saved climbers' lives.

crosslinks covalent bonds that form between polymer chains

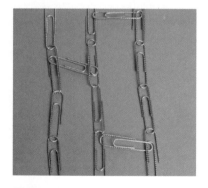

Figure 2
Crosslinks produce a more rigid structure.

Figure 3

Ethene molecules that have other substituted groups produce other polymers. For example, polyvinyl chloride (commonly known as PVC) is an addition polymer of chloroethene, C_2H_5Cl. PVC is used as insulation for electrical wires and drain pipes, and as a coating on raincoat and upholstery fabrics. You may recall performing a test for chlorine in Activity 3.17: Classifying Plastics.

$$\underset{\substack{\text{Cl} \qquad\quad \text{Cl} \qquad\quad \text{Cl}}}{\text{C}=\text{C}+\text{C}=\text{C}+\text{C}=\text{C}+} \longrightarrow \underset{\substack{\text{Cl} \qquad \text{Cl} \qquad \text{Cl}}}{-\text{C}-\text{C}-\text{C}-\text{C}-\text{C}-\text{C}-}$$

vinyl chloride monomers   polyvinyl chloride (PVC)

Properties of Addition Polymers

Plastics are addition polymers that are chemically unreactive. This property makes them ideal for use as containers for chemicals, solvents, and foods. Plastics are stable because the *double* bonds of the monomers have been changed to less reactive *single* bonds. In effect, the *unsaturated* alkene monomers have been transformed into less reactive *saturated* alkanes.

Plastics are generally flexible and mouldable. The long polymer chains are held together by numerous, but weak, intermolecular forces. These forces allow the polymer chains to slide along each other, making them flexible and stretchable. Heating increases the flexibility of plastics because the motion of molecules is increased. The increased motion further disrupts the forces between the molecules. Therefore, plastics can be softened and moulded by heating.

Some monomers have two double bonds. These monomers can therefore link to two separate polymer chains at the same time. As well as forming their regular polymer chain, they form strong covalent bonds, called **crosslinks**, between adjacent polymer chains. The crosslinks hold the adjacent polymer chains firmly together. The more crosslinks there are, the more rigid the plastic is (**Figure 2**).

> ▶ **TRY THIS** activity   ***Skewering Balloons***
>
> The polymers in synthetic rubber are strong, stretchy, and flexible. Just how much can you push them around? Try this!
>
> **Materials:** balloon, thin knitting needle or long bamboo skewer
>
> 1. Inflate a large round balloon. Release a little of the air until the balloon is not taut, and tie a knot at the neck.
> 2. Place the tip of the needle or skewer at the thick part of the balloon, directly opposite the knot. Slowly rotate the needle or skewer and push it into the balloon. Continue pushing the needle through the balloon and out through the thick area around the knot.
>
> (a) Explain your observations. **Figure 3** might help!

Condensation Polymers

Recall that carboxylic acids react with alcohols to form esters (section 3.13), and with amines to form amides (section 3.15). These reactions are called *condensation* reactions. When monomers join, end to end, to form ester or amide linkages, polymers called **polyesters** and **polyamides** are produced. Because polyesters and polyamides result from condensation reactions, these polymers are called **condensation polymers**.

To form a polyester or a polyamide, the monomer molecule must have *two functional groups*, one at each end of the molecule. The functional groups that meet end to end must be a carboxyl group (–COOH) and either a hydroxyl group (–OH) or an amine group (–NH$_2$). For example, nylon is a polyamide formed from two different monomers: one with a carboxyl group at each end, and the other with another amine group at each end. This arrangement allows an amide to form at each junction of monomers, producing long chains of nylon.

polyester a polymer that is formed by condensation reactions resulting in ester linkages between monomers

polyamide a polymer that is formed by condensation reactions resulting in amide linkages between monomers

condensation polymer a polymer that is formed when monomer units are linked through condensation reactions

$$HO - \overset{\overset{\displaystyle O}{\|}}{C} - C - C - C - C - \overset{\overset{\displaystyle O}{\|}}{C} - \boxed{OH + H}\overset{\overset{\displaystyle H}{|}}{N} - C - C - C - C - C - C - NH_2 + \longrightarrow$$

adipic acid   1,6-diaminohexane

$$-\left[\overset{\overset{\displaystyle O}{\|}}{C} - C - C - C - C - \overset{\overset{\displaystyle O}{\|}}{C} - N - C - C - C - C - C - C - N \right]_n - + H_2O_n$$

nylon 6,6

Nylon was designed to be a substitute for silk, so its structure is similar to the structure of silk. Nylon production was increased during the Second World War, when nylon was used to make parachutes, ropes, cords for aircraft tires, and even shoelaces for army boots.

You probably encounter many polyesters in your daily life. One of the most familiar polyesters is Dacron, which is found in clothing fabrics (**Figure 4**). One of the monomers that is used to produce Dacron has a carboxyl group at each end. The other monomer has a hydroxyl group at each end. An ester linkage is formed at each junction of the monomers, and long polyester chains are produced.

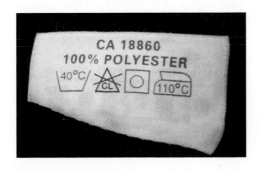

Figure 4
This clothing label indicates that the article of clothing is composed of polyester fabric.

Vinyl

Some unsaturated hydrocarbon groups have special names. For example, the ethene group is sometimes called a vinyl group. Many synthetic products that are commonly called "vinyl" are addition polymers of vinyl monomers with a variety of substituted groups.

vinyl

monomer of Saran wrap
(with vinyl chloride)

monomer of acrylic

monomer of instant glue

Nylon

Nylon was designed in 1935 by Wallace Carothers, a chemist who worked for DuPont. The name *nylon* is a contraction of **N**ew **Y**ork and **Lon**don, where research on nylon was done.

▶ **TRY THIS** activity

Making Nylon (Teacher Demonstration)

One of the early synthetic polymers, nylon, can be made in a lab. It is formed at the boundary of two *immiscible reactants* (reactants that do not mix), so it can be drawn out as a continuous thread.

Materials: eye protection; lab apron; protective gloves; 0.25-mol/L adipyl chloride dissolved in cyclohexane; 0.5-mol/L 1,6-diaminohexane dissolved in 0.5-mol/L sodium hydroxide solution, $NaOH_{(aq)}$; 15-cm length of copper wire; two 50-mL beakers; water

1. Place 10 mL of the 1,6-diaminohexane solution in one beaker.

   (!) 1,6-diaminohexane is toxic and an irritant to the eyes, skin, and respiratory system. Handle with caution, and wash hands thoroughly after this activity.

2. Tilting the beaker at an angle, slowly add 10 mL of the adipyl choride solution so that two distinct immiscible layers are formed.

   Cyclohexane is flammable and irritating to the respiratory system. Keep it away from flames, and do not inhale it.

   Sodium hydroxide is corrosive. Wear eye protection, a lab apron, and protective gloves. Avoid contact with bare skin.

3. Twist one end of the copper wire into a small hook. Use the hook to pull the nylon film that is formed at the boundary of the two liquid layers. Continue pulling to extract the nylon strand until it breaks.

4. Place the nylon in the other clean beaker. Rinse the nylon gently with water.

5. Examine the properties of the nylon: for example, its strength, flexibility, and solubility in different solvents.

6. Dispose of the remaining reactants and products as directed by your teacher.

Properties of Condensation Polymers

Intermolecular bonds play an important part in the properties of condensation polymers. Polyamide chains, such as nylon, have amine groups that can hydrogen-bond with the –C=O groups on other chains. As a result, polyamide chains form exceptionally strong fibres. Similarly, the strong attractive forces between polar groups in polyesters, such as Dacron, hold the separate polymer chains together, giving them considerable strength.

A polymer called Kevlar illustrates the effect of intermolecular bonds in condensation polymers. Kevlar has very special properties. It is stronger than steel and heat-resistant, yet it is lightweight enough to wear. Kevlar is used to make products such as aircraft parts, sports equipment, protective clothing for firefighters, and bulletproof vests for police officers.

What gives Kevlar these special properties? The polymer chains form a strong network of hydrogen bonds, which hold adjacent chains together in a sheet-like structure. The sheets are stacked together to form extraordinarily strong fibres. When woven together, these fibres are resistant to damage, even the damage caused by a speeding bullet (**Figure 5**).

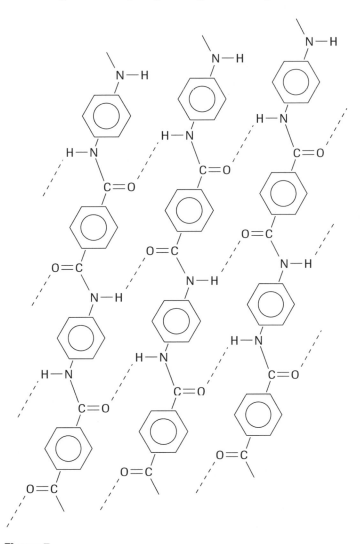

Figure 5
Kevlar is a synthetic fibre that is silky, soft, and light, yet stronger than steel—an extraordinary combination of properties. Chemical engineers have used their imagination and expertise to combine these properties to make products where lightweight strength is essential. Kevlar is used in bullet-proof vests, protective gloves, tires, racing shoes, fibre optics, and aircraft.

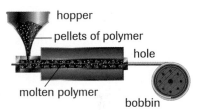

Understanding Concepts

1. What functional groups, if any, must be present in a monomer that undergoes an addition polymerization reaction?

2. (a) Describe the intermolecular forces of attraction among long addition polymer chains.
   (b) Why are these polymers more useful, as materials, than their monomers?
   (c) Why are these polymers more chemically stable than their monomers?

3. Describe a structural feature that is necessary in a monomer added for crosslinking between polymer chains. Include a structural formula to illustrate your answer.

4. What functional group(s) must be present in a monomer of a condensation polymer?

5. Describe the type of chemical bonding within a polyamide chain and the type of chemical bonding between adjacent polyamide chains.

6. (a) What are the typical properties of a plastic?
   (b) What types of bonding would you expect to find within and between the long polymer molecules?
   (c) Explain the properties of plastics by referring to their bonding.

7. State the difference in the structures of a polyester and a polyamide. Give an example of each.

8. Describe the role of each type of chemical bond in a polyamide:
   (a) covalent bonds
   (b) amide bonds
   (c) hydrogen bonds

Making Connections

9. (a) Research the molecular structures of the reactants that are used to make nylon in the activity on page 240. Use these molecular structures to write the reaction equation for the formation of nylon. Circle the reacting functional groups.
   (b) The name "nylon" is often followed by two numbers, as in "nylon 6,6." Find out what these numbers represent.

 www.science.nelson.com

Applying Inquiry Skills

10. Suppose that two new polymers have been designed and synthesized for use as potting materials for plants.
    (a) List and discuss the properties of two ideal polymers that could be used to hold and supply water for a plant over an extended period of time.
    (b) Design an experiment to test and compare the two polymers for the properties you listed. Write a brief description of the procedure you would follow, as well as possible interpretations of experimental results.

11. (a) Contact your community recycling facility to find out what types of plastic products are accepted for recycling in your area. If there are some plastics that are not accepted, find out the reasons.
    (b) Research the molecular structures and functional groups of at least five of these plastics.
    (c) Summarize your findings, including structural formulas, in a well-organized table.

 www.science.nelson.com

12. When household waste is deposited in a landfill site, some of it decomposes. The products of this decomposition may seep into the ground and contaminate water supplies. Work with a partner to complete the following tasks:
    (a) Brainstorm and list properties that would be ideal for a plastic liner for a landfill site, to contain the potentially toxic seepage.
    (b) Describe the general structural features of a polymer that would provide the properties you listed.

13. Natural rubber is made from a resin that is produced by the rubber tree, *Hevea brasiliensis*. Research the commercial production and use of natural rubber, and the circumstances that stimulated the development of synthetic rubber. Write a brief report on your findings.

 www.science.nelson.com

14. The "superabsorbency" of water by sodium polymethylacrylate is ideally suited to its use in baby diapers and other hygiene products. Suggest other applications for which this polymer would be useful.

3.19 The Nicotine Patch

The nicotine patch is one of the best-known and most effective aids available to help people quit smoking. It supplies the body with small doses of the addictive drug nicotine, to ease the body's adjustment to withdrawal from nicotine. This method for taking medication, through contact with the skin, is called transdermal drug delivery. The prefix *trans* means "across," and *dermal* means "the skin."

One of the first patches that was developed was the nitroglycerin patch, in 1981. Before then, scientists generally believed that, although the skin allows sweat and other secretions to leave the body, it acts as a barrier against the *entry* of any substances into the body. During the Second World War, however, soldiers who were prone to angina (health problems associated with the circulatory system) and worked with the explosive nitroglycerin found that the frequency of their angina attacks decreased. This observation prompted the development of nitroglycerin ointments and, later, a patch that is worn like a small adhesive bandage.

The design of the patch itself is very simple. A measured dose of the medication is either held in a reservoir or embedded in a polymer matrix. The bottom of the patch has an adhesive layer that allows the medication to pass through, to the skin. The outside of the patch is a waterproof layer to keep the medication in place (**Figure 1**).

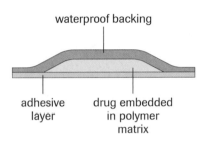

waterproof backing

adhesive layer

drug embedded in polymer matrix

Figure 1
A skin-controlled transdermal drug delivery patch

With advances in organic chemistry, new polymers can be designed to act as a strong yet porous adhesive, as a matrix to control the release of the medication, and as a comfortable and protective outer backing.

Studies of skin cells reveal that, because the skin has both water-soluble and fat-soluble components, any drug molecule that travels through the skin must be a suitable size and must be soluble in both polar and nonpolar solvents.

The major advantage of a patch is the immediate entry of the medication into the bloodstream, without passing through the stomach where it may be broken down by digestion. A patch is easy to apply and pain-free—a welcome alternative to injections. On the other hand, the dosage that is absorbed varies, depending on the individual skin type. As well, the adhesive and the moist conditions under the patch often cause skin irritations over an extended period of use, and some people may have allergic reactions to one of the components of the patch.

Other medications that are available in patches include scopolamine for motion sickness and testosterone for hormone therapy, as well as painkillers, vitamins, and hormonal contraceptives.

▶ **Tech Connect 3.19** *Questions*

Understanding Concepts

1. (a) What properties are necessary for a drug to be a candidate for transdermal delivery?
   (b) What functional groups are present in the nicotine molecule that allows it to be used in a patch (**Figure 2**)? Explain.

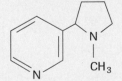

Figure 2
Nicotine molecule

Making Connections

2. Research other examples of organic chemistry leading to the development of useful new medical products, such as gel capsules, artificial skin, artificial heart valves, synthetic joints, absorbable sutures, and denture materials. Prepare a one-page report on the structure, properties, and development of each product you research.

 www.science.nelson.com

3. Several other medications are available in the form of patches. (See examples above.) Research the advantages or disadvantages of a skin patch as a delivery method for *one* of these medications. Discuss any other preferable delivery methods for this medication, and give reasons.

 www.science.nelson.com

3.20 Activity

Making Polymers

Part 1: Making Guar Gum Slime—A Crosslinked Polymer

Guar gum, a vegetable gum that is derived from the guar plant, has a molar mass of about 220 000 to 250 000 g/mol. It has many uses: as a stabilizer, thickener, and film-forming agent for cheese; in salad dressings, ice creams, and soups (**Figure 1**); as a binding and disintegrating agent in tablet formulations; and in suspensions, emulsions, lotions, creams, and toothpastes. In short, guar gum is a very useful polymer.

In this activity, you will make "slime" by creating a reversible crosslinked gel made from guar gum (**Figure 2**). The crosslinking is accomplished by adding sodium borate (commonly called borax), $Na_2B_4O_7 \cdot 10\ H_2O_{(s)}$.

Materials

eye protection
lab apron
protective gloves
guar gum
water
saturated sodium borate (borax) solution,
$\quad Na_2B_4O_7 \cdot 10\ H_2O_{(aq)}$
vinegar, $CH_3COOH_{(aq)}$
100-mL graduated cylinder or measuring spoons
balance or measuring spoons
Popsicle stick or glass rod, for stirring
glass or disposable cup, or beaker
food colouring (optional)
sealable plastic bags for storing slime
small funnel
funnel support

Procedure

1. Measure 80 mL of water into the cup.

2. Add one to two drops of food colouring, if desired.

3. Measure 0.5 g of guar gum ($\frac{1}{8}$ tsp). Add it to the water, and stir until dissolved. Continue stirring until the mixture thickens (approximately 1 to 2 min).

Figure 1
The properties of polymers are useful in many different fields, including the food sciences.

Figure 2
Choose your own colour, and have fun!

 Guar gum, if not food grade, is not safe to taste.

4. Add 15 mL (3 tsp) of saturated borax solution, and stir. The mixture will gel in 1 to 2 min. Let the slime sit in the cup for a few minutes to gel completely. (To store the slime for more than a few minutes, put it in a plastic bag and seal the top of the bag.)

Sodium borate is moderately toxic in quantities of more than 1 g/1 kg of body mass. Wear protective gloves. Wash your hands after handling the slime.

Slime may stain or mar clothing, upholstery, or wood surfaces. Clean up spills immediately by wetting with vinegar, followed by soapy water.

(a) Describe what happens to the slime when you pull it slowly.

(b) Describe what happens to the slime when you pull it quickly.

5. Put some slime on a smooth, hard surface, and hit it with your hand.

(c) Describe what happens to the slime.

6. Place a funnel on a funnel support. Put some slime into the funnel. Push the slime through the funnel.

(d) Describe what happens as the slime comes out of the hole in the funnel.

7. Take about 20 mL of the slime, and add about 5 mL of vinegar.

(e) Describe any changes in the properties of the slime.

8. Dispose of the slime as directed by your teacher.

Analysis

(f) Use your knowledge of the general structure and bonding of polymers to explain each of your observations.

Synthesis

(g) Other substances that are used to thicken food include cornstarch, gelatin, and carrageenan. Research the sources of these substances and their molecular structures. Suggest reasons why these substances have similar properties.

 www.science.nelson.com

Part 2: Making Glyptal—A Polyester

The reaction between an alcohol that has two or more hydroxyl groups (such as glycerol) and an acid that has two or more carboxyl groups produces a polyester

known as an alkyd resin. One alkyd resin is Glyptal, which is commonly used to make paints and enamels.

Glyptal is a thermoset plastic, meaning that it solidifies or "sets" irreversibly when it is heated and cooled. The heating causes a crosslinking reaction between the polymer molecules. In Glyptal resin, orthophthalic acid (from phthalic anhydride) forms crosslinks with several glycerol molecules, holding them together in the polymer (**Figure 3**).

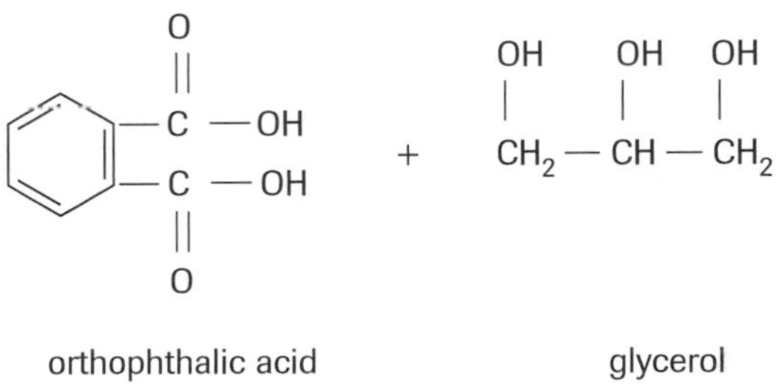

orthophthalic acid   glycerol

Figure 3
Phthalic anhydride is similar to orthophthalic acid, with one H_2O group removed.

Question

What are the properties of a polyester made from the condensation reaction between two monomers: one with three hydroxyl groups and one with two carboxyl groups?

Materials

eye protection
lab apron
protective gloves
2 g glycerol
3 g phthalic anhydride powder
5 mL nonpolar solvent (paint thinner or nail polish remover)
two 100-mL beakers
glass stirring rod
beaker tongs
watch glass to fit beaker
hot plate
small metal container (such as an aluminum pie dish or a glass filter funnel lined with aluminum foil)

Procedure

9. Place 2 g of glycerol and 3 g of phthalic anhydride in a 100-mL beaker. Cover the beaker with a watch glass. Heat the beaker gently on a hot plate until the mixture boils, and then boil gently for 5 min.

   (T) Phthalic anhydride is toxic and a skin irritant. Handle it with care, and wear eye protection, a lab apron, and protective gloves.

10. Using beaker tongs, carefully pour the solution into a metal container. Let the plastic cool completely at room temperature.

(h) Observe and record the properties of the plastic formed. The trade name for this plastic is Glyptal.

11. Place about 5 mL of cold water in a beaker. Try to dissolve a piece of the plastic in the water.

12. Remove the plastic from the water, and dry it.

13. Heat approximately 100 mL of water in a beaker until it is boiling. Remove the beaker from the heat. Place the plastic in the hot water for 2 min.

14. Remove the plastic with the tongs, dry it, and examine it for flexibility and elasticity. Record your observations.

15. Place about 5 mL of paint thinner or nail polish remover in a beaker. Try to dissolve a piece of the plastic in the solvent.

   (flame icon) The solvent is flammable, so it must be kept well away from any open flame.

16. In a fume hood, allow the solvent containing any dissolved plastic to evaporate. Observe any residue that is formed.

17. Dispose of the materials according to your teacher's instructions.

Analysis

(i) Summarize your observations of the properties of the polyester Glyptal.

Synthesis

(j) What makes this plastic suitable for use in household paints?

(k) What other types of consumer products would this plastic be suitable for?

(l) Using molecular structures, explain how each of the properties you observed can be explained by the functional groups of the plastic.

Have you ever felt the irritation of having a small eyelash or a tiny dust particle in your eye? Now imagine placing a circular plastic disk, nearly a centimetre in diameter, onto the cornea of each eye (**Figure 1**). You may be a contact lens wearer who does so daily, hardly aware of the intrusion except for the fact that you can see much better (**Figure 2**). Since the 1960s, when contact lenses became widely available, there have been many improvements in the materials used, and the developments are continuing. Patent lawyers submit many new patent applications each month, but only a few materials pass the rigorous tests of health agencies and manufacturers.

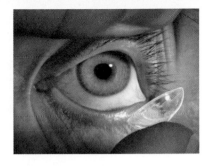

Figure 1
The rapid development of polymer technology has produced a large object that sits on the surface of your eyeball.

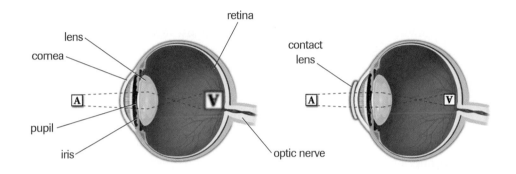

Figure 2
If the natural lens cannot bring the images of distant objects into focus on the retina, a contact lens can solve the problem.

Hard Contact Lenses

The first contact lenses were made in 1887. They were manufactured from glass and were designed to cover the entire eye. These contact lenses were replaced by plastic lenses in 1938. Over the next 10 years, the size of the lenses was changed to cover only the cornea.

The material that was used in these hard lenses was a plastic called PMMA (polymethylmethacrylate). This plastic was firm and uncomfortable to wear. A greater concern was the finding, in the mid-1970s, that the lenses did not allow much oxygen to reach the eye. The cornea does not contain blood vessels, and all of its oxygen is obtained from a film of tears on its surface. Without oxygen, the cornea swells and may develop small cysts, becoming more susceptible to infection. The old PMMA hard lenses are now obsolete. They have been replaced by soft contact lenses and by new hard lenses that allow oxygen to enter the eye.

Soft Contact Lenses

More comfortable soft contact lenses were introduced in 1971. Soft contact lenses are made of a material known as polyHEMA: a synthetic polymer of a monomer called HEMA (**Figure 3**). This polymer swells in water to form a soft and flexible hydrogel. Crosslinking between polymer strands gives the lens elasticity, a property that provides a comfortable fit for the lens wearer.

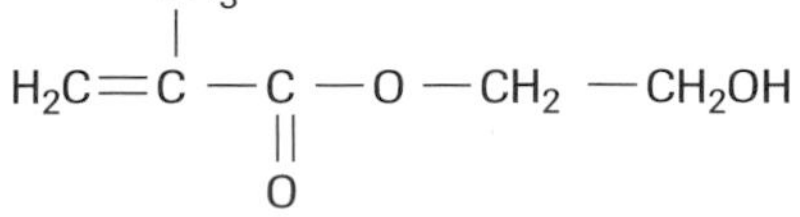

$$H_2C = C - \underset{\underset{O}{\|}}{C} - O - CH_2 - CH_2OH$$

with CH_3 on the second carbon.

Figure 3
Structure of HEMA
(2-hydroxyethylmethylacrylate)

Even polyHEMA isn't ideal, though, when made into contact lenses. Its elasticity allows the lenses to change their shape each time the eyelids blink, for better comfort. Because they cannot fully recover from these numerous daily deformations, the lenses perform poorly in correcting vision. Another problem is the water content. The higher the water content in the lenses, the more oxygen that becomes available to the cornea, which is good. With more water, however, the degree of bending of light decreases, so thicker lenses are needed to achieve the prescription requirements.

Rigid Gas-Permeable Lenses

While soft lenses were being developed, the race began for the best material to make hard lenses that were gas permeable (that allowed gases, such as oxygen, to pass through). In 1978, new lenses, commonly called rigid gas permeables (RGPs), were introduced. The first material that was used was cellulose acetate butyrate, a polymer whose structure allowed more gases to reach the eyes. This material was not highly stable, however, and also tended to form deposits of proteins and lipids on the surface of the lenses. Manufacturers of the newer RGPs claim that, since their lenses do not contain water, they are easier to handle, are not prone to buildup of deposits, and last longer than soft lenses. The goal for the future is to develop lenses that are gas permeable, rigid, and easy to maintain, have good optical properties, and can be worn for long periods of time. Research is ongoing in this field, and the world of polymer chemistry offers endless possibilities.

▶ Case Study 3.21 *Questions*

Understanding Concepts

1. Explain how crosslinking gives the material used to make a soft contact lens elastic properties.

Making Connections

2. Do you agree that "the world of polymer chemistry offers endless possibilities"? Make a list of your reasons for agreeing or disagreeing.

3. (a) Compare the features, good and bad, of each type of contact lenses: hard lenses, soft lenses, and rigid gas-permeable lenses.

   (b) Use your comparison to illustrate how organic chemistry has contributed to improvements in the field of vision correction and eye care.

4. Speculate on future developments for contact lenses. What properties might be desirable for contact lenses? How might these properties be achieved?

5. (a) With a partner or in a small discussion group, brainstorm and list all the desired properties of a material that could be used to manufacture contact lenses.

   (b) From what you have learned in this unit, suggest some types of molecules that may provide some of the properties you listed in (a).

Extension

6. Research and provide examples of the use of organic chemistry to solve technical problems in the medical field. Write a report on one of the problems you find, including a description of the problem to be solved, existing technical solutions, and the role of organic chemistry in an improved solution. As a starting point for your research, consider:
   - drug delivery systems, such as estrogen patches and gel capsules for timed release
   - artificial flexible joints
   - medical textiles, such as adhesives
   - medical equipment, such as materials for angioplasty
   - polymers as UV blockers

 www.science.nelson.com

Key Understandings

3.1 Hydrocarbons
- Demonstrate an understanding of the characteristics of the carbon atom in the formation of long chains.
- Draw Lewis structures to represent covalent bonding in organic molecules.
- Identify double and triple bonds, and describe how they explain the properties of compounds.
- Experimentally determine some of the products of burning hydrocarbons.

3.2 Activity: Building Molecular Models
- Build molecular models of some alkanes, alkenes, and alkynes, and represent these compounds using structural formulas.

3.3 Fractional Distillation and Cracking
- Describe how organic compounds can be separated by distillation.
- Explain how distillation and cracking enable useful substances to be obtained from petroleum.
- Experiment to discover some physical properties of motor oils.

3.4 Investigation: Separating a Mixture by Distillation
- Apply the principle underlying the use of distillation to separate organic compounds.
- Select and use apparatus safely to separate a mixture of liquids by distillation.

3.5 Functional Groups
- Describe the relationship between functional groups and the observable properties of compounds.

3.6 Explore an Issue: The Cost of Your Cold Drink
- Identify organic compounds that are used for refrigeration, and collect information on the benefits and drawbacks of using these substances.

3.7 Alcohols and Ethers
- Identify the functional groups of alcohols and ethers.
- Explain the properties of alcohols and ethers by referring to their functional groups.

3.8 Investigation: Properties of Alcohols
- Investigate some physical and chemical properties of alcohols, and identify patterns and trends in your observations.
- Identify some of the products of the combustion of alcohols and alkanes, and write balanced chemical equations to represent the combustion reactions.

3.9 Aldehydes and Ketones
- Identify the functional groups of aldehydes and ketones.
- Explain the properties of aldehydes and ketones by referring to their functional groups.
- Observe the effects of propane (acetone) as an organic solvent, and be aware of the necessary precautions to take when handling such a solvent.

3.10 Safe Use of Organic Solvents
- Explain the dangers associated with the use of organic solvents, and identify the precautions that need to be taken.
- Research the career of firefighter, including personal characteristics, necessary training, and starting salary.

3.11 Carboxylic Acids
- Identify the functional group of carboxylic acids.
- Explain the properties of carboxylic acids by referring to their functional group.
- Safely use a carboxylic acid to make a bath bomb.

3.12 Investigation: Properties of Carboxylic Acids
- Determine, through experimentation, the physical and chemical properties of acetic acid and stearic acid, and identify patterns and trends in your observations.

3.13 Esters
- Identify the functional group of esters.
- Explain the properties of esters by referring to their functional group.
- Use structural formulas to describe the condensation reaction by which esters are formed.

3.14 Activity: Synthesis of Esters

- Synthesize an ester by a condensation reaction, and determine the ester's properties.

3.15 Amines and Amides

- Identify the functional groups of amines and amides.
- Explain the properties of amines and amides by referring to their functional groups.

3.16 Explore an Issue: Regular or Diet?

- Identify artificial sweeteners as organic compounds synthesized for a specific purpose.
- Analyze the risks and benefits of using a sugar substitute.

3.17 Activity: Classifying Plastics

- Experimentally investigate some physical and chemical properties of a variety of plastics.
- Appreciate that many new organic compounds, such as plastics, have been developed for specific purposes.
- Describe how the development of plastics has brought both benefits and drawbacks.
- Research plastic recycling.

3.18 Polymers

- Use structural formulas to describe the reactions that result in the formation of a variety of polymers.
- Describe several polymers and the purposes for which they were developed.

3.19 Tech Connect: The Nicotine Patch

- Explain how the characteristics of various polymers make them suitable for transdermal drug delivery.

3.20 Activity: Making Polymers

- Synthesize two polymers: guar gum slime and Glyptal.
- Account for the properties of these polymers by referring to their molecular structures.

3.21 Case Study: Contact Lenses

- Describe how polymers with various properties have been developed as contact lens materials.
- Understand that the development of a product is often an ongoing process, with each new compound bringing a new balance of benefits and drawbacks.

Problems You Can Solve

3.1

- Draw structural diagrams of alkanes, alkenes, and alkynes.
- Write combustion and addition reactions of hydrocarbons.

3.7

- Name and draw alcohols.

Key Terms

3.1
hydrocarbon
alkane
alkene
alkyne
structural isomer
addition reaction
unsaturated
saturated

3.3
petroleum
fractional distillation
cracking

3.5
organic family
functional group
hydrogen bond

3.7
alcohol
ether
alkyl group

3.9
ketone
aldehyde
carbonyl group

3.10
flammable liquid
combustible liquid
flashpoint
spontaneous combustion

3.11
carboxylic acid
carboxyl group

3.13
ester
esterification
hydrolysis

3.15
amine
amide

3.16
carbohydrate
amino acid

3.18
plastic
polymer
addition polymer
monomer
crosslinks
polyester
polyamide
condensation polymer

▶ *MAKE* a summary

(a) If you have not already done so as you studied this unit, prepare a large index card for each organic family. On the front of each card,

- write the name and general formula of the organic family;
- give an example, including the IUPAC name, common name, and structural formula.

On the back of each card,

- list the characteristic properties, such as relative boiling points and solubility in polar and nonpolar solvents;
- name the characteristic functional groups;
- list the types of intermolecular forces.

(b) Arrange your index cards in different sequences or groupings: for example, by general trends of increasing boiling points or increasing solubility in polar solvents; by functional groups; by presence or absence of oxygen atoms; or by reaction type. (The families of organic compounds are listed in **Table 1** on the following page.)

Table 1 Families of Organic Compounds

Family name	General formula	Example	Structural formula
alkanes	$-\overset{\mid}{\underset{\mid}{C}}-\overset{\mid}{\underset{\mid}{C}}-$, C_nH_{2n+2}	propane	$CH_3 - CH_2 - CH_3$
alkenes	$-\overset{\mid}{C}=\overset{\mid}{C}-$, C_nH_{2n}	propene (propylene)	$CH_2 = CH - CH_3$
alkynes	$-C\equiv C-$, C_nH_{2n-2}	propyne	$CH \equiv C - CH_3$
alcohols	$R - OH$	1-propanol	$CH_3 - CH_2 - CH_2 - OH$
ethers	$R - O - R'$	methoxyethane (ethyl methyl ether)	$CH_3 - O - CH_2 - CH_3$
aldehydes	$R - \overset{O}{\overset{\|}{C}} - H$	propanal	$CH_3 - CH_2 - \overset{O}{\overset{\|}{C}} - H$
ketones	$R - \overset{O}{\overset{\|}{C}} - R'$	propanone (acetone)	$CH_3 - \overset{O}{\overset{\|}{C}} - CH_3$
carboxylic acids	$R - \overset{O}{\overset{\|}{C}} - OH$	propanoic acid	$CH_3 - CH_2 - \overset{O}{\overset{\|}{C}} - OH$
esters	$R - \overset{O}{\overset{\|}{C}} - O - R'$	methyl ethanoate (methyl acetate)	$CH_3 - \overset{O}{\overset{\|}{C}} - O - CH_3$
amines	$R - \overset{R'}{\overset{\mid}{N}} - R''$	propylamine	$CH_3 - CH_2 - CH_2 - \overset{H}{\overset{\mid}{N}} - H$
amides	$R - \overset{O}{\overset{\|}{C}} - \overset{R''}{\overset{\mid}{N}} - R'$	propanamide	$CH_3 - CH_2 - \overset{O}{\overset{\|}{C}} - \overset{H}{\overset{\mid}{N}} - H$

Note: R refers to alkyl groups or H.

Making Soap

Fats and oils are esters formed from the esterification reaction between long-chain carboxylic acids (fatty acids) and a three-carbon alcohol (glycerol). When the ester linkages in a fat or oil are broken, and a sodium salt of the fatty acid is formed, soap is the product (**Figure 1**, on the next page). The other product is glycerol. The reaction requires heating and the presence of a strong base, such as sodium hydroxide. The word equation for the reaction is

| fat or oil | + | sodium hydroxide | → | soap | + | glycerol |
| (ester) | + | (base) | → | (Na⁺ salt of the acid) | + | (alcohol) |

As you carry out this activity, you will be assessed on your laboratory techniques, your safety procedures in handling and disposing of materials, and your use of equipment. You will also be evaluated on the ability of your soap to produce a lather in distilled water.

Materials

eye protection
lab apron
15 g fat (lard or vegetable shortening)
15 g oil (cooking oil, such as corn oil, canola oil, or olive oil)
18 pellets of sodium hydroxide, $NaOH_{(s)}$
15 mL ethanol
vinegar
4 g sodium chloride, $NaCl_{(s)}$
distilled water
food colouring and perfume (optional)
250-mL beaker
two 100-mL beakers
forceps
glass stirring rods
beaker tongs
filter funnel and paper
ring stand
ring clamp
hot plate
balance
mould for soap (optional)

Procedure

1. Refer to **Table 1**, on the next page, for a summary of the contents of the beakers used in this activity.

2. Set up a 100-mL beaker, and label it "A." Using forceps, add 18 pellets of solid sodium hydroxide to beaker A. Do not allow the pellets to touch your skin. Add 10 mL of distilled water to the sodium hydroxide pellets, and stir with a glass rod to dissolve. Set this beaker aside.

Criteria

Assessment
Your completed task will be assessed according to the following criteria:

Process
- Follow safe laboratory practices for handling and disposing of materials.
- Use appropriate personal protection in the laboratory.
- Select appropriate instruments, and use them effectively and accurately.

Product
- Hand in your soap, which should produce a lather when tested.
- Write answers to (a) to (j).

Table 1 Contents of Beakers A, B, and C

Beaker	A (100 mL)	B (250 mL)	C (100 mL)
Contents	18 pellets of NaOH$_{(s)}$ 10 mL distilled water	15 g fat or oil 15 mL ethanol	4 g NaCl 20 mL cold distilled water
How to dissolve contents	Stir to dissolve.	Warm gently on a hot plate to dissolve.	Stir to dissolve.

 Sodium hydroxide pellets, NaOH$_{(s)}$, are extremely corrosive to eyes and skin. They must be handled with forceps. If they come in contact with skin, rinse with copious amounts of cold water. If the solution is splashed in the eyes, flush with water at an eyewash station for at least 10 min, and then get medical attention.

3. Set up a 250-mL beaker, and label it "B." Add 15 g of a fat (such as lard or shortening) or an oil (such as corn oil or olive oil) to beaker B. Add 15 mL of ethanol to the fat or oil. Warm the mixture very gently on a hot plate, stirring with a glass rod to dissolve.

 Ethanol is highly flammable. Ensure that there are no open flames.

4. Pour the contents of beaker A into beaker B. Heat the mixture gently on the hot plate (low setting). Stir the mixture continuously for at least 25 min. If the mixture bubbles or splatters, use tongs to remove the beaker from the hot plate. Then, when the beaker has cooled slightly, return it to the hot plate.

5. When the reaction is complete, the mixture should thicken and have the appearance and consistency of creamy pudding. Remove the beaker from the heat, and allow it to cool. You may add a drop or two of food colouring to the mixture at this stage.

6. Set up another 100-mL beaker, and label it "C." Add 4 g of sodium chloride and 20 mL of cold distilled water. Stir to dissolve.

7. Add the cold sodium chloride solution from beaker C to the soap mixture in beaker B. Sodium chloride solution should cause the soap to precipitate from the solution.

8. Add 10 mL of vinegar to the mixture to neutralize any excess sodium hydroxide. Pour off any liquid into the sink.

9. Add 10 mL of distilled water to wash the excess vinegar off the soap. Pour off any liquid into the sink.

10. If desired, add a few drops of perfume or scent to the soap at this stage.

11. Set up a filter funnel and filter paper. Pour the soap mixture into the funnel, taking care not to puncture the filter paper. The soap will remain on the filter paper in the funnel.

Figure 1
Soap-making was a yearly event undertaken by early Canadian settlers. Waste cooking grease and wood ashes (which contain bases such as sodium hydroxide) were collected year-round. When enough had been collected, they were boiled together in quantities determined by trial and error, with uncertain results. Advances in the industrial production of soaps led to great improvements in personal hygiene and public health.

12. The remaining soap can be left on the filter paper to dry, or it can be taken out of the filter paper, shaped, and left to dry on a paper towel.

13. Hand in your soap to your teacher for evaluation. Your teacher will test the ability of your soap to lather in distilled water.

 Do not use this soap for washing because it may contain residual sodium hydroxide.

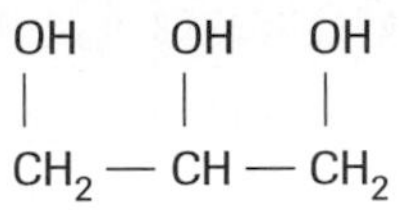

Analysis

(a) The structure of glycerol is shown in **Figure 2**. Identify its functional groups, the organic family to which it belongs, and its intermolecular forces.

(b) The structure of stearic acid, one of the fatty acids in lard, is also shown in **Figure 2**. Identify its functional groups, and explain its low solubility in water.

(c) One molecule of glycerol reacts with three molecules of stearic acid to produce one molecule of ester. What other product is formed? What is the name of this type of reaction? Write a balanced structural formula equation for this reaction.

(d) Vinegar was used to rinse excess aqueous sodium hydroxide off the soap. Draw a structural formula for the active ingredient in vinegar. Write a balanced chemical equation for the reaction between this ingredient and aqueous sodium hydroxide.

Evaluation

(e) How successful was this Procedure for making soap? Suggest any necessary improvements.

Synthesis

(f) Predict whether the boiling point of glycerol is higher or lower than the boiling point of 1-propanol. Explain your answer in terms of intermolecular forces.

(g) Refer to the structure of glycerol to predict whether it is soluble in polar and/or nonpolar solvents. Give reasons for your prediction.

(h) What products would be formed if glycerol were allowed to burn? Write a balanced chemical equation for the combustion reaction of glycerol.

(i) Ethanol, a flammable solvent, was used in this activity to dissolve the fat or oil. Describe five important precautions to take when handling, storing, and disposing of flammable organic solvents.

(j) Soaps are made from natural fats and oils, but detergents are synthetic compounds made from petroleum products. Research the composition and use of detergents. Write a paragraph to present arguments for and against the use of detergents.

 www.science.nelson.com

$$OH \quad OH \quad OH$$
$$| \qquad | \qquad |$$
$$CH_2 — CH — CH_2$$

glycerol

$$CH_3 — (CH_2)_{16} — \overset{\displaystyle}{\underset{\displaystyle O}{C}} — OH$$

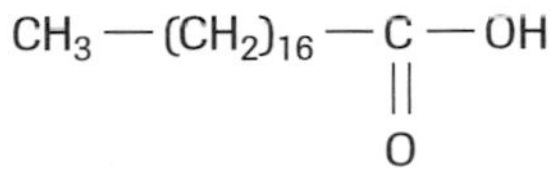

stearic acid

Figure 2
The two major components of soap

Understanding Concepts

1. Draw a structural formula for each of the following organic compounds:
   (a) acetylene (ethyne)
   (b) rubbing alcohol (2-propanol)
   (c) acetone (propanone)
   (d) acetic acid (ethanoic acid)   (3.1)

2. (a) Explain how the two technologies of fractional distillation and cracking are both important for producing sufficient quantities of gasoline to meet demand.
   (b) Identify three useful fuels, other than gasoline, that come from petroleum.   (3.3)

3. Arrange the following compounds in order of increasing boiling points. Give reasons for your order.

CH_3COOH $CH_3CH_2CH_2CH_3$
   A   **C**

CH_3CH_2CHO $CH_3CH_2CH_2OH$
   B   **D**   (3.9)

4. Give the IUPAC name and draw the structural formula for the compound with each of the following common names:
   (a) acetone
   (b) acetic acid
   (c) formaldehyde
   (d) glycerol
   (e) diethyl ether   (3.11)

5. For each of the following pairs of compounds, predict the relative boiling points of the two compounds by arranging the compounds in order of increasing boiling point. Give reasons for your order.
   (a) CH_3CH_2OH and $CH_3CH_2CH_2CH_2CH_2OH$
   (b) $C_2H_5-O-C_2H_5$ and $C_2H_5-\underset{\underset{O}{\|}}{C}-C_2H_5$

   (c) $\underset{\underset{CH_3CH}{\|}}{O}$ and $\underset{\underset{CH_3COH}{\|}}{O}$ (3.12)

6. For each of the following pairs of compounds, predict the relative solubility of the two compounds in water by arranging the compounds in order of increasing solubility in water. Give reasons for your order.
   (a) propane and 1-propanol
   (b) CH_3COOH and CH_3COOCH_3
   (c) 2-butanol and 2-butanone   (3.13)

7. Write the structural formula for each of the following compounds:
   (a) 2-butanol
   (b) ethoxyethane
   (c) 2-butanone
   (d) ethanoic acid
   (e) methyl methanoate   (3.13)

8. When esters are prepared in the reaction of an alcohol with a carboxylic acid, the product formed can often be separated from the reactants by cooling the reaction mixture. Is the solid (formed when the reaction mixture is cooled) the alcohol, the acid, or the ester? Explain your answer with reference to the molecular structure of each reactant and product.   (3.14)

9. Identify the functional group of each of the following compounds and the organic family to which the compound belongs:
   (a) $CH\equiv CCH_2CH_3$
   (b) 1-propanol
   (c) CH_3CH_2COOH
   (d) hexanal
   (e) $CH_3CH_2OCH_2CH_2CH_3$
   (f) CH_3NH_2
   (g) 2-pentanone
   (h) propyl ethanoate
   (i) $CH_3CH_2CONHCH_3$
   (j) $CH_3CH(CH_3)\underset{\underset{O}{\|}}{C}CH_2CH_3$

   (k) $H-\underset{\underset{OH}{|}}{C}=O$

   (l) $CH_2=CHCH_2CH_3$
   (m) $CH_3-O-CH_2CH_3$
   (n) $CH_3CH_2\underset{\underset{CH_3}{|}}{CH}CH_2OH$

   (o) $CH_3C\equiv CH$
   (p) $CH_3CH_2CH_2COOCH_3$

(q)

$$CH_3CH_2\overset{\displaystyle O}{\overset{\displaystyle \|}{CH}}$$

(r)

$$CH_3\overset{\displaystyle O}{\overset{\displaystyle \|}{C}}CH_2CH_3$$

(s) $CH_3CH_2CH_2CH_2NH_2$

(t) $CH_2 = CHCHCHCH_2CH_3$
     $\underset{OH}{|}\ \underset{OH}{|}$

(u)

$$CH_3CH_2CH_2\overset{\displaystyle O}{\overset{\displaystyle \|}{C}}NHCH_2CH_3$$

(v) $CH_3–CH = CH–CH_2–CH =$
      $CH–CH_2–CH = CH_2$ (3.15)

10. Many organic compounds have more than one functional group in a molecule. Copy each of the following structural diagrams. Circle and label the following functional groups: hydroxyl, carboxyl, carbonyl, ester, amine, and/or amide.

(a)

(b)

(c)

(d)

$$H_2N-\overset{\displaystyle O}{\overset{\displaystyle \|}{C}}-NH_2$$

(e)

$$HO-\overset{\displaystyle O}{\overset{\displaystyle \|}{C}}-CH_2-\underset{\underset{\displaystyle OH}{\overset{\displaystyle |}{\underset{|}{C=O}}}}{\overset{\displaystyle OH}{\overset{\displaystyle |}{C}}}-CH_2-\overset{\displaystyle O}{\overset{\displaystyle \|}{C}}-OH$$

            (3.15)

11. Write a complete structural formula equation for each of the following reactions. Name and classify the chemicals. Where possible, classify the reaction.
   (a) $C_3H_6 + Cl_2 \rightarrow C_3H_6Cl_2$
   (b) $C_3H_7COOH + CH_3OH \rightarrow C_3H_7COOCH_3 + HOH$
   (c) $C_6H_5CH_3 + O_2 \rightarrow CO_2 + H_2O$
   (d) $NH_3 + C_4H_9COOH \rightarrow C_4H_9CONH_2 + H_2O$
             (3.15)

12. Write an equation to illustrate each of the following reactions, using a simple example:
   (a) the complete combustion of ethanol
   (b) an addition reaction of an alkene to produce an alcohol
   (c) a condensation reaction of an amine (3.15)

13. Intermolecular bonds between polymer strands contribute to the elastic properties of a polymer.
   (a) Explain briefly why intermolecular bonds increase the elasticity of a polymer.
   (b) What structural features of a monomer are needed for intermolecular bonds to form between polymer chains? (3.18)

Applying Inquiry Skills

14. Design a procedure to separate a mixture that contains methanol, ethanol, and 1-pentanol. Explain, with reference to intermolecular forces, why your procedure is effective for separating this mixture. (3.8)

15. Suppose that you are the safety technician in an organic chemistry laboratory. Write a list of safety guidelines for handling, storing, and disposing of organic solvents. In your list, refer to combustibility and toxicity. (3.10)

16. Suppose that you are given an unlabelled colourless liquid and an unlabelled white solid. Each compound may be an alcohol, a short-chain carboxylic acid, or a long-chain carboxylic acid. Describe several tests that you could perform to classify the two compounds. Explain how the results of your tests would lead to the classification. (3.12)

17. Suppose that you are a laboratory technician. You have been asked to synthesize the ester ethyl ethanoate using readily available materials and equipment in the laboratory.
    (a) Name the alcohol that you would use, and write its structural formula.
    (b) Name the acid that you would use, and write its structural formula.
    (c) Write a balanced chemical equation to show the reaction that would take place to synthesize the ester.
    (d) What is the name of this type of reaction?
    (e) List the equipment you would select for the reaction.
    (f) Describe the steps that you would follow in the procedure.
    (g) Describe the safety procedures that you would use for handling and disposing of the materials in the reaction. (3.14)

18. Use a molecular model kit to build models of the following compounds. Use your models to show the elimination of a small molecule in a condensation polymerization reaction.
    (a)

    $$HO - \overset{\overset{\displaystyle O}{\|}}{C} - \overset{\overset{\displaystyle O}{\|}}{C} - OH \quad \text{and} \quad H_2NCH_2CH_2NH_2$$

    (b)

    $$H_2N - CH_2 - \overset{\overset{\displaystyle O}{\|}}{C} - OH \qquad (3.18)$$

19. Here is a simple kitchen recipe for making glue slime: First, mix a volume of white glue with an equal volume of water. Next, make a saturated solution of laundry borax. Finally, mix equal volumes of the two solutions.
    (a) What is the role of the borax in the slime?
    (b) Design an experiment to change the elasticity of the slime, by varying one or more steps in the recipe. Predict the effect of the change, and give reasons for your prediction. (3.20)

Making Connections

20. (a) Using print or electronic resources, find and draw the structures of four different organic compounds used as coolants in the air conditioning or refrigeration industry.
    (b) What do the compounds have in common?
    (c) List some environmental impacts of each compound.
    (d) Suggest why the environmental impacts of these compounds may differ. (3.6)

 www.science.nelson.com

21. (a) Glucose, glycerol, and ethylene glycol all have a sweet taste, although ethylene glycol is very toxic. Research the molecular structures for these compounds and rubbing alcohol.
    (b) Predict, with reasons, the relative melting points and boiling points of rubbing alcohol, glucose, glycerol, and ethylene glycol.
    (c) Predict, with reasons, the solubility of each of these compounds in water and in gasoline.
    (d) Suggest a reason why ethylene glycol (car antifreeze) must be stored safely and spills must be cleaned up.
    (e) Do the structures of these four compounds support the hypothesis that taste receptors respond to functional groups in the compounds tested? Explain. (3.8)

 www.science.nelson.com

22. The distinction between "natural" and "synthetic" is usually based on the source of a product: whether the product is made by living organisms or by a laboratory procedure. Sometimes the product is even the same. For example, when bananas are dissolved in a solvent and the flavouring is extracted, the pentyl ethanoate obtained is labelled "natural flavour." When pentyl ethanoate is synthesized by esterification of ethanoic acid and pentanol, it is labelled "artificial flavour."
    (a) Write an equation for the synthesis of pentyl ethanoate.
    (b) In your opinion, what criteria should be used to distinguish a "natural" product from a "synthetic" product?
    (c) Research the differences in the source and processing methods of vanilla flavouring. Write a report on your findings. (3.13)

 www.science.nelson.com

23. There are several artificial sweeteners in use today, besides aspartame. Research some of the sweeteners that are used in sugarless gum. Draw their structural formulas. Investigate their advantages and disadvantages over sugar and aspartame. Which sweetener do you think is the best? Defend your decision. (3.16)

 www.science.nelson.com

24. There is ongoing research into the development of polymers for use in dentistry and related fields. For example, polymers are used to make moulds of patients' teeth, so they require certain properties, such as low solubility, high tensile strength, and a high softening temperature. They must also be nontoxic, of course. Research current practice and recent advances in the use of organic polymers in dentistry. Write a report on your findings. Make sure that you cover the following topics:
- the desired properties of dental polymers and the shortfalls of available materials
- the key structural features of the monomers in current and prospective dental polymers
- the rate and degree of the formation of crosslinks between polymer chains, and the properties of the molecule formed (3.20)

 www.science.nelson.com

25. Give examples of ten different organic compounds that are significant in your life. Indicate whether each compound is natural or synthetic, whether it is a polymer, and how you use it. Also indicate whether you could easily live without it, and give reasons. (3.21)

26. Give three examples of different synthetic polymers that are important in our lives. For each example, identify the monomer and the type of polymerization reaction involved. (3.21)

27. Research examples of how organic chemistry is used to address health, safety, or environmental problems. Examples of topics include leaded and unleaded gasoline, dry cleaning solvents, aerosol propellants, and pesticides and fertilizers. Write a report, or present a case study. (3.21)

 www.science.nelson.com

Extension

28. Why are some organic halides toxic, while others are not? Why are some organisms affected more than others? Use the Internet to find out, using the following key words in your search: bioaccumulation, fat-soluble, food chain. Report on your findings by writing a short article for a popular science magazine or web site. (3.6)

 www.science.nelson.com

29. The compound *p*-aminobenzoic acid (PABA) is the active ingredient in some sunscreen lotions.
   (a) Research and draw a structural formula for PABA.
   (b) Predict some properties of PABA, such as solubility, melting point, and chemical reactivity.
   (c) Research the properties of PABA, its role in protecting against ultraviolet (UV) radiation, and any possible hazards in its use (to the wearer or to the environment). (3.11)

 www.science.nelson.com

30. What makes bubble bath bubble? Research bubble bath to find the key ingredient. How is this ingredient made? What properties enable it to make bubbles in a bathtub? (3.13)

 www.science.nelson.com

31. Research each of the following polymers of glucose. Explain its importance to the organism, and describe how its structure is related to its function.
   (a) starch
   (b) glycogen
   (c) cellulose (3.19)

 www.science.nelson.com

Chemistry in the Environment

Until recently, most Canadians probably never thought much about the contamination of their drinking water. Instead, they chose to live by the motto "out of sight, out of mind." Events in Walkerton, Ontario, and elsewhere in Canada changed this attitude.

Water is essential for life. Every organism is between 50% to 90% water. As well, every organism requires a continuous supply of water to live. Humans must drink about 2 L of water daily to replace water lost through excretion, exhalation, and evaporation from the skin.

A view of Earth, taken from space, shows that a significant portion of Earth is covered with water. Only 2.5% of Earth's water is fresh water, however. Of this 2.5%, less than 0.4% is directly available for use.

Human activities are affecting the quality of freshwater sources. Natural processes, such as the water cycle, that clean and purify water can no longer offset the damage caused by human activities. Human activities also add pollutants to the air we breathe. These pollutants contribute to acid rain, climate change, and ozone depletion. To protect the environment, we need a better understanding of the chemistry of water and air, and the impact of our activities on the environment.

▶ Overall Expectations

In this unit, you will be able to

- demonstrate an understanding of the nature and role of elements and compounds, such as acids and bases, in the environment and gases in the atmosphere;

- accurately and effectively use the techniques involved in the quantitative analysis of solutions;

- assess the effects and implications, for society, of the levels of various substances in the environment;

- demonstrate an awareness of the need for both governments and individual citizens to take measures that will ensure a healthy environment.

ARE YOU READY?

Knowledge and Understanding

1. Describe the evidence that might lead you to believe that a chemical change has taken place in each of the following situations (**Figure 1**).

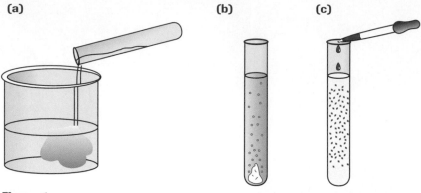

(a) (b) (c)

Figure 1
Chemical reactions

2. Copy and complete the following diagram of a classification of matter (**Figure 2**).

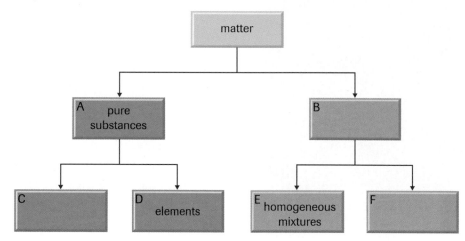

Figure 2
A classification of matter

3. Match each substance in **Table 1** to the classification categories shown in **Figure 2**.

Table 1 Classification of Matter

Substance	A or B	C or D or E or F
vinegar		
pure water		
sulfur		
air		
drinking water		
smog		
lye (sodium hydroxide)		

4. A small child leaves a birthday party with a helium-filled balloon. The string slips from the child's hand, and the balloon rises into the sky. What do you think will eventually happen to the balloon? Explain your reasoning.

5. How is the volume of air in a hot-air balloon maintained as the balloon rises away from the surface of Earth? In your answer, include an analysis of two factors that affect the volume of air in the balloon.

6. **Figure 3** shows a series of diagrams of a crystal of solid sucrose. The sucrose is suspended in an aqueous solution of sucrose over a period of 11 days.

day 1   day 3   day 5   day 7   day 11

Figure 3
A sucrose crystal is suspended in a sucrose solution for 11 days.

(a) List the states of matter that you see in the first container (day 1).
(b) Is the solution saturated or unsaturated from day 1 to day 5? Explain.
(c) Why did the sucrose crystal stay the same size from day 5 to day 7?
(d) Is the solution saturated or unsaturated from day 8 to day 11? Explain.

7. Magnesium metal reacts with hydrochloric acid to produce magnesium chloride, $MgCl_{2(aq)}$, and hydrogen gas.
(a) Write a balanced chemical equation for this reaction.
(b) Classify the type of reaction that your equation represents.

8. Balance each of the following equations:
(a) $KClO_{3(s)} \rightarrow KCl_{(s)} + O_{2(g)}$
(b) $Fe_{(s)} + O_{2(g)} \rightarrow Fe_2O_{3(s)}$
(c) $N_{2(g)} + H_{2(g)} \rightarrow NH_{3(g)}$

9. Write a sentence for each chemical equation in question 8. Include amounts (in moles), names, and states for all reactants and products.

10. (a) Define the terms "acid" and "base."
(b) Why is vinegar (5% acetic acid solution) safe to consume, while hydrochloric acid (5% solution) is not safe to consume?

11. Carbonic acid, $H_2CO_{3(aq)}$, is one of the acids that is present in normal rain. It is formed when carbon dioxide in the air reacts with water. Write a balanced chemical equation to represent this reaction.

12. (a) Write a balanced chemical equation for the reaction between sodium hydroxide, $NaOH_{(aq)}$, and hydrochloric acid, $HCl_{(aq)}$.
(b) Classify the type of reaction that your equation represents.

Inquiry and Communication

13. A piece of limestone chalk is placed in a container of dilute hydrochloric acid. The chalk slowly reacts, producing a colourless gas. Suggest as many ways as you can to make this chemical change go faster.

14. Design an experiment to answer the following question: "How does altitude affect the boiling point of pure water?" Include a brief plan. Identify dependent, independent, and controlled variables in your plan.

Making Connections

15. As a result of several complaints from citizens regarding sour-tasting tap water, Summerville's water commissioner ordered the city water utility to measure and record the pH of the tap water for one year. **Table 2** shows the data collected.

Table 2   Acidity of Summerville's Tap Water

Month	Jan.	Feb.	March	April	May	June	July	Aug.	Sept.	Oct.	Nov.	Dec.
Average pH	6.1	6.3	5.9	6.1	5.8	5.7	5.3	4.8	5.0	5.5	5.8	6.1

Based on these data, the commissioner recommended that no action be taken since the average pH was 5.7. When asked to justify this decision, the commissioner said, "People consume salad dressing at a pH of approximately 2.9 and soft drinks at a pH of approximately 3.4 all the time. Surely tap water, with an average pH of 5.7, is no cause for concern." Comment on the commissioner's point of view.

16. In the 1930s, chemists produced a series of synthetic chemicals called chlorofluorocarbons, or CFCs. Common CFCs include $CFCl_3$, CF_2Cl_2, $C_2F_3Cl_3$, and $C_2F_4Cl_2$.
    (a) What were some of the initial uses of these gases?
    (b) Why has the production of CFCs been banned in many countries, including Canada?

17. For many years, individuals, industries, and governments believed that "the solution to pollution is dilution." Even if pollutants are diluted by water or air, however, they can again become concentrated in the same or a different geographic area. For example, municipal waste that is discharged into a river can concentrate in pools downstream as river water levels drop in the summer. Identify a solid or liquid pollutant that has become concentrated in the environment. Describe the circumstances under which it was originally released, with the expectation that it would become diluted.

Math Skills

18. In many jobs, the salary earned is directly related to the time spent on the job.
    (a) Draw a graph for the direct variation between salary and time spent on the job. Use time (in hours) for the horizontal axis and salary (in dollars) for the vertical axis.
    (b) Like many relationships in science, the slope of your graph has a specific meaning. What does the slope of your graph represent?
    (c) Suppose that the salary varied inversely with the time spent on the job. Draw a graph for this inverse relationship. Would you want a job that pays this way? Why or why not?

Technical Skills and Safety

19. In this unit, you will work with chemical substances in solution.
    (a) What should you do immediately if a solution spills onto exposed skin?
    (b) Draw or describe the WHMIS symbol for a corrosive substance.

20. Flammable and combustible materials require special attention in the science laboratory. What do the following WHMIS symbols (**Figure 4**) represent?

(a)   **(b)**   **(c)**

Figure 4
WHMIS symbols

21. Copy and complete **Table 3** by using check marks to indicate the classes of fire extinguishers that are suitable for the three classes of fires.

Table 3   Types of Fire Extinguishers

| Class of fire | Class of fire extinguisher | | |
	Class A water	Class B carbon dioxide	Class C dry chemical
A (wood, paper, cloth)			
B (flammable liquids)			
C (live electrical equipment)			

Look at the photograph of the Great Lakes, taken from space (**Figure 1**), and think about the following questions:

- Is life on Earth sustainable?
- Will the biosphere—the part of Earth where life exists—continue to provide the resources that support life?

Now look at the photograph on the first page of this unit. You can see that Earth is mostly covered with water, which appears pure and clean, and able to support life. Walking through your community and carefully observing the environment around you, would you have a different perspective? Would you answer the two questions differently?

You are fortunate to live in North America, where water is abundant and accessible (**Figure 1**). For example, the Great Lakes contain 25% of the world's fresh water. How much of this seemingly abundant supply, however, is safe to use?

The demand for fresh water is increasing. The water cycle has always served as nature's principal water purification process. Unfortunately, evaporated water is now exposed to pollutants in the air. These pollutants are produced by human activities, such as the burning of fossil fuels. For example, combustion in automobiles produces nitrogen oxides, $NO_{x(g)}$, which react with water to form nitric acid, $HNO_{3(aq)}$. Similarly, the burning of high-sulfur coal produces sulfur dioxide, $SO_{2(g)}$, which reacts with water to form sulfurous acid,

Figure 1
People living near the Great Lakes seem to have access to an abundant supply of fresh water.

$H_2SO_{3(aq)}$. Nitric acid and sulfurous acid enter the water cycle and return to Earth as acid precipitation.

The burning of fossil fuels also produces pollutants that affect the quality of the atmosphere. One of these pollutants is carbon dioxide, $CO_{2(g)}$, a greenhouse gas that is linked with global climate change. Air pollution and smog are major problems in many parts of Canada. They contribute to several thousand deaths per year.

Is there a solution? What actions can we take as individuals to improve the quality of the water and air in Canada? Before deciding on a possible course of action, you need to understand the chemistry of the environment—the chemistry of water and air.

▶ TRY THIS activity   *Simulated Water Treatment*

In this activity, you will see what happens during the settling part of the water treatment process. You will use alum, which is widely available at pharmacies and grocery stores. Alum refers to a hydrate of one of the following compounds: aluminum sulfate, $Al_2(SO_4)_{3(s)}$; sodium aluminum sulfate, $NaAl(SO_4)_{2(s)}$; potassium aluminum sulfate, $KAl(SO_4)_{2(s)}$; or ammonium aluminum sulfate, $NH_4Al(SO_4)_{2(s)}$. For the purposes of this activity, it is not important to know which compound is used. Alum is relatively harmless, and it may be safely washed down the sink with plenty of water.

Materials: alum, soil, household ammonia, 3 drinking glasses, 3 teaspoons

1. Label three glasses with the numbers 1, 2, and 3. Fill each glass nearly full of tap water at room temperature.

2. Add 5 mL (a level teaspoon) of alum to glass 1. Stir gently until solid crystals are no longer visible. Add a teaspoon of household ammonia, and stir. What do you observe?

3. Repeat step 2 with glass 2, but add 5 mL of soil before adding the alum.

4. To glass 3, add only 5 mL of soil to the water.

5. Place the glasses where they will not be disturbed. Record your observations when you return to class the next day.

(a) What differences can you see among the three glasses after several hours? Explain these differences.

(b) Why is the gelatinous precipitate allowed time to settle, rather than separating it by filtration?

(c) Based on your observations, would a water sample treated with alum and household ammonia be safe to drink? Why or why not?

💡 REFLECT on your learning

1. List ten substances that can dissolve in water.

2. Give three examples of substances that do not dissolve in water. Explain why these substances do not dissolve in water.

3. (a) What are some properties of acids?
   (b) How can you explain these properties of acids?

4. (a) List some properties of bases.
   (b) How can you explain these properties of bases?

5. What is the difference between a strong acid and a weak acid?

6. Since many gases are invisible, how can you observe their properties?

7. Explain how a hot-air balloon works.

8. What affects the quality of air in the atmosphere?

Figure 1
This water is generally unavailable for drinking.

Access to clean water is essential for life on Earth. Without water, a person will only survive for a maximum of ten days. Yet, more than one billion people in the developing world do not have access to safe drinking water. Another three billion people do not have access to adequate sanitation systems, which would reduce exposure to water-related diseases. As a result, an estimated 14 to 30 thousand people, mostly young children and elderly people, die every day from water-related diseases. How can this happen when 70% of Earth's surface is covered with water?

About 97% of the water on Earth is in the oceans and is, therefore, not useful for drinking. About 2% is locked in ice caps and glaciers, and is not available for drinking (**Figure 1**). Less than 1% of Earth's water is fresh water or water in the liquid state (**Table 1**).

In Canada, we are very fortunate because we have access to an abundant supply of water. We simply turn on the tap to get the water we need, without worrying about its quality or the quantity we are using. When there is a disruption in the water supply, it is usually temporary and due to a break in a water line.

Suppose, however, that your community experienced a drought that lasted for several years, or that your local source of fresh water became contaminated. You would be asked to conserve water or to boil the water before using it. In fact, you may have already experienced lawn-watering regulations during a dry summer when nonessential water uses were restricted or even banned. Water use restrictions are usually voluntary. If you were asked to restrict your water uses, which uses would you give up first?

Table 1   Sources of Water on Earth

Source	Percentage of total water	Percentage of fresh water
Salt water (97.44%)		
oceans	96.5%	
saline/brackish ground water	0.93%	
saltwater lakes	0.006%	
Fresh water (2.53%)		
glaciers	1.74%	68.7%
ground water	0.76%	30.06%
ice, permafrost	0.022%	0.86%
freshwater lakes	0.007%	0.26%
soil moisture	0.001%	0.05%
atmospheric water vapour	0.001%	0.04%
marshes, wetlands	0.001%	0.03%
rivers	0.0002%	0.006%

▶ *TRY THIS* activity   *How Much Water Is Essential?*

Do you know how much water is used in your home every day? In this activity, you will determine the amount of water that is used over a three-day period.

1. Prepare a table, similar to **Table 2**, for each person in your home. Keep a record of water use for each person over a three-day period.

2. (a) Calculate the quantity of water used by each person for each activity over three days. Record the amount in **Table 2**.

(b) Determine the total volume of water used by each person over the three-day period.

(c) Determine the average volume of water used by each person in one day.

(d) How much water is used in your home in one day?

(e) Which water use would you give up if you had to conserve water?

Table 2   Water-Use Journal

Water use	Number of times in one day	Total for three days	Average amount of water used (L)	Amount of water used (L)
Bathroom				
flushing toilet			× 18 L	
showering (10 min)			× 100 L	
bathing in tub			× 60 L	
brushing teeth*			× 10 L	
shaving			× 20 L	
washing hands*			× 8 L	
Kitchen				
cooking			× 20 L	
washing dishes by hand			× 35 L	
using dishwasher			× 40 L	
Laundry				
using washing machine			× 225 L	
Outside				
washing car*			× 400 L	
watering lawn			× 35 L/min	
Other				
Total water used				

*while water is running

Although we have access to fresh water in Canada, there are growing concerns about the safety of our water supply. Historically, people drank water from the nearest source, usually a river or stream, without too many harmful effects. The water was purified by natural processes, such as the water cycle.

Recall what you know about the water cycle. Energy from the Sun causes water to evaporate and rise. When water evaporates, contaminants that may have dissolved in the water are left behind. As the water vapour rises, it cools

DID YOU KNOW ?

Conservative Showers
A 5-min shower with a low-flow showerhead uses only 35 L of water.

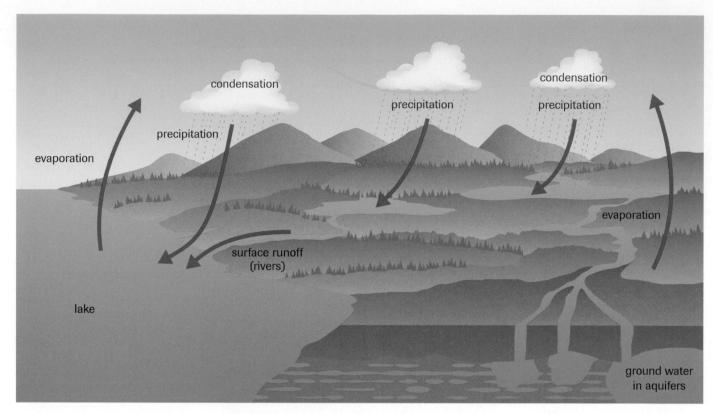

Figure 2
The water cycle

and condenses into mist, fog, and clouds. When the water vapour condenses into a liquid, it returns to Earth as precipitation (rain, snow, sleet, and hail). This sequence repeats itself endlessly in a cycle called the water cycle (**Figure 2**).

Most of the precipitation soaks directly into the soil, seeping downward until it becomes **ground water**. Ground water collects in porous rock structures called **aquifers**. As the water makes its way to the aquifers, it dissolves small amounts of soil and rock, which add impurities to the water in the form of iron, zinc, magnesium, and calcium.

The precipitation that finds its way into lakes, ponds, rivers, and streams becomes **surface water**. Surface water is the primary source of drinking water for most Canadians, although ground water is also used.

As populations have increased and communities have grown, the contamination of surface water and ground water has increased. The natural processes of water purification cannot keep up. Water's physical and chemical properties have resulted in many substances being dissolved in and contaminating water.

The Physical Properties of Water

Water has unique properties. Water is called "the universal solvent" because of its unique ability to dissolve a large number of substances. Pure water is clear, colourless, odourless, and tasteless. Tap water often has a characteristic taste,

ground water water from precipitation that seeps underground and collects in aquifers

aquifer an underground formation of porous rock that collects or holds ground water

surface water water in lakes, ponds, rivers, and streams

and sometimes an odour, caused by substances that have been dissolved in it. While some of these substances are present in water as a result of natural processes, such as the erosion of soil and rocks, others are contaminants or pollutants.

Whether another substance floats or sinks in water is related to the density of water. You may recall that density is the ratio of the mass of a substance to its volume. The density of pure water is 1.0 g/mL. The density of seawater is slightly higher and varies with the amount of salt that is dissolved in it. When crude oil spills from a leaking tanker, it usually floats on the surface of the water, where it is subjected to wind and ocean currents. Crude oil floats because its density is generally less than the density of seawater. Some of the heavier (more dense) components of crude oil sink to the bottom, often causing significant ecological damage to the ocean floor.

Water is unique because of how its density changes with its state. Generally, the density of a substance decreases as the substance changes from a solid to a liquid, and eventually to a gas. Solid water (ice), however, is less dense than liquid water. As liquid water freezes, it expands to occupy a larger volume than it did as a liquid. As a result, the density of solid ice is 0.9 g/mL. Since ice is less dense than water, it floats on the surface of the water. This property is very important. Floating ice on a lake, for example, acts like an insulating blanket, keeping the water underneath from freezing solid. If water did not have this property, all the fish in the lake would die.

Heat capacity is a measure of how much energy a substance absorbs to increase its temperature or how much heat a substance releases when its temperature decreases. Water has a very high heat capacity, which means that it can absorb a lot of heat with only a small increase in temperature (**Figure 3**). This property means that large bodies of water, such as the oceans or the Great Lakes, can act as heat reservoirs, or heat sinks. The Great Lakes, for example, can absorb a large amount of heat in the summer before the

The Exxon Valdez
In March, 1989, the *Exxon Valdez* oil tanker ran aground in Prince William Sound, spilling approximately 257 000 barrels, or 38 800 metric tonnes, of oil—equal to 125 Olympic-size swimming pools. While the *Valdez* oil spill is not the largest oil spill worldwide, it is considered to be the number one spill in terms of environmental damage. Cleanup efforts continued for four summers. Some beaches were never cleaned and remain oiled today. Despite the massive cleanup efforts, wave action from winter storms did more to clean the beaches than all the human effort.

Today, the *Exxon Valdez*, renamed the *Sea River Mediterranean*, hauls oil across the Atlantic Ocean. Under law, it can never return to Prince William Sound.

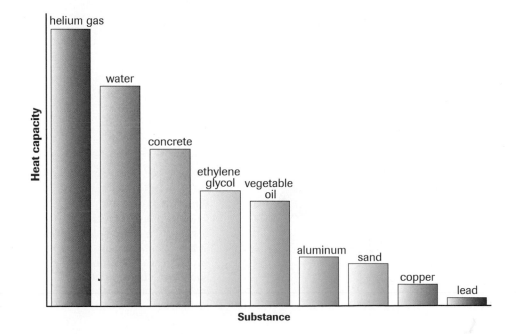

Figure 3
This bar graph shows the heat capacities of common substances. To calculate the heat capacities of different substances, scientists use the same mass of each substance, yielding what is called the specific heat capacities, measured in J/(kg•°C). Water has a very high heat capacity compared with other substances.

(a)

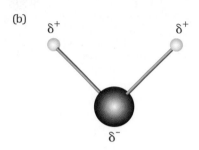

(b)

Figure 4
(a) The oxygen atom in a water molecule has two single polar covalent bonds and two lone pairs of electrons. This bonding results in a bent (V-shaped) molecule.
(b) A water molecule has a slightly negative charge on the oxygen atom and a slightly positive charge on each hydrogen atom.

water temperature increases significantly. Prevailing winds that move across the cool lake water help to moderate warmer temperatures over adjacent land masses. When the temperature of the air decreases, the heat that is stored in the water helps to moderate the colder air temperatures over adjacent land masses. Southwestern Ontario benefits from the Great Lakes heat sink, often having cooler summer temperatures and warmer winter temperatures than other parts of the province.

The Chemical Behaviour of Water

You learned about hydrogen bonding in section 3.5. Many of the unique physical properties of water can be explained by looking at the forces between water molecules.

Water is a polar molecule that consists of an oxygen atom bound to two hydrogen atoms with covalent bonds (**Figure 4(a)**). The large difference between the electronegativity of the hydrogen atom and the electronegativity of the oxygen atom produces a highly polar bond. (You learned about electronegativity in section 1.12.) The small hydrogen atom has a slight positive charge, while the oxygen atom has a slight negative charge (**Figure 4(b)**). As a result, the more positively charged hydrogen atoms of one water molecule exert a very strong attractive force on the more negatively charged oxygen atom of a neighbouring water molecule. This intermolecular force is called a hydrogen bond (**Figure 5**).

The concept of hydrogen bonding can be used to explain many of water's physical properties. For example, water has a high boiling point (100°C) because large amounts of energy are required to break the hydrogen bonds between water molecules in the liquid state. In contrast, a compound with a similar structure, hydrogen sulfide, $H_2S_{(g)}$, has a very low boiling point (−61°C) because it does not form hydrogen bonds between its molecules.

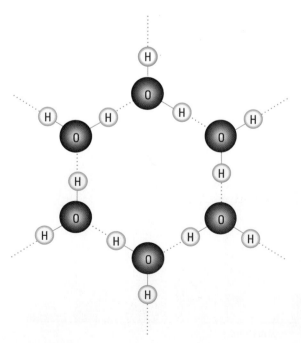

Figure 5
In ice, hydrogen bonds (···H−) between the water molecules result in a regular hexagonal crystal structure. Ice is less dense than liquid water because of the increased space between its molecules.

Since water is a polar molecule, it dissolves other polar molecules ("like dissolves like"). Ionic compounds also dissolve in water. To understand why, you need to look at electrolytes (**Figure 6**).

In 1887, a Swedish scientist named Svante August Arrhenius (1859–1927) proposed that, when dissolving, particles of a substance separate from each other and disperse throughout the solution. Recall, from section 1.11, that some substances, such as sucrose, separate into electrically neutral molecules in water. **Figure 7** shows an electrically neutral sucrose molecule surrounded by water molecules. The resulting solution does not conduct electricity. Sucrose is a *nonelectrolyte*.

When dissolved in water, ionic compounds, such as sodium chloride, $NaCl_{(s)}$, form solutions that conduct electricity. According to Arrhenius, electrically charged particles must be present for a solution to conduct electricity. When sodium chloride dissolves, it *dissociates* into sodium ions and chloride ions (**Figure 8**). The positively charged sodium ions are surrounded by the negatively charged oxygen ends of water molecules. The negatively charged chloride ions are surrounded by the positively charged hydrogen ends of water molecules. The dissociation equation can be written as

$$NaCl_{(s)} \rightarrow Na^+_{(aq)} + Cl^-_{(aq)}$$

Notice that water, $H_2O_{(l)}$, is not a reactant and does not appear in the equation. You will look at dissociation reactions again in section 4.7.

Understanding how water can dissolve ionic solids is the first step in understanding how water can dissolve many other substances, such as sugar.

Figure 6
Some examples of electrolytes

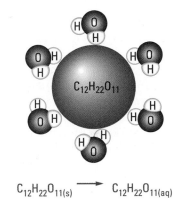

$$C_{12}H_{22}O_{11(s)} \longrightarrow C_{12}H_{22}O_{11(aq)}$$

Figure 7
This model of sucrose dissolved in water shows that a nonelectrolyte does not conduct electricity.

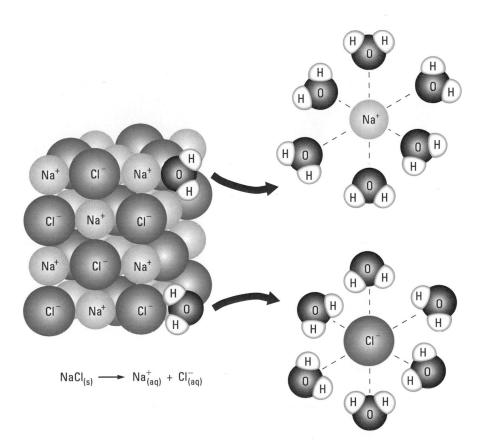

$$NaCl_{(s)} \longrightarrow Na^+_{(aq)} + Cl^-_{(aq)}$$

Figure 8
This model represents the dissociation of sodium chloride into ions.

Chemistry in the Environment **273**

Water's success as a solvent, however, means that both surface water and ground water can easily become contaminated with a great variety of substances. You will learn more about substances that dissolve in water in the next section.

▶ *Section 4.1* *Questions*

Understanding Concepts

1. Briefly describe two ways in which water is stored on Earth.

2. The total supply of water on Earth has probably been the same for billions of years. Explain why.

3. (a) Describe the bonding in a water molecule.
   (b) Describe the shape of a water molecule.

4. (a) Explain why water has a high boiling point compared with hydrogen sulfide, $H_2S_{(g)}$, a compound with a similar structure. Use an illustration in your explanation.
   (b) Use your answer to (a) to explain why water has such a high heat capacity.
   (c) How do bodies of water moderate extremes in climate?
   (d) What property of water causes roadways to deteriorate more in winter than they do in summer?

5. (a) Explain the term "polar" in relation to a water molecule as a whole.
   (b) How does this property of water account for the large number of substances that dissolve in water?

6. **Figure 9** shows the oil tanker *Prestige* spilling its cargo into the ocean. Seawater is denser than pure water. The density of the oil that is spilling from the ship is 0.9 g/mL. Describe the implications of this density difference for
   (a) aquatic life
   (b) oil-spill cleanup operations
   (c) the possibility of an oil fire

Making Connections

7. In Samuel Taylor Coleridge's classic poem, "The Rime of the Ancient Mariner," a sea captain describes the desperation of his crew who have run out of fresh water to drink:

   Water, water, everywhere,
   Nor any drop to drink.

   Describe what these lines mean in terms of what you have learned in this section.

8. Test results indicate that your local source of drinking water contains a toxic substance at levels that are higher than acceptable. You are given 500 L of safe water as your only source of water for three days.
   (a) What steps would you take to reduce your water consumption?
   (b) Identify any consequences of your actions.

Figure 9
In November, 2001, the tanker *Prestige* sank off the coast of Spain, spilling 77 000 t of oil into the ocean.

Today, many people carry bottled water (**Figure 1**). Why is bottled water so popular? Is it safer than tap water, or is it simply more accessible and convenient? How is bottled water similar to or different from tap water?

The Canadian Bottled Water Association recognizes two sources for its products: underground springs and wells, and municipal supplies. When municipal water is used, regulations require bottled water companies to reprocess the water by distillation. Distillation is a two-step purification process in which the water is vaporized and then condensed. Is the water that comes from underground springs and wells as pure as distilled water?

To answer this question, you need to consider the water cycle. As you learned earlier, water that falls back to Earth as precipitation seeps through the soil and porous rock, and collects in aquifers. On its journey through the soil, water dissolves a variety of substances from the surrounding rocks and minerals. Low concentrations of some substances give water flavour and promote human health. For example, iron ions, $Fe^{2+}_{(aq)}$ and $Fe^{3+}_{(aq)}$, and small amounts of magnesium ions, $Mg^{2+}_{(aq)}$, are constituents of blood and tissues.

Other substances that dissolve in water have negative effects. Compounds that contain sulfur give water a bad odour. High concentrations of iron ions in drinking water can give water a bad taste. Compounds that contain metal ions of mercury, lead, cadmium, and arsenic (heavy metal ions) can be harmful to human health.

Hard Water

When washing with soap, do you find that the soap does not produce a good lather or leaves a floating scum on the surface of the water? If you do, you may have **hard water**. While hard water is an inconvenience, it does not pose a health risk.

Some rocks, such as limestone, contain minerals that will dissolve in water. These minerals include calcium ions, $Ca^{2+}_{(aq)}$, magnesium ions, $Mg^{2+}_{(aq)}$, iron(II) ions, $Fe^{2+}_{(aq)}$, iron(III) ions, $Fe^{3+}_{(aq)}$, and manganese ions, $Mn^{2+}_{(aq)}$. When slightly acidic water flows through limestone, higher than normal amounts of these ions may dissolve in the water.

Table 1 shows an approximate grading system that is used for quantifying water hardness. The hardness index is the total concentration of calcium and magnesium ions expressed in milligrams per litre of water, or parts per million (1 mg/L = 1 ppm).

When soap mixes with soft water, it forms a cloudy solution with a layer of suds on top. In hard water, soap reacts with calcium ions and magnesium ions to form insoluble compounds, or *precipitates*. These precipitates appear as scum on the surface of bath water. Scum is the source of the "bathtub ring" that remains behind after the bath water is drained.

Figure 1
There are 49 different brands of bottled water produced by members of the Canadian Bottled Water Association. Is bottled water safer than tap water?

DID YOU KNOW ?

Bottled Water
Europeans drink far more bottled water than Canadians do. In 2000, Canadians consumed 27.2 L of bottled water per person, while Europeans consumed 93 L of bottled water per person. Can you explain the difference?

hard water water that contains dissolved calcium, magnesium, and iron ions

Table 1 Water Hardness

Hardness index (mg/L or ppm)	Water classification
<50	soft
50–200	slightly hard
200–400	moderately hard
400–600	hard
>600	very hard

If you have noticed a hard scale deposit on the inside of a kettle, the hard water in your home likely contains the hydrogen carbonate ion, $HCO_{3(aq)}^-$. Calcium hydrogen carbonate, $Ca(HCO_3)_{2(aq)}$, is formed when CO_2-laden water comes into contact with chalk or limestone, $CaCO_{3(s)}$, according to the equation

$$CaCO_{3(s)} + H_2O_{(l)} + CO_{2(g)} \rightarrow Ca(HCO_3)_{2(aq)}$$

As a result of this reaction, water has relatively high concentrations of both $Ca_{(aq)}^{2+}$ and $HCO_{3(aq)}^-$ ions. When hard water is boiled, the reaction proceeds in the reverse direction, depositing calcium carbonate according to the equation

$$Ca(HCO_3)_{2(aq)} \xrightarrow{heat} CaCO_{3(s)} + H_2O_{(l)} + CO_{2(g)}$$

Boiling the water causes solid calcium carbonate to remove the calcium ions from the water, which helps to soften the water. Unfortunately, the calcium carbonate forms a scale on the bottom of the kettle. This scale reduces the efficiency of heat transfer from the element to the water in the kettle (**Figure 2**).

Figure 2
Scale deposits caused by hard water can reduce the efficiency of some kitchen appliances and impede the flow of water through water pipes.

▶ *TRY THIS* activity   *Testing for Hard Water*

Boiling water is not an effective way to soften water. Adding sodium carbonate, $Na_2CO_{3(s)}$, softens water by removing calcium ions as an insoluble precipitate, solid calcium carbonate, $CaCO_{3(s)}$.

$$Ca_{(aq)}^{2+} + CO_{3(aq)}^{2-} \rightarrow CaCO_{3(s)}$$

What would you observe if you compared the layer of soapsuds in a sample of hard water with the layer of soapsuds in a sample of hard water mixed with sodium carbonate and filtered?

Materials: hard water sample; liquid hand soap; 3 test tubes; filter funnel; filter paper; ruler; sodium carbonate, $Na_2CO_{3(s)}$

1. Half-fill a test tube with hard water. Add one or two drops of liquid hand soap. Place your thumb over the top of the test tube, and shake well.

2. Measure and record the thickness of the soapsuds on the surface of the water.

3. Half-fill a second test tube with hard water. Add a pinch of sodium carbonate to the test tube. Cover the top of the test tube, and shake well.

4. Filter the contents of the test tube. Pour the filtrate into a clean, dry test tube.

5. Add one or two drops of liquid soap to the test tube. Place your thumb over the top of the test tube, and shake well.

6. Measure and record the thickness of the soapsuds at the surface.

(a) How does the thickness of the soapsuds in the two test tubes compare?

(b) Is sodium carbonate effective for softening water? Explain your answer.

Softening Hard Water

Hard water can be softened for laundry use by removing calcium and magnesium ions. One of the earliest methods of softening water was to add sodium carbonate, known as washing soda, to the water along with the laundry soap. Hard-water ions precipitate as calcium carbonate, $CaCO_{3(s)}$, and magnesium carbonate, $MgCO_{3(s)}$. These precipitates are rinsed away with the rinse water. Other water softeners that are directly added to water include commercial products, such as Calgon. Calgon contains both sodium carbonate and sodium hexametaphosphate, which react with calcium and magnesium ions to produce complex ions that do not precipitate. Calcium ions react with hexametaphosphate ions according to the following net ionic equation:

$$2\,Ca^{2+}_{(aq)} \quad + \quad P_6O_{18}{}^{6-}_{(aq)} \quad \rightarrow \quad [Ca_2(P_6O_{18})]^{2-}_{(aq)}$$

calcium ions + hexametaphosphate ion → calcium hexametaphosphate ion

Home water softeners that are linked directly into the plumbing system use an ion exchange process to soften water. These water softeners consist of a large tank filled with ion exchange resin (plastic) grains through which the water passes. The resin is highly concentrated with sodium ions, Na^+, attached to sulfonate ions, $SO_3{}^-$ (**Figure 3(a)**). When water passes through the resin, the dissolved calcium ions, $Ca^{2+}_{(aq)}$, and the magnesium ions, $Mg^{2+}_{(aq)}$, attach to the sulfonate group in the resin, displacing the sodium ions (**Figure 3(b)**). The sodium ions dissolve in the water. In this way, water that passes through the resin is "softened." The sodium ions, which have been exchanged for calcium and magnesium ions in the hard water, do not form a precipitate with soap or a hard scale inside water pipes or electrical appliances.

After a period of use, all the sodium ions are replaced by hard water ions, so the resin requires regeneration. The resin is rinsed with a concentrated brine (sodium chloride, $NaCl_{(aq)}$) solution from a salt tank that is attached to the water softener (**Figure 4**). Sodium ions from the brine replace the calcium and magnesium ions on the resin. The hard water ions are washed down the drain,

Figure 4
A home water-softening unit attached to a brine tank

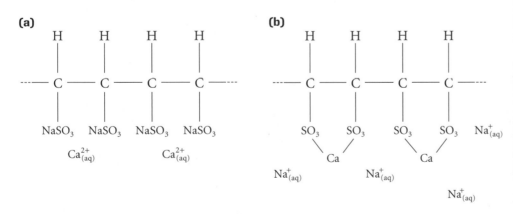

Figure 3
(a) Each resin molecule consists of hundreds of thousands of these units. Each sulfonate group acts as a site where positive (metal) ions may be attached and held.
(b) Two aqueous calcium ions have been attached and held by the resin surface, in exchange for four aqueous sodium ions.

Chemistry in the Environment **277**

along with excess sodium ions, and the resin is ready to exchange ions with incoming hard water.

In parts of Ontario, water hardness is such a problem that municipalities use the **soda-lime process** to soften water as part of the water purification process. In the soda-lime process, both sodium carbonate (washing soda), $Na_2CO_{3(s)}$, and calcium hydroxide (slaked lime), $Ca(OH)_{2(s)}$, are added to the hard water. The washing soda causes calcium ions to precipitate as calcium carbonate (equations 1 and 2 below). Although adding calcium hydroxide to hard water that contains calcium ions may seem strange, after a series of reactions, the calcium ions in the lime and the calcium ions in the hard water precipitate, as do the magnesium ions (equations 3 and 4):

$$Ca(OH)_{2(s)} \rightarrow Ca^{2+}_{(aq)} + 2\,OH^-_{(aq)} \qquad (1)$$

$$2\,OH^-_{(aq)} + 2\,HCO^-_{3(aq)} \rightarrow 2\,CO^{2-}_{3(aq)} + 2\,H_2O_{(l)} \qquad (2)$$

$$CO^{2-}_{3(aq)} + Ca^{2+}_{(aq)} \rightarrow CaCO_{3(s)} \qquad (3)$$

$$CO^{2-}_{3(aq)} + Mg^{2+}_{(aq)} \rightarrow MgCO_{3(s)} \qquad (4)$$

Calcium carbonate and magnesium carbonate are much less soluble than the hydrogen carbonate (bicarbonate) compounds of these ions in hard water.

Dissolved Oxygen

So far, you have only considered solids that dissolve in water. What about gases that dissolve in water? Think about what happens when you open a can or bottle of a carbonated soft drink. You may hear a "whoosh," and see a rush of bubbles as dissolved carbon dioxide gas escapes from the beverage. Are there other gases that dissolve in water?

Every animal requires oxygen gas, $O_{2(g)}$, either by inhaling it or by absorbing it from water. Some oxygen gas dissolves directly into the water from the air above the surface of the water. Additional oxygen mixes into the water by **aeration**. Aeration occurs when water flows over a dam or rocks in a stream, or breaks as waves on a beach. Finally, oxygen is added to water in the process of photosynthesis, as aquatic plants produce food from sunlight and carbon dioxide.

Aquatic organisms compete for the available oxygen in water. Oxygen-consuming bacteria feed on natural wastes (dead plant and animal material) and on human and industrial wastes in the water. If too much waste is dumped into the water, bacteria populations explode and, in the process, reduce the amount of dissolved oxygen in the water. Decreased oxygen levels in the water can negatively affect aquatic populations of fish and other animals. For example, fish cannot live in water that has a dissolved oxygen level that is less than 4 ppm. If the dissolved oxygen content of a freshwater source decreases, game fish that require higher oxygen concentrations, such as walleye (pickerel) and trout, will move to other regions or die.

Therefore, as the populations of pollution-tolerant organisms, such as bacteria, increase, the amount of dissolved oxygen in the water decreases and the diversity of aquatic organisms decreases. This inverse relationship is the basis for a method of assessing water quality, called the Benthic macro-invertebrate analysis, in which pollution-tolerant organisms are monitored. The Benthic macro-invertebrate analysis is often part of a comprehensive water-quality monitoring program.

▶ **TRY THIS** activity

Determining the Concentration of Dissolved Oxygen in Water

The concentration of dissolved oxygen (DO) in a freshwater source is an indicator of water quality. In this activity, you will measure the DO in local freshwater samples and analyze your data.

Materials: DO test kits, sample bottles, freshwater samples

1. Obtain a sample of fresh water and a DO test kit from your teacher. Follow the instructions in the test kit to measure the DO in the sample you are given.

2. Perform three tests on one freshwater sample, and average your results.

3. Repeat the tests using a different sample of fresh water.

(a) Why is it important to perform three tests on each water sample, and average your results?

(b) What do your results tell you about the quality of the sources of water from which your samples were taken?

▶ **Section 4.2 Questions**

Understanding Concepts

1. What are the two ions that are most commonly responsible for hard water?

2. What dissolved substance in ground water might cause the water to have an unpleasant odour?

3. Identify two substances that are considered unpleasant or unsafe in drinking water. Explain your choices.

4. Why is hard water a problem for Ontario and for states that border the Great Lakes?

5. (a) Identify two problems that hard water can cause.
   (b) Describe the more serious problem you identified in (a).

6. How does hard water reduce the efficiency of an appliance that is used to heat water?

7. (a) Write a complete, balanced chemical equation to show how sodium carbonate removes calcium ions and magnesium ions from hard water.
   (b) Explain how this process softens hard water.

8. Why does a home water-softening unit require regeneration?

9. High concentrations of iron(III) ions, $Fe^{3+}_{(aq)}$, in drinking water contribute to water hardness and may also give water a bad taste. Write a balanced equation to show how these ions can be removed using sodium carbonate.

10. (a) How does oxygen gas become dissolved in water?

(b) Describe two ways that the concentration of oxygen in water can be decreased in a natural environment.

Applying Inquiry Skills

11. Write a complete procedure that you could use to test the effectiveness of each of the following methods for removing calcium ions and magnesium ions from hard water:
    (a) sand filtration
    (b) Calgon
    (c) sodium ion exchange resin

Making Connections

12. Would an ion-exchange water-softening process be appropriate for large-scale municipal water softening? Give as many reasons as possible to support your decision.

13. Suppose that a water-softening unit is used in a home that draws water from a well.
    (a) Does soft water come out of all the taps in the home?
    (b) What health concerns are related to drinking water with a high sodium content?
    (c) Research the health issues that are associated with drinking soft water (water that has had all the metal ions removed).

 www.science.nelson.com

Canada has more access to sources of fresh water within its boundaries than any other country in the world. So it may seem strange that many Canadians are concerned about the safety of their drinking water. What do you think are the reasons for this concern? Most Canadians live in the southern part of the country and in urban centres. The increased levels of human activities in these areas invariably have some effect—usually negative—on the water supply (**Figure 1**).

There are three types of contaminants that contribute to water pollution: physical contaminants, biological contaminants, and chemical contaminants.

Physical Contaminants

Physical contaminants are objects that do not dissolve in water. Examples of physical contaminants are oil and petroleum products, garbage, floating debris, tree branches, and silt, clay, and other soil particles.

The removal of physical contaminants is the first step in the water purification process. Large particles and debris, such as algae, sticks, and garbage, are removed by coarse screens. Oil and some liquid petroleum products, which do not dissociate in water, generally float on the surface and can be skimmed away fairly easily. The water then passes through a sand filter or is pumped into settling tanks to allow the small particles to sink to the bottom. You may recall removing small soil particles from a water sample in the Try This Activity in "Getting Started." You added alum and household ammonia, which combined to form a sticky precipitate of aluminum hydroxide, $Al(OH)_{3(s)}$. The precipitate settled to the bottom of the glass, taking the fine soil particles with it and, thus, clearing the water.

Biological Contaminants

Water can also be polluted by biological contaminants, which include bacteria and viruses. For example, many of the beaches along the Great Lakes are closed each summer because the water is contaminated with fecal coliform bacteria.

In the spring of 2000, the drinking water in the town of Walkerton, Ontario, became contaminated with *E. coli* bacteria (**Figure 2**). Heavy rainfall in southwestern Ontario had saturated the ground to the extent that surface contaminants were quickly carried to and mixed with water in underground aquifers. One of these aquifers, located under Walkerton, serves as the source of water for the town. Contamination of the wells that draw water from this aquifer resulted in the deaths of seven people and caused hundreds of people to become seriously ill. Residents had to boil water used for drinking and add bleach to water used for washing.

Subsequent investigations established that the contaminant was a particularly deadly strain of *E. coli* bacteria, traced to agricultural runoff from cattle manure. Normally these impurities are filtered out as rainwater passes

Figure 1
This river is polluted with physical, biological, and chemical contaminants.

DID YOU KNOW ?

Water Pollution
One litre of oil can pollute more than two million litres of water.

Figure 2
The contamination of Walkerton's water supply raised concerns about the processes that are used to ensure a high quality of drinking water in Canada.

through the soil, long before they reach the level of ground water. During the wet spring, however, the soil was quickly saturated with water and the runoff containing *E. coli* easily found its way into the three wells that tapped the Walkerton aquifer.

This tragedy could have been averted if the town's chlorination system, designed to administer chlorine to kill bacteria in the water, had been working properly. An investigation revealed that the chlorination system had not been working properly and that staff had falsified reports to indicate that the system was working.

Circumstances surrounding the Walkerton tragedy have raised concerns about the reliability of our water purification processes. For some Canadians, the solution may be to buy bottled water or to use filtration systems in their homes. However, most Canadians recognize the need to regulate and enforce safe drinking water standards.

Chemical Contaminants

Chemical contaminants are substances that are soluble in water. They include manufactured chemicals, metal ions and compounds, pesticides, fertilizers, and soluble petroleum products. People are producing, transporting, using, and dumping larger quantities and a greater variety of chemicals than ever before. Even when we incinerate chemicals or put them where we think they will do no damage, such as buried in a landfill, the chemicals may leak, run off, or leach out, and find their way into our sources of water (**Figure 3**).

The environment has a natural capacity for handling some chemical contaminants. Certain chemicals, such as fertilizers and some industrial wastes, are naturally degradable. They can be broken down by natural

Figure 3
Can you identify any chemical contaminants in this photograph?

DID YOU *KNOW* ?

Drinking Water and Chlorine
Drinking water contains chlorine, which is added to disinfect the water and to inhibit bacterial growth. Chlorine can be removed from drinking water by letting the water sit in an open container for 48 h. Passing the water through an activated charcoal filter, such as a Brita filter, will also effectively remove the chlorine and its associated taste.

DID YOU *KNOW* ?

What's Wrong with Lake Erie?
Thousands of migratory water birds die around Lake Erie from eating fish infected with Type E botulism. Bacteria that naturally live in lake sediment produce this toxin.

DID YOU *KNOW* ?

Follow the Flush
Each year, septage haulers in Ontario remove about 500 million litres of human waste from septic tanks. In 2002, only 12 of more than 60 companies properly disposed of the waste. Violations included spreading untreated waste on crops too close to harvest time or too near waterways.

DID YOU *KNOW* ?

Great Lakes Chemical Soup
Over 360 chemical compounds have been identified in the Great Lakes. Persistent toxic chemicals destroy aquatic ecosystems and are potentially dangerous to humans.

Table 1   Sources and Effects of Common Chemical Contaminants in Water

Contaminant	Source	Effect
heavy metals (mercury, cadmium)	landfill containing batteries	• interfere with brain and nerve development in vertebrates
salts (sodium chloride, potassium chloride)	road-salt runoff	• kill freshwater organisms • make water unsuitable for drinking
nitrates, phosphates, sulfates	fertilizers, naturally occurring minerals	• encourage plant growth, which results in algal blooms and depletes oxygen in water
pesticides	farms, gardens	• are toxic to invertebrates • can accumulate to toxic levels in vertebrates
organic compounds, petroleum products	gasoline, oil spills	• are poisonous or carcinogenic • can interfere with oxygen diffusion into water

Figure 4
In some school laboratories, waste chemicals are disposed of in specially designated containers.

Figure 5
Hazardous household product symbols (HHPSs) identify both the nature and the degree of the hazard.

chemical processes or by bacteria. Other chemical contaminants persist in the environment, either degrading very slowly or not at all. These contaminants include pesticides (such as DDT), petroleum and petroleum products, PCBs, and heavy metals (such as lead, mercury, and cadmium). **Table 1** lists the sources and effects of some common chemical contaminants.

The Solution

How can you, as an individual, help the water pollution problem? In school laboratories, most chemical wastes from school science investigations are washed down the drain, while only some of the wastes require special disposal (**Figure 4**). Do these actions safeguard the quality of our freshwater sources?

The school laboratory is only one place where hazardous chemical products may be used. Many consumer products, such as household cleaners, paints, and batteries, contain hazardous chemicals (**Figure 5**). Products labelled as corrosive, flammable, reactive, or toxic should be properly disposed of at a designated hazardous waste facility. Although government regulations control the disposal of industrial hazardous wastes, the same regulations do not apply to the disposal of household hazardous wastes. Instead, municipalities rely on educated and committed citizens to protect the environment by taking special care when purchasing, using, and disposing of potentially hazardous substances.

All levels of government in Canada have some responsibility for ensuring the safety of our drinking water. The provinces and territories are responsible for setting and enforcing standards to ensure that adequate treatment facilities for drinking water are available. Municipal governments are responsible for supplying safe drinking water to residents as an essential public service. Federal–provincial water quality guidelines list the maximum acceptable concentrations (MACs) of chemicals that can be in drinking water after treatment (**Table 2**). A water treatment plant operator is responsible for ensuring that local drinking water meets these federal–provincial guidelines. ◆■ Lab technicians at water-treatment plants and testing labs assess the quality of the water samples that are sent to them.

Table 2 Maximum Acceptable Concentrations (MACs) of Selected Chemicals in Canada's Drinking Water

Substance	Typical source	MAC (ppm or mg/L)
arsenic	mining waste, industrial effluent	0.025
benzene	industrial effluent, spilled gasoline	0.005
cadmium	leached waste from landfill	0.005
chloride	natural sources, runoff	250*
cyanide	mining waste, industrial effluent	0.2
fluoride	natural sources, additive for control of tooth decay	1.5
lead	leached waste from landfill, old plumbing	0.01
mercury	industrial effluent, agricultural runoff	0.001
nitrate and nitrite	agricultural runoff	10.0
sulfate	natural sources, agricultural runoff	500*
tetrachloroethylene	dry cleaners	0.030
trihalomethanes (THMs)	water chlorination	0.100

* Aesthetic objective, not health related: Aesthetic objectives have been established for certain chemicals that impair the taste, odour, or colour of water. These chemicals do not pose a risk to human health.

▶ **SAMPLE** problem

Determining the Safety of Water

A sample of the drinking water in a mining town in northern Ontario has arsenic levels of 0.03 ppm. Is this water safe to drink? Explain.

Step 1: Express Concentration in mg/L or ppm
The arsenic level in the water sample is 0.03 ppm.

Step 2: Determine Maximum Acceptable Concentration (MAC) of Substance
In **Table 2**, the MAC of arsenic is 0.025 ppm.

Step 3: Compare MAC with Given Value
Since the sample of drinking water has a higher level of arsenic than the MAC, the water is not safe to drink.

Example
A 10-mL water sample in a southern Ontario town is tested and found to contain 6.8 mg of fluoride. Is the water safe to drink?

Solution
According to **Table 2**, the MAC of fluoride is 1.5 ppm, or 1.5 mg/L. Therefore, 6.8 mg in a 10-mL sample of drinking water far exceeds the allowable limit. The water is not safe to drink.

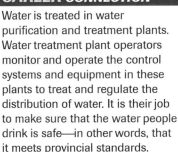

CAREER CONNECTION

Water is treated in water purification and treatment plants. Water treatment plant operators monitor and operate the control systems and equipment in these plants to treat and regulate the distribution of water. It is their job to make sure that the water people drink is safe—in other words, that it meets provincial standards.

Water treatment plant operators collect water samples daily for chemical and bacterial analyses. These samples are analyzed on site or at an independent test facility. As well, operators keep a detailed record of results and changes that are made to the system. Failing to complete testing procedures or falsifying results can have grave consequences, as the Walkerton tragedy showed. A diploma from a Community College program is an essential requirement for this position.

(i) What would you identify as the most important aspect of a water treatment plant operator's job?
(ii) Research the Community College program(s) that would qualify a graduate for employment as a water treatment plant operator.
(iii) Determine the starting salary for a water treatment plant operator.
(iv) How much demand is there for this position?

 www.science.nelson.com

Understanding Concepts

1. (a) Name the three types of contaminants that contribute to water pollution.
   (b) Prepare a table with three columns, one for each type of contaminant. Think of ten water pollutants, and place each pollutant in the correct column of your table.

2. Briefly describe two ways in which pollutants can enter surface water and ground water.

3. A 10-mL sample of drinking water was analyzed and found to contain 5.4 mg of nitrate ions. Is the water safe to drink? (*Hint:* Refer to **Table 2**.)

4. Why is it important that the maximum allowable concentration of contaminants in drinking water not be exceeded?

Making Connections

5. The maximum acceptable concentrations of the chemicals listed in **Table 2** are based on an average daily intake of 1.5 L of drinking water.
   (a) If a community's drinking water contained the maximum acceptable levels of lead, how much lead would a person in the community consume in a year?
   (b) Would you willingly drink this water? Give reasons for your decision.

6. (a) Examine the labels of different brands of bottled water to see what information is provided about the source of the water and the treatment process.
   (b) Research the safety standards that apply to suppliers of bottled water.
   (c) Do you think these standards are adequate? Give your reasons.

 www.science.nelson.com

7. (a) Research how and where your community's drinking water is treated.
   (b) What are the possible sources of water contaminants?
   (c) What testing standards are used to ensure the safety of drinking water?
   (d) What happens to the drinking water if tests indicate that the MAC of a substance has been exceeded?

 www.science.nelson.com

8. Drinkable water is important when hiking or camping in wilderness areas. Any natural water you encounter may look clear and clean, but there may be dissolved substances that could cause you to become ill. Research some portable technologies that can be used to purify water. Include examples of technologies that involve both physical and chemical treatments.

 www.science.nelson.com

Testing for Ions in Water

Now that you have learned about the contaminants that may affect water quality, it is time to investigate the quality of the drinking water in your community.

In the first part of this investigation, you will use qualitative, or diagnostic, tests to check for the presence of certain ions in tap water. A positive test—a change in colour or the formation of an insoluble precipitate—indicates that an ion is present. (Refer to **Table 1**.) A negative test (no colour or precipitate), however, does not mean that the ion is absent. The ion may be present in such small amounts that it does not produce enough colour or solid precipitate to be seen.

Table 1   Confirming Tests

Ion	Expected product
Fe^{3+}	red with $SCN^-_{(aq)}$
Ca^{2+}	white precipitate with $C_2O_4^{2-}_{(aq)}$
SO_4^{2-}	white precipitate with $Ba^{2+}_{(aq)}$

In the second part of this investigation, you will quantitatively determine the concentration of sulfate ions in tap water. You will test a water sample that contains sulfate ions using excess barium chloride to produce barium sulfate, a precipitate.

$$2\,Na^+_{(aq)} + SO_4^{2-}_{(aq)} + Ba^{2+}_{(aq)} + 2\,Cl^-_{(aq)} \rightarrow$$
$$BaSO_{4(s)} + 2\,Na^+_{(aq)} + 2\,Cl^-_{(aq)}$$

You will then calculate the concentration of sulfate ions in the water sample, based on the amount of precipitate formed. You will use the calculation method that is shown in the Sample Problem.

Questions

Part 1: What ions are present in a sample of drinking water?

Part 2: What is the concentration of sulfate ions, in ppm, in a sample of tap water?

Prediction

(a) Predict the ions that are present in a sample of drinking water.

▶ SAMPLE problem

Calculating Concentration from Mass of Precipitate

When a 100-mL sample of drinking water is tested for sulfate ions with barium chloride solution, $BaCl_{2(aq)}$, a white precipitate forms. The precipitate is collected in a pre-weighed filter and dried in an oven. The dried filter and precipitate have a combined mass of 0.056 g. When the mass of the original filter is subtracted, the mass of the precipitate collected is 0.053 g. What is the concentration of sulfate ions in the drinking water sample (in mg/L)?

The net ionic equation for this reaction is
$$Ba^{2+}_{(aq)} + SO_4^{2-}_{(aq)} \rightarrow BaSO_{4(s)}$$

Step 1: Calculate Mass of Required Substance

$$m_{SO_4^{2-}} = 0.053 \text{ g } \cancel{BaSO_4} \times \frac{1 \cancel{\text{ mol BaSO}_4}}{233.39 \text{ g } \cancel{BaSO_4}} \times$$

$$\frac{1 \cancel{\text{ mol SO}_4^{2-}}}{1 \cancel{\text{ mol BaSO}_4}} \times \frac{96.06 \text{ g } SO_4^{2-}}{1 \cancel{\text{ mol SO}_4^{2-}}}$$

$$m_{SO_4^{2-}} = 0.0218 \text{ g } SO_4^{2-}$$

Step 2: Convert Units of Sample to mg and L

You know that 0.0218 g of sulfate ions is collected from 100 mL of water. You need to convert to mg/L.

First, determine the mass of precipitate in mg.

$$0.0218 \text{ g} \times 1000 \text{ mg/g} = 21.8 \text{ mg}$$

Now determine the volume in litres (L).

$$100 \text{ mL} \times \frac{1 \text{ L}}{1000 \text{ mL}} = 0.1 \text{ L}$$

Step 3: Determine Concentration in mg/L

$$\frac{21.8 \text{ mg}}{0.1 \text{ L}} = 218 \text{ mg/L}$$

The concentration of sulfate ions in the water sample is 218 mg/L, or 218 ppm.

Materials

eye protection
lab apron
protective gloves
microtray
microdroppers
filter paper
filter funnel
distilled water
tap water
reference solutions:
  0.1-mol/L iron(III) nitrate, $Fe(NO_3)_{3(aq)}$
  0.1-mol/L calcium chloride, $CaCl_{2(aq)}$
testing solutions:
  0.5-mol/L potassium thiocyanate, $KSCN_{(aq)}$
  0.1-mol/L sodium oxalate, $Na_2C_2O_{4(aq)}$
  0.1-mol/L barium chloride, $BaCl_{2(aq)}$
dilute acetic acid, $HC_2H_3O_{2(aq)}$

stirring rod
two 250-mL beakers
150-mL beaker
retort stand
iron ring clamp
water sample
graduated cylinder
electronic balance

Part 1: Qualitative Analysis of a Water Sample

Procedure

1. Rinse a microtray with distilled water, and shake it dry.

2. Make an observation table.

3. Using a microdropper, add one drop of distilled water to three wells in the first row of the microtray (**Figure 1**).

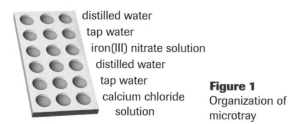

distilled water
tap water
iron(III) nitrate solution
distilled water
tap water
calcium chloride solution

Figure 1
Organization of microtray

 Avoid cross-contamination of the microdroppers and solutions. Use each dropper for one solution only.

4. Using another microdropper, add one drop of tap water to three wells in the second row of the microtray.

5. Add one drop of iron(III) nitrate (reference solution—a source of iron ions) to three wells in the third row of the microtray.

6. To test for iron ions, add one drop of potassium thiocyanate (testing solution) to each of the nine wells. Record your observations in your table. Note the appearance and colour of the starting solutions and any product.

7. Using a microdropper, add one drop of distilled water to three wells in the fourth row of the microtray.

8. Using another microdropper, add one drop of tap water to three wells in the fifth row of the microtray.

9. Add one drop of calcium chloride (reference solution—a source of calcium ions) to three wells in the sixth row of the microtray.

10. To test for calcium ions, add one drop of dilute acetic acid and one drop of sodium oxalate (testing solution) to each of the nine wells you filled in steps 7 to 9. Record your observations in your table.

 Oxalates are toxic and should be treated with care.

Analysis

(b) Answer the Question for Part 1. Refer to **Figure 2**.

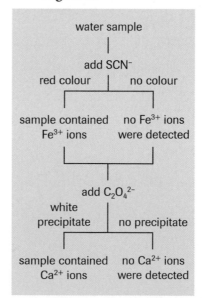

Figure 2
Flow chart for Part 1

(c) Why are three samples tested for each ion?

(d) What is the purpose of a control? Which solution was the control in these tests?

(e) Write balanced chemical equations to represent each of the reactions between the ions and the test chemicals used in Part 1.

Evaluation

(f) Evaluate your Prediction.

(g) Suggest any necessary improvements to the Procedure for Part 1.

(h) How might your observations have changed in Part 1 if you had not used a clean microdropper for each solution? Would this omission have introduced error into your experimental results for Part 1? Explain.

Synthesis

(i) Suggest why the tests you performed in Part 1 do not absolutely confirm the absence of an ion.

Part 2: Quantitative Analysis of a Water Sample

Procedure

11. Obtain a water sample from your teacher.

12. Measure 40 mL of the sample, and add it to a clean, dry 250-mL beaker.

13. Add 50 mL of barium chloride solution (testing solution) to the beaker. Thoroughly mix the solutions using a stirring rod.

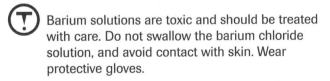

Barium solutions are toxic and should be treated with care. Do not swallow the barium chloride solution, and avoid contact with skin. Wear protective gloves.

14. Determine the mass of a clean, dry piece of filter paper. Fold the filter paper, and place it in a filter funnel suspended over a clean, dry 250-mL beaker (**Figure 3**).

15. Carefully pour the mixture from step 13 into the filter funnel. Make sure that the mixture does not rise above the level of the filter paper and spill down the outside.

16. Rinse the beaker several times with distilled water. Pour the rinse into the filter paper.

17. Carefully remove the filter paper. Place it in a clean, dry 150-mL beaker for drying. When the

Figure 3
Laboratory setup

filter paper and its contents are dry, determine the mass of the precipitate collected.

18. Wash your hands thoroughly after the experiment. Follow your teacher's instructions for disposing of waste materials.

Analysis

(j) Calculate the concentration of sulfate ions in the water sample, using the mass of the precipitate and the appropriate balanced chemical equation.

(k) Answer the Question for Part 2.

Evaluation

(l) Identify any potential sources of error in Part 2.

(m) Suggest any necessary improvements to the Procedure for Part 2 that would reduce the sources of error.

Synthesis

(n) Does the sulfate ion concentration in the water sample exceed the aesthetic objective for this ion, given in **Table 2** in section 4.3? Explain.

Tech Connect

Water is a precious resource (**Figure 1**). Many Canadians are concerned about the potential health risks associated with contaminants in fresh water and, therefore, about the safety of our drinking water. As a result, water-treatment devices that disinfect and remove chemicals from water are becoming increasingly popular in homes. Manufacturers of these devices also build larger models for treating municipal water supplies.

Figure 1
Since less than 1% of the world's water is fresh water available for drinking, we need to protect this resource for current and future generations.

Chlorination is the most widely used method for disinfecting municipal water. Although chlorine leaves a taste in the water, it is very effective against a wide range of biological contaminants. There are safety issues, however, associated with the use of chlorine. When it is used in water that contains decaying organic matter, trihalomethanes (THMs) are produced and remain dissolved in the water. Studies in the Great Lakes Basin have shown that exposure to THMs in drinking water may be associated with an increased risk of bladder and colon cancer. For this reason, the MAC of THMs has been set at 0.100 mg/L.

A Canadian-based company, called Trojan Technologies, produces water disinfection systems that use ultraviolet (UV) radiation instead of chemicals. Trojan makes both large units for treating municipal water supplies and smaller units for home use. During the disinfection process, water flows past UV lamps. The lamps deliver a lethal dose of UV energy to the biological contaminants (such as bacteria, viruses, moulds, and algae) in the water (**Figure 2**).

Figure 2
Advantages to using ultraviolet radiation rather than chlorine to disinfect water include no odour or taste added to the water, and no chemicals used.

While both chlorination and UV radiation of water effectively remove most biological contaminants, neither method by itself is very effective at removing dissolved chemicals from drinking water.

Distillation is another water purification method. It removes both biological contaminants, such as microorganisms, and chemical contaminants, such as heavy metals. However, it does not remove all organic chemical compounds.

Ozone can also be used to oxidize and inactivate biological contaminants in water. Ozone is generated using air and high-voltage electricity. There is concern that treating polluted waters with ozone leads to the formation of small amounts of THMs. As well, the disinfection of water is temporary since water that is left standing in pipes can become re-contaminated.

For best results, distillation and ozone methods are often combined with carbon filtration systems to disinfect water. Unfortunately, the carbon filters can become saturated with contaminants very quickly. At this point, the filters can no longer effectively remove any more contaminants from the water.

Most disinfection methods require relatively clear water to ensure maximum efficiency. As a result, at least two methods are often used in combination: one to clear the water and another to remove unwanted organisms and dissolved chemicals. Ultimately, the best method involves a multi-barrier device, in which the water is first filtered and then disinfected. One such device, designed by a Canadian company, Pure Water Corporation, uses a combination of ozone, UV radiation, and filtration to purify water (**Figure 3**). This device is designed to imitate the natural water purification process.

Figure 3
A home water purification and disinfection device uses ozone, UV radiation, and filtration.

▶ TRY THIS activity   *A Simple Distillation*

During distillation, biological contaminants and most dissolved chemicals are removed from water. Boiling the water at 100°C kills most of the biological contaminants. The water vapour is then collected and allowed to condense. Is pure water produced during distillation?

Materials: eye protection, retort stand, iron ring clamp, wire gauze, heat source, watch glass, ice cube, salt-water solution, 150-mL beaker, microscope, microscope slide

1. Assemble the apparatus as illustrated in **Figure 4.** Add 100 mL of the salt-water solution to the beaker.

2. Use the heat source to bring the solution in the beaker to a slow, rolling boil.

3. Place an ice cube on the upper surface of the watch glass. Observe what happens on the underside of the watch glass.

(a) Describe the liquid that collects on the underside of the watch glass.

(b) What causes the water vapour to form in the beaker?

(c) What happens to the water vapour when it comes into contact with the cool watch glass?

Figure 4
Laboratory setup

4. Allow the contents of the beaker to cool to room temperature. Carefully tilt the watch glass to allow a drop of the liquid from the underside to fall off onto a clean, dry microscope slide. Place the slide on the stage of a microscope. After the liquid droplet has evaporated to dryness, observe the slide for any signs of a residue.

(d) Has the process of distillation removed the dissolved salt from the water? What evidence do you have to support your answer?

▶ Tech Connect 4.5 Questions

Understanding Concepts

1. Which contaminants are Canadians most concerned about when they consider the safety of their drinking water? Why?

2. Create a table with three columns. In the first column, list four water purification methods. In the second column, list the advantages of each method.

In the third column, list the disadvantages of each method.

Extension

3. Research other home water treatment methods. Incorporate this additional information into your table.

 www.science.nelson.com

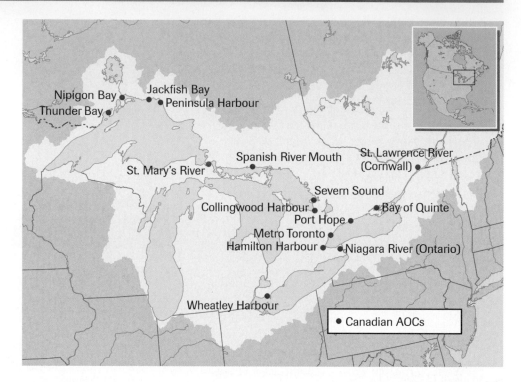

Figure 1
This map shows the Canadian designated Areas of Concern in the Great Lakes Basin. They are pollution "hot spots" for which Canada has responsibility.

Thirty years ago, Canada and the United States signed the Great Lakes Water Quality Agreement in an effort to restore and protect the Great Lakes Basin. The Great Lakes Action Plan 2001–2006 describes activities and actions for the Great Lakes Basin that are being undertaken jointly by the Government of Canada, the Government of Ontario, and United States federal and state agencies.

The main goals of the Great Lakes Action Plan are to establish a healthy environment, to protect the health of citizens, and to sustain communities. These goals are to be met through the implementation of seven specific objectives:

- to restore areas of concern
- to conserve ecologically important areas
- to control the introduction of exotic species
- to assess and manage ecosystem health
- to protect and promote human health
- to reduce harmful pollutants
- to advance sustainable use of the Great Lakes Basin

Target results have been identified for each objective, along with specific actions to be carried out over a five-year period.

There are 42 Areas of Concern (AOCs) that have been identified in the Great Lakes Basin (**Figure 1**). In these areas, pollutants from industrial sources, sewage treatment plants, landfills, and agricultural runoff have entered the waterways, destroyed the quality of the water, and created unsafe situations for people and wildlife.

Remedial Action Plans (RAPs) are being developed and implemented for each AOC. A RAP is designed to restore the ecological health of the area and to make recreational uses possible. The RAP identifies the contaminants in the water and their sources, the treatment needed to clean the water, and the preventative measures that need to be taken to relieve future contamination. Public involvement in the development and implementation of an area's RAP is essential for the successful rejuvenation of the area.

Hamilton Harbour

One example of a successful RAP is the RAP for Hamilton Harbour. Hamilton Harbour is a small, enclosed bay on Lake Ontario. It is surrounded by a large population and many industries. Over the last century, Hamilton Harbour became contaminated from sewage and industrial wastes that flowed directly into it. The addition of landfill for a variety of building projects eliminated 65% of the wetland areas around the harbour. Only 7% of the shoreline was accessible to the public. In 1980, Hamilton Harbour had the reputation of being the most polluted area in the Great Lakes (**Figure 2**).

Figure 2
Surrounded by industries, Hamilton Harbour became a dumping ground for municipal, industrial, and agricultural wastes.

Today, people are swimming in Hamilton Harbour because of improvements made to storm-water management and sewage treatment. Public access to the harbourfront has been improved. Families can enjoy the parks around the harbour or cycle along its shoreline (**Figure 3**). Fish populations are improving because of habitat restoration. In part of the harbour known as Cootes Paradise, a carp barrier has been constructed to keep out carp, but not other species of fish. These changes have occurred because governments, environmentalists, business people, people from industries, and concerned individuals have worked together to solve a shared problem and to create a cleaner, safer harbour.

Figure 3
A view of Hamilton Harbour today

The Yellow Fish Marking Program

Despite the success of the cleanup of Hamilton Harbour so far, the work continues. A "Yellow Fish" program, initiated in British Columbia, has been implemented in Hamilton. To raise awareness about the impact of toxic substances on local waters, participants paint yellow fish symbols beside storm drains to remind members of the community not to pour harmful, hazardous materials down the drains. Hazardous materials include oil, cleaners, paints, pesticides, products with warning labels, and even soap from washing cars. Hazardous materials that are poured down storm drains go directly into the closest creek, and eventually into the harbour. There they can harm fish and other aquatic life, as well as vegetation, wildlife, and people. Yellow, fish-shaped flyers are distributed to Hamilton homes to inform residents about the proper way to dispose of hazardous materials. Information about alternatives to toxic products is also made available.

The Success of Hamilton Harbour

Although all the targets of the RAP for Hamilton Harbour have not yet been met, the remaining targets are attainable. One of the reasons that this RAP has been so successful is that the citizens of Hamilton and the neighbouring region have played a pivotal role. The RAP was developed as a community-based program. By focusing on public involvement, people recognized the role they played in the harbour's problems and, thus, became more interested in participating in a solution.

The cleanup of Hamilton Harbour has not only benefited fish and wildlife, but it has also had positive effects on human health by reducing exposure to contaminants. Habitat restoration, improvement of water quality (**Figure 4**), rehabilitation of shorelines, and the creation of shoreline trails have contributed to the health and well-being of the people of Hamilton.

Figure 4
A secchi disc is lowered into the water until it reaches a depth at which it can no longer be seen. This depth is the measure of the water's clarity. Before Hamilton Harbour was cleaned up, the clarity was 1 to 2 m. The RAP target for Hamilton Harbour is 3 m.

▶ *Case Study 4.6* *Questions*

Understanding Concepts

1. (a) Identify the goals and objectives of the Great Lakes Action Plan that have been addressed in the restoration of Hamilton Harbour.
   (b) Do you think that all the goals and objectives in a Remedial Action Plan should be met before an AOC can be removed from the list?

2. Identify and explain one reason for the success of the RAP for Hamilton Harbour.

3. Describe two examples of evidence of the successful restoration of Hamilton Harbour.

4. How has human health benefited from the restoration of the harbour?

Extension

5. The focus of the Great Lakes Action Plan is the restoration of 42 Areas of Concern in the Great Lakes Basin. The cleanup of Hamilton Harbour is only one success story. Choose another AOC, and research the actions that are being taken to restore it. Prepare a report that explains how the objectives of the Great Lakes Action Plan are being met.

 www.science.nelson.com

Much of what we know about solutions today originated with the theories of chemist Svante Arrhenius. As you learned in section 4.1, Arrhenius developed a theory that the particles of a substance, when dissolving, separate from each other and disperse into the solution. Recall, from Unit 1, that compounds are electrolytes if their aqueous solutions conduct electricity. Compounds are nonelectrolytes if their solutions do not conduct electricity. Most ionic compounds are electrolytes, while most molecular compounds are nonelectrolytes. For example, a solution of sodium chloride (table salt) in water conducts electricity because sodium chloride *dissociates* in water to form positive sodium ions and negative chloride ions. The presence of positive and negative ions in the solution accounts for the flow of electricity.

$$NaCl_{(s)} \rightarrow Na^+_{(aq)} + Cl^-_{(aq)}$$

Acids and bases are water-soluble compounds that share some properties with ionic compounds and some properties with molecular compounds. As well, they have properties that are unique. The properties of acids and bases (and their solutions) are summarized in **Table 1**.

Table 1 The Properties of Bases, Acids, and Their Solutions

Bases	Basic Solutions	Acids	Acidic Solutions
• are water soluble • are electrolytes	• taste bitter* • feel slippery* • turn red litmus blue • neutralize acidic solutions	• are water soluble • are electrolytes	• taste sour* • do not feel slippery • turn blue litmus red • neutralize basic solutions • react with active metals to produce hydrogen gas

* For reasons of safety, it is not appropriate to use taste or touch as diagnostic tests in the laboratory.

Arrhenius eventually extended his theory to explain some of the properties of acids and bases. According to Arrhenius, bases are ionic hydroxide compounds that dissociate in water to form positive metal ions and negative hydroxide ions, $OH^-_{(aq)}$. For example, sodium hydroxide dissociates into sodium ions and hydroxide ions in solution (**Figure 1**).

$$NaOH_{(s)} \rightarrow Na^+_{(aq)} + OH^-_{(aq)}$$

Arrhenius proposed that the hydroxide ions are responsible for the properties of basic solutions, in particular turning red litmus paper blue. The dissociation of bases in water is similar to the dissociation of ionic compounds in water.

Many acids, in their pure form, are a special group of compounds containing at least one hydrogen atom. Examples include hydrogen chloride, $HCl_{(g)}$, hydrogen sulfide, $H_2S_{(g)}$, and hydrogen nitrate, $HNO_{3(l)}$. While salts (such as sodium chloride, $NaCl_{(s)}$) and bases (such as sodium hydroxide, $NaOH_{(s)}$) are composed of ions held together by ionic bonds, the atoms in an acid molecule are held together by relatively strong covalent bonds. Nevertheless, when acids dissolve in water, they form solutions that are good conductors of electricity, just like the solutions of table salt and bases.

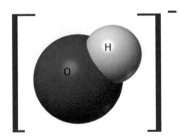

Figure 1
A model of a hydroxide ion, $OH^-_{(aq)}$

According to the Arrhenius theory, acid solutions must, therefore, contain ions. Since acids are electrically neutral molecules, how can their aqueous solutions contain ions? Arrhenius' explanation was that acid molecules produce hydrogen ions, $H^+_{(aq)}$, and anions when they dissolve in water, and that the hydrogen ions are responsible for the properties of the acidic solutions listed in **Table 1**. He described the formation of hydrochloric acid as follows:

$$HCl_{(g)} \rightarrow H^+_{(aq)} + Cl^-_{(aq)}$$

Arrhenius believed that although water acts a solvent, it does not participate in the reaction that produces the ions. In fact, he did not explain exactly how the ions form. After conducting many experiments with acids, scientists now believe that water molecules indeed react with acid molecules, causing the acid molecules to form ions. The process in which ions are formed from electrically neutral molecules (or atoms) is called **ionization**.

When an acid molecule, such as hydrogen chloride, reacts with water, a proton (commonly called an H^+ ion) is transferred from the acid molecule to a water molecule, leaving behind the rest of the acid molecule as an anion. (Remember that a hydrogen atom is composed of one proton and one electron.) The following equation and **Figure 2** describe the ionization of hydrogen chloride when it reacts with water:

$$HCl_{(g)} + H_2O_{(l)} \rightarrow H_3O^+_{(aq)} + Cl^-_{(aq)}$$

ionization a reaction in which electrically neutral molecules (or atoms) produce ions

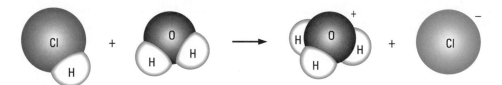

Figure 2
When gaseous hydrogen chloride dissolves in water, the hydrogen chloride molecules are thought to collide and react with water molecules to form hydronium ions and chloride ions.

hydronium ion a hydrogen ion covalently bonded to a water molecule, represented as H_3O^+

The $H_3O^+_{(aq)}$ ion that forms from the combination of a water molecule (H_2O) and the H^+ ion (from the acid molecule) is called a **hydronium ion** and is responsible for the properties of acidic solutions listed in **Table 1**.

Since the hydronium ion may be viewed as a water molecule to which a proton (H^+ ion) is attached by a covalent bond, chemists frequently simplify the ionization equation of an acid by disregarding the water molecules involved in the reaction. Cancelling a water molecule from both sides of the equation produces a simplified ionization equation as follows (for $HCl_{(g)}$):

Ionization equation: $HCl_{(g)} + H_2O_{(l)} \rightarrow H_3O^+_{(aq)} + Cl^-_{(aq)}$

Simplified ionization equation: $HCl_{(g)} \rightarrow H^+_{(aq)} + Cl^-_{(aq)}$

We will use the simplified ionization equation throughout the rest of this unit. However, always keep in mind that, while the simplified ionization equation describes the essential change that takes place in the ionization reaction, it does not communicate the important role of water in causing the acid to ionize, or the fact that the proton, $H^+_{(aq)}$, most likely exists as a hydronium ion, $H_3O^+_{(aq)}$, in solution.

▶ **SAMPLE** problem 1

Writing Dissociation and Ionization Equations

Write a dissociation or ionization equation (as appropriate) for each of the following chemical compounds dissolving in water. State whether the equation represents an ionization reaction or a dissociation.

(a) potassium chloride

> ***Step 1: Write Balanced Equation***
>
> $KCl_{(s)} \rightarrow K^+_{(aq)} + Cl^-_{(aq)}$
>
> ***Step 2: Determine if Equation Describes Ionization Reaction or Dissociation***
> This equation describes a dissociation because $KCl_{(s)}$ is an ionic compound (a salt).

(b) lithium hydroxide

> ***Step 1: Write Balanced Equation***
>
> $LiOH_{(s)} \rightarrow Li^+_{(aq)} + OH^-_{(aq)}$
>
> ***Step 2: Determine if Equation Describes Ionization Reaction or Dissociation***
> This equation describes a dissociation because $LiOH_{(s)}$ is an ionic hydroxide.

(c) hydrogen bromide

> ***Step 1: Write Balanced Equation***
>
> $HBr_{(aq)} \rightarrow H^+_{(aq)} + Br^-_{(aq)}$
>
> ***Step 2: Determine if Equation Describes Ionization Reaction or Dissociation***
> This equation describes an ionization reaction because $HBr_{(aq)}$ is an acid.

Strong and Weak Acids

Most acids are highly soluble in water. The solutions of different acids may vary in their ability to conduct electricity, however, even when their concentrations are the same. For example, a 1.0-mol/L solution of hydrochloric acid, $HCl_{(aq)}$, conducts electricity much better than a 1.0-mol/L solution of acetic acid, $HC_2H_3O_{2(aq)}$. Hydrochloric acid is a better conductor than acetic acid because virtually every molecule of hydrogen chloride, $HCl_{(g)}$, that dissolves in water (100% at 25°C) reacts with water molecules to produce $H^+_{(aq)}$ and $Cl^-_{(aq)}$ ions that conduct electricity.

$$HCl_{(g)} \xrightarrow{\ 100\% \ } H^+_{(aq)} + Cl^-_{(aq)}$$

strong acid an acid that completely ionizes in water to form ions and, therefore, is a good conductor of electricity

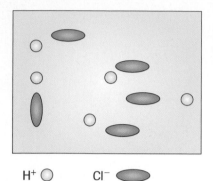

H⁺ ◯   Cl⁻ ⬭

Figure 3
Hydrochloric acid, $HCl_{(aq)}$, is a strong acid and has high electrical conductivity. In solution, every HCl molecule ionizes into $H^+_{(aq)}$ and $Cl^-_{(aq)}$ ions.

weak acid an acid that partially ionizes in water to form ions and, therefore, is a poor conductor of electricity

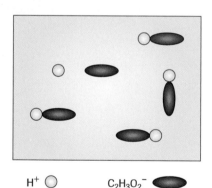

H⁺ ◯ $C_2H_3O_2^-$ ⬭

Figure 4
Acetic acid, $HC_2H_3O_{2(aq)}$, is a weak acid and has poor electrical conductivity. Only a few $HC_2H_3O_{2(aq)}$ molecules ionize into $H^+_{(aq)}$ and $C_2H_3O_{2(aq)}^-$ ions.

Thus, a 1.0-mol/L solution of hydrochloric acid contains virtually no hydrogen chloride molecules, but only hydrogen ions, chloride ions, and water. Acidic solutions like hydrochloric acid, which exhibit high electrical conductivity because virtually all of the acid molecules ionize in water, are called **strong acids** (**Figure 3**). Other examples of strong acids include nitric acid, $HNO_{3(aq)}$, and sulfuric acid, $H_2SO_{4(aq)}$. In the case of acetic acid, only a small percentage (approximately 1.3% at 25°C) of the dissolved hydrogen acetate molecules, $HC_2H_3O_{2(l)}$, ionize to form hydrogen ions and acetate ions, $C_2H_3O_{2(aq)}^-$.

$$HC_2H_3O_{2(l)} \xrightarrow{1.3\%} H^+_{(aq)} + C_2H_3O_{2(aq)}^-$$

In other words, a 1.0-mol/L acetic acid solution contains 0.013 mol $H^+_{(aq)}$, 0.013 mol $C_2H_3O_{2(aq)}^-$, and 0.987 mol $HC_2H_3O_{2(aq)}$. The vast majority of the entities in solution are electrically neutral hydrogen acetate molecules (and water molecules) that do not conduct electricity. Acidic solutions like acetic acid, which exhibit low electrical conductivity because a small percentage of the acid molecules ionize in water, are called **weak acids** (**Figure 4**). Many common acids, such as carbonic acid (in soda pop), $H_2CO_{3(aq)}$, and boric acid (in eye-wash solutions), $H_3BO_{3(aq)}$, are weak acids. A comparison of strong and weak acids, at the same concentration and temperature, is shown in **Table 2**.

Table 2   Electrical Conductivity and Strength of Acids

Acid solution	Formula	Electrical conductivity	Strength	Percentage ionization
hydrochloric acid	$HCl_{(aq)}$	high	strong	100%
nitric acid	$HNO_{3(aq)}$	high	strong	100%
sulfuric acid	$H_2SO_{4(aq)}$	high	strong	100%
carbonic acid	$H_2CO_{3(aq)}$	low	weak	<50%
acetic acid	$HC_2H_3O_{2(aq)}$	low	weak	<50%
nitrous acid	$HNO_{2(aq)}$	low	weak	<50%

In this textbook, you can assume that all bases are ionic hydroxides and, therefore, are strong bases. Although ionic hydroxides have varying solubilities in water, when they do dissolve, they dissociate completely to form ions (100% dissociation). For example, sodium hydroxide, $NaOH_{(s)}$, is a strong base and highly soluble in water. A sodium hydroxide solution contains only sodium ions, $Na^+_{(aq)}$, hydroxide ions, $OH^-_{(aq)}$, and water, resulting in high electrical conductivity (**Figure 5**).

Group 2 elements form the ionic hydroxides magnesium hydroxide, $Mg(OH)_{2(s)}$, calcium hydroxide, $Ca(OH)_{2(s)}$, and barium hydroxide, $Ba(OH)_{2(s)}$. When these bases dissolve in water, two moles of hydroxide ions are produced for every mole of ionic hydroxide that dissolves. These bases are only slightly soluble in water, however.

Concentration and Strength

When you look at an acidic solution, you cannot tell how concentrated the solution is unless it is labelled with this information. Recall, from section 2.5, that the ratio of the quantity of solute to the quantity of solution is the concentration of the solution. A **dilute** solution has a relatively small amount of solute per unit volume of solution (**Figure 6(a)**). A **concentrated** solution has a relatively large amount of solute per unit volume (**Figure 6(b)**).

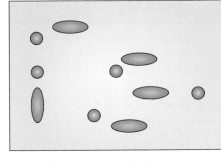

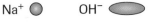

Na$^+$ ⬤   OH$^-$ ⬭

Figure 5
Sodium hydroxide, NaOH$_{(s)}$, is a strong base, which means that it completely dissociates into Na$^+_{(aq)}$ and OH$^-_{(aq)}$ ions in water.

(a)   **(b)**

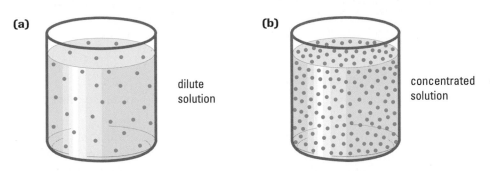

dilute solution

concentrated solution

Figure 6
Notice that the model of the dilute solution **(a)** shows fewer particles per unit volume than the model of the concentrated solution **(b)**.

dilute a relatively small amount of solute per unit volume of solution

concentrated a relatively large amount of solute per unit volume of solution

Note that both strong and weak acids can be concentrated or dilute. For example, a large amount of hydrogen chloride gas dissolved in a small volume of water produces concentrated hydrochloric acid, and a small amount of hydrogen chloride gas dissolved in a large volume of water produces dilute hydrochloric acid. Whether concentrated or dilute, hydrochloric acid is the solution of a strong acid because all of the acid molecules are ionized.

The concentration of a solution can be expressed in several ways. In consumer applications, for example, the concentration of acetic acid is often expressed as a percentage. The label on a bottle of vinegar may read "5% acetic acid," which means that there are 5 mL of pure acetic acid dissolved in every 100 mL of vinegar solution.

In chemistry, the concentration of a solution is most often expressed as a molar concentration. The molar concentration is the amount, in moles, of solute dissolved in 1 L of solution (mol/L). For example, concentrated nitric acid, HNO$_{3(aq)}$, has a molar concentration of 15.4 mol/L when purchased from a chemical supplier.

In certain situations, such as environmental analysis, very low, or dilute, concentrations are often measured in mg/L, or parts per million (ppm). For example, the water in public swimming pools contains about 1 ppm (or 1 mg/L) of chlorine to inhibit bacterial growth.

Hydrogen Ion and Hydroxide Ion Concentrations

The properties of acids are determined by the hydrogen ions that are present in solution. The higher the concentration of hydrogen ions, the more acidic the solution is. How can you determine the molar concentration of hydrogen ions in an acid solution?

As mentioned earlier, the high electrical conductivity of solutions of strong acids and strong bases means that there is essentially 100% ionization of acid molecules in water and essentially 100% dissociation of bases in water. Thus, we are able to determine the concentrations of the ions in solution. For example, consider a 1.0-mol/L solution of hydrochloric acid, $HCl_{(aq)}$. You know that hydrogen chloride is a strong acid, so you can assume that it is 100% ionized in water.

$$HCl_{(aq)} \rightarrow H^+_{(aq)} + Cl^-_{(aq)}$$

initial	1 mol	0	0
final	0	1 mol	1 mol

In this solution, the concentration of hydrogen ions, $[H^+_{(aq)}]$, and the concentration of chloride ions, $[Cl^-_{(aq)}]$, are each 1.0 mol/L because one mole of $HCl_{(aq)}$ ionizes into one mole of $H^+_{(aq)}$ and one mole of $Cl^-_{(aq)}$.

The properties of bases are determined by the hydroxide ions that are present in solution. The higher the concentration of hydroxide ions, the more basic the solution is. How can you determine the molar concentration of hydroxide ions in a basic solution?

Consider a 2.0-mol/L solution of sodium hydroxide, $NaOH_{(aq)}$, a strong base. You can assume that the base completely dissociates in water.

$$NaOH_{(s)} \xrightarrow{100\%} Na^+_{(aq)} + OH^-_{(aq)}$$

One mole of $NaOH_{(s)}$ dissociates into one mole of $Na^+_{(aq)}$ and one mole of $OH^-_{(aq)}$. Therefore, the concentration of sodium ions, $[Na^+_{(aq)}]$, will be 2.0 mol/L and the concentration of hydroxide ions, $[OH^-_{(aq)}]$, will be 2.0 mol/L.

▶ **SAMPLE** problem **2**

Calculating Concentrations of Hydrogen and Hydroxide Ions

Determine the concentration of hydrogen or hydroxide ions in each of the following solutions of strong acids or bases:

(a) 0.5 mol/L $HNO_{3(aq)}$

Step 1: Write Balanced Ionization Equation

$$HNO_{3(aq)} \rightarrow H^+_{(aq)} + NO^-_{3(aq)}$$

Step 2: Determine Molar Ratios of Reactants and Products
From the equation, you know that 1 mol $HNO_{3(aq)}$ ionizes to produce 1 mol $H^+_{(aq)}$ ions and 1 mol $NO^-_{3(aq)}$ ions.

Step 3: Use Molar Ratios to Determine Concentration of Hydrogen or Hydroxide Ions
Therefore, 0.5 mol/L $HNO_{3(aq)}$ ionizes to produce 0.5 mol/L $H^+_{(aq)}$ ions and 0.5 mol/L $NO^-_{3(aq)}$ ions.

(b) 0.25 mol/L $HCl_{(aq)}$

Step 1: Write Balanced Ionization Equation

$$HCl_{(aq)} \rightarrow H^+_{(aq)} + Cl^-_{(aq)}$$

Step 2: Determine Molar Ratios of Reactants and Products

From the equation, 1 mol $HCl_{(aq)}$ ionizes to produce 1 mol $H^+_{(aq)}$ ions and 1 mol $Cl^-_{(aq)}$ ions.

Step 3: Use Molar Ratios to Determine Concentration of Hydrogen or Hydroxide Ions

Therefore, 0.25 mol/L $HCl_{(aq)}$ ionizes to produce 0.25 mol/L $H^+_{(aq)}$ ions and 0.25 mol/L $Cl^-_{(aq)}$ ions.

(c) 0.10 mol/L $LiOH_{(aq)}$

Step 1: Write Balanced Dissociation Equation

$$LiOH_{(aq)} \rightarrow Li^+_{(aq)} + OH^-_{(aq)}$$

Step 2: Determine Molar Ratios of Reactants and Products

1 mol $LiOH_{(aq)}$ dissociates to produce 1 mol $Li^+_{(aq)}$ ions and 1 mol $OH^-_{(aq)}$ ions.

Step 3: Use Molar Ratios to Determine Concentration of Hydrogen or Hydroxide Ions

Therefore, 0.10 mol/L $LiOH_{(aq)}$ dissociates to produce 0.10 mol/L $Li^+_{(aq)}$ ions and 0.10 mol/L $OH^-_{(aq)}$ ions.

(d) 0.435 mol/L $KOH_{(aq)}$

Step 1: Write Balanced Dissociation Equation

$$KOH_{(aq)} \rightarrow K^+_{(aq)} + OH^-_{(aq)}$$

Step 2: Determine Molar Ratios of Reactants and Products

1 mol $KOH_{(aq)}$ dissociates to produce 1 mol $K^+_{(aq)}$ ions and 1 mol $OH^-_{(aq)}$ ions.

Step 3: Use Molar Ratios to Determine Concentration of Hydrogen or Hydroxide Ions

Therefore, 0.435 mol/L $KOH_{(aq)}$ dissociates to produce 0.435 mol/L $K^+_{(aq)}$ ions and 0.435 mol/L $OH^-_{(aq)}$ ions.

(e) 2.0 mol/L $HI_{(aq)}$

Step 1: Write Balanced Ionization Equation

$$HI_{(aq)} \rightarrow H^+_{(aq)} + I^-_{(aq)}$$

Step 2: Determine Molar Ratios of Reactants and Products

1 mol $HI_{(aq)}$ ionizes to produce 1 mol $H^+_{(aq)}$ ions and 1 mol $I^-_{(aq)}$ ions.

Step 3: Use Molar Ratios to Determine Concentration of Hydrogen or Hydroxide Ions

Therefore, 2.0 mol/L $HI_{(aq)}$ ionizes to produce 2.0 mol/L $H^+_{(aq)}$ ions and 2.0 mol/L $I^-_{(aq)}$ ions.

Understanding Concepts

1. What evidence shows that solutions of bases and solutions of acids are composed of separate ions?

2. Based on the chemical formula, predict whether each of the following solutions is acidic, basic, or neutral:
   (a) $HNO_{2(aq)}$
   (b) $LiOH_{(aq)}$
   (c) $Ba(OH)_{2(aq)}$
   (d) $NaCl_{(aq)}$
   (e) $CH_3OH_{(aq)}$

3. Write a balanced dissociation or ionization equation to represent each of the following compounds dissolving in water:
   (a) potassium bromide
   (b) aluminum hydroxide
   (c) magnesium chloride
   (d) hydrogen iodide

4. Calculate the concentrations of all the ions in a solution of each of the following compounds:
   (a) 0.45 mol/L potassium hydroxide
   (b) 0.50 mol/L sodium chloride
   (c) 0.375 mol/L hydroiodic acid

Answers

4. (a) $[K^+_{(aq)}] = [OH^-_{(aq)}]$
       $= 0.45$ mol/L
   (b) $[Na^+_{(aq)}] = [Cl^-_{(aq)}]$
       $= 0.50$ mol/L
   (c) $[H^+_{(aq)}] = [I^-_{(aq)}]$
       $= 0.375$ mol/L

After many experiments on acids and bases, chemists realized that acidic solutions contain hydrogen ions *and* hydroxide ions, but more hydrogen ions than hydroxide ions. In contrast, basic solutions contain more hydroxide ions than hydrogen ions. These experiments caused chemists to wonder whether pure water contains hydrogen ions and hydroxide ions as well. Careful testing with extremely sensitive electrical conductivity equipment shows that pure water has very slight electrical conductivity. Therefore, there must be a very small number of ions in pure water. This evidence, and other experimental evidence, indicate the presence of both hydrogen ions and hydroxide ions in pure water. The explanation is that a very small number of water molecules (two of every billion) react with each other to form hydronium ions and hydroxide ions as follows:

$$H_2O_{(l)} + H_2O_{(l)} \xrightarrow{\text{slight}} H_3O^+_{(aq)} + OH^-_{(aq)}$$

As before, this ionization equation can be simplified to

$$H_2O_{(l)} \xrightarrow{\text{slight}} H^+_{(aq)} + OH^-_{(aq)}$$

Pure water is neutral because both the hydrogen ion concentration, $[H^+_{(aq)}]$, and the hydroxide ion concentration, $[OH^-_{(aq)}]$, are equal to 1×10^{-7} mol/L at 25°C. An acidic solution has a hydrogen ion concentration of more than 1×10^{-7} mol/L and a hydroxide ion concentration of less than 1×10^{-7} mol/L. A basic solution has a hydrogen ion concentration of less than 1×10^{-7} mol/L and a hydroxide ion concentration of more than 1×10^{-7} mol/L.

Aqueous solutions can have a wide range of hydrogen ion concentrations. For example, a concentrated solution of hydrochloric acid may have a hydrogen ion concentration of more than 10 mol/L, while a concentrated solution of sodium hydroxide may have a hydrogen ion concentration of less than 10^{-14} mol/L.

pH

The extremely wide range of hydrogen ion concentrations in acidic and basic solutions led to the development of a convenient shorthand method for communicating these concentrations. This method, called "pH" for "power of hydrogen," was invented in 1909 by Danish chemist Sören Sörenson (1868–1939). The **pH** of a solution is defined as the negative of the exponent to the base 10 of the hydrogen ion concentration. This definition is not as complicated as it sounds. For example, a solution with a hydrogen ion concentration of 10^{-5} mol/L has a pH of 5. A solution with a hydrogen ion concentration of 10^{-2} mol/L has a pH of 2. Notice that the pH value does not have units associated with it.

An acidic solution has a hydrogen ion concentration of more than 10^{-7} mol/L, and so has a pH value of less than 7. A basic solution has a hydrogen ion concentration of less than 10^{-7} mol/L and a pH of more than 7. For example, a concentrated solution of a strong base with a hydrogen ion concentration of 10^{-14} mol/L has a pH of 14. An acidic solution with a hydrogen ion concentration of 10^{-4} has a pH of 4. A neutral solution has a hydrogen ion concentration of 10^{-7} mol/L and a pH of 7.

Using pH, the wide range of hydrogen ion concentrations can be expressed as a simple set of positive numbers, as shown in **Figure 7**. Notice that the difference between any two consecutive pH units in the scale corresponds to a ten-fold difference in hydrogen ion concentration. Thus, the hydrogen ion concentration of a solution with a pH of 4 is ten times greater than the hydrogen ion concentration of a solution with a pH of 5.

pH a measure of the acidity of a solution; negative of the exponent to the base 10 of the hydrogen ion concentration

LEARNING *TIP*

Comparing pH Values
A hydrogen ion concentration of 10^{-6} mol/L is ten times greater than a hydrogen ion concentration of 10^{-7} mol/L. The hydrogen ion concentration increases as the pH decreases.

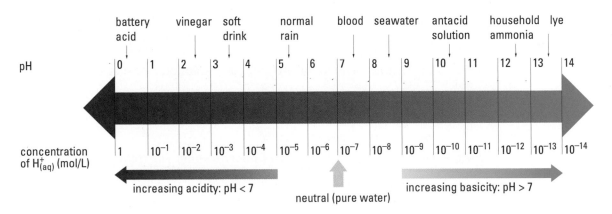

Figure 7
The pH scale can communicate a wide range of hydrogen ion concentrations for a variety of substances.

DID YOU KNOW?

Alkaline
A solution with a pH of more than 7 is sometimes referred to as an alkaline solution. Alkaline comes from the Arabic word *al-qili*, which means "ashes from the saltwort plant." It refers to the potassium hydroxide and potassium carbonate from the plant ashes that dissolved in water.

You can use the following equation to determine the concentration of hydrogen ions, $[H^+]$, when the pH of a solution is known:

$$[H^+] = 10^{-pH}$$

For example, a solution with a pH of 6 has $[H^+_{(aq)}] = 10^{-6}$ mol/L. A solution with a pH of 5, which has $[H^+_{(aq)}] = 10^{-5}$ mol/L, has ten times more hydrogen ions than the solution with a pH of 6:

$$10^{-6} \times 10 = 10^{-5} \text{ mol/L}$$

SAMPLE problem 3

Determining the pH from the Hydrogen Ion Concentration

What is the pH of an acidic solution that has a hydrogen ion concentration of 10^{-3} mol/L?

Step 1: Identify Exponent of Hydrogen Ion Concentration
The exponent is -3.

Step 2: To Find pH, Determine Negative Value of Exponent

$$pH = -(-3)$$
$$pH = 3$$

The pH of the acidic solution is 3.

Example
What is the pH of an antacid solution that has a hydrogen ion concentration of 10^{-11} mol/L?

Solution

$$pH = -(-11)$$
$$pH = 11$$

The pH of the solution is 11.

SAMPLE problem 4

Calculating the Hydrogen Ion Concentration from the pH

What is the hydrogen ion concentration of a solution that has a pH of 10?

Step1: Write Equation for Determining $[H^+]$, Given pH

$$[H^+] = 10^{-pH}$$

Step 2: To Determine Hydrogen Ion Concentration, Substitute pH Value into Equation

$$[H^+] = 10^{-10} \text{ mol/L}$$

The solution has a hydrogen ion concentration of 10^{-10} mol/L.

Example

What is the hydrogen ion concentration of a window-cleaning solution that has a pH of 11?

Solution

$[H^+] = 10^{-pH}$

$[H^+] = 10^{-11}$ mol/L

The window-cleaning solution has a hydrogen ion concentration of 10^{-11} mol/L.

▶ **Practice**

Understanding Concepts

5. Express each of the following concentrations as a pH value:
   (a) $[H^+] = 10^{-8}$ mol/L
   (b) $[H^+] = 10^{-6}$ mol/L
   (c) $[H^+] = 10^{-4}$ mol/L

6. What is the hydrogen ion concentration in each of the following household solutions?
   (a) ammonia, pH = 11
   (b) vinegar, pH = 2
   (c) soda water, pH = 4

Answers

5. (a) pH = 8
   (b) pH = 6
   (c) pH = 4

6. (a) 10^{-11} mol/L
   (b) 10^{-2} mol/L
   (c) 10^{-4} mol/L

Many consumer products specify the pH on the label. Have you ever tried or used commercial products that are promoted as being "pH balanced"? In light of what you have just read, does this description make more sense?

The setting of pH levels in many consumer products is important. For example, baby shampoo has the same pH as a baby's tears. This pH ensures that the shampoo will not sting if it gets into a baby's eyes. To ensure that a product has the proper pH, companies test the pH during manufacturing and before shipping. Some companies even buy samples of their product from stores and then test the pH. These steps are taken to ensure consumer satisfaction with the product.

Understanding Concepts

1. What types of solutes are electrolytes when they dissolve in water to form solutions?

2. Write a balanced chemical equation to describe what happens when each of the following compounds is placed in water:
   (a) $NaI_{(s)}$
   (b) $HCl_{(g)}$
   (c) $Sr(OH)_{2(s)}$

3. (a) State and explain, in your own words, the Arrhenius theory of acids.
   (b) State and explain, in your own words, the Arrhenius theory of bases.

4. Each of the following compounds is added to water and mixed thoroughly. List the formulas of the ions, atoms, or molecules that are present in the resulting solution. (Do not write ionization or dissociation equations.)
   (a) potassium hydroxide
   (b) hydrogen nitrate
   (c) hydrogen acetate
   (d) sodium bromide

5. Write a dissociation equation to explain the electrical conductivity of each of the following solutions:
   (a) lithium hydroxide
   (b) aluminum hydroxide
   (c) barium hydroxide

6. What is the evidence for strong and weak acids? Explain the difference between strong and weak acids in terms of degree of ionization.

7. Explain the terms "concentrated solution" and "dilute solution."

8. What is the hydrogen ion concentration in each of the following solutions?
   (a) pure water, pH = 7
   (b) household ammonia, pH = 11
   (c) vinegar, pH = 2
   (d) carbonated drink, pH = 4
   (e) drain cleaner, pH = 14

9. Express the hydrogen ion concentration in each of the following solutions as a pH value:
   (a) grapefruit juice, $[H^+] = 10^{-3}$ mol/L
   (b) rainwater, $[H^+] = 10^{-5}$ mol/L
   (c) milk, $[H^+] = 10^{-7}$ mol/L
   (d) soap, $[H^+] = 10^{-10}$ mol/L

10. What is the pH of pure water or any neutral aqueous solution at 25°C?

11. A water sample has a pH of 5. By what factor does the hydrogen ion concentration change to achieve a pH of 7?

12. What is the molar concentration of each ion in a 0.15-mol/L solution of each of the following chemicals? (Assume strong acids and strong bases.)
    (a) hydrobromic acid, $HBr_{(aq)}$
    (b) potassium hydroxide, $KOH_{(aq)}$
    (c) fertilizer, $NH_4Cl_{(aq)}$

13. Which solution has a higher concentration of hydrogen ions: a 0.2-mg/100-mL solution of hydrochloric acid or a 3-ppm solution of hydrochloric acid? Explain.

Applying Inquiry Skills

14. Describe a diagnostic test or a simple procedure that would distinguish between a solution of an acid and a solution of a base.

15. During an investigation, a chemist adds equal amounts of five different chemicals to equal volumes of water. The chemist then tests the chemicals for their solubility, and the aqueous solutions for their electrical conductivity and effect on litmus paper. Complete the Analysis in the following lab report:

    Question
    Which of the chemicals, numbered 1 to 5 in **Table 3**, is which of the following compounds: $KCl_{(s)}$, $NaOH_{(s)}$, $Mg_{(s)}$, $HC_2H_3O_{2(l)}$, $C_{12}H_{22}O_{11(s)}$?

    Analysis
    Based on the observations in **Table 3**, identify each chemical. Explain your reasons.

Table 3 Observed Properties of Substances

Chemical	Solubility in water	Electrical conductivity of solution	Effect of solution on litmus
1	high	low	blue to red
2	high	high	red to blue
3	none	none	no change
4	high	none	no change
5	high	high	no change

Hazardous products are found not only in industries, but also in homes (**Table 1**). Some of these products require special precautions for safe use and disposal. Instructions for environmentally safe disposal are not required on the labels, however, despite the toxic or corrosive chemicals that many of these products contain (**Figure 1**). The manufacturer and the consumer of a hazardous product both have a responsibility for properly disposing of the product's container and residue. Unfortunately, safe disposal methods do not exist for some hazardous products, such as herbicides and pesticides. Other products, such as leftover paints and old batteries, should be dropped off at special collection facilities, although some communities do not have such facilities.

Table 1 Categories of Household Hazardous Products

Category	Examples
household products	cleaners, disinfectants, household batteries, pharmaceuticals
paint products	latex paint, oil-based paint, stains and finishes, thinners, solvents
automobile products	motor oil, antifreeze, transmission fluid, brake fluid, lead-acid batteries
garden products	fungicides, herbicides, poisons, fertilizers

Figure 1
Corrosive substances, such as acids and bases, are found in many household products, including cleaners.

What happens if hazardous household products are not disposed of properly? Hazardous wastes that are thrown out with regular garbage can contaminate landfill sites. Residue wastes that are poured down the drain may contaminate surface water, while wastes that are buried can contaminate ground water.

Disposing of hazardous household products properly can be expensive. The cost of disposing of one can of paint at a household hazardous waste depot has been estimated to be $80. Typically, 75% of the hazardous wastes collected at a waste depot are paints.

Along with paint, household batteries represent a significant portion of hazardous waste that finds its way into municipal landfills. Batteries are used to power large and small devices, from watches, pagers, and cell phones to clocks, computers, power tools, and vehicles. All batteries contain toxic substances. Some batteries contain heavy metals (such as mercury, cadmium, and lead), which are toxic and bioaccumulative. Batteries that end up in municipal landfills and trash incinerators can disperse significant amounts of heavy metals and other toxic substances into the air and water. For this reason, battery waste prevention and recycling strategies are essential.

Figure 2
How can you safely dispose of products with these symbols on their labels?

Does your community have a recycling depot for batteries? What about other household products (**Figure 2**)? How can you safely dispose of these household products? Is the safe disposal of these household products solely the responsibility of the waste treatment engineer at the waste treatment facility?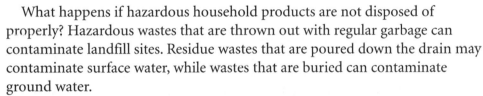

What happens to hazardous waste products once they are dropped off at a collection depot? The products are separated and sent to a waste treatment facility, where they are treated or destroyed. Waste treatment engineers design equipment and develop processes, based on engineering and chemistry principles, to treat the waste in treatment facilities. They also assess existing facilities to see how the facilities can be adapted to treat the increasing amounts of waste produced by a growing population.

Waste treatment engineers promote environmental sustainability by promoting good waste management practices. They encourage the "3 Rs" (reduce, reuse, and recycle) and help to develop residential composting programs.

In addition to understanding mathematics and science, waste treatment engineers need drawing and design skills. They also need excellent decision-making skills and communication skills because they work closely with government officials, contractors, architects, and clients. They must be able to analyze data, review calculations, and prepare cost estimates. Computer skills are imperative, especially the ability to work with specialized software. A successful waste treatment engineer should enjoy being innovative, doing work that requires precision, and making decisions about the methods of treatment needed for wastes that are brought to the plant.

(i) List the types of work that a waste treatment engineer would typically do.
(ii) Research the educational requirements that are needed to become a waste treatment engineer.
(iii) Research the different ways that waste can be treated. What are the different waste categories? Provide two examples of substances in each category.

 www.science.nelson.com

Making Choices

You make choices every day about what to wear and what to eat. Do you make choices that affect the environment: for example, how to dispose of hazardous household products that require special handling and disposal? What choices do you have for disposing of these products? Do you take them to the appropriate municipal waste disposal facility, or do you put them in household garbage for curbside collection? Statistics show that less than 10% of Canadian households participate in household hazardous waste disposal programs offered in their communities. How do you choose the best way to dispose of household products?

You can follow a decision-making model to help you make appropriate choices for any issue. A decision-making model is explained in detail in Appendix A2.

Risk–Benefit Analysis

In section 4.1, you examined your consumption of water. Did you apply any of the steps of the decision-making model when you did so? Perhaps a risk–benefit analysis would have helped you determine which water uses you would give up if necessary. What is a risk–benefit analysis? How do you do one?

The risk that is associated with an action is the potential for harm associated with the action. The harm can be financial, medical, or social. The risk is related to both the likelihood of the event occurring and the severity of its consequences. People have different perceptions of risk, however. Therefore, a risk to one person may be a benefit to another person. A benefit is an action that is advantageous. For example, let's look at the risks and benefits associated with producing an electric car. A risk to the manufacturer is that the car may not be accepted by consumers and may not sell. The benefit is that the car will not produce as many air pollutants as a car that exclusively burns fossil fuels. Does the benefit outweigh the risk involved in producing an electric car? (You will have an opportunity to find out more about electric vehicles in Unit 5.)

An analysis of the risks and benefits that are associated with a decision can help you determine the best course of action. For example, consider the risk–benefit analysis done by an individual who used an asthma inhaler that contained a CFC propellant (**Table 2**). Assigning cost and probability values are very subjective, since they are based on the individual's values and perspectives on the issue. In this case, the risks outweighed the benefits. The individual lobbied for the manufacturers of asthma inhalers to eliminate the use of CFCs in the inhalers. Canada, in fact, has established a phase-out schedule for CFC inhalers, which will be prohibited as of 2005.

Table 2 Risk–Benefit Analysis of Using Ozone-Depleting CFCs

Risks				Benefits			
Possible result	Cost of result (scale of 1 to 5)	Probability of result occurring (%)	Cost × probability	Possible result	Benefit of result (scale of 1 to 5)	Probability of result occurring (%)	Benefit × probability
CFCs reduce the ozone in the stratosphere.	very serious (5)	conclusive research (100%)	500	CFCs are cost-effective propellants in inhalers.	high (4)	somewhat likely (25%)	100
CFCs remain in the atmosphere for more than 10 years.	very serious (5)	likely (80%)	400	CFC inhalers have always worked for me.	high (3)	likely (40%)	120
Total risk value			**900**	**Total benefit value**			**220**

▶ **Understanding the Issue**

Understanding the Issue

1. Look around your home at the variety of commercial products. How can you determine which, if any, of these products are hazardous?

2. (a) Identify the four broad categories of hazardous household products.
   (b) Give two examples of products in each category.

3. For most hazardous products, there are no instructions for proper disposal given on the labels. Why do you think this is the case?

4. Suggest three strategies for reducing the quantity of waste batteries that are appearing in municipal landfill sites.

 www.science.nelson.com

5. Summarize the decision-making process. Briefly explain the steps that are involved in the process.

▶ **Take a Stand**

How Will I Dispose of a Hazardous Household Product?

1. Use the decision-making model outlined in Appendix A2 to decide how you will dispose of common household products.

(a) Work in small groups to identify two hazardous commercial products that are found in many Canadian homes. Identify both the hazard and the level of the hazard from the product label. Research the component(s) of each product that require careful handling and disposal.

 www.science.nelson.com

(b) Brainstorm the various disposal options that are available for each household product. Examine each option from an economic perspective, a social perspective, and an environmental perspective.

Decision-Making Skills

- Define the Issue
- Defend a Decision
- Analyze the Issue
- Identify Alternatives
- Research
- Evaluate

(c) Conduct a risk–benefit analysis of two possible solutions: disposal in the normal waste stream (curbside collection) or disposal at an identified waste disposal facility.

(d) Decide which solution you will follow. Be prepared to defend your decision in a position paper that you present to your class, or in a short multi-media presentation that is suitable for a school assembly.

Reflection

2. Think about the data you collected on household water uses in section 4.1 and any decisions you made about your water uses. Re-examine the data, and use the decision-making model to re-evaluate your decisions. Would you still make the same decisions?

Dilution and pH

Many consumer products, such as juices and cleaners, are purchased in a concentrated form and diluted with water before use. For example, water is added to make orange juice from frozen concentrate. Water is also added to many floor cleaners. A concentrated product is often less expensive to buy because there are less costs involved in transporting and shipping the product, and less packaging is required. Less packaging also means that the product is more environmentally friendly.

Sometimes acids are diluted with water. Unlike diluting concentrated orange juice, however, care must be taken because heat is often produced when acids are diluted with water. If a large volume of concentrated acid solution is added to a small volume of water, the water may boil and splatter some of the acid out of the container. For this reason, you should always add the acid slowly to the water (**Figure 1**), while constantly stirring the solution to dissipate the heat.

In Part 1 of this investigation, you will sequentially dilute a 0.1-mol/L acid solution by withdrawing 10-mL samples and diluting them with water to a

Inquiry Skills

○ Questioning	○ Planning	● Analyzing
○ Hypothesizing	● Conducting	● Evaluating
● Predicting	● Recording	● Communicating

volume of 100 mL. You will then measure the pH values of the resulting solutions using a pH meter and estimate the pH values using universal pH paper. Finally, you will establish a set of standard indicator colours that correspond to the pH values. In Part 2, you will determine the pH of an acidic solution by colorimetric analysis.

Question

What effect does the dilution of an acidic solution have on the pH of the solution?

Prediction

(a) Predict the effect that the dilution of a solution has on the pH of the solution. Consider how the number of hydrogen ions in a solution changes as the solution is diluted.

Materials

eye protection
lab apron
wash bottle of pure water
distilled water
10-mL graduated or volumetric pipette (**Figure 2**)
pipette bulb
0.1-mol/L hydrochloric acid, $HCl_{(aq)}$
five 100-mL volumetric flasks with stoppers
250-mL beaker
five 100-mL beakers
universal indicator
6 test tubes
test-tube rack
solution of unknown concentration
pH meter

Figure 1
When diluting all concentrated acids, always add the acid to water.

Part 1: Dilution

Procedure

1. Obtain 50 mL of 0.1-mol/L hydrochloric acid in a 250-mL beaker.

    Hydrochloric acid, even when dilute, is corrosive. If any of the solution is spilled on your skin or clothing, wash the area with plenty of cold water. Report spills immediately to your teacher.

2. Label the volumetric flasks with the numbers 1 to 5.

3. Add 40 mL of distilled water to flask 1. Rinse a pipette first with distilled water and then with the solution.

4. Use the pipette to add 10 mL of 0.1-mol/L hydrochloric acid to the flask (**Figure 3(a)**). Add distilled water to the 100-mL mark (**Figure 3(b)**). Stopper the flask, and invert it several times to thoroughly mix the contents (**Figure 3(c)**).

5. Rinse the pipette with distilled water and then with a small amount of the solution from flask 1.

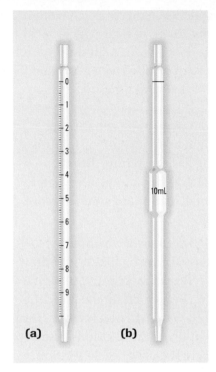

(a)   **(b)**

Figure 2
(a) A graduated pipette measures a range of volumes.
(b) A volumetric pipette is calibrated to give a fixed volume.

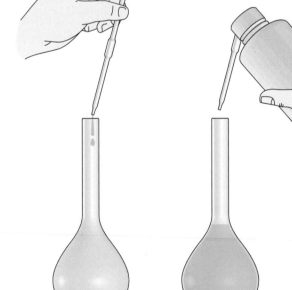

(a)   **(b)**   **(c)**

Figure 3
(a) A pipette is used to add 10 mL of hydrochloric acid to a flask of distilled water.
(b) The initial amount of hydrochloric acid is not changed by adding water to the flask.
(c) In the final dilute solution, the initial amount of hydrochloric acid is still present, but it is distributed throughout a larger volume. It is diluted.

6. Add 40 mL of distilled water to flask 2. Use the pipette to withdraw 10 mL of hydrochloric acid from flask 1 and add it to flask 2. Fill the flask to the 100-mL mark with distilled water. Stopper the flask, and invert it several times to mix the contents.

7. Repeat steps 5 and 6 three more times using flasks 3, 4, and 5, and taking 10 mL of solution from the preceding flask each time. Be sure to rinse the pipette each time with distilled water and then with a small amount of solution from the preceding flask.

8. Number five small beakers with the numbers 1 to 5, to match each flask. Using a pipette, withdraw 50 mL from each volumetric flask and place it into the appropriate beaker. Be sure to rinse the pipette with distilled water after each withdrawal to avoid contamination of the solutions. Use a pH meter to measure the pH of each solution.

(b) Record the pH for each solution in a table that is similar to **Table 1**.

9. Label five test tubes with the numbers 1 to 5, to match each flask. Using a clean pipette, withdraw 50 mL from each flask and place it in the appropriate test tube. Rinse the pipette with distilled water each time to avoid contamination of the solutions. Add two to three drops of universal indicator solution to each test tube. Place the test tubes in a rack to compare the colours (**Figure 4**).

10. Compare the colours of the solutions with the colours shown on the label of the universal indicator. Assign a pH value to each solution.

(c) Record your observations in your table.

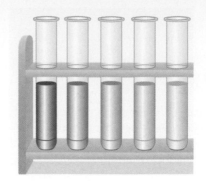

Figure 4
The colour of the solution in each test tube matches the universal indicator colours for pH values from 2 to 6.

Analysis

(d) Calculate the $[H^+]$ in each flask, using the pH values from the indicator. Record your results in your table.

(e) What effect does dilution have on the pH of the solution? Be as specific as possible in your answer.

(f) How does the pH meter reading for each solution compare with the pH value you determined using the indicator scale?

(g) Why was 40 mL of distilled water added to the flask before the hydrochloric acid was added? Explain.

(h) Why did you not put 100 mL of water into the flask in step 3?

Table 1 pH Values of Dilute Acid Solutions

Solution	1	2	3	4	5	unknown
pH from meter						
Universal indicator colour						
pH from indicator						
Concentration of H+ (mol/L)						

Evaluation

(i) What are some sources of error in this part of the investigation? Evaluate the Materials and Procedure for Part 1. What improvements could you make to the Materials and Procedure to reduce your sources of error?

(j) How accurate was your Prediction?

(k) Which method gives the best pH value: the indicator scale or the pH meter? Explain your answer.

Synthesis

(l) For many years, people disposed of liquid wastes into lakes and rivers. They believed that the large volume of water could dilute the wastes. This belief is summarized as the "solution to pollution is dilution." Based on what you have learned in this investigation, should acid waste be dumped into rivers and lakes? Explain your answer.

(m) If a 0.1-mol/L solution of sodium hydroxide, $NaOH_{(aq)}$, had been used in this investigation instead of 0.1-mol/L hydrochloric acid, would you expect the same result? Explain.

(n) Why is the dilution of concentrated solutions commonly performed?

Part 2: pH of a Water Sample
Procedure

11. Pour 50 mL of the solution of unknown concentration into a clean, dry test tube. Add two to three drops of universal indicator to the test tube. Compare the colour of this solution with the colours of the solutions from step 10 to determine the closest match.

(o) Record your observations in your table.

12. Use a pH meter to measure the pH of the unknown sample. Record the pH in your table.

13. Wash your hands thoroughly. Follow your teacher's instructions for disposing of waste materials.

Analysis

(p) Calculate the $[H^+]$ in the water sample. Record your result in your table.

Evaluation

(q) What are some sources of error in this part of the investigation? Evaluate the Materials and Procedure for Part 2. What improvements could you make to the Materials and Procedure to reduce your sources of error?

Synthesis

(r) Colorimetric analysis is the process of determining the concentration and pH of a solution of unknown concentration by matching the colour to solutions of known concentration and pH. A colorimeter mechanically performs the same analysis. Research how a colorimeter works. How might a colorimeter improve the accuracy of your results in this investigation? How do you think the results obtained using a colorimeter would compare with the results obtained using a pH meter?

 www.science.nelson.com

4.10 Investigation

Acid–Base Reactions

A significant number of the chemical substances you encounter daily are acids and bases. For example, vinegar is a dilute solution of acetic acid, many fruit juices contain citric acid, and soap contains lye, or sodium hydroxide.

In Part 1 of this investigation, you will observe some general reactions of acids and bases. In Part 2, you will demonstrate the acid–base character of solutions of oxides formed from metals and nonmetals.

Questions

Part 1: What are some of the characteristic chemical reactions of acids and bases?

Part 2: How do aqueous oxide solutions compare with the substances that are present in acidic precipitation? How can you use the characteristic chemical reactions of acids to help you explain acid precipitation and the damage it causes?

Prediction

(a) Read the Procedure. Identify the reactant(s) in each step, and, where possible, predict the products of the reaction.

(b) Identify the acids that contribute to acid precipitation. Do any of these acids appear as products in this investigation?

Materials

eye protection
lab apron
100-mL beaker
4 test tubes
mossy zinc
3.0-mol/L hydrochloric acid, $HCl_{(aq)}$
wooden splints
sodium hydrogen carbonate (baking soda), $NaHCO_{3(s)}$
marble chips
3.0-mol/L sodium hydroxide solution, $NaOH_{(aq)}$
aluminum pieces
filtered limewater, $Ca(OH)_{2(aq)}$

red and blue litmus paper
glass droppers
piece of charcoal (carbon)
piece of sulfur
piece of magnesium
steel wool (iron)
deflagrating spoon
Bunsen burner
4 gas bottles
4 glass plates
distilled water
oxygen source
aluminum foil

Part 1: Reactions of Acids and Bases

Procedure

1. Obtain 10 mL of hydrochloric acid in a beaker. Place a small piece of mossy zinc in the bottom of a clean, dry test tube. Use a glass dropper to add just enough hydrochloric acid to cover the zinc. Observe the reaction, and record your observations in a table.

    Hydrochloric acid, even when dilute, is corrosive. If any of the solution is spilled on your skin or clothing, wash the area with plenty of cold water. Report spills immediately to your teacher.

2. Bring a lighted splint up to the mouth of the test tube, and observe what happens. Record your observations.

3. Place a pinch of sodium hydrogen carbonate (baking soda) in the bottom of a clean, dry test tube. Use a glass dropper to add ten drops of hydrochloric acid to the baking soda. Use another dropper to add one to two drops of limewater to the inside of the test tube near the top. Observe any changes in the limewater as it trickles down the inside of the test tube. Record your observations.

4. Place a marble chip in the bottom of a test tube. Use a dropper to add just enough hydrochloric acid to cover the marble chip. Use another dropper to add one to two drops of limewater to the inside of the test tube near the top. Observe any changes in the limewater. Record your observations.

5. Add a few pieces of aluminum to the bottom of a clean, dry test tube. Add enough sodium hydroxide solution to cover the aluminum pieces. Gently shake the test tube. Feel the test tube, and observe the contents. Record your observations.

 Sodium hydroxide is corrosive, even in dilute solutions. Any spills on skin or clothing, or in the eyes, should be washed immediately with lots of cold water and reported to your teacher.

6. If a gas is produced in step 5, test the gas with a lighted splint. Record your observations.

Analysis

(c) What did you observe when the lighted splint was placed at the mouth of the test tube in step 2? What does your observation tell you about one of the products of the chemical reaction?

(d) The reaction of zinc metal with hydrochloric acid is an example of a single displacement reaction. Write a complete word equation and a balanced chemical equation to describe this reaction.

(e) What changes did you observe in the limewater in step 3? Based on your observations, what is one of the products? The other products of the reaction in step 3 are sodium chloride and water. Write a complete word equation and a balanced chemical equation to describe this reaction.

(f) What changes, if any, did you observe in the limewater in step 4? Based on your observations, what is one of the products? The other products

of the reaction in step 4 are calcium chloride and water. Write a complete word equation and a balanced chemical equation to describe this reaction.

(g) Describe the chemical test for carbon dioxide gas.

(h) What did you observe in step 6 when you brought a lighted splint to the mouth of the test tube? What do your observations tell you about one of the products of the chemical reaction in step 5?

(i) Answer the Question for Part 1 at the beginning of the investigation.

Evaluation

(j) Did your observations in this part of the investigation enable you to answer the Question for Part 1?

(k) Evaluate your first Prediction.

(l) Identify any sources of uncertainty or error in this part of the investigation. What improvements could you make to the Procedure for this part of the investigation to reduce your sources of error?

Synthesis

(m) The chemical reaction in step 5, between aluminum and a basic solution, is the same reaction that is used to open a plugged drain. Commercial drain cleaners are a mixture of aluminum and sodium hydroxide. These substances react as follows when dissolved in water:

$$Al_{(s)} + NaOH_{(aq)} \rightarrow Na_3AlO_{3(aq)} + H_{2(g)}$$

Balance the skeletal equation. Then explain why this chemical reaction can clear a drain that is clogged with fat.

(n) Why do you think baking soda is used in most cake and bread recipes? Why is milk also an ingredient in many of these recipes?

Part 2: Acidic and Basic Oxides

Procedure

7. Create a table that is similar to **Table 1** to record your observations.

8. Add 5 to 10 mL of distilled water to each of the four gas bottles. Fill each bottle with oxygen from the source (**Figure 1**). Cover the mouth of each bottle with a glass plate to prevent the oxygen from escaping.

   Oxygen supports combustion. Keep it well away from open flames at all times.

9. Line a deflagrating spoon with a double thickness of aluminum foil. Place a pea-sized piece of sulfur on the foil. Heat the sulfur with a Bunsen burner flame until it starts to burn. Place the burning sulfur into the first bottle of oxygen.

   Fumes from burning sulfur are extremely irritating and can trigger an asthma attack. This part of the activity should be performed under a fume hood, if one is available.

10. When the reaction is complete, remove the spoon. Cover the bottle with the glass plate, and shake. Test the resulting solution with red and blue litmus paper. Record your observations in your table.

11. Repeat steps 9 and 10 using a small piece of charcoal. Record your observations in your table.

12. Curl a small piece of magnesium around a pen or pencil to make a coil. Remove the coil from the pen, and hook it to the bottom of the deflagrating spoon. Heat the magnesium in a Bunsen burner flame. Once ignited, be careful not to let the burning magnesium come into contact with the sides of the gas bottle. Slide open the glass plate on one of the bottles, and plunge the deflagrating spoon and its contents into the oxygen. When the reaction is complete, remove the deflagrating spoon, cover the bottle with the glass plate, and shake the contents. Test the resulting solution with red and blue litmus paper. Record your observations.

   Do not look directly at the burning magnesium. The deflagrating spoon is very hot. Do not let it touch the glass plate.

Figure 1
Filling a bottle with oxygen

Table 1 Properties of Oxides

| Element burned | | Observations during burning | | Oxide produced | Indicator colour of oxide solution | pH of oxide solution | Nature of oxide in water (acid, base, or neutral) |
Name	Symbol	In air	In oxygen				
sulfur							
carbon							
magnesium							
iron							

13. Place a small ball of steel wool in a deflagrating spoon, and heat in a flame. Slide open the glass plate on one of the bottles, and plunge the deflagrating spoon and its contents into the oxygen. When the reaction is complete, remove the deflagrating spoon, cover the bottle with the glass plate, and shake the contents. Test the resulting solution with red and blue litmus paper. Record your observations.

14. Dispose of chemical wastes as directed by your teacher. Wash your hands.

Analysis

(o) How did the way the elements burned in air compare with the way they burned in pure oxygen?

(p) What type of solution is formed when metal oxides react with water? What type of solution is formed when nonmetal oxides react with water?

(q) The iron in steel wool burns in oxygen to produce an insoluble oxide. Does the pH of the water change? Why or why not?

(r) Write balanced chemical equations to represent the burning of the elements carbon, sulfur, and magnesium in oxygen, and the reaction of the oxide of each element with water.

(s) Answer the Questions for Part 2 at the beginning of the investigation.

Evaluation

(t) Identify any sources of uncertainty or error in this part of the investigation. What improvements could you make to the Procedure for this part of the investigation to reduce your sources of error?

Synthesis

(u) What might happen if you spilled a soft drink or lemon juice on the metal shelves of a refrigerator and did not clean up your spill immediately?

(v) Oven cleaners are basic. Why should you never use an oven cleaner on aluminum pots and pans?

(w) Buildings, marble statues, and metal bridges all suffer from the effects of acid precipitation. Explain why, in terms of what you have learned in this investigation.

Figure 1
This statue has been affected by acids in the environment.

In your community, have you noticed any marble or limestone statues or buildings that are showing signs of deterioration (**Figure 1**)? Metal window frames in older buildings may also show the same kind of surface breakdown. This damage to limestone and marble, as well as to some metals, is caused by the acids that are present in the environment.

How do acids form in the environment? In section 4.10, you learned that some combustion reactions involving nonmetals produce oxides. The oxides react with water to form acids. For example, carbon dioxide is produced during the combustion of carbon compounds, including petroleum and wood. It is also produced by living organisms during respiration. Carbon dioxide is a greenhouse gas. You will look more closely at greenhouse gases in section 4.15.

Carbon dioxide reacts with water to form carbonic acid:

$$CO_{2(g)} + H_2O_{(l)} \rightarrow H_2CO_{3(aq)}$$

Carbonic acid occurs naturally in the atmosphere and contributes to the slightly acidic nature of precipitation.

Sulfur dioxide is produced during the combustion of coal because coal generally contains sulfur impurities. Sulfur dioxide reacts with water to form sulfurous acid:

$$SO_{2(g)} + H_2O_{(l)} \rightarrow H_2SO_{3(aq)}$$

Sulfurous acid is a component of acid rain, which you will look at in section 4.13.

Characteristic Chemical Reactions of Acids and Bases

Acids and bases participate in a number of characteristic chemical reactions. These reactions are described as "characteristic" because you can get clues about an unidentified substance by seeing how it participates in a reaction and what products are formed. For example, imagine the scene at a train derailment. Emergency response workers see that a tank car has begun to leak (**Figure 2**). They notice that the leaking liquid is reacting with metal. They also notice that bubbles are being produced where the liquid contacts the metal. They know that all acids react with active metals, such as magnesium, $Mg_{(s)}$, zinc, $Zn_{(s)}$, and aluminum, $Al_{(s)}$, to form hydrogen gas, $H_{2(g)}$. They deduce that the spilled chemical is an acid and take appropriate measures.

The acid–metal reaction is a single displacement reaction in which the metal displaces the hydrogen from the acid. For example, iron reacts with hydrochloric acid in the following reaction to produce hydrogen gas and a salt solution:

$$Fe_{(s)} + 2\,HCl_{(aq)} \rightarrow H_{2(g)} + FeCl_{2(aq)}$$

Figure 2
At a train derailment, there is the potential for acid from a leaking tank car to react with metal to produce hydrogen gas.

Another characteristic chemical reaction is the double displacement reaction between a carbonate, such as sodium carbonate, $Na_2CO_{3(s)}$, or calcium carbonate, $CaCO_{3(s)}$, and an acid. The unstable carbonic acid that is produced quickly decomposes to form a salt, carbon dioxide gas, and water. To clean up a spill of hydrochloric acid, for example, emergency workers could add sodium carbonate to the acid:

$$2\,HCl_{(aq)} + Na_2CO_{3(s)} \rightarrow 2\,NaCl_{(aq)} + CO_{2(g)} + H_2O_{(l)}$$

This reaction is called an acid neutralization. The active ion, $H^+_{(aq)}$, has been removed from the spill and converted into "neutral" water, $H_2O_{(l)}$, during the chemical reaction.

Acids also react with bases in another double displacement reaction, called **neutralization**. The products of this neutralization reaction are a salt and water. Neutralization of a strong acid by a strong base causes the pH of the solution to move toward the neutral value of 7. For example, the neutralization of hydrochloric acid with sodium hydroxide is

$$HCl_{(aq)} + NaOH_{(aq)} \rightarrow NaCl_{(aq)} + H_2O_{(l)}$$

In this reaction, the hydrogen ions, $H^+_{(aq)}$, combine with the hydroxide ions, $OH^-_{(aq)}$, to form water, according to the following balanced equation:

$$H^+_{(aq)} + OH^-_{(aq)} \rightarrow H_2O_{(l)}$$

The chloride ions, $Cl^-_{(aq)}$, in hydrochloric acid and the sodium ions, $Na^+_{(aq)}$, in sodium hydroxide remain in solution. If the water is evaporated, these ions will combine to form solid sodium chloride, $NaCl_{(s)}$. Thus, in general, when an acid reacts with a base, water and a salt are formed.

neutralization a reaction between an acid and a base that yields a salt and water

> **LEARNING TIP**
>
> **Salts**
> Recall that a salt is an ionic compound consisting of a positive ion, or cation, bonded to a negative ion, or anion. Two examples of salts are sodium chloride (table salt), Na^+Cl^-, and potassium bromide, K^+Br^-.

Similarly, the neutralization of nitric acid with potassium hydroxide is

$$HNO_{3(aq)} + KOH_{(aq)} \rightarrow KNO_{3(aq)} + H_2O_{(l)}$$

When a concentrated solution of a strong acid is neutralized by a concentrated solution of a strong base, the reaction is highly exothermic (energy is produced) and can result in an explosion. Emergency workers must be aware of this hazard when neutralizing a spill of a concentrated solution of either a strong acid or a strong base.

Characteristic reactions of acids are summarized in **Table 1**.

Table 1   Characteristic Reactions of Acids

Acids React with Metals
acid + active metal → salt + hydrogen gas
$2\,HCl_{(aq)} + Fe_{(s)} \rightarrow FeCl_{2(aq)} + H_{2(g)}$
Acids React with Carbonates
acid + carbonate → salt + carbon dioxide gas + water
$2\,HCl_{(aq)} + Na_2CO_{3(s)} \rightarrow 2\,NaCl_{(aq)} + CO_{2(g)} + H_2O_{(l)}$
Acids React with Bases
acid + base → salt + water
$HCl_{(aq)} + NaOH_{(aq)} \rightarrow NaCl_{(aq)} + H_2O_{(l)}$

titration a laboratory procedure that involves the carefully measured and controlled addition of a solution, usually from a burette, into a measured volume of the sample being analyzed

Figure 3
An initial volume reading is taken before any titrant is added to the sample solution. Then the titrant is added until the reaction is complete (when a small drop of titrant changes the colour of the sample). The final burette reading is taken.

▶ **Practice**

Understanding Concepts

1. (a) Explain what is meant by the expression "characteristic chemical reaction."
   (b) Give an example, different from those given in **Table 1**, of each of the three characteristic reactions of acids.

2. A spill of hydrochloric acid can be "neutralized" by treating the spill with aluminum, sodium hydroxide, or sodium carbonate.
   (a) Write a balanced chemical equation for each reaction.
   (b) List one advantage and one disadvantage of each method of neutralization.

3. Write a balanced chemical equation to represent the neutralization reaction for each of the following acid–base combinations:
   (a) potassium hydroxide and hydrochloric acid
   (b) lithium hydroxide and nitric acid
   (c) potassium hydroxide and sulfuric acid

Acid–Base Titration

A **titration** is a common method of quantitative chemical analysis. It is used to determine the concentration of a substance in solution. A known volume of the sample to be analyzed is usually transferred into a flask (**Figure 3**). The burette contains a solution of an accurately known concentration. This

▶ TRY THIS activity *A Neutralization Reaction*

Acids undergo neutralization reactions with bases. In this activity, you will observe the reaction between hydrochloric acid and sodium hydroxide solution.

Materials: eye protection; lab apron; evaporating dish or 100-mL beaker; 10-mL graduated cylinder; 0.5-mol/L hydrochloric acid, $HCl_{(aq)}$; 0.5-mol/L sodium hydroxide solution, $NaOH_{(aq)}$; phenolphthalein indicator; microdropper; stirring rod; pH paper or universal indicator solution

1. Measure 10 mL of sodium hydroxide solution, and put it into an evaporating dish or beaker. Add one or two drops of phenolphthalein indicator.

2. Use the microdropper to add the hydrochloric acid drop by drop. Stir the solution after each drop. Continue adding drops until the solution just turns colourless.

3. Use a universal indicator or pH paper to determine the pH of the colourless solution.

(a) The neutralization of sodium hydroxide solution by hydrochloric acid is a double displacement reaction. Write a balanced chemical equation for this reaction.

(b) If the neutralized solution were evaporated to dryness, what would you expect to see in the bottom of the container? How could you determine the identity of what you saw?

---

solution is called the **standard solution**. During the titration, the solution in the burette, called the **titrant**, is added drop by drop to the sample. Alternatively, the standard solution can be in the flask, so the solution of unknown concentration is the titrant. The titrant is added until the reaction between the two chemicals is judged to be complete. To identify this point, an indicator is used. The indicator changes colour when the reaction is complete (**Table 2**).

standard solution a solution of precisely and accurately known concentration

titrant the solution in a burette during a titration

Table 2 Indicator Colour Change as the Endpoint of Titration

Indicator	Acidic	Basic
litmus	red	blue
methyl orange	red	yellow
bromothymol blue	yellow	green
phenolphthalein	colourless	red

The **endpoint** is reached when one drop of the titrant changes the colour of the indicator. At the endpoint, no more titrant is added and the volume that has been used is recorded. The mole ratio of reactant and product, from the balanced chemical equation, is then used to determine the concentration of the solution in the flask.

endpoint the point in a titration at which a sharp change in property, such as a colour change, occurs

At least three trials should be performed during a titration analysis to improve the reliability of the concentration calculated. Sometimes more than three trials are necessary to get three consistent volumes (usually ± 0.01 mL). The three most consistent volumes are averaged before calculating the concentration.

Titration is used to analyze acids in the environment, such as the acids that may be present in freshwater sources. Titration is also used for quality control in industrial and commercial operations and for scientific research.

Quantitative Analysis Using Titration

The concentration of hydrochloric acid can be analyzed by titration with sodium hydroxide solution. Three 10.0-mL samples of hydrochloric acid are titrated with a standardized 0.200-mol/L solution of sodium hydroxide. The results for the three trials are shown in Table 3. What is the concentration of hydrochloric acid?

Table 3   Titration of $HCl_{(aq)}$ with $NaOH_{(aq)}$

Trial	1	2	3	Average
Final burette reading	13.85 mL	26.95 mL	39.85 mL	
Initial burette reading	0.70 mL	13.90 mL	26.90 mL	
Volume of $NaOH_{(aq)}$ added	13.15 mL	13.05 mL	12.95 mL	13.05 mL

Step 1: Write Balanced Chemical Equation for Reaction, and List Given Values

$$HCl_{(aq)} + NaOH_{(aq)} \rightarrow NaCl_{(aq)} + H_2O_{(l)}$$

$v_{HCl} = 10.0 \text{ mL} = 0.0100 \text{ L}$

$v_{NaOH} = 13.05 \text{ mL} = 0.01305 \text{ L}$

$c_{NaOH} = 0.200 \text{ mol/L}$

Step 2: Calculate Amount in Moles of Standard Solution Required for Complete Reaction

$n_{NaOH} = v_{NaOH} c_{NaOH}$

$\qquad = 0.01305 \text{ L} \times 0.200 \text{ mol/L}$

$n_{NaOH} = 0.002\,61 \text{ mol}$

Step 3: From Balanced Equation, Determine Amount in Moles of Titrant Required for Complete Reaction

From the balanced equation, you know that 1 mol of $NaOH_{(aq)}$ requires 1 mol of $HCl_{(aq)}$ for complete reaction. Therefore, at the endpoint, the amount in moles of $NaOH_{(aq)}$ added is equal to the amount in moles of $HCl_{(aq)}$.

$n_{NaOH} = n_{HCl} = 0.002\,61 \text{ mol}$

Step 4: Calculate Concentration of Titrant

Since you know that there was 0.002 61 mol of HCl in the 0.0100-L sample, you can calculate the concentration of hydrochloric acid.

$$c_{HCl} = \frac{n_{HCl}}{v_{HCl}}$$

$$= \frac{0.002\,61 \text{ mol}}{0.0100 \text{ L}}$$

$$c_{HCl} = 0.261 \text{ mol/L}$$

The concentration of hydrochloric acid is 0.261 mol/L.

Example

Hydrochloric acid, $HCl_{(aq)}$, is labelled as having a concentration of 0.235 mol/L. To check that the concentration on the label is correct, the solution is analyzed by titration. A 10.0-mL standard solution of 0.150-mol/L sodium carbonate, $Na_2CO_{3(aq)}$, is used for the titration. The results for three trials are shown in **Table 4**. What is the concentration of hydrochloric acid?

Table 4 Titration of $Na_2CO_{3(aq)}$ with $HCl_{(aq)}$

Trial	1	2	3	Average
Final burette reading	13.35 mL	26.05 mL	38.85 mL	
Initial burette reading	0.55 mL	13.30 mL	26.00 mL	
Volume of $HCl_{(aq)}$ added	12.80 mL	12.75 mL	12.85 mL	12.80 mL

Solution

$$2\,HCl_{(aq)} + Na_2CO_{3(aq)} \rightarrow H_2CO_{3(aq)} + 2\,NaCl_{(aq)}$$

$$v_{HCl} = 12.80\ mL = 0.01280\ L$$
$$v_{Na_2CO_3} = 10.0\ mL = 0.0100\ L$$
$$c_{Na_2CO_3} = 0.150\ mol/L$$

$$n_{Na_2CO_3} = v_{Na_2CO_3}\,c_{Na_2CO_3}$$
$$= 0.0100\ L \times 0.150\ mol/L$$
$$n_{Na_2CO_3} = 0.001\ 50\ mol$$

$$n_{HCl} = 0.001\ 50\ mol\ Na_2CO_3 \times \frac{2\ mol\ HCl}{1\ mol\ Na_2CO_3}$$

$$n_{HCl} = 0.003\ 00\ mol$$

$$c_{HCl} = \frac{n_{HCl}}{v_{HCl}}$$
$$= \frac{0.003\ 00\ mol}{0.01280\ L}$$
$$c_{HCl} = 0.234\ mol/L$$

The concentration of hydrochloric acid is 0.234 mol/L, which is close to the concentration on the label.

Understanding Concepts

Answers

4. 0.0225 mol/L

5. 0.460 mol/L

6. 2.20 mol/L

7. 0.024 mol/L

8. 2.92 mol/L

4. A 25.0-mL water sample is analyzed by titration to determine its sulfurous acid, $H_2SO_{3(aq)}$, content. A standard 0.105-mol/L solution of sodium hydroxide, $NaOH_{(aq)}$, is used for the titration. The results for three trials are shown in **Table 5**. What is the concentration of sulfurous acid in the water sample?

Table 5 Titration of $H_2SO_{3(aq)}$ with $NaOH_{(aq)}$

Trial	1	2	3	Average
Final burette reading	11.15 mL	21.75 mL	32.40 mL	
Initial burette reading	0.30 mL	11.15 mL	21.70 mL	
Volume of NaOH added	10.85 mL	10.60 mL	10.70 mL	10.72 mL

5. An average volume of 57.50 mL of 0.200-mol/L standardized lithium hydroxide solution, $LiOH_{(aq)}$, is required to neutralize 25.0-mL samples of hydrobromic acid, $HBr_{(aq)}$, in a titration reaction. What is the concentration of the hydrobromic acid solution?

6. An average of 86.19 mL of a 0.765-mol/L standard sodium hydroxide solution, $NaOH_{(aq)}$, is used in a titration analysis of 30.0-mL samples of a hydrochloric acid solution, $HCl_{(aq)}$. What is the concentration of hydrochloric acid?

7. Analysis shows that 9.44 mL of 0.050-mol/L potassium hydroxide, $KOH_{(aq)}$, is needed to titrate a 10.0-mL water sample that is suspected of containing sulfuric acid, $H_2SO_{4(aq)}$. Determine the molar concentration of the sulfuric acid in the water.

8. Hydrochloric acid, $HCl_{(aq)}$, is used in a product that removes hard-water scale from electrical kettles. An average of 150.01 mL of 0.0974-mol/L barium hydroxide solution, $Ba(OH)_{2(aq)}$, is required to react completely with 10.0 mL of the kettle-scale remover in a titration. What is the molar concentration of the acid in the scale remover?

► Section 4.11 Questions

Understanding Concepts

1. Describe three characteristic reactions of acids.

2. Write a complete, balanced chemical equation to represent the neutralization of each of the following pairs of solutions:
   (a) potassium hydroxide and hydrochloric acid
   (b) lithium hydroxide and hydrobromic acid
   (c) sodium hydroxide and nitric acid
   (d) calcium hydroxide and sulfuric acid
   (e) barium hydroxide and hydrochloric acid
   (f) magnesium hydroxide and acetic acid

3. Why are several trials done in a titration?

4. Four samples of sulfuric acid were titrated consecutively with 0.484-mol/L sodium hydroxide solution to the endpoint. The initial volume reading and the final volume reading after each titration are shown in **Figure 4**. Calculate the concentration of the sulfuric acid.

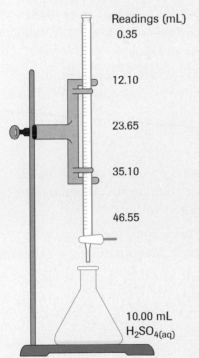

Readings (mL)
0.35

12.10

23.65

35.10

46.55

10.00 mL
$H_2SO_{4(aq)}$

Figure 4
Sodium hydroxide titrant is added to samples of sulfuric acid in successive trials.

5. Phenolphthalein indicator was used in the titration of 25 mL of acid solution of unknown concentration. The acid was titrated with a standard solution of sodium hydroxide, $NaOH_{(aq)}$. The results are given in **Table 6**. What volume of $NaOH_{(aq)}$ would you use to calculate the concentration of the acid? Explain your reasons.

Table 6 Titration of an Acid of Unknown Concentration with $NaOH_{(aq)}$

Trial	Volume of $NaOH_{(aq)}$	Indicator colour
1	20.53 mL	light pink
2	21.55 mL	dark pink
3	20.30 mL	light pink
4	20.65 mL	light pink

6. A chemical spill of hydrobromic acid, $HBr_{(aq)}$, can be neutralized with sodium hydroxide, $NaOH_{(aq)}$, or sodium carbonate, $Na_2CO_{3(aq)}$. Write a balanced chemical equation for each reaction.

Applying Inquiry Skills

7. Complete the Experimental Design for the following investigation to determine the relative strengths of some acids.

   Question
   What is the order of the acids, from fastest to slowest, in their reaction with a fixed mass of magnesium, $Mg_{(s)}$?

   Experimental Design
   Write a description of an experimental design that would allow you to complete **Table 7**. Identify the independent, dependent, and controlled variables.

Table 7 Acidity of 0.10-mol/L Acid Solutions

Acid solution	Chemical formula	Time for reaction
hydrochloric acid	$HCl_{(aq)}$	
acetic acid	$HC_2H_3O_{2(aq)}$	
hydrofluoric acid	$HF_{(aq)}$	
formic acid	$HCHO_{2(aq)}$	
nitric acid	$HNO_{3(aq)}$	
hydrocyanic acid	$HCN_{(aq)}$	

Making Connections

8. Lime (calcium oxide, $CaO_{(s)}$) is often added to soil to make it more basic. Lime, however, is not very soluble in water. Why is this property an advantage for a soil additive?

Quality Control of Vinegar

Every consumer product is required, by law, to have the minimum quantity of the active ingredient listed on the label. Manufacturers and government consumer affairs departments employ analytical chemists to monitor and check these products, using a concept called quality control. Analytical chemists and technicians ensure that production standards, such as the quantity of the active ingredient, are met.

Imagine that you are working for a consumer affairs department. You have received a complaint that a school cafeteria has been serving watered-down vinegar to the students. You need to test the acetic acid concentration of the vinegar to determine whether it matches the concentration on the label: 5.0% acetic acid.

Question

What is the molar concentration of acetic acid in a sample of vinegar from the school cafeteria?

Materials

eye protection
lab apron
vinegar sample of unknown concentration
0.10-mol/L standard sodium hydroxide solution, $NaOH_{(aq)}$
phenolphthalein indicator
distilled water
25-mL volumetric pipette
50-mL burette
ring stand with burette clamp
250-mL volumetric flask with stopper
250-mL Erlenmeyer flask
25-mL graduated pipette

Procedure

1. Rinse and fill the burette with 0.10-mol/L sodium hydroxide solution, $NaOH_{(aq)}$. Record the starting volume to the nearest 0.01 mL.

 Sodium hydroxide is corrosive, even in dilute solutions. Any spills on skin or clothing, or in the eyes, should be washed immediately with lots of cold water and reported to your teacher.

2. Vinegar is too concentrated to be titrated directly. To dilute the vinegar solution, pipette 25 mL of vinegar into a clean 250-mL flask. Add distilled water to bring the final volume up to the 250-mL mark. Stopper the flask, and invert it several times to mix the contents.

3. Pipette 25 mL of your diluted vinegar solution into an Erlenmeyer flask. Add two to three drops of phenolphthalein indicator to the diluted vinegar sample.

4. Titrate the diluted vinegar sample with the standard sodium hydroxide solution from the burette until a single drop produces a change from colourless to faint pink. This point is the endpoint. Record the final volume of the burette reading to the nearest 0.01 mL.

5. Repeat steps 3 and 4 at least two more times until you get three consistent results. Record all your observations and data.

6. Dispose of surplus reactants and all product solutions as directed by your teacher. Wash your hands.

Analysis

(a) Determine the average volume of 0.10-mol/L sodium hydroxide solution that is required to neutralize the diluted vinegar.

(b) Answer the Question. Since the vinegar was diluted by a factor of 10, your answer will have to be multiplied by 10 to get the concentration of the vinegar in the cafeteria sample.

(c) Compare your answer with the 5% acetic acid concentration listed on the label.

Evaluation

(d) Evaluate your observations. Is the method of titration as a quantitative analysis tool adequate for solving quality control issues such as this one?

(e) Is someone in the cafeteria diluting the vinegar?

Have you ever visited northern Ontario and observed how clear and clean the water in some lakes appears (**Figure 1**)? You may have even thought about how little pollution there was in the water. However, a lake that appears this clean and pollution-free may be dead. When the blue sky fills with grey clouds and rain starts to fall in the lake, the rain may be as acidic as vinegar!

Acid-Forming Pollutants

Normal rain is usually slightly acidic, with a pH of 5.6, because carbon dioxide in the air dissolves in moisture in the atmosphere to form dilute carbonic acid, $H_2CO_{3(aq)}$.

$$CO_{2(g)} + H_2O_{(l)} \rightarrow H_2CO_{3(aq)}$$

Carbon dioxide is naturally present in the atmosphere. All living organisms produce carbon dioxide during respiration. Carbon dioxide is also produced during volcanic eruptions and forest fires. Human activities, such as burning fossil fuels in automobiles and burning wood for heat, cause levels of carbon dioxide to increase.

Nitrogen oxides, which are collectively represented by the formula $NO_{x(g)}$, are naturally produced by lightning strikes and plant decay. Nitrogen oxides react with water in the atmosphere to form a weak acid, nitrous acid, $HNO_{2(aq)}$, and a strong acid, nitric acid, $HNO_{3(aq)}$. Both acids contribute to the natural acidity of rain.

$$2\,NO_{2(g)} + H_2O_{(l)} \rightarrow HNO_{3(aq)} + HNO_{2(aq)}$$

Nitrogen oxides are also produced during the combustion of fossil fuels in an internal combustion engine.

Volcanic eruptions produce sulfur oxides, $SO_{x(g)}$, which also contribute to the formation of acids in the atmosphere. Sulfur dioxide, $SO_{2(g)}$, is a product of the combustion of coal and oil, and the smelting of sulfur-containing ores. These two activities release more sulfur dioxide into the atmosphere than all the sulfur dioxide produced from natural sources.

Sulfur dioxide reacts with water in the atmosphere to form sulfurous acid, $H_2SO_{3(aq)}$, a weak acid.

$$SO_{2(g)} + H_2O_{(l)} \rightarrow H_2SO_{3(aq)}$$

Both oxygen and ozone, $O_{3(g)}$, which makes up smog, convert some sulfur dioxide to sulfur trioxide, $SO_{3(g)}$. Sulfur trioxide reacts with water to form sulfuric acid, $H_2SO_{4(aq)}$, a strong acid.

$$SO_{3(g)} + H_2O_{(l)} \rightarrow H_2SO_{4(aq)}$$

Nitrogen oxides and sulfur oxides, in the form of gases, and other airborne pollutants, in the form of particles (solids), can be carried long distances in the atmosphere from their point of origin. These distances give the oxides

Figure 1
An area of blue sky and clear lakes

Chemistry in the Environment **325**

acid precipitation any form of natural precipitation that has an unusually high acidity (pH less than 5.6)

acid deposition acid-forming pollutants that fall to Earth as wet deposition (such as rain, hail, drizzle, fog, and snow) and as dry deposition (such as dust and other particulate matter)

Figure 2
An Environment Canada acid precipitation collector

time to react with water vapour to form the acids that are responsible for decreasing the pH of the precipitation to below 5.6 and forming **acid precipitation**, commonly called "acid rain." Acid precipitation is also called **acid deposition** because the acids can fall to Earth wet (as rain, snow, hail, or sleet) or dry (as dust or other particulate matter) (**Figure 2**).

Neutralizing Acid Precipitation

Acid precipitation can be neutralized by basic substances, such as limestone (calcium carbonate), $CaCO_{3(s)}$, deposits in bedrock, or by dilution.

$$2\,HNO_{3(aq)} + CaCO_{3(s)} \rightarrow Ca(NO_3)_{2(aq)} + CO_{2(g)} + H_2O_{(l)}$$

Fortunately, much of the bedrock in the area around the Great Lakes is limestone. As a result, the Great Lakes are able to neutralize the acid precipitation that collects in them. Even if the Great Lakes did not have a lakebed of limestone, however, it would take years for the pH of the water to decrease significantly. Several millimetres of acid precipitation would have little effect on the pH of a large volume of water, since the water dilutes the acid precipitation. You observed this effect in section 4.9. Over time, though, dilution loses its effectiveness, even in large lakes.

When the environment cannot neutralize acid deposition, damage occurs to lakes, forests, wildlife, and even people. The problem of acid deposition is particularly severe in central and northern Ontario, where the rock does not contain limestone. When limestone is not present to neutralize acid precipitation, lakes can become acidified. Acidified lakes do not support the same variety of life (**Table 1**) as healthy lakes, which generally have a pH higher than 6.0. Acidified lakes cause normally insoluble aluminum hydroxide to react with hydrogen ions to form soluble aluminum ions. Aluminum can accumulate in fish to a toxic level.

$$Al(OH)_{3(s)} + 3\,H^+_{(aq)} \rightarrow Al^{3+}_{(aq)} + 3\,H_2O_{(l)}$$

In a similar chemical reaction, acid precipitation causes an increase in levels of toxic metals, such as copper, lead, and mercury. Toxic metals in sources of drinking water can rise to levels where the metals can affect human health.

Table 1   Effect of Water pH on Aquatic Life in Lakes and Rivers

pH	Effect
6.0	• Crustaceans, insects, and some plankton species begin to disappear.
5.0	• Changes occur in the plankton community. • Less desirable species of mosses and plankton appear. • Some fish populations die out. Smallmouth bass, walleye, brook trout, and salmon are the least tolerant of acidity. • Most fish eggs cannot hatch.
<5.0	• There are no fish. • The lake bottom is covered with undecayed material. • The surrounding shoreline may be dominated by mosses. • Terrestrial animals that depend on aquatic ecosystems are affected.

Acid rain and the damage it causes were first noticed in the late 1970s. Abnormally high acid levels were detected in precipitation and lakes in Nova Scotia. Severe losses in fish populations were also detected in acidified lakes southwest of Sudbury, Ontario. Sudbury, located among the granite and igneous rocks of the Canadian Shield, has minimal ability to neutralize acid precipitation.

Initial attempts were made to neutralize the acids that had accumulated in the lakes and surrounding soil in the Canadian Shield, using a process called liming. These attempts resulted in varying degrees of success. *Liming* refers to the addition of limestone, primarily in the form of calcium carbonate, $CaCO_{3(s)}$, to neutralize acidic lake water and surrounding soil. For example, calcium carbonate added to sulfuric acid produces calcium sulfate and carbonic acid:

$$CaCO_{3(s)} + H_2SO_{4(aq)} \rightarrow CaSO_{4(s)} + H_2CO_{3(aq)}$$

Carbonic acid, $H_2CO_{3(aq)}$, is a weak acid that does not ionize to the same extent that sulfuric acid does. Much of it remains in its molecular form. Calcium sulfate, on the other hand, dissociates slightly in water to produce aqueous calcium ions, $Ca^{2+}_{(aq)}$, an essential plant nutrient. Aqueous calcium ions help to restore plant life on lake bottoms and in the surrounding soils. Calcium ions are also used by shellfish for the development of their protective shells, and by young and adult fish for scale formation and bone development.

There are many advantages to using lime for neutralizing acidic water and soil. Lime is inexpensive, since it occurs naturally. It is also easy to distribute, and it dissolves easily in water. The short-term effects of liming include improved water clarity and restoration of habitat for many aquatic organisms, both plant and animal. After an acidic lake has been limed, the populations of aquatic organisms increase and return to normal levels.

The problem with liming acidic lakes is that applications need to be repeated to keep the water from returning to acidic conditions. Liming is now viewed as a short-term solution for use in specific lakes. The focus on solving the problem of acid precipitation has shifted to the reduction or elimination of acidic oxides at their sources.

Solutions to the Problem of Acid Rain

Acid-forming pollutants can move across the United States from the Ohio Valley, Cleveland, and Detroit areas into Ontario and Quebec (**Figure 3**). For example, about 50% of the sulfur oxides deposited in Canada come from sources in the United States. Similarly, air masses that carry pollution from central Canada drift southeastward to northern New York, Vermont, and Maine.

As more became known about the flow of acid deposition back and forth between the United States and Canada, the problem became a source of political tension between the two countries. Environmental groups worked to educate the media and the public about acid rain, and to encourage political action. Political lobbying has achieved a number of significant agreements between the two countries, to reduce the cross-border flow of acid-forming pollutants.

> ### DID YOU KNOW ?
>
> **Who Survives the Rain?**
> Frogs have the greatest tolerance for acidic conditions. They can survive in water with a pH of 4.0.

> ### DID YOU KNOW ?
>
> **A Towering Solution?**
> One solution to the problem of acid rain was to build very tall smokestacks, which released the acidic oxides higher into the atmosphere. The idea was to disperse or dilute the acid-forming gases before they could react with water to form acid rain. Unfortunately, the taller smokestacks only sent the problem elsewhere, since airborne acidic pollutants can travel hundreds of kilometres from their source.

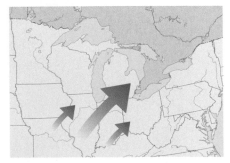

Figure 3
Map of eastern North America showing prevailing summer and winter winds that carry polluting air masses

In 1997, Canada and the United States signed the first agreement to develop a Joint Plan of Action on Transboundary Air Pollution. Since then, several agreements have been signed, each one reducing the targets for emission of airborne pollutants from the previous targets. Although we have been told that previous targets were met, the problem still exists. Perhaps the solution lies in developing better methods to measure emissions at their source.

Canada's ratification of the Kyoto Protocol means that, by 2012, Canada must reduce its emissions of pollutant gases to 6% below 1990 levels (**Figure 4**). To achieve these goals, the federal government must develop a plan that involves all Canadians. Success depends on each person's commitment to reduce automobile use (by car pooling or using public transit), to drive fuel-efficient vehicles, to retrofit older homes to eliminate air leaks, and to plant trees. These are just some of the ways that individual Canadians can help to reduce acid-forming pollutants and solve the problem of acid rain. Can you think of any other ways?

(a)

(b)

Figure 4
(a) Industries and power plants are the major producers of sulfur dioxide.
(b) When fitted with a scrubber that absorbs sulfur dioxide, emission levels of this gas are significantly decreased.

▶ *TRY THIS* activity   *A Marble Reaction*

You may have noticed buildings or statues that have been affected by acid rain. Buildings and statues that are made of marble (a form of limestone, $CaCO_{3(s)}$) are particularly vulnerable to acid rain.

Materials: electronic balance; marble chips; two 250-mL beakers; dilute sulfuric acid, $H_2SO_{4(aq)}$; 50-mL graduated cylinder; filter and filter paper; distilled water in a squeeze bottle

1. Obtain approximately 5.0 g of marble chips (limestone rock), and measure the mass accurately. Place the chips in a 250-mL beaker.

2. Add 50 mL of dilute sulfuric acid to the beaker. Observe the mixture for several minutes, and record your observations.

3. Stir the mixture vigorously until it stops bubbling.

4. Pour the mixture through a piece of filter paper. Rinse the marble chips with distilled water from a squeeze bottle.

5. Allow the marble chips to dry. Determine their mass.

(a) Has the mass of the marble chips changed?

(b) What do you think happened to the marble chips? Write a balanced chemical equation to explain your answer.

▶ *Section 4.13* Questions

Understanding Concepts

1. (a) What is the approximate pH of normal rain?
   (b) Why is the pH of normal rain not 7.0? Explain.

2. (a) Identify two gases that are largely responsible for acid precipitation.
   (b) Describe the sources of these gases.

3. Summarize the steps involved in the formation of acid precipitation.

4. Describe one chemical method that is used to reverse the process of acidification in a lake.

5. Write a balanced chemical equation for the formation of an acid when each of the following compounds dissolve in water:
   (a) carbon dioxide
   (b) nitrogen dioxide
   (c) sulfur dioxide

6. List four ways in which human activities have contributed to the production of gases that dissolve in water to form acid precipitation.

7. Why is acid rain less of a problem in the Great Lakes Basin than in the lakes of northern Ontario?

Making Connections

8. Acid rain has effects on the environment and the economy. Research either the environmental or economic consequences of acid rain. Prepare a report for the class. Be prepared to present your report.

 www.science.nelson.com

9. How would planting trees help to solve Canada's acid rain problem? Explain.

Extension

10. Choose one of the following roles. Prepare to respond, as part of your community, to a news report that a Canadian pulp and paper company has been given permission by the Ontario government to open a new plant in northwestern Ontario. The plant will produce sulfur dioxide emissions, but these emissions will fall within the guidelines outlined in the latest Canadian–American agreement on the transboundary flow of acid-producing emissions. The plant is going to be built on the bank of a river, upstream from an Aboriginal community that relies on hunting and fishing for food. Your community is located southeast of the proposed plant. Your response will contribute to your community's position paper, which will be released to the press. Prepare an argument based on the role you have chosen, and be prepared to contribute to the position paper.

Roles:
local provincial politician
spokesperson for the company
environmental activist
Aboriginal person, who has family in the community upstream
average citizen
municipal politician

 www.science.nelson.com

Figure 1
From Earth's atmosphere, we get the gases we breathe.

kinetic molecular theory All substances contain particles that are in constant, random motion.

Although you may live *on* the surface of Earth, you live *in* Earth's atmosphere (**Figure 1**). You breathe gases and, through your activities, you create gases that affect the quality of Earth's atmosphere. You probably do not think about the approximately 14 kg of air you breathe in daily. Before you can understand the nature and role of gases in your environment, you must understand the properties of gases.

Gases are one of the three states of matter. Solids, liquids, and gases have particular properties (**Table 1**). These properties are largely explained by the forces between the particles that make up the substance. The particles can be atoms, molecules, or charged ions. Solids have a definite shape, which suggests that there are strong forces between the particles. Liquids take the shape of a container, which suggests that the forces between the particles are relatively strong, but weak enough to allow the particles to move past one another. Gases, which have no definite shape or volume, appear to have little or no forces between particles.

The relationship between particles, the forces that bind them, and the speed at which they move is called the **kinetic molecular theory**. This theory proposes that all states of matter are composed of tiny particles that are constantly in motion. In solids, there is very little particle motion because of the strong forces of attraction between the particles. These strong forces explain why many solids are rigid. In liquids, the particles remain together but move around in a jumbled, less orderly, state. Gas particles move independently of one another in a straight-line motion. This movement results in large spaces between the particles, which makes gases highly compressible.

As gas particles move about, they collide with each other and with the walls of the container in which they are stored. Every moving object, even an atom or a molecule, has energy called kinetic energy. The faster the motion of the object, the greater is its kinetic energy. Because the particles of a substance are always colliding, at any instant some particles are moving faster than others.

Table 1   Properties of States of Matter

State	Properties	Examples
solid	• has a definite shape and volume • is virtually incompressible • does not flow easily	table salt, $NaCl_{(s)}$; ice, $H_2O_{(s)}$; magnesium oxide, $MgO_{(s)}$
liquid	• assumes the shape of the container, but has a definite volume • is virtually incompressible • flows readily	liquid water, $H_2O_{(l)}$; acidic solutions; basic solutions
gas	• assumes the shape and volume of the container • is highly compressible • flows readily	air; carbon dioxide, $CO_{2(g)}$; nitrogen oxides, $NO_{x(g)}$

Therefore, in any sample of matter, there is a range of speeds at which the particles are moving and a corresponding range of kinetic energies.

The temperature of a substance is an indirect measure of the average kinetic energy of its particles. If the average speed of the particles increases, the average kinetic energy of the particles increases. This relationship is shown by the increase in the temperature of the substance.

Gas Pressure

As you know, Earth's gravity exerts a constant downward force on you, which keeps you on the ground. This force can be distributed over a large or small area. When you lie down, your weight, or force of gravity acting on you, is spread out over the area of your body. When you stand, your weight is concentrated over the smaller area of your feet. Pressure is the force per unit area. Therefore, you exert less pressure on the ground when you are lying down than when you are standing. The greater the area that is occupied by an object on Earth's surface, the lower is the pressure that the object exerts over this area. For example, if you wear boots when walking through deep snow, you sink into the snow. If you wear snowshoes, the force is distributed over a greater surface area. You exert less pressure on the ground and, therefore, are able to walk on top of the snow.

Air is a mixture of mainly nitrogen gas, $N_{2(g)}$, and oxygen gas, $O_{2(g)}$. Air also contains much smaller amounts of carbon dioxide, $CO_{2(g)}$, and other gases. Each gas exerts its own pressure, which contributes to the pressure of the mixture. This pressure is called air pressure.

You rarely notice air pressure until it changes. For example, you may notice a change in air pressure when you are riding in an elevator to the top of a tall building. If you have flown in an airplane, you probably noticed changes in air pressure when taking off and landing. Mountain climbers notice a change in air pressure at high altitudes because the air pressure at higher altitudes is much lower. The gases in the air are able to expand, or spread out, so that each breath of air contains fewer oxygen molecules. Scuba divers (**Figure 2**) need to be aware of both air and water pressure changes. Survival underwater depends on the external pressure of the water being offset by the pressure of the air entering a diver's lungs.

The pressure that is exerted by a gas is created when moving air particles collide with the surface of Earth, with objects on the surface, or with each other. The pressure of a gas that is stored in a container, such as a compressed gas cylinder, results from the moving gas particles colliding with the walls of the container and with each other.

The SI unit for pressure is the pascal (Pa). The pascal represents a force of one newton (N) on an area of one square metre ($1 m^2$): $1 Pa = 1 N/m^2$. Atmospheric pressure and the pressure of many gases are often measured in kilopascals (kPa), where $1 kPa = 1000 Pa = 1 kN/m^2$.

At sea level, the average **atmospheric pressure** is about 101.3 kPa. Scientists used this value as a basis to define one standard atmosphere (1 atm) as exactly 101.325 kPa. For convenience, standard ambient pressure (SAP) is defined as exactly 100 kPa.

Figure 2
Scuba divers must understand both air and water pressure, and how they affect the body.

atmospheric pressure the force per unit area that is exerted by air on all objects within the atmosphere

For many years, experiments with gases were done at a temperature of 0°C and a pressure of 1 atm (101.325 kPa). These conditions were known as standard temperature and pressure (STP). However, 0°C is not a convenient temperature because laboratory temperatures are more likely to be closer to 25°C. As a result, the new standard is called standard ambient temperature and pressure (SATP). SATP is defined as 25°C and 100 kPa.

Boyle's Law: The Relationship between Pressure and Volume

As you know, solids and liquids cannot be compressed to any appreciable extent. Despite any pressure that is exerted on a solid or a liquid, the volume will remain essentially the same. Gases, including air, are very compressible. An applied pressure will squeeze a gas into a smaller volume.

In 1662, British chemist Robert Boyle (1627–1691) determined that as pressure increases, the volume of a gas decreases. This relationship, known as **Boyle's law**, states that *the volume of a gas decreases as the surrounding pressure increases, provided that the temperature and the amount of gas remain constant* (**Figure 3**). Boyle's law can be conveniently written as

$$p_1 v_1 = p_2 v_2$$

where p_1 and v_1 represent the initial pressure and volume measurements and p_2 and v_2 represent the second set of pressure and volume measurements.

This equation can be rearranged to solve for any one of the four variables.

How does Boyle's law relate to what happens to a scuba diver? The human body, while made up mostly of solids and liquids, also contains air spaces in the lungs, ears, sinuses, stomach, and intestines. These air spaces respond to the pressure of the air in the diver's air tank and to the pressure exerted by the water as the diver descends. As the pressure exerted by the water on a diver increases during descent, the volume of the air spaces in the diver's body decreases and the diver experiences a "squeeze." Divers equalize the pressure on descent by inhaling more air from the tank so that the air spaces in the body are filled with air at the same pressure as the surroundings. Failure to do so will result in pain, damage to tissue, or even death.

If the air spaces in the diver's body were filled with water, pressure would have no effect (as shown by the effect of pressure on the water-filled container in **Figure 3**). However, because the spaces in the diver's body are filled with air, the pressure exerted by the water during descent has the same effect on these air spaces as it does on the air-filled container in **Figure 3**. Collapse of the air spaces can only be prevented by inhaling more air from the air tank.

On ascent, the reverse happens. As the pressure on the diver decreases, the volume of air in the spaces increases according to Boyle's law. Since the spaces are finite, the air must be released. Divers release the air by breathing out and "clearing" their ears.

Failure to recognize and compensate for these changes in pressure and volume can have devastating effects. Beginning divers must spend time practising safe descent and ascent procedures.

Boyle's law As the pressure on a gas increases, the volume of the gas decreases proportionally, provided that the temperature and amount of gas remain constant.

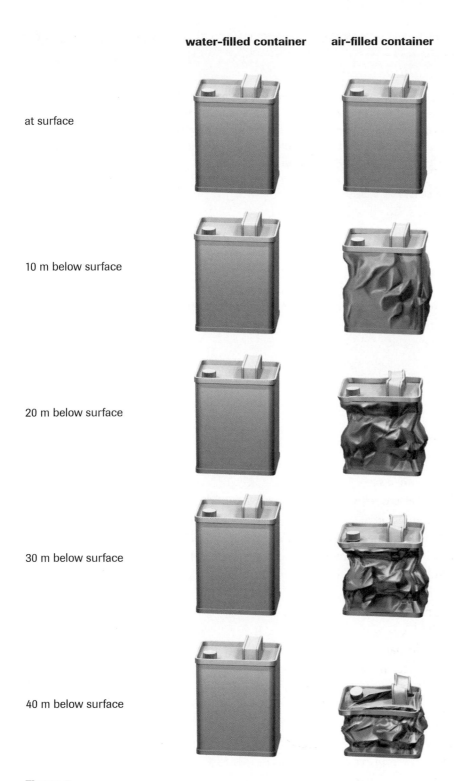

water-filled container   **air-filled container**

at surface

10 m below surface

20 m below surface

30 m below surface

40 m below surface

Figure 3
The effects of increasing pressure on a water-filled container (left) and an air-filled container
(right)

Calculations Using Boyle's Law

A 550-L weather balloon at 98 kPa is released from the ground and rises into the atmosphere. At a relatively low altitude, it is captured by a weather airplane. Instruments in the plane indicate that the air pressure is 75 kPa. Assuming that the temperature is the same as on the ground, what is the volume of the captured balloon?

Step 1: List Given Values

$v_1 = 550$ L

$p_1 = 98$ kPa

$p_2 = 75$ kPa

$v_2 = ?$

Step 2: Rearrange Equation to Solve for Unknown, Then Substitute Given Values

$$p_1 v_1 = p_2 v_2$$

$$v_2 = \frac{p_1 v_1}{p_2}$$

$$= \frac{98 \text{ kPa} \times 550 \text{ L}}{75 \text{ kPa}}$$

$$v_2 = 720 \text{ L}$$

The volume of the captured balloon is 720 L.

Example

A weather balloon that contains 35.0 L of helium gas, at 98 kPa, is released and rises into the atmosphere. Assuming that the temperature remains constant, determine the volume of the balloon when it rises to an altitude where the atmospheric pressure is 25.0 kPa.

Solution

$v_1 = 35.0$ L

$p_1 = 98$ kPa

$p_2 = 25.0$ kPa

$v_2 = ?$

$$p_1 v_1 = p_2 v_2$$

$$v_2 = \frac{p_1 v_1}{p_2}$$

$$= \frac{98 \text{ kPa} \times 35.0 \text{ L}}{25.0 \text{ kPa}}$$

$$v_2 = 140 \text{ L}$$

The volume of the balloon is 140 L.

▶ **Practice**

Understanding Concepts

1. List the three main properties of gases, and write a brief explanation of each property.

2. Define the term "atmospheric pressure."

3. What is the theoretical explanation for the observed pressure of a gas inside a closed container?

4. A bicycle pump contains 0.650 L of air at SATP. Assuming that the temperature remains constant, what pressure is required to change the volume to 0.250 L?

5. 110 mL of oxygen gas at a pressure of 300 kPa is released into a balloon under a pressure of 200 kPa. What is the final volume of the balloon, assuming constant temperature?

Answers

4. 260 kPa

5. 0.16 L

Charles's Law: The Relationship between Volume and Temperature

More than 100 years after Robert Boyle had determined the relationship between the pressure and volume of a gas, French physicist Jacques Charles (1746–1823) determined the relationship between the volume and temperature of a gas after observing hot-air balloons. He observed that a very small volume of air, blown into a hot-air balloon, would expand to fill the balloon when the air was heated. He found that this relationship existed for all gases. Furthermore, when you look at the volume–temperature graphs for several gases (**Figure 4**, on the next page), you can see that the temperature that corresponds to a volume of zero can be extrapolated to −273°C. This temperature, called **absolute zero**, is the lowest possible temperature.

Absolute zero is the basis for another temperature scale, called the **Kelvin temperature scale**. On this scale, absolute zero (−273°C) is zero kelvin (0 K). This scale has the same size divisions as the Celsius temperature scale. Using the Kelvin temperature scale, STP is 273 K and 101 kPa, and SATP is 298 K and 100 kPa.

The direct relationship between the volume of a gas and the temperature of the gas (on the Kelvin temperature scale) is known as **Charles's law**. According to this law, *as the temperature of a gas increases, the volume increases proportionately, provided that the pressure and the amount of gas remain constant.* Charles's law can be conveniently written, comparing any two sets of volume and temperature measurements, as

$$\frac{v_1}{T_1} = \frac{v_2}{T_2}$$

where v_1 and T_1 represent the initial volume and temperature measurements and v_2 and T_2 represent the second set of volume and temperature measurements. Note that T must be in kelvin.

DID YOU KNOW ?

Hydrogen Balloons
In 1873, Jacques Charles made and released the first hydrogen balloon. The balloon ascended 3 km and flew for 2 h, covering 43 km.

absolute zero 0 K or −273°C; believed to be the lowest possible temperature

Kelvin temperature scale a temperature scale that has 0 K at absolute zero and degrees with the same magnitude as the degrees on the Celsius temperature scale

Charles's law As the temperature of a gas increases, the volume increases proportionately, provided that the pressure and the amount of gas remain constant.

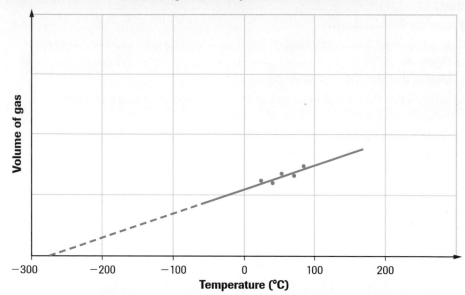

(a) Cooling a Gas Sample at Constant Pressure

Volume of gas

Temperature (°C)

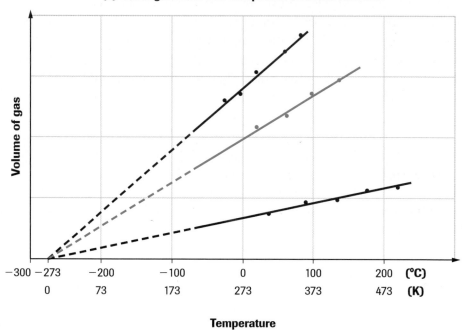

(b) Cooling Several Gas Samples at Constant Pressure

Volume of gas

Temperature

Scuba Tanks and the Gas Laws

Divers must be careful when storing full scuba tanks on a hot summer day. If a tank is left in a hot car trunk or outside in the sunlight, the temperature increase can cause the volume of the air in the tank to increase to the point where the metal of the tank could rupture. To prevent this problem, most tanks are fitted with a "burst disk." A burst disk causes a tank to release all its air if the temperature of the air in the tank gets too high.

Figure 4
When the graphs of several careful volume–temperature experiments are extrapolated, all the lines meet at absolute zero, −273°C, or 0 K.

Charles's law does not have any effect on a scuba diver in the water. It becomes very important, however, when filling and storing a diver's scuba tank. When a scuba tank is filled with compressed air, the motion of the air molecules and the collisions they undergo cause the tank to heat up. Less air can be pumped into a hot tank than into a cooler tank. As a result, scuba tanks are often filled in cold-water baths to ensure that they are full.

▶ *SAMPLE* **problem** *2*

Calculations Using Charles's Law

The volume of a gas inside a cylinder with a movable piston (Figure 5) is 0.30 L at 25°C. The gas in the cylinder is heated to 315°C. What is the final volume of the gas when the temperature is 315°C?

Step 1: List Given Values, and Convert Temperature from Celsius to Kelvin

$V_1 = 0.30$ L

$T_1 = 25° + 273 = 298$ K

$T_2 = 315° + 273 = 588$ K

$V_2 = ?$

Step 2: Rearrange Equation to Solve for Unknown, Then Substitute Given Values

$$\frac{V_1}{T_1} = \frac{V_2}{T_2}$$

$$V_2 = \frac{V_1 T_2}{T_1}$$

$$= \frac{0.30 \text{ L} \times 588 \text{ K}}{298 \text{ K}}$$

$$V_2 = 0.59 \text{ L}$$

The final volume of the gas at 315°C is 0.59 L.

Example

A 100-L sample of oxygen gas at 25°C is cooled at constant pressure in preparation for storage. To what value must the temperature of the sample be lowered before the volume is reduced to one-quarter of its original size?

Solution

$V_1 = 100$ L

$V_2 = \frac{1}{4}(100 \text{ L}) = 25$ L

$T_1 = 25° + 273 = 298$ K

$T_2 = ?$

$$\frac{V_1}{T_1} = \frac{V_2}{T_2}$$

$$T_2 = \frac{V_2 T_1}{V_1}$$

$$= \frac{25 \text{ L} \times 298 \text{ K}}{100 \text{ L}}$$

$$T_2 = 74 \text{ K}$$

The temperature must be lowered to 74 K, or −200°C.

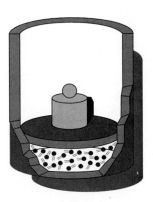

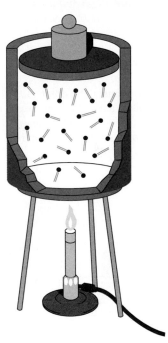

Figure 5
The volume of a gas in a movable piston increases as the temperature of the gas increases.

Understanding Concepts

Answers

6. 1.2 L

7. (a) 2.1 L
   (b) 0.12 L

9. 0.1 L

6. When Jacques Charles observed hot-air balloon flights, burning straw was used to heat the air inside the balloons. Today, propane burners are used to warm the air inside the balloons from an average temperature of 20°C to about 80°C. What volume does every initial 1.0 L of air become at the higher temperature?

7. The cap is removed from an empty 2.0-L plastic bottle. The empty bottle is placed in a refrigerator at 5°C. After a while, the bottle is removed from the refrigerator and allowed to warm to room temperature (21°C).
   (a) What volume will the air in the bottle expand to at the warmer temperature?
   (b) How much air will escape from the bottle?

8. (a) 12.7 mL of butane gas is released from a lighter before it sparks, when the temperature is 22°C. What volume is released when the temperature is −11°C? (*Hint:* The boiling point of butane is −0.5°C.)
   (b) Based on this information, explain why butane lighters do not work well outdoors at very cold temperatures.

9. Cooking pots have loose-fitting lids to allow air and water vapour to escape while the food is cooking. A 1.5-L pot is being used to boil vegetables. The vegetables and water occupy 1.0 L of the total volume of the pot. They are heated from 22°C to 100°C. What volume of gas will escape from the pot?

Dalton's law of partial pressures
The total pressure of a mixture of nonreacting gases is equal to the sum of the partial pressures of the individual gases.

partial pressure the pressure, p, that a gas in a mixture would exert if it were the only gas in the same volume, at the same temperature

Dalton's Law of Partial Pressures

John Dalton (1766–1844) is best known for his development of atomic theory (Unit 1). However, he also studied the properties of gases and atmospheric air. He hypothesized that gas particles behaved independently and that the pressure exerted by a particular gas is the same whether it exists by itself or in a gas mixture (assuming constant temperature). Dalton's hypothesis was later verified by a number of experiments, and it is now known as **Dalton's law of partial pressures**. This law states that *the total pressure of a mixture of nonreacting gases is equal to the sum of the* **partial pressures** *of the individual gases.* The equation for Dalton's law is

$$p_{total} = p_1 + p_2 + p_3 + \dots$$

where p_{total} is the total pressure of the mixture and
p_1, p_2, and p_3 are the partial pressures of each gas in the mixture.

Dalton's law of partial pressures can be explained using the kinetic molecular theory. Gases consist of a large number of particles that are constantly moving and colliding with each other and with the walls of the container. The particles in a mixture of nonreacting gases behave in the same way. It makes no difference whether the container holds one kind of gas molecule or several kinds of gas molecules (**Figure 6**).

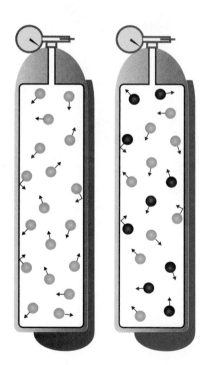

Figure 6
In both cylinders, the frequency of the collisions of the gas particles with each other and with the walls of the container is the same. Therefore, the pressure is the same in both cylinders. One cylinder contains one kind of gas particle, while the other cylinder contains two kinds of gas particles.

> ▶ **SAMPLE** problem *3*

Calculations Using Dalton's Law

A scuba diver's compressed air tank is filled with an air mixture that has an oxygen partial pressure of 28 atm and a nitrogen partial pressure of 110 atm. What is the total pressure in the tank?

Step 1: List Given Values

p_{O_2} = 28 atm

p_{N_2} = 110 atm

p_{total} = ?

Step 2: Substitute Given Values into Equation, and Solve

$$p_{total} = p_{O_2} + p_{N_2}$$
$$= 28 \text{ atm} + 110 \text{ atm}$$
$$p_{total} = 138 \text{ atm}$$

The total pressure in the tank is 138 atm.

Example

A tank of compressed air that is used by a firefighter holds nitrogen at a partial pressure of 300 kPa. The tank has a total pressure of 385 kPa. What is the partial pressure of the oxygen in the tank?

Solution

p_{total} = 385 kPa

p_{N_2} = 300 kPa

$$p_{O_2} = p_{total} - p_{N_2}$$
$$= 385 \text{ kPa} - 300 \text{ kPa}$$
$$p_{O_2} = 85 \text{ kPa}$$

The pressure of the oxygen in the tank is 85 kPa.

> ▶ **Practice**

Understanding Concepts

10. Use the kinetic molecular theory to explain how a gas exerts pressure.

11. Suppose that you have 100 molecules of oxygen in a container at a certain temperature.
  (a) If you added another 100 molecules of oxygen to the same container, while keeping the temperature and volume the same, what would happen to the pressure in the container? Explain your answer.
  (b) If you added 100 molecules of nitrogen to the 100 molecules of oxygen in the same container at the same temperature, what would happen to the pressure?
  (c) How does your answer to (b) explain Dalton's law of partial pressures?

12. A 1.0-L cylinder of air at SATP contains carbon dioxide with a partial pressure of 2 kPa. If all the other gases were removed from the cylinder and the temperature remained the same, what would be the pressure of the carbon dioxide in the cylinder?

Applying Dalton's Law

How does Dalton's law of partial pressures help to explain respiration and the diffusion of gases into and out of body tissues? During respiration, you breathe in oxygen and breathe out carbon dioxide. The air that you exhale contains more carbon dioxide and less oxygen compared with the air that you inhale.

Oxygen, which makes up approximately 21% of the air you breathe, exerts a partial pressure of approximately 21 kPa at SATP. The partial pressure of oxygen determines how much oxygen is absorbed by your lungs. At the top of Mount Everest, the air pressure may only be about 33 kPa compared with 100 kPa at SATP. Even though the air at the top of the mountain has the same amount of oxygen (21%), the partial pressure exerted by the oxygen is 21% of 33 kPa, or 6.3 kPa. This pressure is not enough to cause the diffusion of oxygen into lung tissue. To survive, most humans require a partial pressure of oxygen of approximately 10 kPa.

Scuba divers, on the other hand, experience the opposite effect. As a diver descends, the pressure on the lungs increases. The air that enters the lungs from the scuba tank is also subject to this increase in pressure. At the surface, the partial pressure of the oxygen component of air is 21 kPa (21% of 100 kPa). At a depth of 10 m and a pressure of 200 kPa, the partial pressure of the oxygen increases to 42 kPa. This increase can result in too much oxygen being absorbed by the lung tissue and entering the bloodstream, which can damage the tissue.

Just as dangerous to a diver is the carbon monoxide component of the air in a scuba tank. If the diver's tank is filled with compressed air near a compressor that is spewing carbon monoxide, as much as 0.5% of the compressed air in the tank is carbon monoxide. Diving to a depth of 40 m increases the partial pressure of carbon monoxide by almost five times, a level that is poisonous. Thus, even though the percentage composition of the air in the tank remains the same as the diver descends, the effect of Dalton's law of partial pressures can make breathing deadly.

Henry's Law: The Relationship between Solubility and Pressure

When you breathe in air, what happens to the gas molecules after they are absorbed into the lung tissue? Oxygen molecules enter the bloodstream and are carried to cells to participate in cellular respiration. Recall that cellular respiration is the process in which cells produce the energy they need to survive. How much oxygen can dissolve in the bloodstream?

British chemist and physician William Henry (1775–1836) determined that the amount of gas that will dissolve in a liquid at a given temperature is directly proportional to the pressure of the gas. This relationship means that the solubility of a gas is higher at higher pressures. Therefore, a liquid under pressure will continue to absorb and dissolve a gas until the pressure that is created by the gas in trying to escape the liquid is the same as the pressure that is being exerted on the gas to keep it dissolved in the liquid. This point is called the *saturation point*.

Gases under pressure in a scuba diver's body dissolve in the bloodstream, in a process called "ingassing." During a diver's descent, the pressure of the air inhaled by the diver increases to equal the pressure of the water around the diver. This increase results in a higher gas pressure in the lungs, which causes an increase in the diffusion of air from the lungs into the blood until the saturation point is reached.

When ascending, a diver must be careful because the blood has become saturated with gases. As the water pressure on the diver and the inhaled air pressure in the diver decreases, the blood can no longer hold the gases (air) in solution. The air is released in a process called "offgassing." The excess gases in the blood must be carried to the lungs and released during the diver's normal breathing process. If the rate of offgassing is too rapid, the gases in the lung tissue and bloodstream cannot escape quickly enough. Large air bubbles can form in the lung tissue and joints. The air bubbles cause decompression sickness, commonly called "the bends." When the air bubbles form in tissue around the mid-section of a diver, they cause severe back and abdominal pain. The victim often "bends" over in an attempt to relieve this pain. In severe cases, skin tissue takes on a "bubble wrap" appearance. If a diver is foolish enough to hold his or her breath during a rapid ascent, the result can be a ruptured lung or an embolism in the brain causing death.

Dive organizations use Henry's law and the factors that affect how quickly gases enter and leave a diver's tissues to create dive tables. Dive tables help divers understand the time that is needed for proper offgassing and safe ascents. Divers must learn about and remember the properties of gases and the gas laws to participate safely in scuba diving.

You may be convinced that the best location to breathe easily is on Earth's surface, at or near sea level. The atmosphere delivers all the oxygen you need and provides a sink to absorb all your exhaled carbon dioxide. However, the atmosphere also absorbs airborne wastes from human activities, and from industries and technologies.

▶ *Section 4.14* *Questions*

Understanding Concepts

1. Solids share some properties with liquids and have some properties that are unique.
   (a) Describe the properties that are the same for both solids and liquids. Explain why these properties are the same.
   (b) Describe the properties that are different for solids and liquids. Explain why these properties are different.

2. (a) Describe the forces between gas molecules at room temperature.
   (b) Describe the motion of gas molecules at room temperature.

3. Explain each of the following situations:
   (a) A partially inflated balloon is released from ground level and becomes larger as it rises in the atmosphere. (Assume that the temperature remains constant.)
   (b) A carbonated drink produces more bubbles when it is at a high altitude than when it is at ground level.

4. In your notebook, complete each of the following temperature conversions:
   (a) 25°C = _____ K
   (b) 30°C = _____ K
   (c) −35°C = _____ K
   (d) 312 K = _____ °C
   (e) 208 K = _____ °C

5. The inside volume of an automobile tire is 27 L at 225 kPa and 18°C.
   (a) What volume would the air inside the tire occupy if it escaped? (Assume that the atmospheric

pressure is 98 kPa and the temperature remains constant.)

(b) How many times larger is the new volume compared with the old volume? How does this ratio compare with the change in pressure?

6. A weather balloon that contains helium gas at 100 kPa is released and rises in the atmosphere. How would the volume of the balloon change if the pressure dropped to 25 kPa? What two assumptions must you make to answer this question?

7. A sealed syringe contains 50 mL of a gas at 100 kPa. The plunger is depressed to compress the gas to a volume of 10 mL. Estimate the new pressure, assuming constant temperature.

8. Carbon dioxide gas is produced by yeast in bread dough and causes the dough to rise (**Figure 7**). During baking, the carbon dioxide expands at constant pressure in the dough. Predict the final volume of the carbon dioxide in a loaf of bread if 0.5 L of carbon dioxide is initially in the dough and the dough is heated from 25°C to 200°C.

Figure 7
The lightness of baked goods, such as bread, is a result of gas bubbles trapped in the dough when it is heated.

Applying Inquiry Skills

9. An investigation is performed to determine the relationship between the temperature and pressure of a gas. A compressed cylinder of gas, with a pressure gauge attached, is lowered into water baths at various temperatures. The data collected are given in **Table 2**. Complete the Prediction, Analysis, and Evaluation in the following lab report.

Question
What effect does the temperature of a gas have on the pressure it exerts?

Prediction
(a) Predict the effect of the temperature of a gas on the pressure that the gas exerts. Explain your reasoning.

Analysis
(b) Analyze the data in **Table 2** to answer the Question.

Table 2   Gas Pressure at Various Temperatures

Temperature (°C)	Pressure (kPa)
0	100
20	106
40	115
60	123
80	129
100	135

Evaluation
(c) Evaluate your Prediction.

10. Design an investigation to determine what would happen if you placed a marshmallow in a container and then gradually reduced the air pressure in the container. Complete a report, including a Prediction and a detailed Procedure.

11. The volume of a flexible container of gas is reduced to one-quarter of its original volume in an experiment. Assuming that the temperature and amount of the gas were constant, what variable must have changed? By how much did this variable change?

Making Connections

12. If you use a meat baster, you squeeze the rubber bulb and insert the tip of the baster into the meat juice. Use what you have learned about air pressure to explain why the juice rises in the baster when the bulb is released.

13. Research the invention and refinement of the barometer to measure the pressure of a gas. Create a chronological flow chart that includes dates, names, and diagrams.

 www.science.nelson.com

Life on Earth is supported by the combination of energy from the Sun and the oxygen component of Earth's atmosphere. As **Figure 1** shows, the atmosphere consists of different layers. The troposphere starts at Earth's surface and extends 8 to 16 km above Earth's surface. The troposphere contains all the gases that we breathe and 75% of all the gases in the atmosphere. This homogeneous solution of gases is continually mixing. (See **Table 1** on the next page.) The stratosphere begins above the troposphere and extends another 50 km. Gaseous ozone, $O_{3(g)}$, in the stratosphere acts as an ultraviolet "shield," protecting us and the environment from some forms of ultraviolet radiation.

Earth's atmosphere serves as a reservoir or sink for airborne wastes from human activities and industrial processes. Human activities are changing the concentrations of some of the components of the atmosphere. Perhaps the two most significant effects of human activities on the atmosphere are global warming and the thinning of the ozone layer, due to the increased concentrations of greenhouse gases.

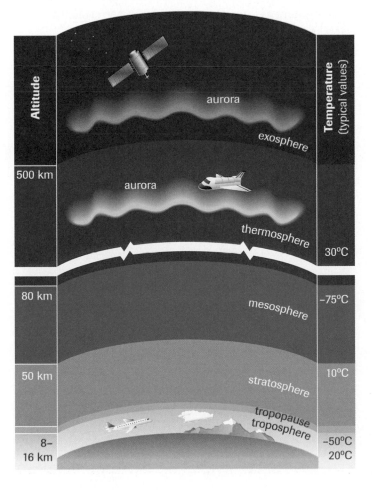

Figure 1
The atmosphere consists of several different layers.

Table 1 Composition of Air in the Troposphere (by Volume)

Substance	Formula	Percentage composition
nitrogen	$N_{2(g)}$	78.08%
oxygen	$O_{2(g)}$	20.95%
argon	$Ar_{(g)}$	0.93%
carbon dioxide	$CO_{2(g)}$	0.033%[+]
neon	$Ne_{(g)}$	0.001 8%[*]
ammonia	$NH_{3(g)}$	0.001 0%[*]
helium	$He_{(g)}$	0.000 5%[*]
methane	$CH_{4(g)}$	0.000 2%[*]
krypton	$Kr_{(g)}$	0.000 11%[*]
others: hydrogen, water vapour, ozone	$H_{2(g)}$, $H_2O_{(g)}$, $O_{3(g)}$	<0.000 1%[*]

[+]Levels of carbon dioxide are predicted to double in the next 100 years, if present trends continue.

[*]This amount is considered to be a trace amount.

Greenhouse Gases and Global Warming

The balance between the absorption of energy from the Sun and the radiation of heat energy from the surface of Earth back into space determines the average annual temperature at any location on Earth. **Figure 2** shows what happens to solar energy as it approaches the atmosphere. Some solar radiation is reflected back into space by clouds and solid particles in the atmosphere. About 25% of incoming solar energy is used to power the water cycle: the continuous cycling of water from Earth's surface through repeated evaporation and condensation cycles. Almost half the energy from the Sun is absorbed by Earth to warm the surface. Objects on Earth's surface, however, re-radiate energy back into space. Much of the re-radiated energy is absorbed by gases in the troposphere. As a result, heat is trapped close to Earth's surface and prevented from being released into space. These gases re-radiate some of the absorbed energy back to Earth's surface. Energy can pass back and forth

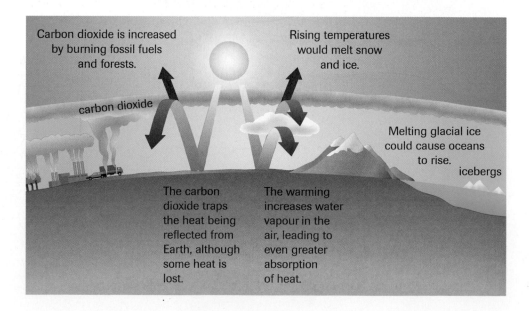

Figure 2

The distribution of the Sun's energy at the surface of Earth several times between atmospheric gases and Earth's surface before escaping out into space.

The trapping of heat energy in Earth's atmosphere is known as the **greenhouse effect** because it is similar to the way that heat is trapped in a greenhouse on a sunny day. The gases that contribute to the greenhouse effect are called **greenhouse gases** (GHGs). The major greenhouse gases include atmospheric water vapour, carbon dioxide (CO_2), methane (CH_4), and nitrous oxide (N_2O). Other GHGs include chlorofluorocarbons (CFCs), hydrochlorofluorocarbons (HCFCs), and hydrofluorocarbons (HFCs).

Most greenhouse gases occur naturally and enter the atmosphere through natural processes (**Table 2**). However, human activities are increasing the concentrations of greenhouse gases.

greenhouse effect a theory describing how heat is trapped near Earth's surface by carbon dioxide, atmospheric water vapour, and other gases

greenhouse gases gases in the atmosphere that trap heat close to Earth's surface, contributing to the greenhouse effect

Table 2 Sources of Major Greenhouse Gases

Greenhouse gas	Natural sources	Sources related to human activities
water vapour, $H_2O_{(g)}$	evaporation and transpiration processes that are part of the water cycle	combustion of fossil fuels and wood
carbon dioxide, $CO_{2(g)}$	plant and animal respiration, decay of organic matter in soil, volcanoes and forest/grass fires, oceans	combustion of fossil fuels, deforestation, industrial processes such as cement production
methane, $CH_{4(g)}$	decay of organic matter in wetlands, chemical reactions in soil	livestock farming, rice cultivation, biomass burning, landfills, coal mining
nitrous oxide, $N_2O_{(g)}$	soil and water denitrification under anaerobic conditions	nitrogen fertilizers, combustion of fossil fuels and wood
CFCs, HCFCs, HFCs		aerosols, refrigeration units, air conditioners

Natural reactions remove some greenhouse gases from the atmosphere. For example, water vapour returns to the surface of Earth as precipitation. Carbon dioxide is removed from the atmosphere during photosynthesis by forests and agricultural crops. These gases have reached their destinations, or "sinks." When natural sources and sinks are in balance, the atmospheric concentrations of water vapour and carbon dioxide are in balance.

The Ozone Layer and CFCs

The ozone layer is the part of the stratosphere that absorbs ultraviolet (UV) radiation from the Sun. Some of the UV radiation still reaches the surface of Earth. While small doses of UV radiation can be healthy, larger doses (especially larger doses of the higher energy UV-B radiation) can cause sunburn and have been linked to the development of cataracts and an increased risk of skin cancer. As well, excessive UV-B radiation inhibits the growth processes of most green plants.

Ozone is formed in the stratosphere. UV radiation from the Sun reacts with oxygen gas, splitting oxygen molecules into separate oxygen atoms.

$$O_{2(g)} + UV \text{ energy} \rightarrow O_{(g)} + O_{(g)}$$

DID YOU KNOW?

How Thick Is the Ozone Layer?
The ozone layer in the stratosphere is several kilometres thick. If all the ozone molecules that make up this layer were taken to sea level where the pressure is greater, the ozone layer would be less than 3 mm thick!

global warming the increase in the average temperature of Earth's atmosphere

Individual oxygen atoms are extremely reactive. They react with oxygen molecules, O_2, to form ozone.

$$O_{(g)} + O_{2(g)} \rightarrow O_{3(g)}$$

Ozone also absorbs UV radiation and decomposes back to oxygen molecules and oxygen atoms. This natural balance of ozone formation and decomposition in the atmosphere has been altered by chlorine atoms, produced from the decomposition of chlorinated hydrocarbon molecules (chlorofluorocarbons, or CFCs). Chlorofluorocarbons (such as Freon-11, CCl_3F, and Freon-12, CCl_2F_2) have been used as propellants in aerosol cans, and as cooling fluids in air conditioners and refrigerators. Since CFCs are relatively inert, they remain unchanged until they reach the stratosphere. There they are exposed to ultraviolet radiation. The UV radiation causes single chlorine atoms to split off from the CFC molecules. The chlorine atoms are now able to react with ozone, converting it into atmospheric oxygen gas. The chlorine atoms are regenerated and, therefore, able to continue destroying ozone molecules.

$$CCl_3F_{(g)} + UV\ energy \rightarrow CCl_2F_{(g)} + Cl_{(g)}$$
$$Cl_{(g)} + O_{3(g)} \rightarrow ClO_{(g)} + O_{2(g)}$$
$$ClO_{(g)} + O_{(g)} \rightarrow Cl_{(g)} + O_{2(g)}$$

Notice that the chlorine atom that reacts in the second reaction is regenerated by the third reaction. The depletion of ozone creates "holes" in the ozone layer, which allow higher levels of UV radiation to reach Earth. These holes are actually areas of decreased ozone concentration where there are so few ozone molecules that the ability to absorb UV radiation is significantly reduced.

Since 1750, the concentrations of greenhouse gases have increased significantly. The concentration of carbon dioxide has increased by 31%, the concentration of methane has increased by 151%, and the concentration of nitrous oxide has increased by 17%. In addition, human-made chemicals, or synthetic greenhouse gases (including CFCs, HCFCs, and HFCs) have been added to the mix.

Trends show that Earth's climate is changing, which may be related to the increase in greenhouse gases. Many scientists believe that the world is getting warmer. This trend is referred to as **global warming**. Extensive data demonstrate that the global average temperature has increased over the past 150 years. This change in temperature, however, has taken place as a series of warming and cooling cycles over intervals of several decades. Some climate experts believe that Earth is currently experiencing a natural warming cycle, and that temperatures will return to normal again. These climate experts do not support the idea of global warming.

Other climate experts predict that Earth's average temperature will increase between 1°C and 3°C over the next century. This increase is related to the expected doubling of carbon dioxide amounts in the atmosphere. For many Canadians, warmer temperatures may seem like a good idea. Scientists warn, however, that severe weather events (such as droughts, severe storms, floods, and tornadoes) will occur (**Figure 3**). A change in climate would also have an impact on Canada's water resources. Clearly, human activities have an enormous effect on the atmosphere.

Solutions to Atmospheric Problems

The obvious solution to ozone depletion is to eliminate the substances that are destroying the ozone layer. In the late 1970s, the United Nations made several attempts to enlist the cooperation of member countries to reduce the use of CFCs. A decade later, a convention in Geneva laid the groundwork for a worldwide agreement on a ban of all substances that deplete the ozone layer. This agreement was signed in Montreal and is called the Montreal Protocol.

Unfortunately, most of the refrigeration equipment that existed when the Montreal Protocol treaty was signed is still in existence today. It is estimated that more than 100 megatonnes of CFCs are still in use today in refrigeration equipment in homes and automobiles in Canada. Some automobile manufacturers have begun using hydrofluorocarbons (HFCs) as alternatives to CFCs in vehicle air-conditioning systems. Although HFCs have less impact on the ozone layer, they contribute to the greenhouse effect. Solving one problem apparently has only contributed to another.

A promising alternative, the product of cooperation between the scientific community and industries, appears to be hydrofluoroether (HFE), $C_4F_9OCH_3$. HFE does not react with ozone. It is not a greenhouse gas, and it is nontoxic. Other hydrocarbons (such as propane, pentane, isobutane, and cyclopentane) are also effective refrigerants and are widely available.

The public often interprets warnings from the scientific community about the degradation of the environment as being exaggerated. Occasionally, advancements in technology validate the warnings. Consider the predictions that increased levels of carbon dioxide would lead to global warming.

In 1958, the first reliable and continuous measurements of atmospheric carbon dioxide levels were taken at an observatory in Hawaii. These measurements confirmed that carbon dioxide levels were increasing. However, more than 10 years passed before climate changes resulting from increased atmospheric levels of carbon dioxide became an issue. Even 20 years later, at an Earth Summit in Rio de Janeiro, Brazil, 150 nations could not reach an agreement to stabilize greenhouse gas emissions voluntarily by the year 2000.

Five years later, in 1997, in Kyoto, Japan, 160 nations finally reached an agreement, called the Kyoto Accord, to set legally binding targets for cutting greenhouse gas emissions. The process by which the Accord is to be implemented, called the Protocol, commits countries to reducing greenhouse gas emissions to levels that are 5.2% below 1990 levels, by 2008 to 2012. Although the United States withdrew from this agreement in 2001, Canada became the 100th country to ratify the Kyoto targets on December 10, 2002. The Protocol is expected to enter into force in mid-2003. To fulfill its commitment under the Kyoto Protocol, Canada must reduce emissions of greenhouse gases to 6% below 1990 levels. Since emissions of these gases have risen since 1990, Canada will have to cut emissions by 20% from current levels.

Figure 3
Severe flooding in Newfoundland during the winter of 2003 is related to climate change.

DID YOU *KNOW* ?

CFCs and Asthma Inhalers
By the year 2005, almost 30 years after signing the Montreal Protocol banning the use of CFCs as aerosol propellants, Canada will have completely eliminated CFCs in asthma inhalers.

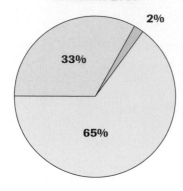

Global Greenhouse Emissions 2000

33%
2%
65%

☐ countries rejecting Kyoto/ no reduction target
☐ countries with Kyoto target
☐ Canada

Source: U.S. Department of Energy

Figure 4
Canada is a small contributor of greenhouse gases (2%) but will pay heavily because of its fossil-fuel based natural resource production.

The surest method to meet these targets is to reduce the use of fossil fuels, which can be done by improving the efficiency of present sources of energy or by shifting to alternative energies, such as wind and solar power. Another solution may be to protect Canada's "carbon sinks"—our forests and farmlands that absorb carbon from the atmosphere in the carbon cycle. In section 4.16, you will look at your role in helping to meet Canada's commitment to the Kyoto Protocol.

Within Canada, Alberta and Ontario are leading the opposition to the Kyoto Protocol. Opponents of the Kyoto Protocol point out that some countries, including the United States and Australia, have opted out of Kyoto. They also point out that some of the developing nations, which are responsible for 65% of global greenhouse gas emissions, are exempt from reduction targets under Kyoto. Countries that have a Kyoto target are responsible for 33% of the total global greenhouse gas emissions. It is interesting to note that Canada, which has committed to the proposed reduction targets, is responsible for only 2% of global greenhouse gas emissions (**Figure 4**). However, Canada is one of the highest per capita emitters, largely because of its resource-based economy, climate, and size.

Other arguments against Canada's participation are given below:

- Predictions about global warming are exaggerated.
- The manufacturing industry will have to modify its technologies, resulting in higher prices for products.
- As many as 450 000 jobs may be lost in the manufacturing sector.
- Higher-priced products will be less competitive in global markets.
- Investors will favour countries that do not have extra costs associated with Protocol compliance. For example, investment in the Alberta tar sands project will dwindle, effectively shutting down the project.
- Jobs will be lost because of the difficulty of remaining competitive with the United States.

Since Canada ratified the Kyoto Protocol, one Alberta Tar Sands investor, TrueNorth Energy, has withdrawn a $3.5 billion investment in the Tar Sands. True North Energy cited rising labour and material costs, jittery capital markets, and uncertainty over Canada's ratification of Kyoto as reasons. Following this announcement, however, Suncor Energy INC., the biggest oil sands investor, announced that its spending on the project will continue.

▶ *Understanding the Issue*

1. Why is Earth's atmosphere sometimes referred to as a "sink"?

2. (a) How is ozone formed in the atmosphere? Use appropriate equations.

   (b) Explain the role of ozone in protecting Earth from UV radiation.

   (c) Explain, using chemical equations, how CFCs cause the depletion of ozone.

   (d) Identify the consequences of ozone depletion.

   (e) Suggest reasons why it has taken 30 years to eliminate the use of CFCs as propellants in asthma inhalers.

3. (a) Why are HFCs, once thought to be a promising alternative, not used to replace CFCs?

   (b) Briefly describe how using alternative hydrocarbon refrigerants can be viewed as a possible solution to the ozone depletion problem.

4. Explain what is meant by the "energy balance," as it applies to the energy from the Sun that reaches Earth's atmosphere.

5. (a) Explain how the greenhouse effect works, relative to Earth's energy balance.

   (b) Name the gases that contribute to the greenhouse effect.

   (c) What will happen if the amount of greenhouse gases added to the atmosphere continues to increase?

6. Explain why the Kyoto Protocol proposes to reduce the amount of greenhouse gases.

▶ *Take a Stand*

Debate: Should Canada Have Ratified the Kyoto Protocol?

Debate continues on whether Canada should have ratified the Kyoto Protocol.

Statement: Canada was right to ratify the Kyoto Protocol, a blueprint for greenhouse gas reductions.

In a small group, prepare to debate the statement. Your group will either be defending or opposing the statement. Research arguments to support your position. Collect scientific evidence from primary sources wherever possible. Topics may include, but are not limited to, the following list:

- activities that are responsible for Canada's contribution to global greenhouse gas emissions (include chemical equations for these activities, showing carbon dioxide gas as a product)

Decision-Making Skills

- ● Define the Issue   ● Analyze the Issue   ● Research
- ● Defend a Decision   ● Identify Alternatives ● Evaluate

- economic forecasts of the costs involved in Canada's participation in Kyoto, in terms of jobs, investment in Canadian business and industry, and taxes

- evidence that these forecasts may be wrong or at least overly negative

- an action plan proposed by Alberta or another province

Assemble evidence to support the position assigned to your group, and prepare to debate the position.

After the debate, discuss, in your group, how you could have improved your group's performance.

 www.science.nelson.com

Do you remember the "bad-air" days of past summers, when a combination of heat, air pollutants, and poor air movement made outdoor activities both unpleasant and unhealthy for some people? Poor air quality, commonly referred to as "bad air," consists of primary air pollutants and secondary air pollutants. Secondary air pollutants, which include ground-level ozone and smog, are formed during chemical reactions between primary pollutants and sunlight or components of air.

Primary Air Pollutants

Primary air pollutants include carbon dioxide, $CO_{2(g)}$, carbon monoxide, $CO_{(g)}$, sulfur dioxide, $SO_{2(g)}$, methane, $CH_{4(g)}$, nitrogen oxides, $NO_{x(g)}$, particulate matter (PM), and the vapours from volatile organic compounds (VOCs). VOCs are the gases or odours given off by solvents and coatings used in industrial processes, as well as by cleaning products, paints, pesticides, and fertilizers. The primary air pollutants and their effects are listed in **Table 1**.

You may not think of carbon dioxide as being a pollutant since it is produced during animal respiration and is required for plant photosynthesis. What is considered to be polluting is the increase in carbon dioxide levels in the air, largely resulting from human activities.

As you know, the combustion of hydrocarbons produces carbon dioxide, carbon monoxide, and nitrogen oxides. When air that contains nitrogen and oxygen is drawn into an internal combustion engine, these gases react at the high temperatures to produce nitrogen monoxide.

$$N_{2(g)} + O_{2(g)} + heat \rightarrow 2\,NO_{(g)}$$

Nitrogen monoxide is released into the atmosphere through automobile exhaust. In the atmosphere, nitrogen monoxide is oxidized to nitrogen dioxide, $NO_{2(g)}$.

$$2\,NO_{(g)} + O_{2(g)} \rightarrow 2\,NO_{2(g)}$$

Nitrogen dioxide plays a major role in the production of ground-level ozone.

Another major contributor to air pollution is particulate matter (PM), which includes all airborne particles. Fine particles get into the air from smoke, construction sites, forest fires, road dust, and vehicle emissions. Fine particles are also formed in the atmosphere during chemical reactions involving gaseous pollutants.

Secondary Air Pollutants

The two main secondary air pollutants are ground-level ozone and smog.

Ground-Level Ozone

Ground-level ozone is a secondary pollutant that is formed near Earth's surface. The formation of ozone is a very complicated process. In simple

Table 1 Primary Air Pollutants and Their Effects

Component	Properties	Effects
carbon dioxide, $CO_{2(g)}$	• has no odour or colour	• is necessary for photosynthesis by plants • increasing levels linked to global warming • can increase the acidity of water
carbon monoxide, $CO_{(g)}$	• has no odour or colour	• reduces body's ability to use oxygen • low-level, short-term exposure decreases athletic performance and aggravates cardiac symptoms • can cause premature death
sulfur dioxide, $SO_{2(g)}$	• has a pungent odour • is fairly soluble in water	• reacts in water vapour to produce acids • aggravates asthma at high levels
nitrogen dioxide, $NO_{2(g)}$	• has an irritating odour • is a reddish-brown colour • reacts with VOCs to form ground-level ozone	• inhibits plant growth (at 0.5 ppm) • causes human respiratory distress (3 to 5 ppm) • reacts in water vapour to produce acids
methane, $CH_{4(g)}$	• is a clear, colourless gas • has no odour • fuels Bunsen burners • is produced by bacterial decomposition of organic waste	• can cause human respiratory distress • can cause death by asphyxiation
particulate matter (PM)	• includes PM_{10} (particles that are < 10 µm in diameter) and $PM_{2.5}$ (particles that are < 2.5 µm in diameter) • contributes to smog	• $PM_{2.5}$ particles get into the respiratory system • causes breathing and respiratory problems • causes irritation, inflammation, and damage to the lungs • can irritate the eyes, nose, and throat • can cause premature death
volatile organic compounds (VOCs)	• evaporate quickly at ordinary temperatures • are present in the atmosphere at very low levels • are found in higher concentrations indoors than outdoors • react with nitrogen oxides to form ground-level ozone	• cause burning eyes • may be irritants • are harmful to humans with heart disease or respiratory conditions • may be carcinogenic (for example, formaldehyde and benzene)

terms, ozone is produced through a series of reactions involving VOCs and nitrogen oxides. First, nitrogen dioxide decomposes in sunlight into nitrogen monoxide, $NO_{(g)}$, and oxygen atoms, $O_{(g)}$.

$$NO_{2(g)} + \text{sunlight} \rightarrow NO_{(g)} + O_{(g)}$$

The atomic oxygen, $O_{(g)}$, then reacts with oxygen in the air to produce ground-level ozone, $O_{3(g)}$.

$$O_{(g)} + O_{2(g)} \rightarrow O_{3(g)}$$

Scientists think that components in the VOCs catalyze the reaction (make the reaction go faster).

Ground-level ozone and stratospheric ozone are the same molecule: O_3. Whereas ozone in the stratosphere blocks harmful solar radiation, ground-level ozone is a pollutant and affects human health and damages vegetation. The effects of ground-level ozone on people and the environment are listed in **Table 2**, on the next page.

Table 2 The Effects of Ground-Level Ozone

Pollutant	Properties	Effects
ozone, $O_{3(g)}$	• has a strong odour • is colourless • is a strong oxidant	• irritates the respiratory tract and eyes • at high levels, results in chest tightness, coughing, and wheezing • is harmful to people with respiratory conditions • causes leaf damage in many plants and trees • cracks rubber and corrodes metals

Figure 1
Smog is generally observed in urban centres.

Smog

The term "smog" originally described a mixture of smoke and fog in the air. Today, smog refers to a bad air condition that results from a mixture of air pollutants seen as a haze in the air (**Figure 1**).

Smog, as we know it today, was first observed in the 1940s in Los Angeles, California. In 1952, a brownish haze that irritated the eyes, nose, and throat, and damaged crops, was identified as *photochemical smog*. Smog requires sunlight (the "photo" in photochemical) and the products of the combustion of fossil fuels in vehicles. The two main ingredients of smog are ground-level ozone and fine particles. Smog also contains nitrogen dioxide, sulfur dioxide, carbon monoxide, and ammonia. Fine particles and the reddish-brown nitrogen dioxide give smog its colour. The effects of smog are listed in **Table 3**.

Table 3 Effects of Photochemical Smog

Pollutant	Effects
photochemical smog	• causes health problems in children and elderly people • affects people with respiratory problems • aggravates asthma and emphysema

Monitoring Air Quality

The Air Quality Index (AQI) is used in Ontario to indicate the air quality. The lower the air quality index, the better the air quality. The air quality index is based on hourly measurements of the six most common air pollutants: sulfur dioxide, ozone, nitrogen dioxide, total reduced sulfur (TRS) compounds, carbon monoxide, and suspended particles (SP). These pollutants are monitored because they have adverse effects on humans and the environment at high levels. A mobile Trace Atmospheric Gas Analyzer, called TAGA (**Figure 2**), along with several air-monitoring stations across the province, provide air quality data from different locations.

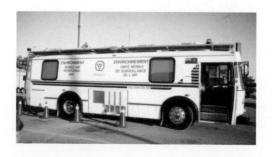

Figure 2
The TAGA (Trace Atmospheric Gas Analyzer)

Regularly throughout the day and night, the concentration of each pollutant is converted into a number using a common scale, or index (**Figure 3**). At each reporting station, the pollutant with the highest number at a given hour becomes the AQI reading for the station. For example, if the AQI for ozone is 20 at the Barrie reporting station, and this number is the highest for the six pollutants, the AQI for Barrie is reported to be 20. Specifically, it would be reported as "Barrie: AQI 20, reason: ozone." AQI data are available on the Ministry of the Environment web site and in the local press.

Substance	Monday readings (ppm)	Tuesday readings (ppm)	Wednesday readings (ppm)	Thursday readings (ppm)	Friday readings (ppm)
sulfur dioxide	153	612	288	351	428
ozone	53	85	32	60	76
nitrogen dioxide	233	771	460	588	697
TRS	17	56	33	24	46
carbon monoxide	63	428	216	412	366
SP	12	65	32	40	62

Figure 3
Sample AQI data

As the air quality changes, the AQI reading also changes. If the air quality value is below 32, the air quality is considered to be relatively good. If the AQI value ranges from 32 to 49, there may be some adverse effects for some people. An index value in the range of 50 to 99 may have some short-term adverse effects on human and animal populations. As well, it may cause significant damage to vegetation and property. An AQI value of 100 or higher may cause adverse effects for a large proportion of those exposed.

Actual concentrations of the pollutants are measured in parts per billion (ppb). AQI levels are estimated using a conversion table that is similar to the table used for ozone (**Table 4**).

Table 4 Relationship between Ground-Level Ozone Concentrations and AQI Levels

Concentration (ppb)	0–24	25–49	50–74	75–199	>200
Index level	0–15	16–31	32–49	50–99	>100
Effects	no harmful effects	some vegetation damage	some vegetation damage	odour and additional vegetation damage	harmful to people with asthma and bronchitis

For example, using the information in **Table 4**, a measured ozone concentration of 78 ppb (which is at the low end of the range of concentrations) falls within the AQI range of 50–99. It is assigned an AQI index level of 52 or 53. At this level, ground-level ozone is detectable by odour and causes visible damage to vegetation.

The concentrations of all six pollutants and the corresponding AQI levels are given in **Table 5**.

Table 5 Relationship between Ground-Level Pollutant Concentrations and AQI Levels

SO_2

Concentration (ppb)	0–99	100–199	200–299	300–799	>800
Index level	0–15	16–31	32–49	50–99	>100

O_3

Concentration (ppb)	0–24	25–49	50–74	75–199	>200
Index level	0–15	16–31	32–49	50–99	>100

NO_2

Concentration (ppb)	0–99	100–199	200–299	300–699	>700
Index level	0–15	16–31	32–49	50–99	>100

TRS (total reduced sulfur)

Concentration (ppb)	0–9	10–19	20–29	30–79	>80
Index level	0–15	16–31	32–49	50–99	>100

CO

Concentration (ppb)	0–49	50–99	100–149	150–499	>500
Index level	0–15	16–31	32–49	50–99	>100

SP (suspended particles)

Concentration (ppb)	0–9	10–19	20–29	30–89	>90
Index level	0–15	16–31	32–49	50–99	>100

A station reading of 25 ppb SP (suspended particles) falls in the middle of the 20–29 ppb range. Therefore, an AQI value of 40 would be assigned to this pollutant, representing the corresponding midpoint in the AQI range.

What Can You Do?

During days when the air quality is bad, residents of the affected areas can help the situation by curtailing activities or behaviours that contribute to the problem. Consider the following suggestions for actions to reduce smog levels and improve air quality:

Reduce Automobile Use

- Emissions from cars, trucks, and buses contribute significantly to the formation of smog. To reduce smog, drivers should leave their cars at home and carpool, take public transportation, walk, or cycle.

- In Ontario, the Drive Clean program (started in 1999) requires mandatory testing of vehicle emissions. To keep vehicles on the road, drivers must maintain their vehicles and meet emission requirements.

- Automobile pollution can be reduced or eliminated in a well-tuned, well-maintained car. Eliminate idling, drive at moderate speeds, and keep tires inflated to proper levels. Waiting until after sundown to refuel, when smog levels decrease, helps to ensure that gasoline vapours do not contribute to smog.

Conserve Electricity

- Stations that burn fossil fuels to generate electricity contribute to smog and increase levels of carbon dioxide, a greenhouse gas. Decreasing the amount of electricity used helps to reduce smog. Turning off the lights when leaving a room and turning down an air conditioner reduces the consumption of electricity.

Limit the Use of Small Engines

- Small gasoline engines in lawn mowers, chain saws, and leaf blowers emit high levels of air pollutants. Not mowing the lawn, or using a manual lawn mower to do the job, on bad air days helps to reduce pollution.

Use Air-Friendly Products

- Aerosol sprays and cleaners, oil-based paints, and other chemical products contribute VOCs to both indoor and outdoor air. Less toxic alternatives, such as latex paints that are low in VOCs, are available.

Do Not Light Up

- Whether lighting a gas barbecue or a cigarette, smoke adds more pollutants and further lowers the air quality in and around your home.

Educate Yourself and Family Members

- Discuss, with your family, what you can do to help improve air quality.

As you examine these suggestions, you might ask if they are only applicable on bad air days. How many of the suggestions make sense all year round?

DID YOU KNOW ?

Car-Free Days
Car-free days are held in over 1400 cities in 38 countries every September 22. These events help to raise awareness and inform people about the long-term environmental effects of smog and air pollution.

Interpreting the Air Quality Index

Suppose that the following data (**Table 6**) were collected at an air-monitoring station close to where you live. Refer to **Table 5**, and estimate the Air Quality Index for each day. Identify the pollutant that has the greatest impact on air quality.

Table 6 Sample Weekly AQI Data

Pollutant	Concentration (ppb)				
	Monday	**Tuesday**	**Wednesday**	**Thursday**	**Friday**
SO_2	153	612	288	351	428
O_3	52	85	32	60	76
NO_2	233	771	460	588	697
TRS*	17	56	33	24	46
CO	63	428	216	412	366
SP	12	65	32	40	62

* Total Reduced Sulfur

▶ Section 4.16 Questions

Understanding Concepts

1. Describe each of the following components of air pollution, and give an example:
   (a) primary pollutants
   (b) secondary pollutants
   (c) particulate matter

2. Explain what is meant by the term "smog."

3. (a) Explain what is meant by the term "photochemical smog."
   (b) List the three components of photochemical smog.

4. (a) How is ground-level ozone the same as stratospheric ozone?
   (b) How is ground-level ozone different from stratospheric ozone?

5. (a) Name the six pollutants that are measured to determine air quality.
   (b) Why are these six pollutants used?

6. (a) Briefly explain how the Air Quality Index is determined.
   (b) What are the consequences associated with each range of values for ozone in the Air Quality Index?

Making Connections

7. Research last week's AQI readings for the station that is closest to where you live. Use **Table 5** to determine the pollutant concentration for the pollutant that is responsible for the AQI. What are the environmental or health implications of these values?

 www.science.nelson.com

8. Review the list of suggestions for actions to help reduce smog levels and improve air quality.
   (a) Which suggestion will likely result in a long-term improvement in air quality?
   (b) Which suggestions, if any, would you identify as impractical and not likely to result in any improvement in air quality? Explain your choice.

9. Do you think the solutions to air quality problems that are outlined in this section will actually work? Explain your reasons. If you answered "no," what do you think must occur for air quality to improve?

Extension

10. Discuss how individuals can contribute to the improvement of air quality through their choice of transportation.

 www.science.nelson.com

The Trace Atmospheric Gas Analyzer (TAGA) is a self-contained mobile laboratory that is able to sample air or emissions from various sources. The TAGA can be used to analyze both outdoor and indoor air pollutants at concentrations in the low parts per billion (ppb). It can also be used to monitor air emissions from waste disposal sites to ensure that operations are within acceptable limits.

Chemical spills and bad odours can also be investigated using the TAGA. Any airborne contamination from a site can be quickly identified and tracked. The TAGA can be used for removal and remediation efforts after the assessment has been done. For example, the TAGA can remain at a site determined to be a source of toxic gases while the material that is producing these gases is removed.

The air monitoring and analysis instruments on the TAGA allow investigations to be done safely. On board the TAGA is equipment for sampling air from various sources (**Figure 1**). Several gas chromatographs can identify most airborne substances, and specialized instruments can analyze organic compounds and inorganic metals.

In a gas chromatograph, a sample of the mixture to be analyzed is injected into a stream of an inert gas, such as helium, that is flowing through a long column. The column is coiled inside a temperature-controlled oven. The carrier gas moves different substances in the mixture through the column at different rates. As each substance exits the column, the chromatograph's detector plots the electric signal it generates against the time it spent in the column. The resulting graph, or chromatogram, shows the composition of the air sample. Each component in the mixture can be identified from the positions of the peaks and the comparative heights of the peaks on the chromatogram (**Figure 2**).

The Ontario Ministry of the Environment currently has a mobile TAGA, which is used to randomly sample air around the province. The mobile TAGA can also be sent to specific locations where air quality may be compromised. For example, it was used at a plant fire in Scarborough to sample the air and determine if the smoke from the fire posed a health hazard for local residents. The TAGA was able to determine that there was no need to evacuate the residents.

Figure 1
Instruments carried by the TAGA are capable of identifying and reporting the concentrations of airborne contaminants.

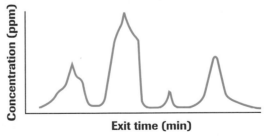

Figure 2
A representative example of a gas chromatogram

▶ **Tech Connect 4.17** *Questions*

Understanding Concepts

1. What general concentration levels of airborne contaminants can the TAGA detect?

2. List four possible uses of this mobile testing unit.

Making Connections

3. A TAGA was used at Ground Zero in the aftermath of the 9/11 tragedy in New York. Research to find out how it helped emergency workers.

 GO   www.science.nelson.com

Key Understandings

4.1 Water: Essential for Life
- Compare essential uses of water with nonessential uses of water.
- Describe the water cycle.
- Describe the physical and chemical properties of water that account for the presence of a variety of substances in environmental water.

4.2 What's in Clean Water?
- Describe how water becomes hard, as well as the methods that are used to soften hard water.
- Describe the relationship between dissolved oxygen in lake or river water and the health of an ecosystem.

4.3 What's in Polluted Water?
- Examine natural and human contributions to water pollution.

4.4 Investigation: Testing for Ions in Water
- Identify and determine the concentration of dissolved ions in environmental water.
- Demonstrate the importance of the quantitative analysis of substances in water samples.

4.5 Tech Connect: Water Disinfection Methods
- Describe different methods (such as chlorine, UV radiation, ozone, and charcoal filters) that are used to disinfect water.

4.6 Case Study: Protecting Canada's Great Lakes Environment
- Describe how governments can work with individuals to improve the water quality in the Great Lakes Basin.

4.7 Acids and Bases
- Explain the terms "acid" and "base" using the Arrhenius theory.
- Write balanced ionization equations for acids and balanced dissociation equations for bases.
- Explain the differences between strong acids and weak acids in terms of the degree of ionization.
- Explain the differences between concentrated and dilute acids.
- Describe the relationship between pH values and hydrogen ion concentrations in acids and bases.

4.8 Explore an Issue: The Disposal of Household Products
- Examine a variety of disposal options for household hazardous products.
- Consider the environmental implications of the disposal of common household products.
- Use a decision-making model to make a personal choice about the disposal of hazardous products.

4.9 Investigation: Dilution and pH
- Use a pH meter and an indicator to demonstrate the effect of dilution on the pH of an acidic solution.

4.10 Investigation: Acid–Base Reactions
- Investigate some general reactions of acids and bases.
- Investigate the acid/base character of the solutions of both metal and nonmetal oxides.

4.11 Reactions of Acids and Bases
- Write balanced chemical equations to represent the reactions of acids and bases, including neutralization reactions.
- Use appropriate scientific vocabulary, such as "neutralization" and "titration," to communicate ideas related to the chemical analyses of acids and bases.

4.12 Activity: Quality Control of Vinegar
- Conduct an acid–base titration to determine the concentration of an acid.

4.13 Acid Rain
- Identify the gases that are responsible for acid rain and the sources of these gases.
- Describe how acid rain is formed.
- Describe the chemistry that is used to neutralize the effects of acid rain.

4.14 Gas Laws: Why Gases Behave the Way They Do
- Examine the properties of gases, including some of the gas laws, to gain a better understanding of gaseous behaviour in the atmosphere.

4.15 Explore an Issue: Canada and the Kyoto Protocol

- Identify the composition of gases in the atmosphere, and explain how human activities are affecting this composition.
- Identify greenhouse gases and their sources.
- Describe the effect of CFCs on stratospheric ozone.
- Investigate government actions, including ratification of the Kyoto Protocol, that are designed to improve air quality, as well as the arguments against such actions.
- Debate Canada's ratification of the Kyoto Protocol.

4.16 Air Quality Solutions—Your Role

- Use data to calculate the Air Quality Index, and recommend actions that would help to reduce smog levels and improve air quality.

4.17 Tech Connect: The TAGA

- Investigate the applications of the Trace Atmospheric Gas Analyzer (TAGA), a self-contained mobile laboratory that samples air emissions.
- Explain the importance of the quantitative analysis of substances in air.

Key Terms

4.1
ground water
aquifer
surface water

4.2
hard water
soda-lime process
aeration

4.7
ionization
hydronium ion

strong acid
weak acid
dilute
concentrated
pH

4.11
neutralization
titration
standard solution
titrant
endpoint

4.13
acid precipitation
acid deposition

4.14
kinetic molecular theory
atmospheric pressure
Boyle's law
absolute zero
Kelvin temperature scale
Charles's law

Dalton's law of partial pressures
partial pressure

4.15
greenhouse effect
greenhouse gases
global warming

Key Equations

4.7
- $[H^+] = 10^{-pH}$

4.14
- $p_1V_1 = p_2V_2$

- $\dfrac{V_1}{T_1} = \dfrac{V_2}{T_2}$

- $p_{total} = p_1 + p_2 + p_3 + \ldots$

Problems You Can Solve

4.3
- Compare the concentrations of chemicals in a sample of drinking water with the Maximum Acceptable Concentrations (MACs) to determine if the sample is safe to drink.

4.4
- Calculate the concentration of a particular ion that is present in a water sample.

4.7
- Calculate the pH of a solution from $[H^+]$.
- Calculate $[H^+]$ from the pH.
- Calculate the concentration of an acid or base using titration data.

4.9
- Calculate the hydrogen ion concentration at each stage of dilution.

4.11
- Balance chemical equations to represent the reactions of acids and bases.

4.14
- Solve problems that involve gases using Boyle's law (pressure and volume), Charles's law (volume and temperature), and Dalton's law of partial pressures.

▶ *MAKE* a summary

Make a concept map that shows your understanding of how the actions and personal choices of an individual help to ensure a healthy environment.

(a) Choose a personal action that will improve the quality of either water or air.

(b) Consider the action from an economic, scientific, and environmental perspective.

(c) Outline three risks and three benefits of your proposed action.

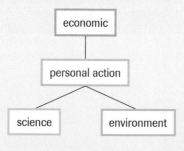

A Microscale Analysis of Hard Water

Calcium ions and magnesium ions are the main culprits in making water hard. These ions precipitate with soap ions to produce an insoluble precipitate that forms a "bathtub ring." Hard water is also a nuisance in swimming pools, where calcium ions and magnesium ions can precipitate with carbonate ions in the water to form grey or brownish deposits on pool walls and fixtures.

The concentrations of dissolved calcium and magnesium ions in a water sample can be determined by titrating the water sample with a chemical called EDTA. Calcium and magnesium combine with an ionic form of EDTA in a 1:1 mole ratio:

$$M^{2+}_{(aq)} + EDTA^{2-}_{(aq)} \rightarrow MEDTA_{(aq)}$$

where M^{2+} is a metal ion with a 2+ charge, such as Ca^{2+} or Mg^{2+}.

For this performance task, you will determine the concentrations of dissolved magnesium ions and calcium ions in different water samples. You will complete a report that includes a Procedure, Analysis, and Evaluation of the experiment.

Question

Which water sample, from a variety of sources, is the hardest?

Experimental Design

(a) Write a complete Experimental Design for an experiment in which you will titrate different water samples with EDTA to determine their relative hardness. Each analysis requires 15 drops of the water sample, 1 to 2 drops of Erichrome Black T indicator, and 2 drops of pH 10 buffer. Include a control test with distilled water.

Materials

eye protection
lab apron
Erichrome Black T indicator
pH 10 buffer
24-well microtrays
3 Berol pipettes
three 150-mL beakers
2 water samples from various sources
dropper bottles of 0.01-mol/L EDTA
distilled or deionized water

Procedure

(b) Write a complete Procedure for your Experimental Design. Describe safety precautions and include disposal instructions, which you can obtain from your teacher. Ask your teacher to approve your Procedure before continuing with your experiment.

▶ **Criteria**

Assessment

Your completed task will be assessed according to the following criteria:

Process

• Develop an appropriate Procedure.

• Safely carry out your Procedure when approved.

• Record observations with appropriate precision.

• Analyze your results.

• Evaluate your Experimental Design, Materials, Procedure, and skills.

• Estimate/identify errors, and evaluate your Prediction.

Product

• Prepare a suitable lab report, including appropriate tables for observations.

• Demonstrate an understanding of the relevant scientific concepts.

• Use appropriate terms, symbols, equations, and SI units correctly.

Wear eye protection and a lab apron. EDTA and Erichrome Black T are toxic if ingested, and pH 10 buffer solution is an irritant.

Analysis

(c) Answer the Question.

(d) According to your observations, rank the water samples from softest to hardest.

Evaluation

(e) Evaluate your observations by evaluating the Experimental Design, the Materials, the Procedure, and your skills as an experimenter. List any sources of error that may have influenced your results.

(f) How confident are you that your techniques and measurements resulted in acceptable evidence? Is there anything that could cast doubt on your evidence?

(g) Are there any changes that you would make to the Experimental Design or to the Procedure? Describe briefly.

Understanding Concepts

1. Describe the physical and chemical properties of water. (4.1)

2. Briefly explain each of the following properties of water:
   (a) the bonding in a water molecule
   (b) the polarity of a water molecule
   (c) why water is such a good solvent
   (d) why water exists as a liquid most of the time (**Figure 1**)
   (e) why bodies of water are able to moderate extremes in climates
   (f) why most oils float on the surface of water (4.1)

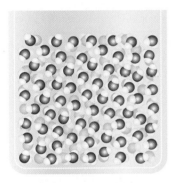

Figure 1
Water is found in the liquid state most of the time.

3. Water is known as the universal solvent. What implication does this property have on drinking water? (4.1)

4. (a) What two metal ions are most responsible for the hardness of water?
   (b) Why is ground water hard in some areas, but not in other areas? (4.2)

5. Describe how hard water can reduce the efficiency of appliances that are used to heat water. (4.2)

6. List the three categories of contaminants in water. Give at least two examples of each contaminant. (4.3)

7. A sample of water contains 1.55 ppm of dissolved nitrate. Calculate the mass of nitrate in a 1.0-L sample of this water. (4.3)

8. A diagnostic test for a certain pollutant is applied to a sample of drinking water (**Figure 2**). Will a negative test result prove that the ion is not present? Explain. (4.4)

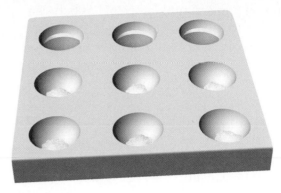

Figure 2
The formation of coloured precipitates is a diagnostic test for certain ions in water. Measuring the mass of a precipitate provides the concentration of one of these ions in a water sample.

9. Distinguish between qualitative analysis and quantitative analysis. (4.4)

10. The Great Lakes Action Plan 2001–2006 identifies 42 Areas of Concern.
    (a) What is an Area of Concern?
    (b) What is a Remedial Action Plan?
    (c) Summarize, in your own words, the seven objectives of a RAP. (4.6)

11. Describe three properties of acids and three properties of bases. (4.7)

12. (a) What is the evidence of a strong acid and a weak acid?
    (b) Explain the terms "strong acid" and "weak acid" using the Arrhenius definition of an acid. (4.7)

13. Classify each of the following solutions as an acid or a base:
    (a) lye, $NaOH_{(s)}$
    (b) vinegar, $HC_2H_3O_{2(aq)}$
    (c) milk of magnesia, $Mg(OH)_{2(s)}$
    (d) slaked lime, $Ca(OH)_{2(s)}$
    (e) window-cleaning solution, $NH_{3(aq)}$ (4.7)

14. A 0.1-mol/L solution of nitric acid, $HNO_{3(aq)}$, has a pH of about 1. A 0.1-mol/L solution of acetic acid, $HC_2H_3O_{2(aq)}$, has a pH of about 3.
    (a) How can you explain this difference in pH?
    (b) Which solution requires greater caution when being used? Why?
    (c) Write the ionization reaction equation for each acid. (4.7)

15. Swimming pools should have a pH of between 7.2 and 7.6 to maintain the correct level of chlorine in the water and to prevent algal growth (**Figure 3**).
    (a) Is pool water acidic, basic, or neutral? Justify your answer.
    (b) A homeowner tests the water in a pool and determines that the pH is 7.9. What could the homeowner add to the water to adjust the pH? Explain. (4.7)

Figure 3
The pH of a swimming pool must be carefully maintained.

16. Drinking water standards in Canada specify that the pH should range from 6.5 to 8.5.
    (a) What household products, if carelessly discarded, might cause the pH of a freshwater source to decrease? List three products.
    (b) What household products, if carelessly discarded, might cause the pH of a freshwater source to increase? List three products. (4.8)

17. A strong acid solution has a pH of 4.0. The acid solution is mixed with water to make a large volume of solution.
    (a) How will the concentration of hydrogen ions be affected?
    (b) How will the pH change?
    (c) How does dilution affect the pH of a basic solution? (4.9)

18. (a) What type of solution forms when soluble metal oxides react with water?
    (b) What type of solution forms when nonmetal oxides react with water? (4.10)

19. Baking soda, $NaHCO_{3(s)}$, is a unique chemical substance because it reacts with both strong acids and strong bases. Write a balanced chemical equation for the reaction between baking soda and hydrochloric acid. (4.11)

20. Three trials in a titration analysis produce the following volumes of a standard base solution delivered from the burette (**Table 1**). From a quality control perspective, is a fourth trial necessary? Why or why not? (4.12)

Table 1   Volume of Standard Base Used during Titration

Trial	Volume of base solution
1	25.55 mL
2	25.45 mL
3	27.55 mL

21. "Just as litmus paper is a chemical indicator, a lake trout is a biological indicator that can be used to indicate the level of acidity of lake water." Explain this statement. (4.13)

22. (a) Name two pollutants that are major contributors to acid precipitation.
    (b) Identify the sources of each pollutant. (4.13)

23. Use the kinetic molecular theory to explain each of the following behaviours of gases:
    (a) The boiling point of a liquid decreases as the altitude increases.
    (b) The volume of a gas in a container can be decreased by increasing the external pressure on the container.
    (c) Gases are more easily compressed than solids or liquids.
    (d) Hot-air balloons are partially inflated with cold air. As the air is heated, the balloon expands and rises. (4.14)

24. Explain how CFCs reduce the number of ozone molecules, resulting in more ultraviolet light reaching the surface of Earth. Use balanced chemical equations in your explanation.   (4.15)

25. What is smog, and how does it form (**Figure 4**)?   (4.16)

Figure 4
Smog contributes to bad air days.

26. Do you think it is important to test for air quality accurately, on a regular basis? Why or why not?   (4.16, 4.17)

Applying Inquiry Skills

27. A wastewater technician wants to determine the concentration of lead(II) ions in a sample of wastewater.
    (a) Suggest an appropriate chemical to add to the wastewater sample to precipitate the lead(II) ions from solution as lead(II) sulfate.
    (b) The technician adds the chemical you suggested in (a) to a volume of the sample, and a precipitate forms. The precipitate is filtered, and the mass of the precipitate is determined. This mass is then used to calculate the concentration of lead(II) ions in the original wastewater sample. Critique the accuracy of the results obtained.   (4.4)

28. A forensic scientist needs to identify four colourless solutions that have been collected at a crime scene. Each solution is tested with red and blue litmus paper. Then it is tested for electrical conductivity. (Note that the four solutions are at the same concentration and temperature.)

Complete the Analysis in the following lab report:

Question
Which of the solutions (labelled A, B, C, and D) contain potassium hydroxide, sugar, sodium chloride, and sulfuric acid?

Observations

Table 2   Results for Litmus and Electrical Conductivity

Unidentified solution	Red litmus	Blue litmus	Conductivity
A	stays red	blue to red	high
B	stays red	stays blue	none
C	red to blue	stays blue	high
D	stays red	stays blue	high

Analysis
Using the Evidence in **Table 2**, answer the Question.   (4.7)

29. Design an experiment to test the Prediction that diluting a solution of a strong acid by a factor of 10 will change the pH by one. Provide a list of Materials and a Procedure that you could follow to carry out your Experimental Design.   (4.7)

30. Each of the elements sodium, $Na_{(s)}$, barium, $Ba_{(s)}$, and sulfur, $S_{(s)}$, has the potential to react with oxygen to form a soluble oxide in water. The oxide may produce either an acidic or basic solution when dissolved in water.
    (a) Design an experiment to produce the oxide of each of these elements and to test the nature of the solution.
    (b) Write a complete chemical equation to show the formation of the oxide of each element.
    (c) Write a complete chemical equation to show the formation of the product when each oxide in (b) dissolves in water.
    (d) Predict whether each solution will be acidic or basic.   (4.10, 4.11)

31. A student is asked to determine the concentration of hydrochloric acid so that the solution can be used as a standard for other class experiments. Samples of a standard sodium carbonate solution are titrated with the hydrochloric acid of unknown concentration.

Methyl orange indicator is used to show the endpoint. Complete the Observations and Analysis in the following lab report:

Question

What is the molar concentration of hydrochloric acid?

Observations

(a) Copy and complete **Table 3**.

Table 3 Titration of 10.0 mL of 0.120-mol/L $Na_2CO_{3(aq)}$ with $HCl_{(aq)}$

Trial	1	2	3	4
Final burette reading (mL)	17.95 mL	35.05 mL	22.95 mL	40.15 mL
Initial burette reading (mL)	0.30 mL	17.90 mL	5.90 mL	22.90 mL
Volume of $HCl_{(aq)}$ added (mL)				

Analysis

(b) Calculate the molar concentration of hydrochloric acid. (4.11)

Making Connections

32. (a) If pure water were tested with classroom electrical conductivity apparatus, would the light bulb glow?
    (b) Can you be electrocuted if you are standing in pure water? Explain. (*Hint:* Pure water contains no dissolved substances.) (4.1, 4.2)

33. Propose a solution to reduce the pollution of ground water. Choose one source of contamination, and suggest ways to minimize or eliminate it. (4.2, 4.3)

34. Why are very precise equipment and technologies needed to test drinking water samples (**Figure 5**)? (4.2, 4.4)

35. The transportation of hazardous wastes from the Nova Scotia tar ponds to a disposal site just outside Sarnia, Ontario, is a contentious issue. Find out whether your community has a hazardous waste disposal depot. What are the skills and qualifications of people who work at these depots? What types of substances do they handle? Do you agree or disagree that

Figure 5
Lives depend on the accuracy of drinking water tests.

Figure 6
Is it safe to transport hazardous wastes over long distances on busy highways?

hazardous wastes from Nova Scotia should be trucked to southwestern Ontario (**Figure 6**)? Why or why not? (4.8)

GO www.science.nelson.com

36. The maximum quantity of oxygen that dissolves in water is 14.7 ppm at 0°C and 8.7 ppm at 25°C.
    (a) What temperature do you think active game fish prefer? Explain your answer.
    (b) If you were trying to catch a fish, where would you cast your lure? Why? (4.14)

37. In movies, acids are often portrayed as dangerous, with the ability to burn or eat through anything. In one movie, for example, an aluminum boat is quickly eaten away by acidic lake water. Is this portrayal accurate? Justify your answer with personal experiences, examples, and explanations. Which acids are the most dangerous? (4.7)

38. Laboratory safety rules require students to wear eye protection and be extremely careful when handling acids, such as hydrochloric acid and sulfuric acid. In contrast, boric acid, $H_3BO_{3(aq)}$, is sold in pharmacies as a soothing eyewash (**Figure 7**). Explain why boric acid is relatively harmless, while hydrochloric acid and sulfuric acid are very corrosive. (4.7)

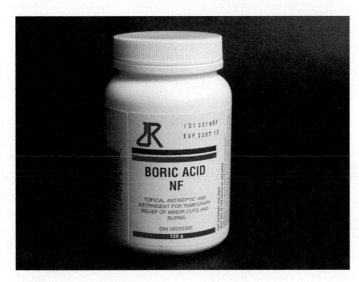

Figure 7
Boric acid is recommended for soothing sore eyes.

39. Gardeners know that coniferous trees like acidic soil. Acid precipitation adversely affects coniferous trees, however. Explain. (4.13)

 www.science.nelson.com

40. Will the information you have learned in this unit affect your attitude toward the materials that you pour down the sink at home or at school? Explain your answer. (4.8)

Extension

41. Reverse osmosis is a water-treatment process that is widely used to produce many brands of bottled water and to purify seawater. Outdoor supply stores sell reverse osmosis kits for purifying water on wilderness trips. Research reverse osmosis to answer the following questions:
    (a) How is pure drinking water obtained from seawater using reverse osmosis?
    (b) Is reverse osmosis commercially viable?
    (c) Research a water-treatment plant that uses reverse osmosis to purify seawater. Present your findings to your class in a report.

 www.science.nelson.com

42. When travelling in the wilderness, people are often advised to take chlorine or iodine tablets to put in their drinking water. Research the purpose of these tablets and how they are used. Prepare a leaflet to be included inside the packets of chlorine or iodine tablets.

 www.science.nelson.com

43. Many chemical substances that are potentially toxic or harmful to the environment have maximum allowable concentrations set by government regulations. Research these regulations.
    (a) If a chemical is dangerous, should the MAC be zero? Explain.
    (b) Is a zero MAC possible? Can it be measured?
    (c) If a nonzero limit is set, how is this limit determined?

 www.science.nelson.com

manganese (engine, brakes, gears, shafts, body panels)

gold (electrical connectors in air bags)

copper (electrical wiring, radiators, instruments)

chromium (exposed trim, engine parts)

mercury (electronics)

nickel (engine, frame, shafts)

tin (solder, lubricants, electrical components)

aluminum (air conditioner parts, trim, body panels, radiator, transmission)

lead (batteries)

tungsten (spark plugs, lights)

platinum and palladium (catalytic converter)

Electrochemistry

Very few of the metals that are used to manufacture this automobile are found in nature in their pure metallic state. Most, in fact, are joined with other elements in compounds or minerals. Chemical processes, which often require a considerable amount of energy, are used to extract these metals from their compounds and purify them.

Once the pure metals have been used to build a car, many start to turn back to their native state as compounds. Preventing this process is one of the many challenges of automobile designers. Although the tendency of metals to become compounds is detrimental to the body of the car, the same process is used in the car's battery to produce useful electricity.

In this unit, you will learn about the reactions of metals, including when they are useful and when they are inconvenient. You will learn how chemical energy is converted into electrical energy (and vice versa), why metals have a tendency to become compounds and how this process can be prevented, how pure metals are made, and how electrochemistry affects the environment.

silver (stereo, mirrors, electrical components)

zinc (battery, electrical components, body panels)

iron (pipes, various engine parts, frame, body)

▶ Overall Expectations

In this unit, you will be able to

- demonstrate an understanding of the chemical processes that take place in galvanic and electrolytic cells;
- investigate, through experimentation, the rate at which metals oxidize;
- build electrochemical cells and describe how they function;
- explain the importance of common electrochemical processes for industry and the consequences of these processes for the environment.

ARE YOU READY?

Knowledge and Understanding

1. The alkaline dry cells in the circuit in **Figure 1** are connected in series.
   (a) If the potential difference across each cell is 1.5 V, what is the potential difference across all three cells?
   (b) What is the potential difference across the light bulb?
   (c) How will the brightness of the bulb change if another cell is added in series?
   (d) Is the voltmeter in series or parallel to the circuit?

Figure 1
Three cells in series

2. Copy and complete each of the following statements:
   (a) Metals tend to _____ electrons to form positively charged ions.
   (b) Nonmetals tend to _____ electrons to form negatively charged ions.

3. What is the charge on the ion that is most commonly formed by each of these metals?
   (a) Na
   (b) Mg
   (c) Al
   (d) Ag
   (e) Zn

4. What is the charge on the ion that is most commonly formed by each of these nonmetals?
   (a) F
   (b) O
   (c) N
   (d) Cl

5. In your notebook, balance each of the following chemical equations:
   (a) ___ $K_{(s)}$ + ___ $Cl_{2(g)}$ → ___ $KCl_{(s)}$
   (b) ___ $Mg_{(s)}$ + ___ $AgNO_{3(aq)}$ → ___ $Ag_{(s)}$ + ___ $Mg(NO_3)_{2(aq)}$
   (c) ___ $C_3H_{8(g)}$ + ___ $O_{2(g)}$ → ___$CO_{2(g)}$ + ___ $H_2O_{(g)}$
   (d) ___ $KClO_{3(s)}$ → ___ $KCl_{(s)}$ + ___ $O_{2(g)}$
   (e) ___ $Al_{(s)}$ + ___ $CuCl_{2(aq)}$ → ___ $Cu_{(s)}$ + ___ $AlCl_{3(aq)}$
   (f) ___ $Fe_{(s)}$ + ___ $O_{2(g)}$ → ___ $Fe_2O_{3(s)}$
   (g) ___ $Fe_2O_{3(s)}$ + ___ $H_{2(g)}$ → ___ $Fe_{(s)}$ + ___ $H_2O_{(g)}$

6. (a) Write the general chemical equation for a synthesis reaction.
    (b) Which of the equations in question 5 represent synthesis reactions?

7. (a) Write the general chemical equation for a single displacement reaction.
    (b) Which of the equations in question 5 represent single displacement reactions?

8. Write the net ionic equation for each of the following chemical equations:
    (a) $Mg_{(s)} + CuCl_{2(aq)} \rightarrow Cu_{(s)} + MgCl_{2(aq)}$
    (b) $Al_{(s)} + 3\ AgNO_{3(aq)} \rightarrow 3\ Ag_{(s)} + Al(NO_3)_{3(aq)}$

9. Predict the product(s) of each of the following chemical reactions. (Assume that a reaction occurs.)
    (a) $Zn_{(s)} + CuSO_{4(aq)} \rightarrow$
    (b) $Zn_{(s)} + AgNO_{3(aq)} \rightarrow$

Inquiry and Communication

10. Write the IUPAC name for each of the following compounds:
    (a) H_2SO_4
    (b) HNO_3
    (c) $NaOH$
    (d) $Ni(OH)_2$
    (e) Na_2SO_4
    (f) $Ca(NO_3)_2$
    (g) $PbSO_4$
    (h) $CuCl_2$
    (i) $FeSO_4$
    (j) $Fe(OH)_3$
    (k) CO_2
    (l) SO_3

11. Write the chemical formula for each of the following compounds:
    (a) magnesium chloride
    (b) silver sulfide
    (c) chromium(VI) oxide
    (d) copper(II) nitrate
    (e) calcium hydroxide
    (f) lead(II) sulfate
    (g) hydrochloric acid
    (h) sulfur dioxide
    (i) carbon monoxide

(a)

(b)

Figure 1
Rust—a nuisance on cars **(a)** but possibly the source of oxygen for the first human colony on Mars **(b)**

What do the old car and the planet Mars in **Figure 1** have in common? The answer is rust. Mars is often called "The Red Planet" because its soil is rich in rust, which is an oxide of iron. Similarly, the iron in this car quickly turns to rust if left unprotected.

Iron, like most metals, is not found naturally in its pure metallic state. It is usually found combined with other elements, such as oxygen and sulfur, in underground mineral deposits. Centuries ago, people learned how to use energy to extract metals from their minerals to craft tools, weapons, and decorative objects (**Figure 2**). This technology became known as metallurgy, and it was crucial to the development of our modern technological society. Most metallurgical processes involve a type of chemical reaction called an oxidation–reduction reaction (also known as a *redox* reaction).

Some reactions of metals and nonmetals occur spontaneously and release energy. As iron rusts, for example, energy is released. This energy has resulted in explosions aboard ships transporting iron. Other reactions, such as extracting pure iron from rust, do not occur spontaneously. These reactions require the input of a great deal of energy to make them happen.

Now let's look into the future. Imagine that you are a member of the first human colony on Mars (**Figure 3**). The Martian atmosphere has far too little oxygen to support life, so how do you produce the oxygen you need? Perhaps you use reactions that extract oxygen from the rust in the soil, just as the early metallurgists did to extract iron on Earth. What is your energy source? Mars does not have organic materials, such as the coal and wood used by the Samurai sword maker. Instead, you collect solar energy and store it as chemical energy in large batteries. When required, this energy is converted into electrical energy and used to extract oxygen from rust. The interconnection between electricity and chemical reactions is called *electrochemistry*.

Figure 2
Making steel requires higher temperatures than a simple wood fire can provide. Only a few cultures developed the technology needed to make steel early in their history. In Japan, steel was used to craft Samurai swords. At a time when there was no written language, the process of sword making became a ritual that was passed on from one generation to the next.

Figure 3
The possible colonization of Mars presents many challenges, not the least of which is the absence of atmospheric oxygen. Is it possible to generate oxygen by reversing the rusting reaction?

Meanwhile, back on Earth…. Electrochemistry is not limited to the realm of "high tech" and science fiction. Many common products, such as aluminum foil, jewellery, batteries, and the chlorine that is used to disinfect pools, are manufactured using the principles of electrochemistry. Many important biological processes, such as photosynthesis and respiration, also involve electrochemical reactions. In this unit, you will explore electrochemistry and its impact on our society.

▶ TRY THIS activity   *Making Money*

Imagine that you are the technician at the Royal Canadian Mint. You are responsible for producing a new penny. How are you going to coat a zinc disk with copper (**Figure 4**)?

This activity provides one possible procedure to coat a zinc strip with copper. The reaction involved is a single displacement reaction. (You learned about single displacement reactions in section 1.14.) Your task is to evaluate the procedure and comment on its effectiveness.

Materials: eye protection; lab apron; zinc strip (approximately the same size as a penny); 20 mL 0.1-mol/L copper(II) sulfate solution, $CuSO_{4(aq)}$; plastic cup; tweezers; sandpaper

1. Clean the zinc strip with sandpaper.
2. Place the zinc strip and 20 mL of the 0.1-mol/L copper(II) sulfate solution in the plastic cup. Allow the cup to stand undisturbed for about 10 min.

 Copper(II) sulfate is toxic. Avoid skin contact. Wear eye protection.

3. Examine the zinc strip, and record your observations.
4. Use tweezers to remove the zinc strip from the solution.
5. Examine the strip to see how well the copper adhered to the surface.
6. Dispose of the contents of the cup as directed by your teacher.

(a) Write a chemical equation for the reaction that occurred to produce copper metal.

(b) Evaluate the procedure you used to coat zinc with copper by identifying specific problems.

Figure 4
In 1997, the Royal Canadian Mint changed the composition of the penny to reduce manufacturing costs. Before 1997, a penny cost 1.5¢ to manufacture. For a few years after 1997, the penny was made primarily of zinc with only a thin coating of copper applied over it. Then the core was changed to steel. These changes cut the cost of making a penny almost in half!

💡 REFLECT on your learning   ▼

1. What does it mean when we say that a substance is "oxidized"?
2. How do batteries work?
3. What changes happen inside a cell or battery when it is recharged?
4. What is corrosion? How can it be prevented?
5. In what ways are chemical reactions and electricity related?

If you place a zinc strip in a solution of copper(II) sulfate, a fuzzy coating of metallic copper forms on the zinc (**Figures 1** and **2**). In this reaction, copper(II) ions pull two electrons from zinc atoms. The net ionic equation for the reaction is

$$\overset{\displaystyle 2e^-}{\overset{\frown}{Zn_{(s)}}} + Cu^{2+}_{(aq)} \rightarrow Zn^{2+}_{(aq)} + Cu_{(s)}$$

Here are the changes that occur:

- Zn becomes Zn^{2+}, a loss of 2 e^-.
- Cu^{2+} becomes Cu, a gain of 2 e^-.

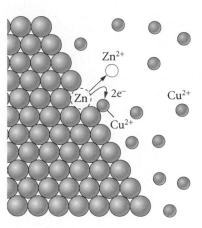

Figure 1
Copper(II) ions, $Cu^{2+}_{(aq)}$, pull two electrons away from zinc, $Zn_{(s)}$, producing copper metal, $Cu_{(s)}$, and zinc ions, $Zn^{2+}_{(aq)}$.

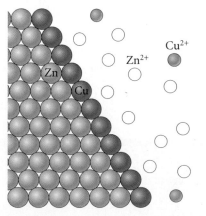

Figure 2
The zinc strip is covered with a heavy coat of copper atoms. The lighter shade of blue, compared with **Figure 1**, suggests that most copper(II) ions, $Cu^{2+}_{(aq)}$, have been changed to atoms of copper metal, $Cu_{(s)}$.

Chemists call the loss of electrons **oxidation** and the gain of electrons **reduction**. In this reaction, zinc atoms, Zn, are oxidized to zinc ions, $Zn^{2+}_{(aq)}$. At the same time, copper(II) ions, $Cu^{2+}_{(aq)}$, are reduced to copper metal, Cu. Reactions in which one reactant is oxidized and the other reactant is reduced are called oxidation–reduction reactions, or **redox reactions** for short.

Note that the electrons gained by one reactant are the electrons lost by the other reactant. Therefore, in any redox reaction, the number of electrons gained and lost must be equal.

oxidation a process in which chemical entities lose electrons

reduction a process in which chemical entities gain electrons

redox reaction a reaction in which one reactant is oxidized and the other reactant is reduced

---

▶ **SAMPLE** problem

Identifying Oxidation and Reduction

Identify the reactant oxidized and the reactant reduced in each of the following reactions:

(a) $Pb_{(s)} + Cu(NO_3)_{2(aq)} \rightarrow Cu_{(s)} + Pb(NO_3)_{2(aq)}$

Step 1: Identify Reacting Entities

First write the total ionic equation.

$$Pb_{(s)} + Cu^{2+}_{(aq)} + 2\,NO^-_{3(aq)} \rightarrow Cu_{(s)} + Pb^{2+}_{(aq)} + 2\,NO^-_{3(aq)}$$

Then write the net ionic equation by eliminating the ions that are common to both sides.

$$Pb_{(s)} + Cu^{2+}_{(aq)} + 2\,\cancel{NO^-_{3(aq)}} \rightarrow Cu_{(s)} + Pb^{2+}_{(aq)} + 2\,\cancel{NO^-_{3(aq)}}$$

$$Pb_{(s)} + Cu^{2+}_{(aq)} \rightarrow Cu_{(s)} + Pb^{2+}_{(aq)}$$

Step 2: Label Charges

$$Pb^0_{(s)} + Cu^{2+}_{(aq)} \rightarrow Cu^0_{(s)} + Pb^{2+}_{(aq)}$$

(*Note:* The charge on the atoms of any uncombined element is always zero.)

Step 3: Identify Loss and Gain of Electrons

In this reaction,

- each Pb becomes Pb^{2+}, a loss of $2\,e^-$ (oxidation)
- each Cu^{2+} becomes Cu, a gain of $2\,e^-$ (reduction)

Therefore, lead atoms are oxidized and copper ions are reduced.

(b) $2\,Mg_{(s)} + O_{2(g)} \rightarrow 2\,MgO_{(s)}$ **(Figure 3)**

Step 1: Identify Reacting Entities

Unlike the reaction in Sample Problem (a), this reaction is between a solid and a gas. There are no ions to consider.

Step 2: Label Charges

$$\overset{0}{2\,Mg_{(s)}} + \overset{0}{O_{2(g)}} \rightarrow \overset{+2\ -2}{2\,MgO_{(s)}}$$

(*Remember:* The charge on the atoms of any uncombined element is always zero.)

Figure 3
The combustion of magnesium in the sparkler is a redox reaction. In addition to the magnesium oxide and bright light, this reaction produces a considerable amount of heat.

Figure 4
Sodium metal and liquid bromine react violently to produce sodium bromide.

Step 3: Identify Loss and Gain of Electrons

In this reaction,

- each Mg becomes Mg^{2+}, a loss of 2 e^- (oxidation)
- each O becomes O^{2-}, a gain of 2 e^- (reduction)

Therefore, magnesium is oxidized and oxygen is reduced.

Example

Identify the reactant oxidized and the reactant reduced in each of the following reaction equations:

(a) $Cu_{(s)} + 2\,AgNO_{3(aq)} \rightarrow 2\,Ag_{(s)} + Cu(NO_3)_{2(aq)}$

(b) $2\,Na_{(s)} + Br_{2(l)} \rightarrow 2\,NaBr_{(s)}$ (**Figure 4**)

Solution

(a) $Cu_{(s)} + 2\,AgNO_{3(aq)} \rightarrow 2\,Ag_{(s)} + Cu(NO_3)_{2(aq)}$

$Cu_{(s)} + 2\,Ag^+_{(aq)} + 2\,\cancel{NO_{3(aq)}} \rightarrow 2\,Ag_{(s)} + Cu^{2+}_{(aq)} + 2\,\cancel{NO_{3(aq)}}$

$Cu^0_{(s)} + 2\,Ag^+_{(aq)} \rightarrow 2\,Ag^0_{(s)} + Cu^{2+}_{(aq)}$

- Each Cu becomes Cu^{2+}, a loss of 2 e^- (oxidation).
- Each Ag^+ becomes Ag, a gain of 1 e^- (reduction).

Therefore, copper atoms are oxidized and silver ions are reduced.

(b) $2\,Na_{(s)} + Br_{2(l)} \rightarrow 2\,NaBr_{(s)}$

$\overset{0}{2\,Na_{(s)}} + \overset{0}{Br_{2(l)}} \rightarrow \overset{+1\ -1}{2\,NaBr_{(s)}}$

- Each Na becomes Na^+, a loss of 1 e^- (oxidation).
- Each Br becomes Br^-, a gain of 1 e^- (reduction).

Therefore, sodium is oxidized and bromine is reduced.

▶ Practice

Understanding Concepts

Identify the reactant oxidized and the reactant reduced in each of the following reaction equations:

(a) $Ca_{(s)} + Sn^{2+}_{(aq)} \rightarrow Sn_{(s)} + Ca^{2+}_{(aq)}$

(b) $CuSO_{4(aq)} + Mg_{(s)} \rightarrow Cu_{(s)} + MgSO_{4(aq)}$

(c) $Zn_{(s)} + 2\,AgNO_{3(aq)} \rightarrow 2\,Ag_{(s)} + Zn(NO_3)_{2(aq)}$

(d) $3\,CuCl_{2(aq)} + 2\,Al_{(s)} \rightarrow 3\,Cu_{(s)} + 2\,AlCl_{3(aq)}$

(e) $Fe_{(s)} + Br_{2(l)} \rightarrow FeBr_{2(s)}$

(f) $2\,Ag_{(s)} + S_{(s)} \rightarrow Ag_2S_{(s)}$

(g) $I_{2(g)} + 2\,K_{(s)} \rightarrow 2\,KI_{(s)}$

(h) $4\,Al_{(s)} + 3\,O_{2(g)} \rightarrow 2\,Al_2O_{3(s)}$

SUMMARY *Redox Reactions*

- The charge on each atom in any uncombined element is always zero.
- In a redox reaction, electrons are transferred from one reactant to another reactant.
- Oxidation refers to the loss of electrons. Reactants that lose electrons are *oxidized*.
- Reduction refers to the gain of electrons. Reactants that gain electrons are *reduced*.

▶ *TRY THIS* activity   **The Pop Can Rip-Off**

Metals vary greatly in their reactivity. In this activity, you will observe the reactivity of aluminum and appreciate why it is useful to protect the inside of an aluminum pop can with plastic.

Materials: eye protection; lab apron; protective gloves; empty soft-drink can; 8- to 10-cm (3- to 4-inch) nail; 150 mL 0.5-mol/L copper(II) chloride solution, $CuCl_{2(aq)}$; plastic tub

1. Insert the nail through the opening of the can. Use the nail to score a smooth circle completely around the inside of the can to cut through the plastic lining inside the can. Be careful not to puncture the can.

2. Pour enough copper(II) chloride, $CuCl_{2(aq)}$, into the can so that the solution is in contact with the exposed aluminum.

 Copper(II) chloride is toxic. Avoid skin contact. Wear eye protection.

3. Place the can in the tub, and allow it to stand undisturbed for about 10 min.

4. Pour out the copper(II) chloride solution into the container provided by your teacher.

5. Carefully pull the can apart.

 The cut edges of the can may be sharp. Wear protective waterproof gloves.

6. Check the inside of the can for evidence of a chemical reaction.

(a) What evidence of a chemical reaction did you observe?

(b) Which reactant was oxidized in this reaction? Which reactant was reduced?

(c) Why is it useful to have a protective plastic lining applied to the inside of an aluminum soft-drink can?

▶ **Section 5.1 Questions**

Understanding Concepts

1. Copy and complete **Table 1** to compare oxidation and reduction.

Table 1   Summary of Oxidation and Reduction

Reaction	Electrons gained or lost	Change in charge on the reactant atom/ion
oxidation		
reduction		

2. State the charge on each ion in each of the following compounds:
   (a) K_2O
   (b) CaO
   (c) Fe_2O_3
   (d) $Cu(NO_3)_2$
   (e) $Al(OH)_3$
   (f) $Fe_2(SO_4)_3$

3. For each of the following reactions, identify the reactant that is oxidized and the reactant that is reduced:
   (a) $Ca_{(s)} + Cl_{2(g)} \rightarrow CaCl_{2(s)}$
   (b) $Zn_{(s)} + S_{(s)} \rightarrow ZnS_{(s)}$

(c) $Mg_{(s)} + Pb^{2+}_{(aq)} \rightarrow Pb_{(s)} + Mg^{2+}_{(aq)}$

(d) $2\,Ag^{+}_{(aq)} + Fe_{(s)} \rightarrow Fe^{2+}_{(aq)} + 2\,Ag_{(s)}$

(e) $NiSO_{4(aq)} + Mg_{(s)} \rightarrow MgSO_{4(aq)} + Ni_{(s)}$

(f) $6\,HCl_{(aq)} + 2\,Fe_{(s)} \rightarrow 2\,FeCl_{3(aq)} + 3\,H_{2(g)}$

(g) $CuCl_{2(aq)} + Mg_{(s)} \rightarrow MgCl_{2(aq)} + Cu_{(s)}$

(h) $3\,AgNO_{3(aq)} + Al_{(s)} \rightarrow 3\,Ag_{(s)} + Al(NO_3)_{3(aq)}$

4. When a metal reacts with a nonmetal, which reactant is oxidized? Which reactant is reduced?

5. Explain why the number of electrons gained and the number of electrons lost in a redox reaction are always equal.

Applying Inquiry Skills

6. A steel (mostly iron) paper clip was placed in a cup of cola (pH 3) and left there for one week. Write the Analysis for the following lab report, including an answer to the Question.

   Question
   Is iron oxidized by cola?

   Observations
   Mass of paper clip at the start = 0.56 g
   Mass of dried paper clip after one week = 0.48 g

Making Connections

7. (a) The dark tarnish that forms on silverware is silver sulfide, $Ag_2S_{(s)}$. **Figure 5** shows a common home remedy for tarnish. The chemical equation for this reaction is

   $$3\,Ag_2S_{(s)} + 2\,Al_{(s)} \rightarrow 6\,Ag_{(s)} + Al_2S_{3(s)}$$

   Identify the reactant oxidized and the reactant reduced.

   (b) Another way to remove tarnish is to scrub and polish the silverware. Which do you think is a better way to clean silverware: the reaction described in (a) or scrubbing and polishing? Give reasons for your answer.

8. Research how photographic images are produced using black and white film (**Figure 6**). Identify the redox reaction that is involved in forming the image.

 www.science.nelson.com

Extension

9. (a) Have you ever felt an ache in your legs while running? It is caused by a buildup of lactic acid in your muscle tissue. The body produces lactic acid during the oxidation of glucose. Research the conditions under which lactic acid is produced.

   (b) Long-distance runners use a value called *lactic acid threshold* (or *lactate threshold*) to help them avoid lactic acid buildup. Research the lactic acid threshold. Explain how runners use this value when training and competing.

 www.science.nelson.com

Figure 5
To remove the tarnish from silverware, soak it in a hot solution of baking soda on aluminum foil in a glass dish.

Figure 6
The magic that takes place in darkrooms can be explained by an understanding of redox reactions.

The redox reactions you have studied so far involve metals. Many important redox reactions, however, do not involve metals. For example, when coal that contains sulfur impurities is burned, sulfur dioxide, $SO_{2(g)}$, is produced (**Figure 1**). The chemical equation for this reaction is

$$S_{(s)} + O_{2(g)} \rightarrow SO_{2(g)}$$

The formation of sulfur dioxide looks very similar to the formation of magnesium oxide (section 5.1). Both are redox reactions.

$$2\,Mg_{(s)} + O_{2(g)} \rightarrow 2\,MgO_{(s)}$$
$$S_{(s)} + O_{2(g)} \rightarrow SO_{2(g)}$$

The main difference is that magnesium oxide is an ionic compound while sulfur dioxide is a molecular compound. As magnesium oxide forms, oxygen atoms pull two electrons from magnesium to form magnesium ions and oxide ions, Mg^{2+} and O^{2-}. In sulfur dioxide, because the difference in electronegativity between sulfur and oxygen is small, sulfur electrons are only partially pulled toward the oxygen atoms. Chemists find it useful to assign "apparent" charges to sulfur and oxygen atoms *as if* the sulfur electrons were completely transferred to oxygen. These "apparent" charges are called **oxidation numbers**. As you will see, oxidation numbers are useful for keeping track of electron changes during redox reactions.

Oxidation Numbers

Consider the example of sulfur dioxide, SO_2. In sulfur dioxide, oxygen is more electronegative than sulfur. Since oxygen gains two electrons to form the oxide ion, O^{2-}, in ionic compounds, the oxygen atoms in sulfur dioxide are assigned an oxidation number of -2. The sum of all the oxidation numbers in a neutral molecule must be zero, since the molecule has no overall charge. You can use this information to calculate the oxidation number of sulfur. In the calculation below, the symbol S represents the oxidation number of sulfur in sulfur dioxide.

$$\begin{array}{cc} S & \overset{O_2}{} \\ \end{array}$$

Oxidation numbers: $\quad S + 2(-2) = 0$
$$S = +4$$

Therefore, sulfur has an oxidation number of $+4$.

Note that the oxidation number of an atom is not fixed. It can change if the atom is involved in a redox reaction. **Table 1**, on the next page, gives a set of general rules for assigning oxidation numbers.

Figure 1
Sulfur dioxide emissions from burning fossil fuels contribute to acid rain.

oxidation number the apparent charge an atom would have if it gained or lost its bonding electrons

Adding Horses

Most reactions involving energy changes are redox reactions. In the dragster, gasoline is oxidized by both oxygen and nitrous oxide, providing a considerable horsepower boost.

Skunk Stink

What's the best remedy for skunk spray? Using tomato juice to get rid of skunk odour is a myth. Tomato juice only masks the odour. To eliminate the odour permanently, you have to eliminate the cause: a group of organic compounds called thiols. Hydrogen peroxide can be used to oxidize thiols in skunk spray into molecules that have little or no odour.

Table 1   Rules for Assigning Oxidation Numbers

Rule	Examples
1. The oxidation number of an atom in an uncombined element is always 0.	K in K, H in H_2, and P in P_4 all have an oxidation number of 0.
2. The oxidation number of a simple ion is the charge of the ion.	The oxidation number of Ca in Ca^{2+} is +2. (Here the oxidation number represents a real charge.)
3. The oxidation number of hydrogen in most compounds is +1.	The oxidation number of H in H_2O, H_2SO_4, and NH_3 is +1.
Exception: metal hydrides	The oxidation number of H in sodium hydride, NaH, is −1.
4. The oxidation number of oxygen in most compounds is −2.	The oxidation number of O in H_2O, MgO, and HNO_3 is −2.
Exception: peroxides	The oxidation number of O in hydrogen peroxide, H_2O_2, is −1.
5. The oxidation number of Group I element ions is +1. The oxidation number of Group II element ions is +2.	The oxidation number of K in K^+ is +1. The oxidation number of Mg in Mg^{2+} is +2.
6. The sum of oxidation numbers in a compound must equal 0.	In H_2O, $2(+1) + (-2) = 0$
7. The sum of oxidation numbers in a polyatomic ion must equal the charge of the ion.	In hydroxide, OH^-, $(-2) + (+1) = -1$

> **SAMPLE** problem 1

Assigning Oxidation Numbers

(a) Determine the oxidation number of the nitrogen atom in potassium nitrate, KNO_3.

Step 1: Write All Known Oxidation Numbers

Use the symbol N to represent the oxidation number of nitrogen in potassium nitrate.

$$
\begin{array}{ccc}
K & N & O_3 \\
+1 & N & -2
\end{array}
$$

(*Note:* Since potassium is a Group 1 element, it has an oxidation number of +1.)

Step 2: Multiply Oxidation Numbers by Number of Each Type of Atom

$$
\begin{array}{ccc}
K & N & O_3 \\
(+1) & N & 3(-2)
\end{array}
$$

Step 3: Solve for Unknown Oxidation Number

$$
(+1) + N + 3(-2) = 0
$$
$$
N - 5 = 0
$$
$$
N = +5
$$

Therefore, the oxidation number of the nitrogen atom in potassium nitrate is +5.

(b) Determine the oxidation number of sulfur in the sulfate ion, $SO_4{}^{2-}$.

Remember that the sum of the oxidation numbers must equal the charge of the ion.

Step 1: Write All Known Oxidation Numbers

Use the symbol S to represent the oxidation number of sulfur in the sulfate ion.

$$S \quad O_4{}^{2-}$$
$$S \quad -2 \quad \text{(from Table 1)}$$

Step 2: Multiply Oxidation Number by Number of Each Type of Atom

$$S \quad O_4{}^{2-}$$
$$S \quad 4(-2)$$

Step 3: Solve for Unknown Oxidation Number

$$S + 4(-2) = -2$$
$$S - 8 = -2$$
$$S = +6$$

Therefore, the oxidation number of sulfur in the sulfate ion is $+6$.

Example

Determine the oxidation number of phosphorus in

(a) sodium phosphate, Na_3PO_4
(b) the phosphite ion, $PO_3{}^{3-}$

Solution

(a)
$$Na_3 \quad P \quad O_4$$
$$3(+1) + P + 4(-2) = 0$$
$$3 + P - 8 = 0$$
$$P = +5$$

The oxidation number of phosphorus in sodium phosphate is $+5$.

(b)
$$P \quad O_3{}^{3-}$$
$$P + 3(-2) = -3$$
$$P - 6 = -3$$
$$P = +3$$

The oxidation number of phosphorus in the phosphite ion is $+3$.

Oxides of Nitrogen
There are a number of nitrogen oxides, each with its own unique properties.

Table 2 Oxides of Nitrogen

Name	Oxidation number of N	Where it can be found
dinitrogen monoxide, $N_2O_{(g)}$	+1	laughing gas
nitrogen monoxide, $NO_{(g)}$	+2	automobile exhaust
nitrogen dioxide, $NO_{2(g)}$	+4	smog

▶ **Practice**

Understanding Concepts

1. Determine the oxidation number of each element in each of the following compounds or ions:
   (a) Cl_2 (c) H_2S (e) Na_2O (g) Fe_2O_3
   (b) S_8 (d) F^- (f) CO_2 (h) H_2O_2

2. Determine the oxidation number of the underlined element in each of the following compounds or ions:
   (a) $H\underline{Cl}O_3$ (c) $\underline{S}O_3{}^{2-}$ (e) $\underline{Cl}O_4{}^-$ (g) $Pb\underline{S}O_4$
   (b) $K\underline{N}O_2$ (d) $Na_2\underline{C}O_3$ (f) $\underline{Fe}(NO_3)_3$ (h) $\underline{Mn}O_4{}^-$

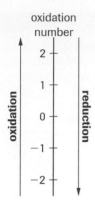

oxidation number

oxidation → reduction

Figure 2
Reduction involves a decrease in oxidation number. Oxidation involves an increase in oxidation number.

Figure 3
The reaction of powdered zinc and sulfur to produce zinc sulfide is rapid and gives off a large quantity of heat.

Figure 4
More sulfuric acid is manufactured annually in North America than any other chemical. One chemical reaction that is used in the industrial synthesis of sulfuric acid is

$$SO_{3(g)} + H_2O_{(l)} \rightarrow H_2SO_{4(aq)}$$

Identifying Redox Reactions Using Oxidation Numbers

All redox reactions, by definition, involve a loss or gain of electrons. Therefore, oxidation numbers must change (**Figure 2**).

- In reduction, oxidation numbers decrease.
- In oxidation, oxidation numbers increase.

The examples in section 5.1 were special cases in which the oxidation numbers were actual ionic charges.

Recognizing changes in oxidation numbers is useful when identifying redox reactions.

> ## SAMPLE problem 2
>
> ### Using Oxidation Numbers to Identify a Redox Reaction
>
> **(a)** Use oxidation numbers to show that the reaction of zinc metal (Figure 3) with sulfur is a redox reaction. The chemical equation for this reaction is
>
> $$Zn_{(s)} + S_{(s)} \rightarrow ZnS_{(s)}$$
>
> **Step 1: Write All Known Oxidation Numbers**
>
> $$Zn_{(s)} + S_{(s)} \rightarrow ZnS_{(s)}$$
> $$\phantom{Zn_{(s)}}0 \quad\ \ 0 \quad\ +2\ -2$$
>
> The oxidation number of zinc changes from 0 to +2.
>
> The oxidation number of sulfur changes from 0 to −2.
>
> Since there is a change in oxidation numbers, the reaction of zinc with sulfur is a redox reaction.
>
> **(b)** Is the following reaction of sulfur trioxide with water (Figure 4) a redox reaction?
>
> $$SO_{3(g)} + H_2O_{(l)} \rightarrow H_2SO_{4(aq)}$$
>
> **Step 1: Write All Known Oxidation Numbers**
>
> $$SO_3 \quad + \quad H_2O \quad \rightarrow \quad H_2SO_4$$
> $$S\ -2 \qquad +1\ -2 \qquad +1\ \ S\ -2$$
>
> **Step 2: Multiply Oxidation Numbers by Number of Each Type of Atom**
>
> $$S \quad O_3 \quad + \quad H_2 \quad O \rightarrow H_2 \quad S \quad\quad O_4$$
> $$S + 3(-2) \qquad 2(+1) - 2 \qquad 2(+1) + S \quad + 4(-2)$$
>
> **Step 3: Solve for Unknown Oxidation Number**
>
> $$S + 3(-2) = 0 \qquad 2(+1) -2 = 0 \qquad 2(+1) + S + 4(-2) = 0$$
> $$S - 6 = 0 \qquad\qquad\qquad\qquad\qquad\quad 2 + S - 8 = 0$$
> $$S = 6 \qquad\qquad\qquad\qquad\qquad\qquad\quad S = 6$$
>
> Since there is no change in oxidation numbers for any of the atoms, the reaction of sulfur trioxide with water is not a redox reaction.

Example

Does the following equation represent a redox reaction?

$$2\ KClO_{3(s)} \rightarrow 2\ KCl_{(s)} + O_{2(g)}$$

Solution

$$\begin{array}{ccccc} 2\,K & Cl & \overline{O_{3(s)}} & \rightarrow 2\,KCl_{(s)} & + O_{2(g)} \\ (+1) + & Cl + & 3(-2) & (+1)(-1) & 2(0) \end{array}$$

$$Cl - 5 = 0$$

$$Cl = +5$$

The oxidation number of chlorine changes from +5 to −1. The oxidation number of oxygen changes from −2 to 0. Therefore, this reaction is a redox reaction.

Practice

Understanding Concepts

3. Which of the following equations represent redox reactions? Which do not represent redox reactions? Justify your answers using oxidation numbers.
   (a) $H_{2(g)} + Cl_{2(g)} \rightarrow 2\ HCl_{(g)}$
   (b) $2\ K_{(s)} + I_{2(g)} \rightarrow 2\ KI_{(s)}$
   (c) $CaCO_{3(s)} \rightarrow CaO_{(s)} + CO_{2(g)}$
   (d) $2\ H_2O_{(l)} \rightarrow 2\ H_{2(g)} + O_{2(g)}$
   (e) $2\ Li_{(s)} + 2\ H_2O_{(l)} \rightarrow 2\ LiOH_{(aq)} + H_{2(g)}$
   (f) $Fe_2O_{3(s)} + 3\ CO_{(g)} \rightarrow 2\ Fe_{(s)} + 3\ CO_{2(g)}$
   (g) $CH_{4(g)} + 2\ O_{2(g)} \rightarrow CO_{2(g)} + 2\ H_2O_{(g)}$
   (h) $2\ NaNO_{3(s)} \rightarrow 2\ NaNO_{2(s)} + O_{2(g)}$
   (i) $Ca_{(s)} + 2\ HCl_{(aq)} \rightarrow H_{2(g)} + CaCl_{2(aq)}$
   (j) $2\ HNO_{3(aq)} + Ca(OH)_{2(aq)} \rightarrow 2\ H_2O_{(l)} + Ca(NO_3)_{2(aq)}$

SUMMARY *Oxidation Numbers*

- An oxidation number is the apparent charge that an atom would have if it gained or lost its bonding electrons.
- You can use oxidation numbers to track the gain or loss of electrons in a reaction.
- You can identify a redox reaction by looking for changes in oxidation numbers.
- Reduction involves a decrease in oxidation number.
- Oxidation involves an increase in oxidation number.

DID YOU KNOW?

Chocolate May Be Good for You!

Free radicals are unusual oxidizing molecules that can damage the walls of blood vessels in the body. Recent research shows that milk chocolate is high in *antioxidants*: substances that limit the oxidation damage caused by free radicals. Eating a reasonable amount of chocolate regularly may prove to be good for your health!

Understanding Concepts

1. Copy and complete **Table 3**.

Table 3 Comparing Oxidation and Reduction

Reaction	Oxidation number increases or decreases?
oxidation	
reduction	

2. Determine the oxidation number of the underlined element in each of the following.
   (a) $\underline{O}_2$
   (b) $\underline{Ag}NO_3$
   (c) $\underline{Mn}O_2$
   (d) $Zn\underline{S}O_4$
   (e) $\underline{Cl}O_3^-$
   (f) $\underline{C}O_3^{2-}$
   (g) $K\underline{Mn}O_4$
   (h) $Mg\underline{S}O_3$
   (i) $Na_2\underline{S}_2O_3$

3. The combustion of hydrogen occurs in a space shuttle's main engines (**Figure 5**).

   $$2\,H_{2(g)} + O_{2(g)} \rightarrow 2\,H_2O_{(g)}$$

   Use oxidation numbers to show that this reaction is a redox reaction.

Figure 5
Redox reactions can be very dramatic, releasing huge quantities of energy.

4. Is the following reaction a redox reaction? Use oxidation numbers to find out.

   $$Ca_{(s)} + 2\,H_2O_{(l)} \rightarrow Ca(OH)_{2(aq)} + H_{2(g)}$$

5. The following reaction is used to extract iron from its ore at a temperature of about 850°C:

   $$Fe_2O_{3(s)} + 3\,H_{2(g)} \rightarrow 2\,Fe_{(s)} + 3\,H_2O_{(g)}$$

   Identify the reactant oxidized and the reactant reduced.

6. The following reactions are involved in the formation of acid rain (section 4.11). Use oxidation numbers to identify which of these reactions are redox reactions.
   (a) $2\,SO_{2(g)} + O_{2(g)} \rightarrow 2\,SO_{3(g)}$
   (b) $3\,NO_{2(g)} + H_2O_{(l)} \rightarrow 2\,HNO_{3(aq)} + NO_{(g)}$
   (c) $SO_{2(g)} + H_2O_{(l)} \rightarrow H_2SO_{3(aq)}$

Making Connections

7. Hydrogen peroxide is an oxidizer that is commonly used in tooth-whitening products (**Figure 6**).
   (a) Research how tooth whitening is performed in a dental office. What is the approximate cost of this procedure?
   (b) Research tooth-whitening products that are designed for use at home. How are they used, and how much do they cost?
   (c) Create a table to compare the dental office procedure for tooth whitening with the home procedure. Use several criteria when making your comparison.

 www.science.nelson.com

Figure 6
Is it better to have your teeth whitened by a dentist or to do it yourself? How would you decide?

8. Ammonium thioglycolate is a chemical used in some hair-straightening products. Research and prepare a poster on the redox changes that occur when ammonium thioglycolate is applied to hair.

 www.science.nelson.com

Developing an Activity Series of Metals

Why does sulfuric acid leaking from a car battery oxidize the battery terminals but not the copper wire coming from the terminals (**Figure 1**)? It is because the metal that the terminal is made from (lead) is more reactive with acid than copper is.

In this activity, you will compare the reactivity of a variety of metals with acid. You will then use your observations to organize the metals in order of reactivity. In other words, you will develop an activity series.

Question

What is the order of reactivity of copper, tin, iron, zinc, and magnesium with hydrochloric acid?

Materials

eye protection
lab apron
equal-sized strips of copper, tin, iron, zinc, and
 magnesium
2.0-mol/L hydrochloric acid, $HCl_{(aq)}$
sandpaper
5 test tubes
test-tube rack

Procedure

1. Clean each metal strip with sandpaper.

Figure 1
Acid corrodes the lead terminals of a leaky battery, but it does not corrode the copper wire.

2. Place each 2-cm strip into a different test tube (**Figure 2**).

3. Carefully pour hydrochloric acid into each test tube, to a depth of 3 cm. Then place each test tube in a test-tube rack. Observe carefully.

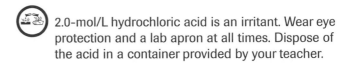
2.0-mol/L hydrochloric acid is an irritant. Wear eye protection and a lab apron at all times. Dispose of the acid in a container provided by your teacher.

(a) Compare the reactivity of each metal in acid, and record your results.

4. Dispose of the acids and metals as directed by your teacher.

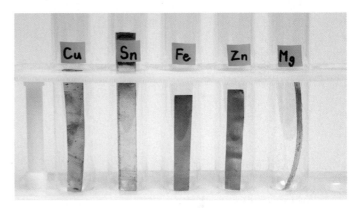

Figure 2
Each test tube contains a sample of one of the metals.

Analysis

(b) Use your observations to answer the Question.

(c) Write a chemical equation for each reaction that took place. Assume that each metal forms an $M^{2+}_{(aq)}$ ion (where M represents the metal) when oxidized by the acid.

Evaluation

(d) Was it difficult to decide on the order of reactivity? Why? Give suggestions for improving the Procedure.

Synthesis

(e) Many older homes still have water pipes that are made of galvanized steel (an alloy of iron coated with a layer of zinc). Use your activity series to explain why copper is now preferred.

activity series of metals a list of metals arranged in order of their reactivity, based on observations gathered from single displacement reactions

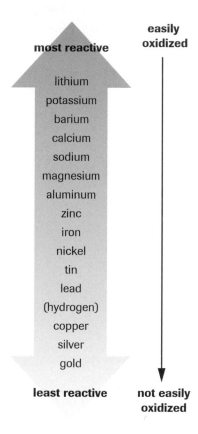

easily oxidized

most reactive

lithium
potassium
barium
calcium
sodium
magnesium
aluminum
zinc
iron
nickel
tin
lead
(hydrogen)
copper
silver
gold

least reactive

not easily oxidized

Figure 1
In the activity series of metals, each metal will displace any metal listed below it in a compound. Hydrogen is usually included in the activity series, even though it is not a metal, because hydrogen can form positive ions, just like metals.

As you know, the stomach contains acid. Some objects, such as paper clips, would react quickly with stomach acid. A gold earring, however, would remain intact in the stomach indefinitely.

How can you predict which metals will react in a single displacement reaction, and which will not? In section 5.3, you compared the reactivity of a few metals in acid. If you experimented using many more metals, you would get the **activity series of metals** given in **Figure 1**. The activity series is a list of metals arranged in order of their tendency to react (become oxidized). The most reactive metals (the most easily oxidized) are at the top of the list. The least reactive metals are at the bottom. For example, the metals from lithium to sodium are so reactive that they even react with relatively unreactive substances, such as water (**Figure 2(a)**). In contrast, copper, silver, and gold are unreactive even in most strong acids (**Figure 2(b)**). Hydrogen is included in the activity series because, like metals, it can be oxidized to form positive ions.

You can use the activity series to predict the products of single displacement reactions. Remember that the general equation for these reactions is

$$A + BC \rightarrow AC + B$$

In general, an element that is higher in the activity series will displace an element that is lower. The lower element is, thus, left as a pure metal. For example, copper will displace silver from silver nitrate, leaving silver metal (**Figure 3**).

$$Cu_{(s)} + 2\,AgNO_{3(aq)} \rightarrow Cu(NO_3)_{2(aq)} + 2\,Ag_{(s)}$$

(a)

(b)

Figure 2
(a) Potassium reacts violently with water in a single displacement reaction. Potassium must be stored in oil to prevent it from reacting with oxygen or water vapour in the air.
(b) Copper in hydrochloric acid does not appear to react.

(a) 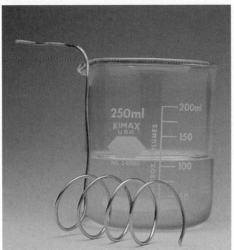 **(b)**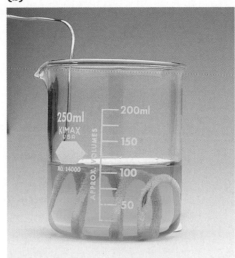

Figure 3
(a) A piece of copper before it is placed into a beaker of silver nitrate solution.
(b) Note the changes after the reaction has occurred. The blue colour of the solution indicates that $Cu^{2+}_{(aq)}$ ions are present.

▶ *SAMPLE* problem

Predicting the Occurrence of a Chemical Reaction

Predict whether a chemical reaction will occur in each of the following situations. Write a balanced chemical equation for each reaction you predict will occur.

(a) Aluminum foil is placed in a solution of silver nitrate, $AgNO_{3(aq)}$.

Step 1: Establish Reactivity of Metals

Aluminum is above silver in the activity series, so aluminum will displace (produce) silver from silver nitrate.

Step 2: Write Unbalanced Equation

Using the general equation,

$$A + B C \rightarrow A C + B$$
$$Al_{(s)} + AgNO_{3(aq)} \rightarrow Al(NO_3)_{3(aq)} + Ag_{(s)} \text{ (unbalanced)}$$

Step 3: Balance Equation

$$Al_{(s)} + 3\,AgNO_{3(aq)} \rightarrow Al(NO_3)_{3(aq)} + 3\,Ag_{(s)}$$

(b) Copper is placed in a solution of iron(II) sulfate, $FeSO_{4(aq)}$.

Step 1: Establish Reactivity of Metals

Copper is below iron in the activity series. Therefore, copper will not displace the iron from iron(II) sulfate. No reaction will occur.

$$Cu_{(s)} + FeSO_{4(aq)} \rightarrow \text{ no reaction}$$

(c) Zinc is placed in a solution of hydrochloric acid, $HCl_{(aq)}$.

Step 1: Establish Reactivity of Metals

Zinc is above hydrogen in the activity series, so a reaction will occur.

LEARNING TIP

Remember …
LEO says GER.

Let It Shine!
Copper and nickel are used to strengthen gold jewellery. Copper and nickel oxidize slowly in air, however. The presence of these metals explains why gold jewellery gradually gets dull, while a piece of pure gold maintains its shine.

Step 2: Write Unbalanced Equation

$$Zn_{(s)} + HCl_{(aq)} \rightarrow ZnCl_{2(aq)} + H_{2(g)}$$

Step 3: Balance Equation

$$Zn_{(s)} + 2\,HCl_{(aq)} \rightarrow ZnCl_{2(aq)} + H_{2(g)}$$

(Notice that hydrogen molecules, $H_{2(g)}$, are produced, rather than H atoms. Hydrogen is always found in nature as diatomic molecules, not separate atoms.)

Example
Use the activity series to predict whether a chemical reaction will occur when lead is placed in a solution of silver nitrate. If you predict that a reaction will occur, write a chemical equation to represent the reaction.

Solution
Lead is above silver in the activity series, so a reaction will occur.

Balanced equation:

$$Pb_{(s)} + 2\,AgNO_{3(aq)} \rightarrow 2\,Ag_{(s)} + Pb(NO_3)_{2(aq)}$$

▶ Practice

Understanding Concepts
Use the activity series to predict whether a chemical reaction will occur when each pair of chemicals is combined. If you predict that a reaction will occur, write a balanced chemical equation. If you predict that a reaction will not occur, write "no reaction."

(a) $Zn_{(s)} + AgNO_{3(aq)} \rightarrow$

(b) $Cu_{(s)} + SnCl_{2(aq)} \rightarrow$

(c) $Mg_{(s)} + HNO_{3(aq)} \rightarrow$

(d) $Ag_{(s)} + H_2SO_{4(aq)} \rightarrow$

(e) $Zn_{(s)} + FeSO_{4(aq)} \rightarrow$

▶ TRY THIS activity   *Coke Corrosion*

Coca-Cola is a fairly acidic liquid, having a pH between 2.5 and 3.0. The acidity is a result of carbonic acid forming when carbon dioxide is dissolved into the Coke, under pressure, during the bottling process. A small amount of phosphoric acid also contributes to the acidity. Although Coke is rumoured to be able to dissolve nails, it will not burn a hole through your stomach. In fact, the fluids that are secreted by your stomach can be over ten times more acidic than Coke. They may have a pH as low as 1.5. In this activity, you will compare the corrosiveness of Coke with the corrosiveness of simulated stomach acid.

(a) Predict which will corrode a paper clip faster: Coke or stomach acid. Include your reason(s).

Materials: eye protection; lab apron; 20 mL Coca-Cola; 20 mL simulated stomach acid (0.030-mol/L $HCl_{(aq)}$); 2 plastic vials; 2 identical paper clips; 125-mL beaker, half-full of acetone; tweezers; electronic balance

1. Determine the mass of each paper clip.
2. Place each paper clip in a plastic vial.
3. To one vial, add enough acid to submerge the paper clip.
4. Add the same volume of Coke to the other vial.
5. Cap the vials, and store them for about one week.
6. Remove the paper clips with tweezers. Visually compare them.
7. Dip each paper clip in a small beaker of acetone. Allow the paper clips to dry on a paper towel. (Keep track of which is which!)

 Acetone is flammable. Keep it away from open flames.

(b) Determine the final mass of each paper clip.

(c) Interpret your results to determine which solution corroded the paper clip faster. Evaluate your reasons in (a).

(d) Assuming a normal stomach, explain why the acidity of Coke is not a health risk.

(e) Now that you've seen what Coke does to a paper clip, is there any truth to the rumour that Coke can dissolve nails? Why?

▶ Section 5.4 Questions

Understanding Concepts

1. Describe the organization of the activity series of metals.

2. Why is hydrogen included in the activity series, even though it is not a metal?

3. For each of the following situations, use the activity series to determine if a reaction will occur. Write a balanced chemical equation for each reaction that will occur. Write "no reaction" if a reaction will not occur.
   (a) $Zn_{(s)} + Cu(NO_3)_{2(aq)} \rightarrow$
   (b) $Au_{(s)} + FeSO_{4(aq)} \rightarrow$
   (c) $Zn_{(s)} + HCl_{(aq)} \rightarrow$
   (d) $Pb_{(s)} + Ni(NO_3)_{2(aq)} \rightarrow$
   (e) $Mg_{(s)} + H_2SO_{4(aq)} \rightarrow$

4. Gold is one of the few metals to be found in nature as an element. Most other metallic elements are found in compounds. Explain why gold is found as an element. Refer to the activity series in your explanation.

Applying Inquiry Skills

5. A lab technician found several unlabelled pieces of silvery gray metals on a lab bench. Describe how the technician could use hydrochloric acid and knowledge of the activity series to identify which metal is silver.

6. Perspiration can be slightly acidic, which explains why the frames of eyeglasses sometimes corrode. Nickel-based frames corrode more than titanium frames, leaving a green stain on the skin.
   (a) Use the activity series to explain why nickel can be corroded by acid.
   (b) What can you infer about the position of titanium, compared with nickel, in an activity series?
   (c) Titanium reacts with oxygen to form a thin surface layer of titanium oxide, which sticks tightly to the titanium below. How does the layer of titanium oxide prevent the titanium from corroding?

7. In an investigation into single displacement reactions, small pieces of silver, lead, nickel, and zinc are placed separately into solutions that contain one of the following ions: $Ag^+_{(aq)}$, $Pb^{2+}_{(aq)}$, $Ni^{2+}_{(aq)}$, $Zn^{2+}_{(aq)}$. Complete the Prediction for the following lab report by copying and completing **Table 1**, then predict an answer to the Question.

Question

Which metal–metal ion combinations result in a reaction?

Table 1 Predictions for Reactions between Metals and Their Ions

Ion / Metal	$Ag^+_{(aq)}$	$Pb^{2+}_{(aq)}$	$Ni^{2+}_{(aq)}$	$Zn^{2+}_{(aq)}$
$Ag_{(s)}$				
$Pb_{(s)}$				
$Ni_{(s)}$				
$Zn_{(s)}$				

8. A chemical technician is faced with the task of disposing of 1 L of a waste copper(II) sulfate solution. Since copper ions are toxic, they must be removed before the solution can be poured down the drain.
   (a) Write a Procedure to describe how steel wool (an alloy of iron) can be used to remove copper from the solution.
   (b) Write a chemical equation for the reaction in (a). (Assume that $Fe^{2+}_{(aq)}$ ions form.)
   (c) Suggest another procedure that could be used to remove the copper ions with a precipitation reaction.
   (d) Comment on which procedure, in your opinion, would be most effective.

Making Connections

9. The Thermite process is a very exothermic reaction that involves aluminum and iron(III) oxide, $Fe_2O_{3(s)}$ (**Figure 4**).
   (a) Write a balanced chemical equation for this reaction.
   (b) Search the Internet for a video clip of this reaction.
   (c) Research and describe at least one potential application of this reaction.

 www.science.nelson.com

Extension

10. Aqueous solutions of chlorine, $Cl_{2(aq)}$, bromine, $Br_{2(aq)}$, and iodine, $I_{2(aq)}$, are placed in different wells of a microtray. A few drops of a solution containing sodium chloride, $NaCl_{(aq)}$, sodium bromide, $NaBr_{(aq)}$, or sodium iodide, $NaI_{(aq)}$, are added to each well, as indicated in **Table 2**. Complete the Analysis for the following lab report:

Question

What is the order of reactivity of chlorine, bromine, and iodine?

Observations

Table 2 Reactions between Halogens and Their Ions

Halogen / Ion	$Cl_{2(aq)}$	$Br_{2(aq)}$	$I_{2(aq)}$
$Cl^-_{(aq)}$	no change	no change	no change
$Br^-_{(aq)}$	colourless to yellow-brown	no change	no change
$I^-_{(aq)}$	colourless to red	colourless to red	no change

Analysis
   (a) Answer the Question by creating an activity series for these three halogens.
   (b) Predict where fluorine would be in your activity series. Justify your prediction.

Figure 4
The Thermite process

Testing the Activity Series

The reactivity of a metal can be predicted by its position in the activity series. The metals toward the top of the series are more reactive (**Figure 1**) than the metals below them. Lead, for example, is more reactive than copper, so it is shown above copper in the activity series. You would therefore expect a single displacement reaction to occur when lead is placed in a solution that contains copper ions. For example, consider the following reaction:

$$Pb_{(s)} + Cu(NO_3)_{2(aq)} \rightarrow Cu_{(s)} + Pb(NO_3)_{2(aq)}$$

The total ionic equation for the reaction is

$$Pb_{(s)} + Cu^{2+}_{(aq)} + 2\,NO^{-}_{3(aq)} \rightarrow Cu_{(s)} + Pb^{2+}_{(aq)} + 2\,NO^{-}_{3(aq)}$$

By eliminating the nitrate ions that are common to both sides, you get the net ionic equation for the reaction:

$$Pb_{(s)} + Cu^{2+}_{(aq)} \rightarrow Cu_{(s)} + Pb^{2+}_{(aq)}$$

In this investigation, you will make a prediction, based on the activity series, about which metals are likely to react with metal ions in solution. You will then observe five metals in five different solutions of metal compounds, to test your prediction.

Figure 1
Since copper is more reactive than silver, it displaces the silver when placed in silver nitrate solution in a single displacement reaction, according to the equation

$$Cu_{(s)} + 2\,AgNO_{3(aq)} \rightarrow Cu(NO_3)_{2(aq)} + 2\,Ag_{(s)}$$

The silver crystallizes around the copper wire.

Inquiry Skills
- ○ Questioning
- ● Hypothesizing
- ● Predicting
- ○ Planning
- ● Conducting
- ● Recording
- ● Analyzing
- ● Evaluating
- ● Communicating

Question

Which metals (silver, copper, iron, zinc, or magnesium) will spontaneously react in which solution (silver nitrate, copper(II) sulfate, iron(II) sulfate, zinc sulfate, or magnesium nitrate)?

Prediction

(a) Use the activity series to predict which metals are likely to react in which solution.

Hypothesis

(b) Explain your Prediction.

Materials

eye protection
lab apron
protective gloves
microtray
pieces of recently sanded silver, copper, iron, zinc, and magnesium
dropper bottles containing 0.1-mol/L solutions of
    silver nitrate, $AgNO_{3(aq)}$
    copper(II) sulfate, $CuSO_{4(aq)}$
    iron(II) sulfate, $FeSO_{4(aq)}$
    zinc sulfate, $ZnSO_{4(aq)}$
    magnesium nitrate, $Mg(NO_3)_{2(aq)}$

Procedure

(c) Place a sheet of paper on the lab bench. Trace the outline of the microtray on the paper, including the position of the rows and columns. Remove the tray. Label the first five columns with the names of the five metals, in the order given in the Materials list. Then label the first five rows with the names of the solutions, in the order given in the Materials list (**Figure 2**, on the next page).

Also, prepare a table like **Table 1**, on the next page, to record your observations. Your table should include a space for notes about each of the 25 wells. (If your table has the same layout as

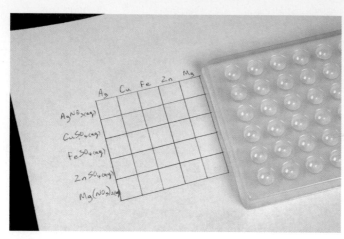

Figure 2
Each well will eventually contain a solution and a small piece of solid metal.

Table 1 Reactions of Metals in Metal Ion Solutions

Metal ion solution \ Metal	$Ag_{(s)}$	$Cu_{(s)}$	$Fe_{(s)}$	$Zn_{(s)}$	$Mg_{(s)}$
$AgNO_{3(aq)}$					
$CuSO_{4(aq)}$					
$FeSO_{4(aq)}$					
$ZnSO_{4(aq)}$					
$Mg(NO_3)_{2(aq)}$					

the microtray, you will not have to label the tray as outlined above.)

Place the microtray back onto the paper.

1. Place a small piece of silver into five wells in the first column of the microtray.

2. Add a similar-sized piece of copper to five wells of the second column of the microtray.

3. Continue this process with iron, zinc, and magnesium.

4. Carefully add three to four drops of silver nitrate to each well in the first row.

 Some of the solutions are toxic and/or irritants. Avoid skin and eye contact. Wear eye protection and a lab apron.

5. Add three to four drops of copper(II) sulfate to each well of the second row.

6. Repeat this process with the remaining solutions.

The silver nitrate solution will stain the skin. Wear protective gloves.

(d) In your observation table, record any evidence of a chemical reaction, such as bubbles of gas, changes in the appearance of the solution or the metal, or precipitates.

7. Dispose of the contents of your microtray as directed by your teacher.

Analysis

(e) Record R in your data table if your observations indicated that a reaction occurred and NR if you noticed no reaction.

(f) Answer the Question.

(g) Write the net ionic equation for each reaction that occurred. (Assume that all metals, except silver, form metal ions with a 2+ charge when they are oxidized. The silver ion, $Ag^+_{(aq)}$, has a 1+ charge.)

Evaluation

(h) Was it difficult to decide whether a reaction was occurring in any of the wells? If so, which metals and solutions were present? Explain why it was difficult to determine whether a reaction was occurring.

(i) Compare your answer in (f), about the reactivity of the metals, with your Prediction in (a).

(j) Comment on the reliability of the activity series for predicting the reactivity of the metals.

Synthesis

(k) Rank the metals in order from *most* reactive (easiest to oxidize) to *least* reactive (hardest to oxidize).

(l) Rank the metal ions in order from *least* reactive (hardest to reduce) to *most* reactive (easiest to reduce).

(m) Compare your answers to (k) and (l).

Conductors and Nonconductors

In sections 5.7 to 5.9, you will learn how batteries work. Since batteries supply electrical energy, many of the substances inside batteries are electrical conductors. In this investigation, you will use a conductivity apparatus to classify a variety of solids, aqueous solutions, and pure liquids as conductors or nonconductors of electricity.

Greenhouse technicians monitor the electrical conductivity of the water that is used in a greenhouse. A high conductivity reading means that the ion concentration in the water is high. If the ion concentration is too high, plant roots stop functioning properly, decreasing their ability to take up water and nutrients. A low conductivity reading may mean that more nutrients should be added.

Question

Which of the test substances in the Materials list are conductors? Which are nonconductors?

Prediction

(a) Predict an answer to the Question.

Materials

eye protection
lab apron
conductivity apparatus
microtray
pH strips or litmus paper
substances to be tested (0.1-mol/L solutions):
  sodium hydroxide, $NaOH_{(aq)}$
  sodium carbonate, $Na_2CO_{3(aq)}$
  sulfuric acid, $H_2SO_{4(aq)}$
  sodium chloride, $NaCl_{(aq)}$
  potassium nitrate, $KNO_{3(aq)}$
  glucose, $C_6H_{12}O_{6(aq)}$
distilled water, $H_2O_{(l)}$
tap water
paste from the inside of a battery (optional)
pieces of sulfur, copper, zinc, and charcoal (carbon)

Procedure

1. Prepare a table to record your observations of conductivity and pH for the substances to be tested.

Inquiry Skills

2. Place two to three drops of each liquid in separate wells of the microtray.

 Many of the solutions, particularly $NaOH_{(aq)}$, are corrosive. Wear eye protection and a lab apron. Avoid skin contact.

3. Test each liquid for electrical conductivity using the conductivity apparatus. Rinse the leads of the tester with distilled water before proceeding to the next solution. Record your observations.

4. Test the four solids for conductivity. Record your observations.

5. Use pH test strips to determine which liquids are acidic, basic, or neutral. Record your observations.

6. Dispose of the contents of the microtray in the sink, and flush with plenty of water.

Analysis

(b) Compare the conductivity of
  (i)   solutions of acids, bases, neutral ionic compounds, and molecular compounds
  (ii) all other liquids
  (iii) metals and nonmetals

(c) Answer the Question.

Evaluation

(d) Compare your Prediction with your Analysis in (b). Account for any differences.

(e) Identify sources of error in the Procedure. Suggest changes to the Procedure that may minimize experimental error.

Synthesis

(f) What do all nonconducting liquids have in common?

(g) In general, where can the elements that are electrical conductors be found in the periodic table?

Have you ever accidentally bitten into a piece of aluminum foil? If you have silver amalgam fillings, you may have experienced a bit of a jolt (**Figure 1**). The aluminium, in contact with a metal filling and your saliva, creates a mini-battery in your mouth, sending electricity into the nerves of your tooth. Believe it or not, this mini-battery works on the same principle as the little button battery in your calculator and the 20-kg lead–acid battery that starts a car. How does a battery work? You'll find out in the next three sections.

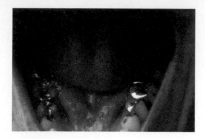

Figure 1
A redox reaction is responsible for the electric shock you may feel if aluminum foil comes into contact with a metal filling in your mouth.

▶ **TRY THIS** activity   *Electrochemical Gizmos*

Potatoes, lemons, bananas, and limes have all been used to power novelty clocks, such as the potato clock in **Figure 2**. The fruits and vegetables really aren't necessary, however! In this activity, you will use simpler materials to power a clock.

Materials: eye protection, 2 copper strips, magnesium strip, sandpaper, connecting wires with alligator clips, LCD clock (battery removed), plastic cup, distilled water, tap water

1. Clean the metals with sandpaper to remove any surface oxides.
2. Attach one alligator clip to a copper strip. Attach another alligator clip to the magnesium strip.
3. Attach the other end of each wire to the "battery" connectors on the LCD clock.
4. Dip both strips into a cup that is half-full of distilled water. Make sure that the metal strips do not touch inside the cup. Does the clock turn on?
5. Replace the distilled water with tap water, and try again.
6. Replace the magnesium strip with another copper strip. Place both copper strips in a cup of tap water.

(a) Which combination of strips and water made the clock turn on?

(b) Why was one combination of strips more successful at running the clock than another combination of strips?

(c) How are distilled water and tap water different? Suggest an explanation for the difference in your observations when using distilled water and tap water.

Figure 2
The energy that is needed to power this digital clock comes from the electrons that are travelling along the wires.

Converting Chemical Energy to Electrical Energy

In section 5.1, you looked at the oxidation of zinc by the copper(II) ion, $Cu_{(aq)}^{2+}$:

$$Zn_{(s)} + Cu_{(aq)}^{2+} \rightarrow Cu_{(s)} + Zn_{(aq)}^{2+}$$

In this reaction, copper ions remove two electrons from zinc atoms. The following redox changes occur:

- Zn becomes Zn^{2+}, a loss of 2 e^- (oxidation).
- Cu^{2+} becomes Cu, a gain of 2 e^- (reduction).

By separating the zinc metal and the solution containing copper ions (**Figure 3**) and by placing a metal conductor (wire) between them, electrons that are lost by zinc are forced to travel through the metal conductor to reach the copper ions. Moving electrons have energy. This energy can be used to power an electrical device, such as a radio or a clock. The apparatus, called a **galvanic cell**, converts chemical energy from a redox reaction into electrical energy. If you think that a galvanic cell sounds like a battery, you're correct! Batteries contain galvanic cells. By learning how the galvanic cell in **Figure 3** works, you will better understand how batteries produce electricity (section 5.9).

The reactions that occur in a galvanic cell are **spontaneous reactions**: they require no outside assistance or energy input to make them occur. In **Figure 3**, the oxidation of zinc and the reduction of copper ions occur in separate compartments, called **half-cells**. Each half-cell contains a solid conductor called an **electrode**. In each half-cell, electron transfers occur between atoms on the surface of the electrode and ions in the solution. The electrode where oxidation occurs is called the **anode**. The electrode where reduction occurs is called the **cathode**. Metals or inert (non-reactive) conductors, such as graphite (the black material inside your pencil), are common electrodes used in galvanic cells.

galvanic cell a device that converts chemical energy from redox reactions into electrical energy

spontaneous reaction a reaction that proceeds on its own without outside assistance

half-cell one of the two compartments of a galvanic cell, composed of an electrode and an electrolyte solution

electrode a solid electrical conductor where, in a galvanic cell, the electron transfers occur

anode the electrode at which oxidation occurs

cathode the electrode at which reduction occurs

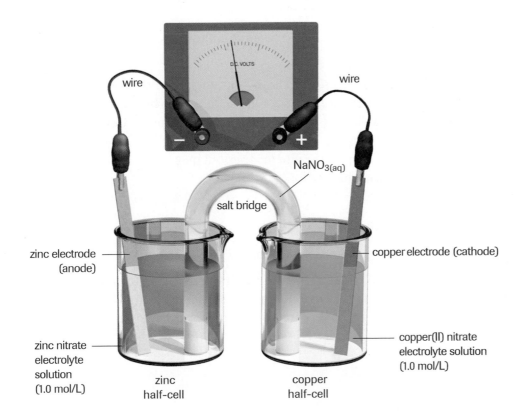

Figure 3
This galvanic cell consists of two half-cells, connected by a metal conductor and a salt bridge.

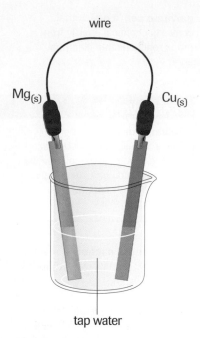

wire

$Mg_{(s)}$

$Cu_{(s)}$

tap water

Figure 4
This simple galvanic cell design produces electricity using two different electrodes and one electrolyte solution.

salt bridge a container of electrolyte solution that connects the two solutions of a galvanic cell

DID YOU KNOW ?

Electrode Charges
The anode and cathode of a galvanic cell are sometimes labelled (−) and (+), respectively. You can think of the anode as being "electron rich" and the cathode as being "electron poor."

In **Figure 3**, each electrode is immersed in an electrolyte solution that contains ions of the same metal as the electrode. The galvanic cell has

- a zinc half-cell consisting of a strip of zinc in a zinc nitrate solution; and
- a copper half-cell consisting of a strip of copper in a copper(II) nitrate solution.

The half-cells are connected to each other using a wire (for electron flow) and a salt bridge (for ion flow). Both connections are necessary for the cell to produce electrical energy. A **salt bridge** is a tube that contains a concentrated solution of an electrolyte, such as sodium nitrate, $NaNO_{3(aq)}$. The electrolyte that is chosen for the salt bridge should not react with the other chemicals in the cell. You will find out more about salt bridges after the next subsection.

Figure 4 shows a simpler galvanic cell design.

Cell Reactions

You can predict that zinc is oxidized in the zinc/copper cell because zinc is higher than copper in the activity series. Therefore, copper(II) ions, $Cu^{2+}_{(aq)}$, are reduced. The chemical equation for this reaction can be divided into two parts, called *half-reactions*. The two half-reactions for the zinc/copper cell are given below.

Anode half-reaction: $\qquad\qquad Zn_{(s)} \rightarrow Zn^{2+}_{(aq)} + 2\,e^-$ (oxidation)

Cathode half-reaction: $\qquad Cu^{2+}_{(aq)} + 2\,e^- \rightarrow Cu_{(s)}$ (reduction)

Notice the following:

- Atoms from the zinc electrode lose electrons and dissolve as $Zn^{2+}_{(aq)}$ ions in the zinc half-cell.
- $Cu^{2+}_{(aq)}$ ions in the copper half-cell gain electrons and become neutral copper atoms.

Therefore, as the cell operates, the mass of the zinc electrode *decreases* and the mass of the copper electrode *increases*. Furthermore, the blue colour of the copper solution fades as more and more of the $Cu^{2+}_{(aq)}$ ions become copper atoms at the copper electrode.

An overall cell reaction for the zinc/copper cell may be obtained by adding the half-reaction of each half-cell:

Anode half-reaction: $\qquad\qquad Zn_{(s)} \rightarrow Zn^{2+}_{(aq)} + \cancel{2\,e^-}$

Cathode half-reaction: $\qquad Cu^{2+}_{(aq)} + \cancel{2\,e^-} \rightarrow Cu_{(s)}$

Overall cell reaction: $\qquad Cu^{2+}_{(aq)} + Zn_{(s)} \rightarrow Cu_{(s)} + Zn^{2+}_{(aq)}$

In summary, electrons that are lost at the anode (where oxidation occurs) flow through the wire to the cathode (where reduction occurs). As with any redox reaction, the electrons that are lost by one reactant are gained by the other reactant.

▶ **SAMPLE** problem

Writing Chemical Equations for Cell Reactions

Write the anode, cathode, and overall cell reactions that occur when each pair of half-cells is combined to form a galvanic cell.

(a) a copper strip in a solution of copper(II) nitrate, $Cu(NO_3)_{2(aq)}$, and a tin strip in a solution of tin(II) chloride, $SnCl_{2(aq)}$

Step 1: Establish Elements Oxidized and Reduced

Tin is higher than copper in the activity series. Therefore, tin atoms are oxidized and copper(II) ions, $Cu^{2+}_{(aq)}$, are reduced.

Anode half-reaction: $\qquad\qquad Sn_{(s)} \rightarrow Sn^{2+}_{(aq)} + 2e^-$

Cathode half-reaction: $\qquad Cu^{2+}_{(aq)} + 2e^- \rightarrow Cu_{(s)}$

Overall cell reaction: $\qquad Sn_{(s)} + Cu^{2+}_{(aq)} \rightarrow Sn^{2+}_{(aq)} + Cu_{(s)}$

(b) an aluminum strip in a solution of aluminum nitrate, $Al(NO_3)_{3(aq)}$, and a silver strip in a solution of silver nitrate, $AgNO_{3(aq)}$

Step 1: Establish Elements Oxidized and Reduced

Aluminum is higher than silver in the activity series. Therefore, aluminum atoms are oxidized and silver ions, $Ag^+_{(aq)}$, are reduced.

Anode half-reaction: $\qquad\qquad Al_{(s)} \rightarrow Al^{3+}_{(aq)} + 3\ e^-$

Cathode half-reaction: $\qquad Ag^+_{(aq)} + e^- \rightarrow Ag_{(s)}$

Notice that the number of electrons lost by each aluminum atom ($3\ e^-$) is not equal to the number of electrons gained by each silver ion ($1\ e^-$).

Step 2: Balance Charges

Multiply both sides of the cathode reaction by 3 so that the number of electrons lost is equal to the number of electrons gained.

Anode half-reaction: $\qquad\qquad Al_{(s)} \rightarrow Al^{3+}_{(aq)} + 3e^-$

Cathode half-reaction: $\qquad 3\ Ag^+_{(aq)} + 3e^- \rightarrow 3\ Ag_{(s)}$

Overall cell reaction: $\qquad Al_{(s)} + 3\ Ag^+_{(aq)} \rightarrow Al^{3+}_{(aq)} + 3\ Ag_{(s)}$

▶ **Practice**

Understanding Concepts

Write the chemical equations for the anode, cathode, and overall cell reactions that occur when a galvanic cell is made using each pair of the following half-cells:

(a) a lead strip in a solution of lead(II) nitrate, $Pb(NO_3)_{2(aq)}$, and a zinc strip in a solution of zinc nitrate, $Zn(NO_3)_{2(aq)}$

(b) a silver strip in a solution of silver nitrate, $AgNO_{3(aq)}$, and a magnesium strip in a solution of magnesium nitrate, $Mg(NO_3)_{2(aq)}$

The Purpose of the Salt Bridge

The salt bridge provides ions to prevent charge buildup from occurring. In **Figure 5**, for example, as $Cu^{2+}_{(aq)}$ ions are removed from the copper nitrate solution on the right, $Na^+_{(aq)}$ ions from the salt bridge move into the cathode half-cell. Similarly, negative nitrate ions flow into the anode half-cell from the salt bridge. As a result, the solution around each electrode remains neutral, even though the concentrations of zinc ions and copper ions are changing.

What happens if you remove the salt bridge? The production of zinc ions, $Zn^{2+}_{(aq)}$, results in a buildup of positive charge around the zinc electrode (the anode). This buildup prevents further production of zinc ions. Similarly, the loss of copper ions, $Cu^{2+}_{(aq)}$, at the cathode leaves the solution negatively charged, which prevents electrons in the copper electrode from being transferred to other copper ions. The end result is that the electrode reactions stop occurring. Thus, removing the salt bridge has the same effect as disconnecting the electrodes.

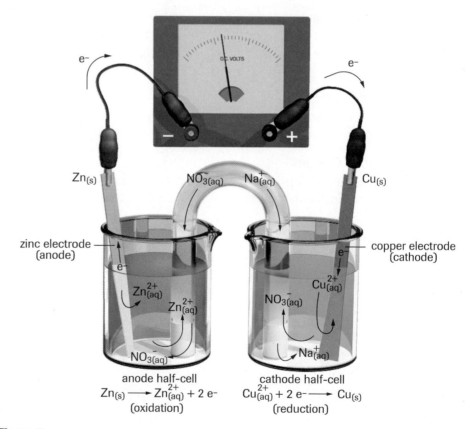

Figure 5
Notice how the electrons flow through the wire (to the right) from anode to cathode. The circuit is completed by negative nitrate ions flowing to the anode through the salt bridge. At the same time, positive sodium ions flow in the opposite direction through the salt bridge.

Cell Potential

The **cell potential** (also called the **voltage**) of the cell in **Figure 5** is 1.20 V (using 1.0-mol/L solutions measured at 25°C). Let's use a water analogy to understand what this measurement means.

Consider the water behind the gates in the lock in **Figure 6**. In **Figure 6(a)**, a kilogram of surface water on the right side of the gate is higher and, therefore, has more gravitational potential energy than a kilogram of surface water on the left side of the gate. Once the water outlet is opened, gravitational potential energy is converted to the kinetic energy of the water flowing from the higher side to the lower side. As the water level drops (**Figure 6(b)**), the difference in gravitational potential energy decreases. Water stops flowing when the difference in gravitational potential energy between the two sides is zero (**Figure 6(c)**).

Similarly, the anode is the "high side" of a galvanic cell, and the cathode is the "low side." Once a connection is made, electrons flow "downhill" from anode to cathode. The cell potential is a measure of the electrical potential energy difference (or potential difference) across the two half-cell electrodes. It is measured using a voltmeter. The units of potential difference are volts (V). As the cell operates, the potential difference gradually decreases until it becomes zero. At this point, electron flow has stopped and the cell is "dead."

Learning about galvanic cell design is useful for understanding how galvanic cells produce electricity. In section 5.9, you will apply your understanding of galvanic cells to the operation of batteries.

cell potential (voltage) a measure of the potential difference across the electrodes

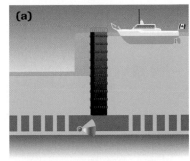
(a)

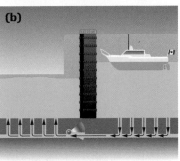
(b)

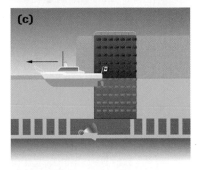
(c)

Figure 6
(a) The water on the right side of the gate has <u>more gravitational potential energy than</u> the water on the left side.
(b) The difference in potential energy decreases as water from the high side flows to the low side through underground pipes.
(c) The flow of water between sides stops when the water levels become equal.

| **SUMMARY** | *Galvanic Cells* |

- The reaction that occurs within a galvanic cell is spontaneous.
- Every galvanic cell consists of two different electrodes and at least one electrolyte.
- Oxidation occurs at the anode.
- Reduction occurs at the cathode.
- Two connections are needed to complete the circuit in a galvanic cell: a wire (for electron flow) and a salt bridge (for ion flow).
- Ions from the salt bridge help to keep each half-cell electrically neutral.
- Positive ions (cations) flow to the cathode.
- Negative ions (anions) flow to the anode.
- The cell potential (or voltage) measures the potential difference across the electrodes.

LEARNING *TIP*

Electron Affinities
Think of the anode as pushing the electrons away, while the cathode is pulling the electrons toward it.

Understanding Concepts

1. What energy change occurs during the operation of a galvanic cell?

2. How are the anode and cathode similar, and how are they different?

3. What are the two "connections" that have to be made before a galvanic cell can produce electricity?

4. A galvanic cell is made using $Mg_{(s)}$ in $Mg(NO_3)_{2(aq)}$ and $Zn_{(s)}$ in $Zn(NO_3)_{2(aq)}$. Write equations for the anode half-cell reaction, the cathode half-cell reaction, and the overall cell reaction.

5. What does cell potential measure?

6. Explain why the cell potential drops to zero when the salt bridge is removed.

7. (a) Identify the anode and cathode in **Figure 7**.
   (b) Write the anode and cathode half-cell reactions.
   (c) Write the overall cell reaction.
   (d) In which direction do electrons flow?
   (e) In which direction do ions flow in the salt bridge?
   (f) Predict which electrode increases in mass and which electrode decreases in mass. Justify your predictions.
   (g) Why is sodium nitrate a more suitable electrolyte for the salt bridge than sodium chloride? (*Hint:* Examine the solubility rules you learned about in Unit 1.)

Applying Inquiry Skills

8. (a) Draw and label a diagram of a galvanic cell that can be constructed using the following materials: beakers, connecting wires, glass U-tube, cotton, distilled water, strip of lead, solid lead(II) nitrate, solid sodium nitrate, strip of nickel, solid nickel(II) nitrate.
   (b) Indicate the direction of electron flow, and identify the anode and cathode in the cell.

Making Connections

9. In the late 1780s, the Italian anatomist Luigi Galvani (after whom the galvanic cell is named) reported a startling discovery. While dissecting the leg muscle of a dead frog, Galvani noticed that the leg suddenly twitched! This effect happened only when the leg muscle came in contact with two different metals. Research and summarize how Galvani interpreted his discovery. Was he correct?

 www.science.nelson.com

Extension

10. Many pH meters contain electrodes that sense the hydrogen ion concentration of a solution and then convert this information into a pH reading. Research and report on how a pH meter works.

 www.science.nelson.com

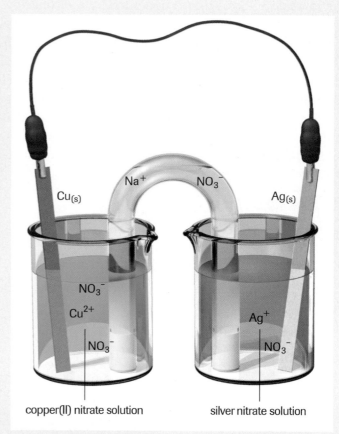

Figure 7
The electrons flow from the copper electrode to the silver electrode.

Designer Cells

The galvanic cell in **Figure 1** was one of the first types of cells to supply a steady output of electrical energy for a long period of time. Although this cell is not very practical as a portable power source, it is still very useful for learning how galvanic cells work.

In this activity, you will construct galvanic cells, measure their cell potentials, and compare them with a more practical galvanic cell—an alkaline dry cell.

Question

How do laboratory galvanic cells compare with alkaline dry cells?

Materials

eye protection
lab apron
0.1-mol/L solutions of
    copper(II) nitrate, $Cu(NO_3)_{2(aq)}$
    iron(II) sulfate, $FeSO_{4(aq)}$
    zinc nitrate, $Zn(NO_3)_{2(aq)}$
    magnesium nitrate, $Mg(NO_3)_{2(aq)}$
    potassium nitrate, $KNO_{3(aq)}$

four 50-mL beakers
100-mL beaker
7 strips of paper towel
8-cm strips of copper, iron, zinc, and magnesium
sandpaper
multimeter or voltmeter and connectors
tweezers
alkaline dry cell
piezo buzzer

 Some of the solutions in this activity are toxic and/or irritants. Wear eye protection and a lab apron. Avoid skin contact.

Procedure
Part 1: Building and Testing Laboratory Galvanic Cells

1. Clean all metals with sandpaper to remove surface oxides.

2. Put seven paper-towel strips in a 100-mL beaker. Add enough 0.1-mol/L potassium nitrate solution to soak the strips. They will be used as salt bridges.

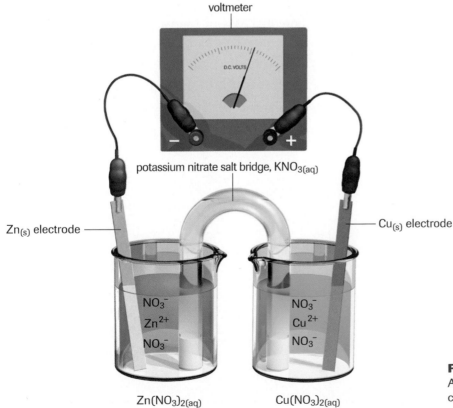

voltmeter

D.C. VOLTS

potassium nitrate salt bridge, $KNO_{3(aq)}$

$Zn_{(s)}$ electrode

$Cu_{(s)}$ electrode

NO₃⁻
Zn²⁺
NO₃⁻

NO₃⁻
Cu²⁺
NO₃⁻

$Zn(NO_3)_{2(aq)}$

$Cu(NO_3)_{2(aq)}$

Figure 1
A galvanic cell consists of two half-cells connected by a wire and a salt bridge.

3. Make a copper half-cell by placing a copper strip in a 50-mL beaker that contains 15 mL of copper(II) nitrate. Prepare the iron, zinc, and magnesium half-cells in the same way.

4. Construct the Zn/Cu galvanic cell as shown in **Figure** 1. Use tweezers to insert the wet paper-towel strips.

5. Connect the multimeter or voltmeter between the two electrodes so that the meter gives a positive reading. (If the reading is negative, reverse the connectors.) Once the reading is positive, the electrode that is attached to the COM connection on the multimeter is the anode of the cell. The other electrode is the cathode. Record which electrode is the anode and which electrode is the cathode. Also, record the maximum voltage of the Zn/Cu galvanic cell.

6. Discard the salt bridge.

7. Repeat steps 4 to 6 for all the other possible galvanic cells that you can make using your half-cells. Always use a clean salt bridge for each new galvanic cell.

Part 2: Comparing Laboratory Galvanic Cells with Alkaline Dry Cells

8. Select the laboratory galvanic cell with the largest observed cell potential. Set the cell up again, using a new salt bridge. Connect this cell to the piezo buzzer, and listen carefully. If you do not hear anything, try reversing the connection.

9. Connect the piezo buzzer to the alkaline dry cell. Again, reverse the connection if you do not hear anything. Compare the intensity of the sound with the intensity of the sound you heard in step 8.

10. Dispose of the metals and solutions as directed by your teacher.

Analysis

(a) (i) Which metal was the anode of the cell most often? Which metal was the cathode of the cell most often?

(ii) Which cell had the largest cell potential (voltage)?

(iii) Compare the positions of the metals you identified in (i) and (ii) in the activity series.

(b) Write the chemical equations for the anode, cathode, and overall cell reactions that took place in each laboratory galvanic cell. (*Note:* In this situation, iron is more likely to form $Fe^{2+}_{(aq)}$ than $Fe^{3+}_{(aq)}$.)

(c) The alkaline dry cell is the modern version of the laboratory cell. Brainstorm, with your lab partner, the criteria you would use to compare the usefulness of these two cells (such as power output, voltage, and portability).

(d) Answer the Question, using the criteria you developed in (c).

Synthesis

(e) Use the activity series to predict how the cell potential of your Mg/Cu cell would compare with the cell potentials of the following galvanic cells:

• galvanic cell 1: Mg/Ag

• galvanic cell 2: Mg/Pb

Explain how you made your predictions.

We live in a portable world. Just think of all the consumer products that require their own portable power sources: power tools, watches, cell phones, and portable CD players, to name just a few (**Figure 1**). We, as consumers, expect that these products will have long-lasting, lightweight power sources. This demand has driven rapid improvements in the design of batteries.

The scientific definition of the word **battery** is two or more galvanic cells connected in series. In everyday language, however, a battery is any portable device that stores power. As you will see, battery or cell design has come a long way from the simple galvanic cells you learned about in section 5.7. Today there are many different kinds of cells and batteries to choose from: cylindrical or rectangular, alkaline or nickel–cadmium (Ni–Cd), low voltage or high voltage. In this section, you will compare some of the cells and batteries that are now available (**Figure 2**).

Cell and Battery Construction

You have likely seen and used many batteries. You know that batteries are usually cylinders or rectangular prisms, sealed with a metal or plastic covering. What is inside, however, and how does the "electricity" get from the inside to the outside?

Outside

The outside casing of a cell has several functions. Perhaps the most important function is to protect the contents from damage, and to protect us from the corrosive chemicals inside. The casing is generally printed to give the voltage of the cell, a code representing its size, and safety information. Because the cell casing is tightly sealed, you should never throw a discharged cell into a fire. An explosion could result as the fluids inside the cell begin to evaporate.

All batteries have two terminals: positive and negative. Most cylindrical and button batteries have the positive terminal on the top. The negative terminal is on the bottom or is sometimes the entire case. Larger batteries, such as the lantern battery in **Figure 2**, have both terminals on the top.

The size of a cell determines the amount of chemicals it can hold and, therefore, the length of time it will last. D cells, for example, last longer than the smaller AA cells under similar operating conditions (**Figure 3**). The AA cell was designed for use in smaller products that do not demand a great deal of electrical energy.

Inside

All cells have three basic internal parts: two electrodes (an anode and a cathode) and an electrolyte. Look at the photo (**Figure 4**, on the next page) and diagram (**Figure 5**, on the next page) of the inside of an alkaline dry cell—one of the most common types of cells. Can you locate these three parts?

battery a group of two or more galvanic cells connected in series

Figure 1
Batteries power many of the products you use every day.

Figure 2
Cells and batteries come in a variety of shapes and sizes, from tiny button watch batteries to large 6-V lantern batteries.

Figure 3
Larger cells last longer because they contain more chemicals.

The centre of the alkaline dry cell in **Figure 4** is a brass pin surrounded by an anode of powered zinc. The cathode chemicals (a mixture of manganese dioxide, MnO_2, and carbon) surround the anode. A thin, porous fabric keeps the anode and cathode chemicals separate while allowing electrolytes to pass through it. The electrolyte in this cell is a thick corrosive paste that contains potassium hydroxide, KOH. Because this electrolyte is a strong base, you should never touch a leaking battery!

Cell Reactions

The chemicals inside a cell affect the cell's properties, such as its voltage. For example, as you can see from the following half-reactions, the chemicals inside an alkaline dry cell (1.5 V) and a Ni–Cd cell (1.2 V) are quite different.

Alkaline Dry Cell

Cathode half-reaction: $$2\,MnO_{2(s)} + H_2O_{(l)} + 2\,e^- \rightarrow Mn_2O_{3(s)} + 2\,OH^-_{(aq)}$$

Anode half-reaction: $$Zn_{(s)} + 2\,OH^-_{(aq)} \rightarrow ZnO_{(s)} + H_2O_{(l)} + 2\,e^-$$

Ni–Cd Cell

Cathode half-reaction: $$2\,NiOOH_{(s)} + 2\,H_2O_{(l)} + 2\,e^- \rightarrow 2\,Ni(OH)_{2(s)} + 2\,OH^-_{(aq)}$$

Anode half-reaction: $$Cd_{(s)} + 2\,OH^-_{(aq)} \rightarrow Cd(OH)_{2(s)} + 2\,e^-$$

The half-cell reactions begin only when a connection is made between the positive and negative terminals. The electrons are like water behind a dam, waiting for the gate to open. An alkaline AA cell can remain unused for over five years and still work! When you press the "on" switch on your flashlight, electrons, produced from the oxidation of zinc, travel through the pin to the

Figure 4

An alkaline dry cell isn't really dry. The paste inside the cell contains the ions that are required to carry the charge and complete the circuit.

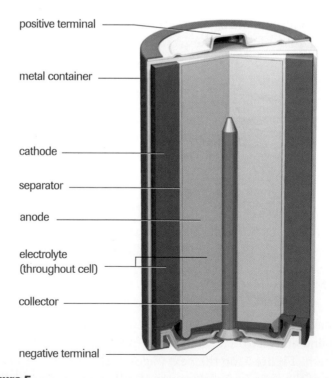

positive terminal

metal container

cathode

separator

anode

electrolyte (throughout cell)

collector

negative terminal

Figure 5

The alkaline dry cell has separated anode and cathode compartments and electrolytes, just like the galvanic cells you learned about in section 5.7.

negative terminal at the bottom of the cell (**Figure 6**). From there, the electrons flow through the connectors to the bulb, where their energy is converted into light energy. The electrons return through the base of the light bulb into the positive terminal, which is connected to the cathode region of the cell.

Cell Arrangements

If larger voltages are required, cells can be connected in series (positive to negative) to become a battery (**Figure 7**). The total voltage of a battery is the sum of the voltages of the individual cells. The 9-V battery in **Figure 7**, for example, contains six 1.5-V alkaline cells. Because the voltage has been increased, a light bulb attached to this battery will glow more brightly than if it were attached to a single alkaline dry cell.

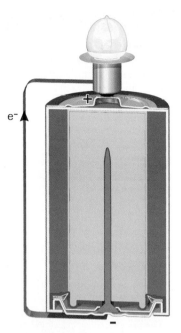

Figure 6
Electrons flow from the negative terminal to the positive terminal.

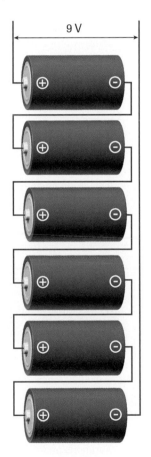

9 V

Figure 7
Six 1.5-V cells connected in series produce a potential difference of 9 V.

Primary and Secondary Cells

Most consumer cells can be grouped into two types. Cells that cannot be recharged are called **primary cells**. These cells are "dead" once the reactants inside them are used up. **Secondary cells** can be recharged using a source of electrical energy, such as a battery recharger. During the recharging process, electrical energy is used to reverse the chemical reactions in the cell, thus regenerating the reactants.

primary cell a cell that cannot be recharged

secondary cell a rechargeable cell

The electric eel contains specialized structures, called electroplates, that can generate electricity. A large eel (over 1 m long) can have more than 6000 electroplates connected in series. The head of the eel has a positive charge, and its tail has a negative charge. Being touched by both the head and the tail of an eel can send a 600-V jolt through the victim, enough to paralyze or kill a human.

Table 1 lists some of the most common primary and secondary cells.

Table 1 Primary and Secondary Cells

Name of cell	Advantages	Disadvantages	Uses
Primary cells			
alkaline dry cell (1.5 V)	• is inexpensive • has a long shelf life • comes in a variety of shapes and sizes • has good energy output for the cost	• has a relatively short lifespan	general household uses
silver oxide cell (1.5 V)	• is very small • has very steady energy output for the life of the cell	• is expensive because of its silver content	hearing aids, watches, calculators
lithium cell (3.0 V)	• is small and lightweight • generates more energy for its mass than most other batteries • lasts three times longer than alkaline batteries of the same size	• is expensive • has a very quick drop in voltage at the end of its life	laptop computers, cell phones, digital cameras, smoke detectors
Secondary cells			
nickel–cadmium (Ni–Cd) cell (1.2 V)	• is the most inexpensive rechargeable battery • recharges quickly • can be recharged over 1000 times • has high energy output • is unaffected by vibrations and shock	• sometimes will not recharge completely because of "memory effect" • is difficult to dispose of, because cadmium is very toxic	power tools, general household uses
lead–acid cell (2.0 V)	• has very high energy output • is reliable for many recharges • is durable • has no memory effect	• is difficult to dispose of, because lead is toxic • is very heavy	automobiles, cordless electric lawn mowers

Burning a Hole in Your Pocket

If coins in your pocket accidentally connect the positive and negative terminals of an unpackaged Ni–Cd cell, enough heat can be produced to burn a hole through your pants.

Table 1   *(continued)*

Name of cell	Advantages	Disadvantages	Uses
Secondary cells (continued)			
nickel metal hydride (NiMH) cell (1.2 V)	• has high energy output • is less toxic than a Ni–Cd cell • can be recharged 300 to 500 times • has an extremely long shelf life (7 to 10 years) • has no memory effect	• is expensive	cell phones, laptop computers, electric vehicles, digital cameras

DID YOU *KNOW* ?

Dead Batteries Come Back to Life!
Have you noticed that primary batteries sometimes seem to recharge themselves after sitting for a while? As the cell operates, hydrogen gas forms on the electrodes, making the cell appear to be "dead." Allowing the cell to sit for a while gives the hydrogen gas time to dissipate.

The problem with both primary and secondary cells is that they can be used for only a relatively short period of time before they have to be either replaced or recharged. There is an alternative! In section 5.11, you will learn about a galvanic cell in which the reactants are continually supplied, so it never runs out.

▶ **TRY THIS** *activity*   **Pickle Power**

Can a pickle generate power? It can't on its own. With two different metals attached to it, however, a pickle can be made into a crude electric cell (**Figure 8**). Two or more pickles together can be made into a battery!

Materials: 2 pickles, two 4-cm magnesium strips, two 4-cm copper wires, connecting wires, digital multimeter, light emitting diode (LED) or piezoelectric buzzer

1. Construct a "pickle cell" by inserting a copper wire and magnesium strip, about 4 cm apart, into a pickle. Make sure that the metals are not in contact inside the pickle.
2. Measure the voltage across the magnesium and copper electrodes.
3. Connect the electrodes to the LED.
4. If the LED does not light up, your pickle cell needs more power. Construct another pickle cell.
5. Connect the two pickle cells so that their combined voltage is greater than the voltage of a single pickle cell.
6. Try to light the LED.
7. Add another pickle cell if necessary.

(a) Sketch a diagram of the arrangement of cells that gave the highest voltage. (In this arrangement, the cells are said to be "in series" with each other.)

Figure 8
A pickle cell

(b) Describe the effect that connecting cells in series has on the overall voltage.
(c) A car battery consists of six 2-V cells connected in series. What is the voltage of a car battery?
(d) At home, find a battery-powered product in which the batteries are arranged in series. What is the total voltage of the batteries in this product?

Understanding Concepts

1. Alkaline AA and C cells from the same manufacturer have the same voltage: 1.5 V. Why are remote-control cars designed to use C cells while portable CD players require AA cells?

2. What is the function of the separator fabric in an alkaline cell?

3. Why is a leak from an alkaline cell dangerous?

4. What determines the voltage of a cell?

5. A 6-V lantern battery consists of 1.5-V cells connected in series. How many cells are required to produce a voltage of 6 V?

6. Use specific examples from your home to explain how primary cells and secondary cells differ.

7. The following half-cell reactions occur in a silver oxide button cell, which is used in watches and cameras.

$$Ag_2O_{(s)} + H_2O_{(l)} + 2\,e^- \rightarrow 2\,Ag_{(s)} + 2\,OH^-_{(aq)}$$
$$Zn_{(s)} + 2\,OH^-_{(aq)} \rightarrow Zn(OH)_{2(s)} + 2\,e^-$$

(a) Identify the chemicals that are oxidized and reduced as this cell operates.
(b) What is the overall cell reaction?
(c) What reactant makes this cell more expensive than an alkaline cell?
(d) Use **Table 1** (Primary and Secondary Cells) to identify two properties that make this cell ideally suited for use in watches.

8. Why is a lithium cell better for digital cameras than an alkaline dry cell?

Making Connections

9. A manufacturer of portable power tools is designing a new cordless electric screwdriver. Which secondary cell from **Table 1** would you recommend for use in this screwdriver? Give reasons for your choice.

10. The lead-acid automobile battery has been so successful that it has not changed much in the past 50 years. Research the answers to the following questions:
(a) What are the anode, cathode, and electrolyte in a lead-acid battery?
(b) How does this battery get recharged?
(c) Why is it particularly important to recycle lead-acid car batteries? Find out where old car batteries can be taken for recycling in your area.

 www.science.nelson.com

Reflecting

11. Sometimes words are used differently depending on the context. The scientific meaning might be quite different from the everyday usage. The scientific definition of a battery is an arrangement of two or more cells connected in series. Therefore, most of the cells in **Figure 2**, at the beginning of the section, cannot be called batteries. Do you think the scientific definition of the word "battery" should be changed?

Exploring

12. Some batteries have colour-changing testers. (**Figure 9**). Research how these testers work.

 www.science.nelson.com

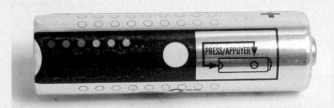

Figure 9
Some batteries incorporate a device that indicates how much charge is left.

13. Battery manufacturers are beginning to print cell capacities on their packaging. An AA nickel metal hydride cell, for example, has a capacity of 1300 mAh. Research what this value means. Report to your classmates on how your findings might affect the choices you make when buying batteries.

 www.science.nelson.com

14. Some batteries suffer from "memory effect." Research how this property affects the life of a cell and how it can be prevented.

 www.science.nelson.com

Until just recently, the nickel–cadmium (Ni–Cd) "battery" (it is actually a cell) was the leading rechargeable battery on the market. The Ni–Cd battery gets its name from its contents. The active material in the cathode is nickel oxyhydroxide, $NiOOH_{(s)}$. The active material in the anode is cadmium, $Cd_{(s)}$.

Ni–Cd batteries (**Figure 1**) became popular because they were relatively inexpensive and, if maintained properly, could be recharged more than 1000 times. Using Ni–Cd batteries instead of disposable (primary) cells regularly was not only cost-effective, but also reduced household waste. Imagine the savings to your wallet and the environment of *not* purchasing 1000 disposable batteries!

Ni–Cd batteries are not ideal, however. If they are not fully discharged before recharging, the amount of charge they will hold (known as their capacity) starts to decrease. This decrease in charge is called "memory effect." It can be caused by crystal growth at the cadmium anode. If the memory effect continues, the capacity of the battery can decrease to the point that it has to be replaced.

Even though most stores encourage recycling of Ni–Cd batteries (**Figure 2**), many of these batteries still end up in landfill sites. Over time, the case of the battery corrodes, allowing the toxic contents to leak out. Scientists estimate that Ni–Cd batteries are responsible for over 50% of the cadmium that leaches from landfills into ground water. Cadmium is extremely toxic to both wildlife and humans. Studies have linked long-term low-level exposure to cadmium to disorders of the lungs, kidneys, and liver.

Recycling is one way to cut back on the negative effects of Ni–Cd batteries. There are alternatives, as well. Rechargeable batteries that contain slightly more environmentally friendly chemicals have already taken over the general-purpose household battery market. These batteries include nickel metal hydride cells, lithium ion cells, and rechargeable alkaline batteries. In fact, finding a general purpose Ni–Cd battery is becoming increasingly difficult. Ni–Cd batteries still rule the portable telephone and power-tool markets, however (**Figure 3**, on the next page). At the time of writing, most of the portable telephones and power tools that are sold at major retailers contain Ni–Cd batteries. Manufacturers are now starting to produce power tools that use nickel metal hydride batteries (**Figure 4**, on the next page).

Figure 1
The nickel–cadmium battery was once a familiar sight in hardware stores.

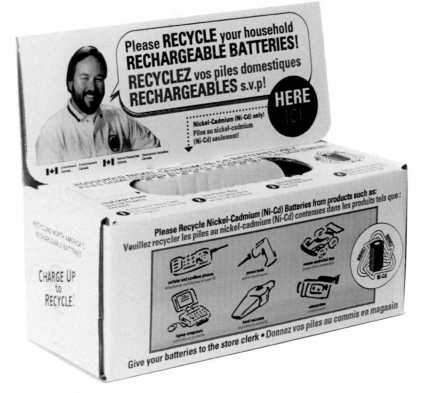

Figure 2
Look for battery recycling boxes in hardware stores.

Figure 3
Open just about any cordless or portable phone, and you are likely to find a Ni–Cd battery inside.

There is no question that banning the production of Ni–Cd batteries will benefit the environment. Those who oppose banning this battery, however, put forth strong arguments:

1. Alternatives to Ni–Cd batteries are often more expensive.

2. The alternatives cannot be recharged as many times as Ni–Cd batteries.

3. There is no alternative battery for some applications. In Canadian homes today, there are thousands of portable telephones and power tools that use Ni–Cd batteries. It is unreasonable to expect Canadians to give up the convenience of using portable devices. It is just as unreasonable to expect Canadians to purchase new devices that do not use Ni–Cd batteries if their Ni–Cd devices work well.

> ## Understanding the Issue

1. Why are Ni–Cd batteries bad for the environment?
2. What is the major technical problem with the use of Ni–Cd batteries?
3. What alternatives are there to the use of Ni–Cd batteries?

Figure 4
This drill comes with a 2.6-V nickel metal hydride battery that the manufacturer claims has greater storage capacity and longer running time than any Ni–Cd battery.

> ## Take a Stand

Should the Production of Ni–Cd Batteries Be Banned?

Decision-Making Skills

- ○ Define the Issue
- ● Analyze the Issue
- ● Research
- ○ Defend a Decision
- ● Identify Alternatives
- ● Evaluate

(a) Ni–Cd batteries are very popular. What are the alternatives to banning them?

(b) Suggest ways in which Canadians could phase out the use of Ni–Cd batteries.

(c) Research the pros and cons of Ni–Cd batteries. Organize your findings in a table. Be sure to record the sources of your information.

(d) Analyze your research findings to decide on an answer to the following question: "Should the production of Ni–Cd batteries be banned?"

(e) Prepare an information leaflet to be handed out at all stores where appliances containing Ni–Cd batteries are sold, or write a letter to a major manufacturer of Ni–Cd batteries, outlining and giving reasons for your position.

GO www.science.nelson.com

Wouldn't it be convenient to have a battery that never runs out? Why not design a battery in which the "used-up" chemicals are continuously replaced with "fresh" chemicals? This replacement is exactly what happens in a fuel cell. As it operates, the cell reactants are continually added and the cell products are continually removed. A fuel cell can run as long as fuel is supplied.

The hydrogen–oxygen fuel cell (more often called a hydrogen fuel cell) is a well-known fuel cell. Perhaps the first car you buy after graduating from college will be completely powered by fuel cells!

As a hydrogen fuel cell operates, hydrogen and oxygen are pumped into separate chambers in the cell: the anode and the cathode. There they are oxidized and reduced. Water is removed from the cell (**Figure 1**).

Anode half-reaction:
$$2\ H_{2(g)} + 4\ OH^-_{(aq)} \rightarrow 4\ H_2O_{(l)} + 4\ e^-$$

Cathode half-reaction:
$$O_{2(g)} + 2\ H_2O_{(l)} + 4\ e^- \rightarrow 4\ OH^-_{(aq)}$$

Overall cell reaction:
$$2\ H_{2(g)} + O_{2(g)} \rightarrow 2\ H_2O_{(l)}$$

A car running on these fuel cells would have only water vapour coming out of its tail pipe. Sounds ideal! So why aren't hydrogen fuel cells being used in all cars?

A few major technical problems need to be solved before hydrogen can be widely used in cars.

- Hydrogen gas is very explosive and, consequently, difficult to store. Engineers need to develop special pressurized tanks to hold hydrogen. To be used in a car, these tanks must be strong enough to withstand a rear-end collision.

- Where would you go for a fill-up? You would need to be able to buy fuel (hydrogen) from a network of hydrogen gas stations across the country. It will take a while before this network is in place, making hydrogen a convenient fuel.

- Is hydrogen really "clean"? Hydrogen is often made by passing electricity through water. Therefore, hydrogen is only as clean as the electricity that is used to make it. For example, if the electricity is generated by burning fossil fuels, are we any better off by using hydrogen fuel cells?

(a)

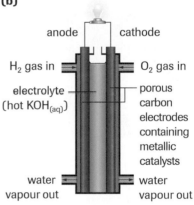

(b)

anode   cathode

H₂ gas in — — O₂ gas in
electrolyte — porous carbon electrodes containing metallic catalysts
(hot KOH_(aq))
water — water
vapour out   vapour out

Figure 1
An alkaline battery **(a)** and a fuel cell **(b)** are both galvanic cells. They both use redox reactions to produce electricity. An alkaline battery is "dead," however, when its anode and cathode chemicals are used up. In a fuel cell, the anode and cathode chemicals are continually replaced as they are used up.

▶ Tech Connect 5.11 Questions

Understanding Concepts

1. How does a fuel cell differ from an alkaline dry cell?

2. Why is the hydrogen–oxygen fuel cell nonpolluting?

3. Is hydrogen a "clean" fuel? Explain your answer.

Making Connections

4. Experimental cell phones that run on miniature hydrogen–oxygen fuel cells exist today. What must be in place before the average consumer can buy one?

5. One solution to the hydrogen storage problem is to make hydrogen as it is needed. A *reformer* is such a device. Research how reformers can be used to produce hydrogen.

 www.science.nelson.com

6. Iceland has ambitious plans to be the first nation to replace all its use of fossil fuels with "clean" hydrogen. Research how Iceland plans to produce the hydrogen it will need.

 www.science.nelson.com

Figure 1
RMS *Titanic* was the most modern luxury liner of its era.

Figure 2
A portion of *Titanic*'s hull showing "rustacles": structures that resemble stalactites in caves. Rusticles are constructed by colonies of microorganisms as the microorganisms consume iron.

corrosion the deterioration of a metal as a result of slow oxidation

On April 10, 1912, the RMS *Titanic* set sail from England on its maiden voyage across the North Atlantic Ocean (**Figure 1**). There were over 2200 people on board. Four days later, *Titanic* hit an iceberg and sank 3800 m to the bottom of the frigid ocean. The wreck of *Titanic* was finally discovered in 1985, about 500 km off the coast of Newfoundland. After examining the hull (**Figure 2**), scientists were surprised to find that the ship was not as badly corroded as they expected.

Corrosion is the deterioration of metals as a result of oxidation. It is common knowledge that salt water promotes corrosion, but does salt water *cause* corrosion? This section will introduce the phenomenon of corrosion, the factors that affect how quickly corrosion occurs, and how corrosion can be prevented.

What Is Corrosion?

With the exception of gold (and occasionally silver), most metallic elements are found naturally only as compounds or minerals. Copper, for instance, is extracted from copper-containing minerals such as malachite (**Figure 3**). Without protection against corrosion, most metals are oxidized by their environment. The colour of a newly installed copper roof (**Figure 4**) does not last long. Exposure to oxygen, moisture, and pollution soon corrodes copper atoms into a green layer of copper compounds that adhere well to the copper metal below. This green layer, sometimes called a patina, forms a tough, protective coating that allows copper roofs to last well over 50 years.

Figure 3
Malachite is a mineral that contains copper.

Figure 4
When copper sheets were first installed on the roofs of the Parliament Buildings, they were the shiny reddish brown of metallic copper. Over time, however, they oxidized until they were coated with a green patina of oxidized copper.

Zinc and aluminum also form protective coatings when they oxidize. These protective coatings are the reason why zinc and aluminum are more corrosion-resistant than iron, despite being above iron in the activity series. Zinc oxidizes to form a hard coating that consists of a mixture of zinc carbonate, $ZnCO_{3(s)}$, and zinc hydroxide, $Zn(OH)_{2(s)}$. Therefore, a galvanized (zinc-coated) chain-link fence can last a long time without deteriorating (**Figure 5**). Aluminum protects itself by reacting with oxygen to form a coating of aluminum oxide, $Al_2O_{3(s)}$, one of the hardest compounds known. Aluminum oxide is hard, chemically stable, and heat-resistant. It stubbornly re-forms as quickly as you can scrape it off, thereby protecting the metal underneath.

Figure 5
The metal in this fence is galvanized steel: steel coated with zinc. As the zinc surface layer corrodes, it forms a protective coating that prevents the steel from corroding.

Rusting of Iron

Rust is the familiar reddish-brown flaky material that is produced when iron-containing metals, such as steel, corrode. Rust is a mixture of oxides and hydroxides of iron. Unfortunately, rust does not stick well to the metal beneath it. Instead, it flakes off easily, exposing fresh metal to further oxidation. This process continues until the metal is completely reacted or "eaten away."

The Redox Chemistry of Rust

In section 5.7, you learned that a galvanic cell consists of two electrodes (an anode and a cathode) in an electrolyte. The electrodes are connected to each other by a conductor. Oxidation occurs at the anode, and reduction occurs at the cathode. A corroding metal is a galvanic cell in which the anode and the cathode are found at different points on the same metal surface. The metal itself is the conducting material that allows electrons to flow from the anode to the cathode (**Figure 6**). The anode is often located where there is a more reactive impurity in the metal, or where the metal has been dented or hit. The cathode can be almost anywhere on the surface of the metal.

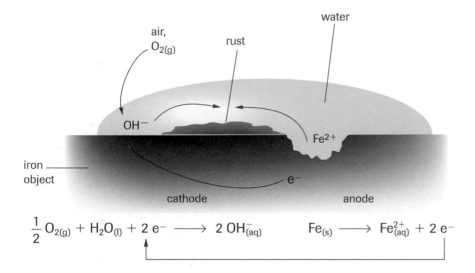

$$\frac{1}{2}\,O_{2(g)} + H_2O_{(l)} + 2\,e^- \longrightarrow 2\,OH^-_{(aq)} \qquad Fe_{(s)} \longrightarrow Fe^{2+}_{(aq)} + 2\,e^-$$

Figure 6
The corrosion of iron occurs through galvanic cell action on the surface of the iron. The oxidation of iron occurs at one location (the anode). The reduction of oxygen occurs at another location (the cathode). The products of these reactions, $Fe^{2+}_{(aq)}$ and $OH^-_{(aq)}$, are involved in additional reactions that lead to the formation of rust.

Figure 7
Rusting of exposed iron is almost negligible when the relative humidity is less than 40%. This iron pillar in Delhi, India, has existed for about 1500 years because of the very dry and unpolluted environment.

Figure 8
Rust around wheel wells is common because of frequent contact with road salt and moisture.

Figure 9
This photo shows the corrosion of aluminum, caused by a steel screw, over a period of six months.

In the presence of moisture and oxygen, iron oxidizes, releasing two electrons:

$$Fe_{(s)} \rightarrow Fe^{2+}_{(aq)} + 2\,e^-$$

The electrons travel through the metal to the cathode, where they are used to reduce oxygen molecules:

$$\frac{1}{2}\,O_{2(g)} + H_2O_{(l)} + 2\,e^- \rightarrow 2\,OH^-_{(aq)}$$

Water is a reactant of corrosion, but it is also a medium through which ions can flow to the anode and the cathode—much like a salt bridge. Iron(II) ions, produced at the anode, are further oxidized by oxygen to form iron(III) ions. The iron(III) ions combine with water molecules and hydroxide ions to form the reddish precipitate that we call rust.

Factors That Affect the Rate of Corrosion

Corrosion is a complex phenomenon that is affected by many variables, including moisture, the presence of electrolytes, contact with less reactive metals, and mechanical stresses.

Moisture

Because water is a reactant, corrosion cannot occur without water. For this reason, steel objects in dry desert climates (**Figure 7**) last much longer than they do in Ontario. In fact, a relative humidity of at least 40% is required for corrosion to take place.

Electrolytes

Road salt (sodium chloride) does not cause corrosion, but it certainly helps! Salt is an electrolyte. It releases sodium and chloride ions as it dissolves in road slush (**Figure 8**). These ions greatly improve the electrical conductivity of water. As iron corrodes, iron(II) ions, $Fe^{2+}_{(aq)}$, are released at the anode (**Figure 6**). Chloride ions from road salt help to offset the buildup of positive charge at the anode. Thus, they are similar to the negative ions (anions) in a salt bridge moving toward the anode of a galvanic cell (section 5.7). Similarly, sodium ions, $Na^+_{(aq)}$, help to offset the buildup of negative hydroxide ions at the cathode. The combination of road salt and salt-water spray off the ocean makes the corrosion of cars worse in the Maritime provinces than in many other parts of Canada.

Contact with Less Reactive Metals

Corrosion can occur when two different metals come into contact with each other (**Figure 9**). The more reactive metal (higher in the activity series) loses electrons to the less reactive metal and becomes oxidized. To prevent corrosion, plumbers must use copper nails and brackets to attach copper pipes to framing lumber in new construction. ▨▮ Steel nails, which contain iron, would corrode quickly since iron is a more reactive metal than copper. Some researchers think that *Titanic* sank faster than it should have because the rivets that were used to join the hull sections were made of a different metal than the hull sections themselves.

Mechanical Stresses

Bending, shaping, or cutting metals can introduce stresses into the structure of the metal. These stresses can become corrosion sites. Have you ever noticed that nails often corrode first at the point, at a bend, or on the head where they are hit by a hammer (**Figure 10**)?

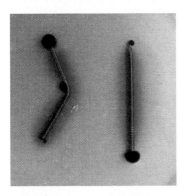

Figure 10
Two nails were placed overnight in a gel containing phenolphthalein and hexacyanoferrate(III) indicator solutions. The blue colour shows where iron is being oxidized (the anode). The pink colour shows where oxygen is being reduced (the cathode). Note that corrosion has occurred at the points and heads of both nails and at the bend in the nail on the left. These are places where the nails had been mechanically stressed.

Figure 11
A dielectric union is used in a plumbing system to connect pipes that are made from different metals. A plastic washer inside the union prevents the galvanized steel and the copper from coming into contact.

CAREER CONNECTION

Most of the water supply pipes that are used in new construction are made of copper. Copper is ideal because it is so unreactive, being near the bottom of the activity series. Many older homes, however, still have galvanized steel pipes. Galvanizing protects a steel pipe from corrosion by coating it with a protective layer of zinc. Attaching a new copper pipe directly to a galvanized pipe would quickly corrode the galvanized pipe. To avoid contact between the two metals, plumbers use a connector known as a dielectric union (**Figure 11**). Inside the large nut on the union is a plastic washer that prevents the copper pipes from coming into direct contact with the galvanized steel pipe (on the left).

(i) Describe another situation in which the contact of two different metals can result in corrosion.

(ii) Lead-based solder was commonly used to join copper piping until the late 1980s. Why was it banned?

GO www.science.nelson.com

(iii) Consult a plumbing contractor in your area to find out about future employment trends, as well as current requirements for apprenticeship programs in this field.

DID YOU KNOW ?

Electron Competition
The corrosion that results from two different metals in contact is sometimes called *galvanic corrosion*.

SUMMARY   *Corrosion*

- Corrosion is the deterioration of metals as a result of oxidation.
- A corroding metal is a galvanic cell in which one part of the metal is the anode and another part of the metal is the cathode.
- In the corrosion (rusting) of iron, oxidation of iron occurs at the anode:

$$Fe_{(s)} \rightarrow Fe^{2+}_{(aq)} + 2\,e^-$$

Reduction of oxygen occurs at the cathode:

$$\frac{1}{2}\,O_{2(g)} + H_2O_{(l)} + 2\,e^- \rightarrow 2\,OH^-_{(aq)}$$

- Factors that affect the rate of corrosion include moisture, dissolved electrolytes, contact with less reactive metals, and mechanical stress.

Understanding Concepts

1. What is corrosion?

2. Why are some reactive metals corrosion-resistant?

3. Which elements are oxidized and reduced during the corrosion of iron in neutral solutions?

4. Write the chemical equation for the reduction of oxygen. Use this equation to explain why an iron object does not rust appreciably in the desert.

5. List four factors that can affect the rate of corrosion.

6. Explain how salt speeds up corrosion.

Applying Inquiry Skills

7. In an investigation into corrosion rate, several identical nails were placed in 0.1-mol/L solutions of different acidity. The nails were observed for evidence of corrosion. Complete the Analysis for the following lab report:

   Question
   What effect does the pH of a solution have on the rate of corrosion?

   Observations

   Table 1   Corrosion of Nails in Various Solutions

Solution	pH	Amount of corrosion
$NaCl_{(aq)}$	7	some
$Na_2CO_{3(aq)}$	10	none
$NaOH_{(aq)}$	13	none
$NaHSO_{4(aq)}$	2	a great deal
$H_2O_{(l)}$	7	some
$H_2SO_{4(aq)}$	1	a great deal
$HNO_{3(aq)}$	1	a great deal

Analysis

(a) Identify the independent variable and two controlled variables.

(b) Which type of solution (acidic, basic, or neutral) showed the greatest amount of corrosion?

(c) Which type of solution showed the least amount of corrosion?

Making Connections

8. Some people claim that it is better to leave a car outside overnight during the winter, rather than in a warm garage, especially if salt is used on the roads. Give reasons for and against this claim.

9. Before a new copper roof is installed, a layer of plastic is applied to the sub-roof to ensure that all the steel nail heads are covered. Why is this step necessary?

10. When RMS *Titanic* was discovered at the bottom of the Atlantic Ocean, scientists were surprised that there was not more corrosion damage.
    (a) Identify two factors that may have slowed the corrosion of *Titanic*.
    (b) Conduct a search of the Internet to check your prediction.

 www.science.nelson.com

Extension

11. During the 1970s, aluminum wiring became quite popular in homes. Since then, however, faulty aluminum wiring has been the cause of many fires. Consult an electrician to find out why aluminum wiring is potentially dangerous.

Is Your Jewellery Poisoning You?

It sounds strange, but it's true: an earring can cause you to develop blisters or break out in a rash. What's in your jewellery that causes the problem? Why can some people who react badly to white gold happily wear 18-karat yellow gold? The answer is nickel—a metal that is present in varying amounts in most metals used for jewellery (**Figure 1**). The symptoms of a nickel allergy (also known as nickel contact dermatitis) can be water blisters and a rash at the site of contact or all over the body.

Each year, an increasing number of Canadians develop allergic reactions to nickel. Why? The main reason is the large number of metallic objects that are made of alloys containing nickel. Earrings, rings, watches and other jewellery, belt buckles, staples, zippers, and even pennies all contain nickel. Because nickel is found in so many objects, and because more people than ever before are having body piercings, the likelihood that someone with a nickel sensitivity will develop a nickel allergy is increasing. Unfortunately, there is no cure and no way to desensitize a sufferer.

DID YOU KNOW?

A Penny Is Nickel
Many metal objects are alloys that contain nickel. Even the current Canadian penny contains about 1.5% nickel.

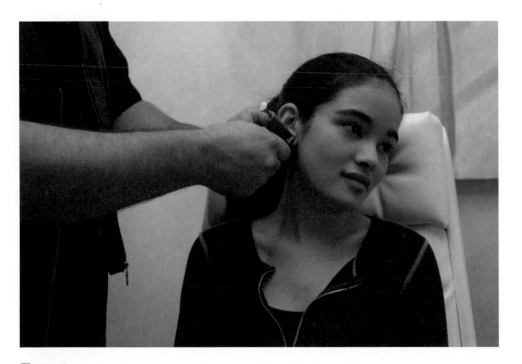

Figure 1
The studs or hoops that are placed in newly pierced ears are stainless steel or yellow gold. Why is this precaution wise?

Figure 2
The stainless steel back of this watch has been "eaten away" by the wearer's sweat.

Figure 3
The fashion for multiple piercings may soon begin to wane.

The Chemistry of a Nickel Allergy

The real cause of a nickel allergy is nickel(II) ions, $Ni^{2+}_{(aq)}$, rather than the metal itself. Nickel ions are produced when nickel metal is oxidized (corroded) by the acidity of body fluids, such as sweat. The reaction involved is

$$Ni_{(s)} + 2\,H^+_{(aq)} \rightarrow Ni^{2+}_{(aq)} + H_{2(g)}$$

Factors That Contribute to a Nickel Allergy

Three factors contribute to a nickel allergy: prolonged, direct skin contact; the presence of electrolytes; and the type of nickel-containing metal involved.

Direct Skin Contact

The skin must be in direct contact with a nickel-containing object, which is why the skin under a watch or ring is a common place for the allergy symptoms to appear.

Presence of Electrolytes

Have you ever noticed that sweat tastes salty? The salty taste is due to the *electrolytes* in sweat. Recall, from Unit 1, that electrolytes are substances that release ions when they are dissolved in water. A watch corrodes much faster in sweat than it would in pure water. Dissolved ions do not cause corrosion, but they make corrosion occur more quickly once it starts. Dissolved ions are the reason why cars rust faster when salt is used on roads in the winter.

Type of Metal

The quality of a metal often determines how quickly it will corrode. Stainless steel, for example, is an alloy of iron, nickel, and chromium. The resistance of stainless steel to corrosion varies, depending on how much chromium it contains. Stainless steel that contains 18% chromium is almost corrosion-proof. Better-quality stainless steel, such as the type used for dental braces and surgical implants, does not corrode at all. The watch in **Figure 2** is badly corroded because it is made of poor-quality stainless steel.

So You Want to Pierce Your ...

The number of nickel allergies has recently increased dramatically because of the popularity of body piercings. The reason is obvious: a nickel-containing object is in constant contact with the body. Piercing any body part results in some bleeding. Blood contains electrolytes, which can oxidize nickel in the studs or hoops, releasing nickel ions. The ions then have easy access, through the new wound, into the body. Soft tissue, like that in the earlobes, heals quickly because it does not have a lot of blood flow.

Problems with Tongue Piercing

Getting your tongue pierced is far riskier than getting your ears pierced (**Figure 3**). Here are three reasons why.

Infection

A tongue is much thicker, and contains more blood vessels, than an earlobe. Consequently, a newly pierced tongue takes much longer to heal than a newly pierced earlobe, and it is more prone to infection than most other body parts. An infected tongue can swell large enough to block the airway to the throat. Some deaths have resulted from complications due to tongue piercing.

Nickel Allergy

Without complications, a newly pierced tongue can take from four to six weeks to heal. Even during a normal healing period, a stud through the tongue is bathed with enough blood and saliva to trigger a nickel allergy.

Nickel and Toxicity

There is enough acidity in saliva to corrode any nickel in the stud. Acidic beverages, such as soft drinks and citric juices, corrode the nickel even faster. Once produced, nickel ions, $Ni^{2+}_{(aq)}$, are swallowed and can be absorbed into the bloodstream. The nickel ions are a known carcinogen (cancer-causing chemical). In fact, compounds that contain nickel are so toxic they have been banned from use in school labs by many school boards across Canada.

Making an Informed Decision

If you are considering a body piercing, it is imperative that you consider carefully the risks involved. Consult with only the most reputable business or health-care professional, and insist that sterile procedures and high-quality surgical steel be used.

DID YOU KNOW ?

A Tale of Two Toonies
The Canadian toonie and the 1-Euro coin are both bimetallic coins. They both have an inner and outer ring made of different metals. A design flaw in the 1-Euro coin makes it corrode when in contact with human sweat. Fortunately, the toonie was designed so that this problem does not occur.

> ### Case Study 5.13 Questions

Understanding Concepts

1. Why are nickel allergies difficult to overcome?
2. Refer to the activity series (section 5.4) to name one metal that would not be affected by the acidity of the skin. Explain your answer.
3. The pH of skin varies from person to person. What effect does a lower pH have on the amount of nickel that is released from a piercing? Explain your answer.
4. If you are wearing a watch, why is it a good idea to wash and dry your wrist and the back of your watch after heavy exercise?
5. Identify two reasons why piercing a tongue is riskier than piercing an earlobe.

Making Connections

6. Should we wear jewellery? Identify some risks and benefits of wearing jewellery. How would you evaluate your use of jewellery?
7. In 2002, neither training nor qualifications were required to work as a body piercer or a tattoo artist. Do you feel that people who are doing these jobs should be trained? In your opinion, what specific training should they have? Explain your answers.

GO www.science.nelson.com

Factors That Affect the Rate of Corrosion

Each year, millions of dollars are spent on replacing equipment and materials that are damaged by corrosion. To better understand how to control this redox reaction, you must first identify the factors that affect how quickly it occurs.

When steel (which is mostly iron) corrodes, iron(II) ions, $Fe^{2+}_{(aq)}$, are produced at the anode. Hydroxide ions, $OH^-_{(aq)}$, are produced at the cathode. The half-reactions are given below.

Anode half-reaction: $Fe_{(s)} \rightarrow Fe^{2+}_{(aq)} + 2e^-$ (oxidation)

Cathode half-reaction:

$$\frac{1}{2} O_{2(g)} + H_2O_{(l)} + 2\,e^- \rightarrow 2\,OH^-_{(aq)} \qquad \text{(reduction)}$$

The presence of $Fe^{2+}_{(aq)}$ and $OH^-_{(aq)}$ indicates that corrosion is taking place. The following chemical tests can be used to detect $Fe^{2+}_{(aq)}$ and $OH^-_{(aq)}$:

- Potassium hexacyanoferrate(III) indicator solution changes colour from yellow to blue if $Fe^{2+}_{(aq)}$ ions are present (**Figure 1(a)**).
- Phenolphthalein indicator changes from colourless to pink if hydroxide ions are present (**Figure 1(b)**).

In this investigation, you will design and conduct four mini-experiments to investigate four factors that may affect the rate of corrosion.

Question

What effect does each of the following factors have on the rate of corrosion of steel?

 (i) type of electrolyte

 (ii) presence of both oxygen and water

(iii) contact with other metals

(iv) mechanical stress

Prediction

(a) Read the list of Materials for each mini-experiment. Predict the effect of each of the factors on the corrosion rate of steel. Explain each prediction.

(a)  (b)

Figure 1
(a) If a solution that contains potassium hexacyanoferrate(III) indicator turns blue, iron(II) ions are present.
(b) If a solution that contains phenolphthalein indicator turns pink, hydroxide ions are present.

 The acid and base solutions used in this experiment are mildly corrosive. Avoid skin contact.

 Phenolphthalein indicator is flammable. Keep it away from open flames.

 Potassium hexacyanoferrate(III) indicator is toxic. Avoid ingestion. Wash your hands thoroughly at the end of the lab.

Part 1: The Type of Electrolyte
Materials

eye protection
lab apron
tweezers
4 cleaned nails
4 test tubes and stoppers
test-tube rack
pH test strips
dropper bottle of 0.1-mol/L potassium hexacyanoferrate(III) indicator
dropper bottle of phenolphthalein indicator
10 mL distilled water
10 mL of each of the following 0.1-mol/L solutions:
 sodium chloride, $NaCl_{(aq)}$
 hydrochloric acid, $HCl_{(aq)}$
 sodium hydroxide, $NaOH_{(aq)}$

Procedure

(b) Write a step-by-step Procedure for your experiment. Include any necessary safety precautions. When your teacher has approved your Procedure, carry out your experiment. Record your observations as you proceed.

(*Hint:* Use only one drop of phenolphthalein indicator per test tube.)

Part 2: The Presence of Both Oxygen and Water
Materials
eye protection   lab apron
tweezers   3 cleaned nails
3 test tubes and 1 stopper
test-tube rack
tap water
freshly boiled tap water (boiling removes dissolved oxygen)
dropper bottle of 0.1-mol/L potassium hexacyanoferrate(III) indicator
dropper bottle of phenolphthalein indicator

Procedure

(c) Write a step-by-step Procedure for your experiment. Include any necessary safety precautions. When your teacher has approved your Procedure, carry out your experiment. Record your observations as you proceed.

Part 3: Contact with Other Metals
Materials
eye protection
lab apron
tweezers
3 cleaned nails
sandpaper
small squares of zinc (1 cm²)
10-cm copper wire
3 test tubes and stoppers
test-tube rack
0.1 mol/L sodium chloride solution, $NaCl_{(aq)}$
dropper bottle of 0.1-mol/L potassium hexacyanoferrate(III) indicator
dropper bottle of phenolphthalein indicator

Procedure

(d) Write a step-by-step Procedure for your experiment. Include any necessary safety precautions. When your teacher has approved your Procedure, carry out your experiment. Record your observations as you proceed.

Part 4: Mechanical Stress
Materials
eye protection
lab apron
tweezers
3 or 4 cleaned nails
pliers
hammer and steel block
3 or 4 test tubes and stoppers
test-tube rack
distilled water
dropper bottle of 0.1-mol/L potassium hexacyanoferrate(III) indicator
dropper bottle of phenolphthalein indicator

Procedure

(e) Write a step-by-step Procedure for your experiment. Consider several different ways of mechanically stressing the nails, such as bending, straightening, and hammering them. Include any necessary safety precautions. When your teacher has approved your Procedure, carry out your experiment. Record your observations as you proceed.

Analysis

(f) Answer the Question, based on your observations in your four mini-experiments.

Evaluation

(g) Evaluate your Procedures. What, if anything, could be done to improve them?

(h) Why was a control included in each mini-experiment?

(i) List some sources of experimental error or uncertainty. How significant do you think they are?

Figure 1
Without corrosion protection, these natural gas pipelines would quickly rust.

There are over 100 000 km of steel pipelines carrying western oil and natural gas to eastern Canada and the United States (**Figure 1**). Underground sections are in constant contact with acidic ground water. Aboveground sections of the pipelines are subject to the harsh conditions of the Canadian climate, including acid rain. Without appropriate corrosion protection, the pipelines would rust quickly, resulting in an ecological disaster.

There are several methods that are currently being used to prevent the corrosion of a wide variety of objects. Some methods are more effective than others, but none are perfect. The methods can be divided into three categories: protective coatings, corrosion-resistant metals, and cathodic protection.

Figure 2
Some bridges are so large that they can take years to paint. Once the job is finished, the painters have to start over again.

Protective Coatings

The simplest method for preventing corrosion is to cover the metal with a rust-inhibiting paint (**Figure 2**), tape, or plastic coating. Paint works well on aboveground structures, effectively protecting them from the corrosive effects of road salt and acid rain. The metal must remain completely covered, however. All it takes for corrosion to begin is a chip or scratch in the coating, exposing bare metal. Rust may be observed only around the chip, but the corrosion damage may extend under the coating for some distance.

Galvanizing involves coating a metal object, such as a steel water tank, with zinc. Galvanizing can be done either by dipping the steel into a hot vat of molten zinc or by electroplating. You will learn more about electroplating in

galvanizing the process of coating iron or steel with a thin layer of zinc

section 5.18. As the zinc oxidizes, it forms a tough, protective coating that adheres well to the zinc below it. The coating is very resistant to further corrosion, so the underlying steel is protected.

Corrosion-Resistant Metals

Improving the corrosion resistance of metals is another method for preventing corrosion. For example, the steel that is used by the auto industry today has fewer undesirable impurities and more corrosion-resistant additives than ever before. New cars should remain rust-free longer, even in Ontario's challenging winter conditions.

Many other corrosion-resistant alloys have been developed, as well. Some alloys of aluminum, for example, are light, corrosion-resistant, and stronger than steel, making them ideal building materials for airplanes. As you learned in section 5.12, aluminum is corrosion-resistant because it forms a tough, protective oxide layer. Alloys of aluminum and magnesium are also being used to manufacture automobile parts.

In some situations, however, the metal must never rust at all. A patient who needs an artificial hip (**Figure 3**), for example, does not want the implant to start rusting once it is in the body! The implant is made from titanium or high-quality stainless steel, which is an alloy of iron, chromium, and nickel.

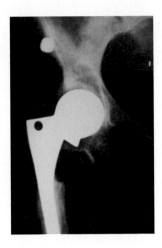

Figure 3
This artificial hip must be made of a corrosion-resistant metal, such as titanium, because it is in constant contact with body fluids that contain dissolved electrolytes.

Cathodic Protection

Cathodic protection works by forcing the metal object you wish to protect into becoming the cathode of a corrosion cell. Cathodic protection can be done in two different ways: with a sacrificial anode or by impressed current.

Sacrificial Anode

Galvanized steel is one common use of a **sacrificial anode**. You have just learned that the surface layer of zinc forms a protective layer of oxide. Even if the galvanizing layer is scratched, however, the steel remains protected because zinc is oxidized more readily than iron (**Figure 4**). Therefore, zinc becomes the anode of the corrosion cell instead of iron. The zinc is very gradually corroded away while protecting the iron, which is why it is called a "sacrificial" anode.

Covering the entire surface of the metal with a sacrificial anode is not always desirable or necessary. A metal object remains protected as long as

cathodic protection a form of corrosion prevention in which the metal being protected is forced to be the cathode of a cell, using either impressed current or a sacrificial anode

sacrificial anode a form of cathodic corrosion protection in which a metal that is more easily oxidized than iron (or some other structural metal) is electrically connected to an iron object

Metal Detector Alert
Passengers with metal implants must notify airline personnel before passing through airport metal detectors. Otherwise, the implants may set off the detectors.

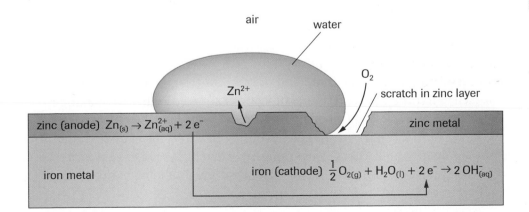

air   water
Zn^{2+}
O_2
scratch in zinc layer
zinc (anode) $Zn_{(s)} \rightarrow Zn^{2+}_{(aq)} + 2\,e^-$ zinc metal
iron metal   iron (cathode) $\frac{1}{2}O_{2(g)} + H_2O_{(l)} + 2\,e^- \rightarrow 2\,OH^-_{(aq)}$

Figure 4
Zinc is a suitable sacrificial anode for iron because it oxidizes more readily. Electrons that are released by the oxidation of zinc protect exposed iron from oxidation.

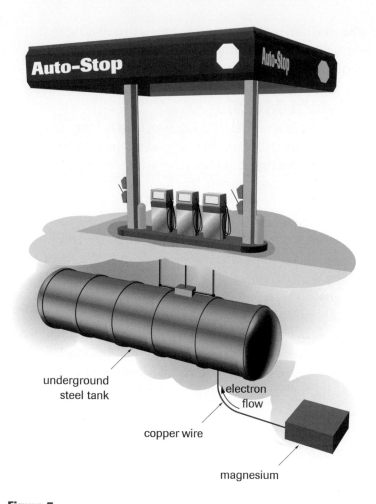

Figure 5
Attaching a block of magnesium to an underground gas tank minimizes corrosion of the metal tank. The magnesium is oxidized, becoming a sacrificial anode.

impressed current a form of cathodic corrosion protection in which the metal object to be protected is attached to the negative terminal of a power source, making the object the cathode in a cell

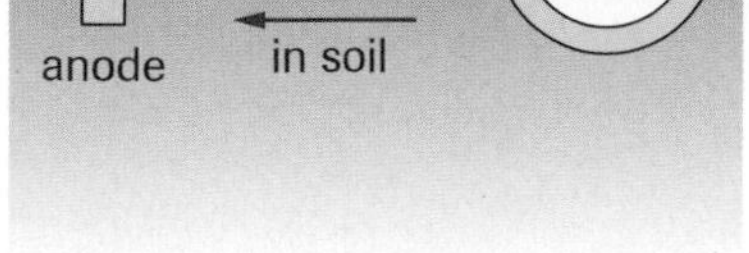

Figure 6
Cathodic protection of a structure, such as a gas pipeline, by impressed current requires a continuous supply of electricity.

there is an electrical connection between it and the sacrificial anode. For example, the block of magnesium in **Figure 5** protects the storage tank because magnesium atoms are easily oxidized, sending their electrons through the connecting wire to the steel tank. Having a continuous supply of electrons prevents the iron atoms in the tank from losing their own electrons to become oxidized. The magnesium block needs to be replaced periodically to maintain protection. Storage tanks, pipelines, and ships are protected using this method.

Impressed Current

Impressed current is another method of cathodic protection. The steel object is attached to the negative terminal of a power source (**Figure 6**), making it the cathode of a cell. Continually pumping electrons into the object prevents the metal from losing electrons and becoming oxidized. The anode is often an inert conductor, such as graphite. Impressed current is used to protect a significant portion of the Canadian oil and natural gas pipeline network.

SUMMARY *Preventing Corrosion*

Corrosion can be prevented by

- protective coatings: for example, paint (isolates the metal from the environment) and corrosion-resistant metal coatings, such as zinc (galvanizing)
- corrosion-resistant metals: for example, stainless steel and titanium
- cathodic protection (making the object to be protected the cathode of a corrosion cell): for example, sacrificial anodes (the metal object is connected to a sacrificial anode—an active metal, such as magnesium or zinc, that oxidizes more readily than the metal object to be protected) and impressed current (a power source is used to pump electrons into the metal object)

▶ Section 5.15 Questions

Understanding Concepts

1. Why does painting prevent corrosion?

2. (a) According to the activity series, which should oxidize more readily, aluminum or iron?
   (b) Describe why iron corrodes much more quickly than aluminum.

3. (a) What is a sacrificial anode?
   (b) Consider the metals gold, silver, and iron. Which metal would be a suitable sacrificial anode for copper? Explain your choice.

4. (a) How are impressed current and sacrificial anodes similar in how they prevent corrosion?
   (b) How are they different?

5. Identify one physical property and one chemical property that steel should have if it is used to make dental braces.

Applying Inquiry Skills

6. (a) Compare the appearance of the nails in **Figure 7**.
   (b) What method of corrosion protection does **Figure 7** illustrate?
   (c) Briefly describe how this method works.
   (d) What are the disadvantages of this method?

7. (a) Compare the appearance of the nails in **Figure 8**.
   (b) What method of corrosion protection does **Figure 8** illustrate?
   (c) Briefly describe how this method works.
   (d) What are the disadvantages of this method?

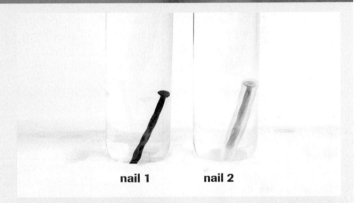

nail 1   **nail 2**

Figure 7
Nail 1 is submerged in salt water. Nail 2 is covered in grease and then submerged in salt water. The photo shows what the nails look like after a few days.

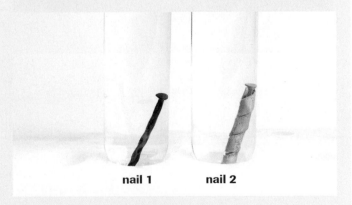

nail 1   **nail 2**

Figure 8
Both nails are submerged in salt water. Nail 1 has no corrosion protection. Nail 2 is wrapped in a strip of zinc. The photo shows the nails' appearance after a few days.

Making Connections

8. (a) Many people claim that applying an underbody oil spray once a year is the best method for preventing corrosion on a car. Describe how the oil prevents corrosion.

   (b) What are some of the disadvantages of oil undercoats?

   (c) Call or visit a service centre that applies "no drip" undercoats. Investigate and describe why these methods may be better than oil undercoats.

9. Galvalume is a new type of steel that has excellent corrosion resistance. Research and prepare a brief summary of

   (a) the ingredients in a Galvalume coating

   (b) what Galvalume is used for

 www.science.nelson.com

10. In order to remain competitive in the global steel market, major Canadian steel manufacturers spend millions of dollars annually on testing and improving the quality of their products. Steel manufacturers employ materials engineering technologists to perform routine tests on the steel being produced (**Figure 9**).

Figure 9
Metals are subjected to extreme conditions during many manufacturing processes.

(a) Research some of the tests that these technologists regularly perform.

(b) Are most of the tests that are conducted by technologists physical or chemical tests? Explain your answer.

(c) Research the career of a materials engineering technologist or a related career in metallurgy. Write a report that includes
   • the personal skills required;
   • the educational background and length of study required to work in this field;
   • two examples of programs offered by educational institutions that could lead to work in this field;
   • salary and working conditions.

 www.science.nelson.com

Extension

11. Many boats spend most of their time in salt water. The metal components of these boats, particularly the propellers, are prime candidates for corrosion. A device called the "Mercathode Protection System" has been developed to reduce the rate of corrosion. Research this device on the Internet, and prepare an illustrated report of your findings.

 www.science.nelson.com

Impressed Current

As you discovered in section 5.15, there are several methods for slowing or stopping corrosion. One of these methods is called cathodic protection. It involves supplying electrons continuously to the metal to be protected, so that the metal becomes the cathode of a corrosion cell. The electron flow prevents the metal from becoming oxidized.

Buried pipelines are very susceptible to corrosion. Leaks in pipes that carry petroleum products or other chemicals would be extremely dangerous. Many pipeline companies hire corrosion engineers and technicians to install and monitor systems that reduce the rate of corrosion. A corrosion technician is involved in designing, installing, and maintaining cathodic protection systems for pipelines and terminal facilities, as well as maintaining corrosion control records. To prepare suitably qualified engineering technicians, several colleges provide courses leading to an Associate Degree in Corrosion Control.

In this activity, you will investigate one method of cathodic protection called *impressed current* (see **Figure 6** in section 5.15). You will submerge two nails in a sodium chloride solution that contains indicators for both hydroxide ions and iron(II) ions (as in section 5.14). One nail will have electrons pumped into it (an impressed current). The other nail will be a control.

Materials

eye protection
lab apron
0.1-mol/L sodium chloride solution, $NaCl_{(aq)}$
dropper bottle of 0.1-mol/L solution of potassium hexacyanoferrate(III) indicator
dropper bottle of phenolphthalein indicator
2 small, identical nails
graphite rod
test tube and stopper
test-tube rack
retort stand
utility clamp
U-tube
funnel
1.5-V battery and holder
connecting wires

Procedure

1. Add two drops of hexacyanoferrate(III) indicator and two drops of phenolphthalein to a test tube.

 The indicators are both poisonous. Wash your hands thoroughly at the end of the experiment.

 Phenolphthalein is also flammable. Keep it away from open flames.

2. Add one of the nails to the test tube.

3. Pour enough salt water into the test tube to submerge the nail. Stopper the test tube, and tip it to mix its contents thoroughly. This test tube will be the control. Place the test tube in the rack, and allow it to stand undisturbed.

4. Carefully clamp the U-tube to the retort stand.

5. Use the funnel to fill the U-tube with the salt solution, to within 2 or 3 cm of the top.

6. Add two drops of hexacyanoferrate(III) indicator to each side of the U-tube.

7. Add two drops of phenolphthalein to each side of the U-tube.

8. Attach the graphite rod to the positive terminal of the battery.

9. Attach a nail to the negative terminal.

10. Add the nail and graphite rod to opposite sides of the U-tube (**Figure 1**, on the next page).

11. Allow the cell to operate for about 1 min. During this time, carefully observe the nail and graphite rod for evidence of chemical reactions. Record your observations.

12. Disconnect the battery.

13. Compare the solutions in the test tube and the U-tube. Record your observations.

14. Dispose of the solutions in the sink, with lots of running water.

Analysis

(a) Interpret your observations. Did either of the nails corrode? If so, what evidence of corrosion do you have?

(b) Account for any difference you observed in the two nails.

(c) Here are two possible reduction reactions for what you observed on the nail side of the U-tube.

$$2\,H^+_{(aq)} + 2\,e^- \rightarrow H_{2(g)}$$

$$\frac{1}{2}\,O_{2(g)} + H_2O_{(l)} + 2\,e^- \rightarrow 2\,OH^-_{(aq)}$$

Use your observations to determine which reaction(s) did occur.

Synthesis

(d) What would you have observed if you had used distilled water instead of salt water? Why?

(e) Suppose that you replaced the graphite rod (the anode) with another nail. What reaction would occur at the anode? What colour change would you expect to see at the anode? Why?

(f) What would happen to the nail in the U-tube if there were a power failure? What colour changes would you expect to see?

(g) Suggest one disadvantage of this method of corrosion protection.

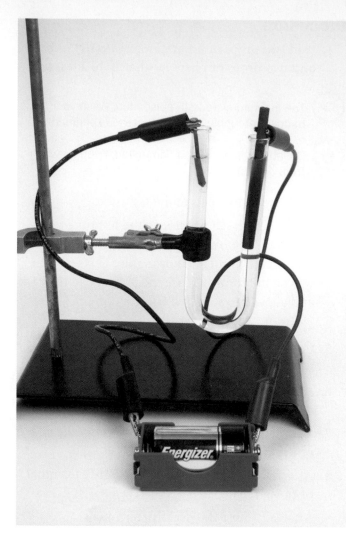

Figure 1
Check that the graphite rod and the nail are attached to the correct terminals of the power supply.

Most of the chemical reactions that you've seen so far in this unit occur on their own. In other words, these reactions occur spontaneously. Some chemical reactions, however, occur only if they are given an energy push. These reactions are called **nonspontaneous reactions**. In the following Try This Activity, you will use electrical energy to "push" a nonspontaneous redox reaction into occurring.

nonspontaneous reaction a reaction that proceeds only with outside assistance

---

▶ **TRY THIS** activity   *Pencil Plating*

Instead of using a chemical reaction to produce electrical energy, you will use electrical energy to make a chemical reaction occur.

Materials: eye protection; lab apron; petri dish; 9-V battery; 2 connecting wires; 2 short pencils, sharpened to a point at both ends; 0.1-mol/L copper(II) sulfate solution, $CuSO_{4(aq)}$

 The copper(II) sulfate solution is toxic and a skin irritant. Wash your hands thoroughly after this activity.

1. Half-fill the petri dish with the copper(II) sulfate solution.
2. Using the connecting wires, connect one graphite end of one pencil to the positive terminal of the battery, and one graphite end of the other pencil to the negative terminal. The graphite in each pencil is now an electrode.
3. Dip just the tip of the free end of each pencil into the solution (**Figure 1**) for about 1 min. Record your observations.
4. Dispose of the contents of your petri dish as directed by your teacher.

(a) What evidence of chemical reactions did you observe?

(b) At which electrode (positive or negative) did you see a colour change?

(c) What do you think the reddish material might be?

(d) What colour change would you expect to observe if you used a zinc sulfate solution instead? Why?

(e) The cell that you made in this activity had two electrodes and an electrolyte—just like a galvanic cell. Describe two ways in which this cell is different from a galvanic cell.

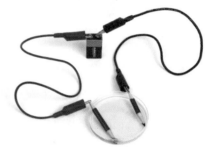

Figure 1
Make sure that the tips of the two pencils do not touch.

In a galvanic cell, redox reactions are used to produce electrical energy. The reverse is also possible. Electrical energy may be used to make a nonspontaneous redox reaction occur. This process is called **electrolysis**.

electrolysis the process of supplying electrical energy to force a nonspontaneous redox reaction to occur

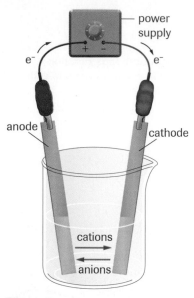

Figure 2
In a typical electrolytic cell, electrical energy is used to push electrons from the anode to the cathode.

Figure 3
The power supply in an electrolytic cell is like this water pump. Effort is required to pull water up from underground. Similarly, energy is needed to push electrons in a direction in which they do not naturally want to go.

"Forcing" a Redox Reaction

A cell in which electrolysis is occurring is called an **electrolytic cell** (**Figure 2**). The power supply acts as an "electron pump," pushing electrons toward the cathode while pulling electrons away from the anode. As a result, the anode becomes positively charged and the cathode becomes negatively charged. Positive ions (cations) in the solution move toward the cathode, where a reduction reaction occurs. Similarly, negative ions (anions) move toward the anode, where oxidation occurs. Without continuous electrical "help" from the power supply, the cell reactions would not occur (**Figure 3**).

As an example, consider what happens during the electrolysis of a copper(II) sulfate solution, $CuSO_{4(aq)}$, using inert (nonreacting) graphite electrodes. Electrons are drawn out of the anode and pumped toward the cathode, making the anode positively charged and the cathode negatively charged. In response, negatively charged sulfate ions, $SO_{4(aq)}^{2-}$, move toward the positively charged anode. Copper(II) ions, $Cu_{(aq)}^{2+}$, are attracted to the negatively charged cathode.

There are two possible reduction reactions that can occur at the cathode:

1. reduction of copper(II) ions: $Cu_{(aq)}^{2+} + 2\,e^- \rightarrow Cu_{(s)}$

2. reduction of water: $2\,H_2O_{(l)} + 2\,e^- \rightarrow H_{2(g)} + 2\,OH_{(aq)}^-$

When the experiment is conducted, the familiar reddish colour of copper is observed at the cathode. Thus, you can assume that reaction 1 occurs. (Although it is possible to predict accurately which reaction occurs in each half-cell, the skills required go well beyond the scope of this textbook.)

Similarly, there are two possible oxidation reactions that can occur at the anode:

1. oxidation of sulfate: $2\,SO_{4(aq)}^{2-} \rightarrow S_2O_{8(aq)}^{2-} + 2\,e^-$

2. oxidation of water: $H_2O_{(l)} \rightarrow \frac{1}{2}\,O_{2(g)} + 2\,H_{(aq)}^+ + 2\,e^-$

Since bubbles of gas were observed at the anode, you can assume that reaction 2 occurs. Therefore, the following two half-reactions occur in this cell:

Cathode half-reaction: $Cu_{(aq)}^{2+} + \cancel{2\,e^-} \rightarrow Cu_{(s)}$ (reduction)

Anode half-reaction: $H_2O_{(l)} \rightarrow \frac{1}{2}\,O_{2(g)} + 2\,H_{(aq)}^+ + \cancel{2\,e^-}$ (oxidation)

Overall cell reaction: $Cu_{(aq)}^{2+} + H_2O_{(l)} \rightarrow Cu_{(s)} + \frac{1}{2}\,O_{2(g)} + 2\,H_{(aq)}^+$

Galvanic and Electrolytic Cells Working Together

A familiar example of a galvanic cell and an electrolytic cell working together is a car battery alternator system. Starting a car draws electricity from the battery, using up some of the chemicals in the battery. The battery would completely discharge if it weren't for the car's built-in recharger: the alternator. Once the engine is running, electrical energy from the alternator is pumped into the battery. There it is used to reverse the cell reaction. These two steps can be summarized by a two-way chemical equation:

$$\text{Pb}_{(s)} + \text{PbO}_{2(s)} + \text{H}_2\text{SO}_{4(aq)} \underset{\text{recharging}}{\overset{\text{discharging}}{\rightleftharpoons}} 2\,\text{PbSO}_{4(s)} + 2\,\text{H}_2\text{O}_{(l)}$$

A car with a discharged battery requires an electrical "boost" from another vehicle to push the battery reaction to the left.

To summarize,

- discharging occurs spontaneously and releases electrical energy;
- recharging is nonspontaneous and requires electrical energy.

Comparing Galvanic and Electrolytic Cells

The similarities and differences between galvanic and electrolytic cells are summarized in **Table 1**.

Table 1   Galvanic and Electrolytic Cells

Type of cell	galvanic cell	electrolytic cell
Diagram	wire; salt bridge, NaNO₃(aq); zinc electrode; copper electrode; zinc nitrate electrolyte solution, Zn(NO₃)₂(aq); copper (II) nitrate electrolyte solution, Cu(NO₃)₂(aq)	power supply; e⁻ e⁻; anode; cathode; copper(II) nitrate electrolyte solution, Cu(NO₃)₂(aq)
Cell reaction	spontaneous	nonspontaneous
Energy	produces electrical energy	requires electrical energy
Energy changes	converts chemical energy into electrical energy	converts electrical energy into chemical energy
Where reduction occurs	cathode	cathode
Where oxidation occurs	anode	anode
Direction of electron flow	anode to cathode	anode to cathode
Direction of ion flow	anions to anode; cations to cathode	anions to anode; cations to cathode

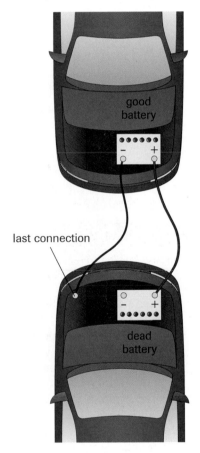

good battery

last connection

dead battery

Understanding Concepts

1. (a) What does it mean when a chemical reaction is described as being nonspontaneous?
   (b) How can a nonspontaneous reaction be forced to occur?

2. Compare the energy changes of a galvanic cell with the energy changes of an electrolytic cell.

3. Describe why recharging a battery is an example of electrolysis.

4. In which type of cell (a galvanic cell or an electrolytic cell) is it necessary to separate the reactants? Why?

Applying Inquiry Skills

5. (a) Draw a diagram of an electrolytic cell that contains a zinc sulfate solution and two graphite (carbon) electrodes. (Assume that the electrodes do not react.) Label the left terminal of the power supply as positive and the right terminal as negative.
   (b) Label the anode and the cathode.
   (c) Show the direction of electron flow in the cell.
   (d) Show the direction in which the zinc ions, $Zn^{2+}_{(aq)}$, and the sulfate ions, $SO_{4(aq)}^{2-}$, move.
   (e) During the operation of the cell, the colour of the submerged portion of the cathode changes to silver. Write the chemical equation for the cathode half-cell reaction.

6. During the electrolysis of tin(II) chloride solution, $SnCl_{2(aq)}$, a silver-coloured metal is deposited at the cathode. A gas with a bleach-like odour is noticed at the anode.
   (a) Name the metal that is produced in this reaction.
   (b) Name the gas that is produced at the anode.
   (c) Use these observations to write the half-reactions that occur at the anode and the cathode of the cell.

7. Consider the diagram of the galvanic cell in **Table 1**, on the previous page. If the positive terminal of a 9-V battery were attached to the copper electrode, and the negative terminal were attached to the zinc electrode, what changes would you expect to see at each electrode? Why?

Making Connections

8. Some restorers of jewellery and antique objects use electrolysis to clean corroded silver (**Figure 4**). Research the process they use. Prepare a brief summary of your findings.

GO www.science.nelson.com

Extension

9. **Figure 5** shows a cell that is used to electrolyze potassium iodide solution.
   (a) Describe the direction of ion flow during the operation of the cell.
   (b) During electrolysis, the solution around the anode becomes a reddish-brown colour. Use this evidence to write a chemical equation for the anode half-reaction.
   (c) The reduction of water occurs at the cathode:
   $$2\,H_2O_{(l)} + 2\,e^- \rightarrow H_{2(g)} + 2\,OH^-_{(aq)}$$
   What changes would you expect to see at the surface of the cathode?
   (d) Describe a chemical test that could be done to prove that hydroxide ions are being produced at the cathode.
   (e) Write an equation for the overall cell reaction in the cell.

Figure 4
Electrolysis can be used to remove the tarnish from silver objects.

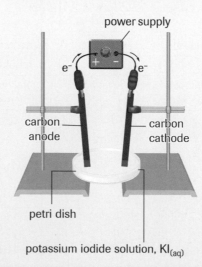

Figure 5
Electrolysis of a potassium iodide solution, $KI_{(aq)}$

As you will see in this section, there are far more applications of electrolysis than recharging batteries. Among other things, electrolysis is used to purify metals and to coat objects with a thin layer of metal. Electrolysis is a multi-billion-dollar industry in Canada, and many Canadian companies are world leaders in the industry.

Aluminum Production

Mention the word "aluminum," and an image of the silvery foil we use to wrap baked potatoes quickly comes to mind. There's more to aluminum, however. Barbecues, car parts, airplanes, and construction materials (**Figure 1**) are just a few of the many products that are made of aluminum.

Here are some of the reasons why aluminum is so versatile:

- It is abundant. Aluminum is the third most abundant element on Earth, behind only oxygen and silicon.
- It can be produced inexpensively on a large scale.
- It is lightweight and can be formed into almost any shape.
- It can be mixed with other metals to form lightweight corrosion-resistant alloys that are stronger than steel.

Figure 1
Aluminum is used to manufacture high-tech panels for building exteriors.

Aluminum is found naturally in a reddish mineral called bauxite. Aluminum production involves processing bauxite into a white powder called alumina (aluminum oxide, Al_2O_3) (**Figure 2**). Alumina is then electrolyzed to aluminum metal.

Because the ionic bond between aluminum and oxygen in aluminum oxide is so strong, separating aluminum from its oxide is extremely difficult. Early attempts to displace aluminum from its oxide using potassium (one of the most reactive metals) met with only limited success. A breakthrough finally came in 1886 when a 22-year-old American, Charles Martin Hall, and a 23-year-old Frenchman, Paul Héroult, independently developed the same electrolytic process for producing aluminum. The Hall-Héroult process, as it became known, is still in use today.

The Hall–Héroult Process

Hall and Héroult's breakthrough came when they discovered that aluminum oxide dissolves in a solvent called cryolite, Na_3AlF_6, at 1000°C, well below its melting point of 2072°C. As it dissolves, aluminum oxide dissociates into its

Figure 2
Processing aluminum involves two steps. First red bauxite is processed into white aluminum oxide (above). Then aluminum oxide is electrolyzed into silver-coloured aluminum.

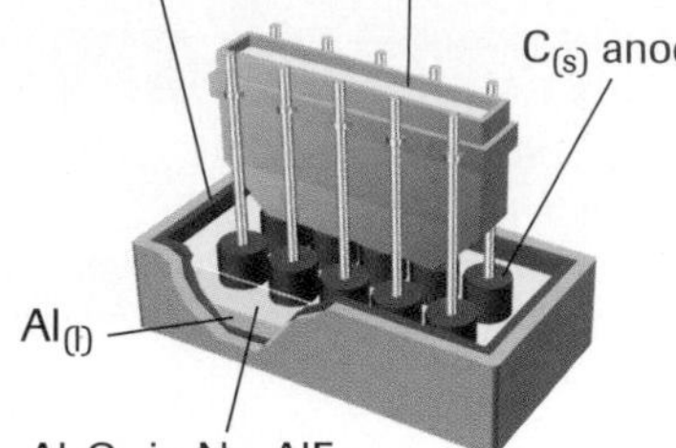

alumina, $Al_2O_{3(s)}$, in hopper

$C_{(s)}$ cathode
(lining of cell)

$C_{(s)}$ anode

$Al_{(l)}$

Al_2O_3 in $Na_3AlF_{6(l)}$
electrolyte

Figure 3
This diagram shows the Hall-Héroult cell for the production of aluminum. Liquid aluminum, produced at the cathode, sinks to the bottom of the cell, where it is drained off.

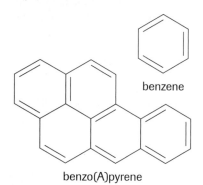

benzene

benzo(A)pyrene

Figure 4
PAHs such as benzo(A)pyrene are made up of several six-sided carbon rings joined together. Each ring is based on an organic molecule called benzene, C_6H_6. Some PAHs, like benzo(A)pyrene, are carcinogenic (cancer-causing).

ions, aluminum, Al^{3+}, and oxide, O^{2-}. The electrolysis cell that is used in the Hall-Héroult process is a steel container lined with carbon, which serves as the cathode (**Figure 3**). Carbon blocks suspended in the electrolyte are the anode. When the power is turned on, aluminum ions are reduced to liquid aluminum. The liquid aluminum sinks to the bottom of the container and is drained away. At the anode, oxide ions are oxidized to oxygen gas, $O_{2(g)}$, and released. The half-cell reactions for this process are given below.

Cathode half-reaction: $\qquad Al^{3+}_{(cryolite)} + 3\ e^- \rightarrow Al_{(l)}$

Anode half-reaction: $\qquad\qquad 2\ O^{2-}_{(cryolite)} \rightarrow O_{2(g)} + 4\ e^-$

Extremely high currents (up to 250 000 A) are required to manufacture aluminum. As a result, the aluminum industry is one of Canada's largest consumers of electricity. For this reason, Canadian manufacturers of aluminum locate their operations near sources of inexpensive hydroelectric power rather than where bauxite is mined. About 14 kW·h of electricity is needed to produce 1 kg of aluminum. This much electricity is enough to operate a 60-W light bulb nonstop for almost ten days!

Aluminum Production and Air Pollution
There are three major air pollutants associated with aluminum production: carbon dioxide, sulfur dioxide, and polyaromatic hydrocarbons.

Carbon Dioxide: A Greenhouse Gas
Oxygen gas, produced at the anode during the Hall-Héroult process, oxidizes the carbon blocks of the anode into carbon dioxide. Consequently, the anodes are consumed and need to be replaced frequently. Carbon dioxide is one the gases that have been linked to global warming through the greenhouse effect.

Sulfur Dioxide
The anodes that are used in the Hall-Héroult cell are made from coke, a pure form of carbon. Coke is produced by burning (oxidizing) coal in specially designed furnaces. During the process, sulfur impurities in the coal are also oxidized to sulfur dioxide. Even with the best sulfur-removal technology, some sulfur dioxide escapes into the atmosphere. There it combines with water vapour to form acid rain (section 4.13).

Polyaromatic Hydrocarbons (PAHs)
The carbon anodes that are used in the Hall-Héroult process are made by pressing large chunks of coke together with a thick, tar-like liquid under heat. The tar helps to bind the coke chunks together. It contains small amounts of a class of organic compounds called polyaromatic hydrocarbons, or PAHs (**Figure 4**). PAHs are commonly found in smoke, soot, and engine exhaust. As the anodes are oxidized, PAHs from the tar are released into the environment. PAH emissions are a concern because some PAHs have been linked to lung cancer in humans. As well, PAHs are harmful to aquatic life. Aluminum production is the major industrial source of PAHs. The largest natural source of PAHs is forest fires.

Electrorefining

The extraction of copper from its ore can yield a product that is up to 99% pure copper. This degree of purity is good enough for household plumbing. It is not good enough for electrical applications, such as the components of computer chips (**Figure 5**), however, because even tiny amounts of impurities decrease the electrical conductivity of copper.

Copper, gold, and nickel are just three of the metals that are refined to a higher degree of purity using a form of electrolysis called **electrorefining**. In a copper electrorefining cell, for example, an impure slab of copper is the anode and a thin sheet of pure copper is the cathode. A laboratory version of this cell is shown in **Figure 6**. The electrolyte is an acidic solution of copper(II) sulfate. When the power is turned on, most of the impure copper anode, including many of its metallic impurities, is gradually oxidized into ions.

Anode half-reactions:

$$Cu_{(s)} \rightarrow Cu^{2+}_{(aq)} + 2\,e^-$$

$$Zn_{(s)} \rightarrow Zn^{2+}_{(aq)} + 2\,e^-$$

$$Fe_{(s)} \rightarrow Fe^{2+}_{(aq)} + 2\,e^-$$

Cathode half-reaction:

$$Cu^{2+}_{(aq)} + 2\,e^- \rightarrow Cu_{(s)}$$

Unreactive metal impurities, such as gold, silver, and platinum, fall to the bottom of the cell as the anode disintegrates. These impurities form a sludge, called *anode mud*, that is periodically drained and processed to extract valuable metals. By carefully controlling the voltage in the cell, only the copper ions are reduced at the cathode (**Figure 7**, on the next page). Other impurities, such as zinc and iron, remain as ions in the solution.

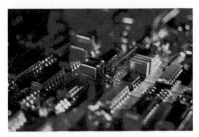

Figure 5
Extremely pure copper is used to manufacture some types of computer chips.

electrorefining the process of increasing the purity of a metal using an electrolytic cell

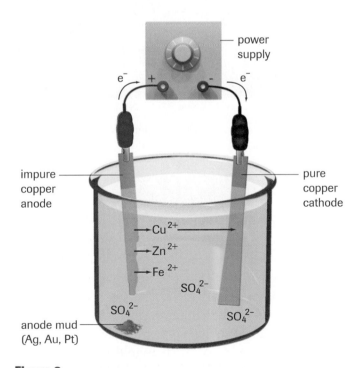

Figure 6
A copper electrorefining cell: As the impure copper anode dissolves away, the copper ions plate onto the pure copper cathode, causing it to become thicker.

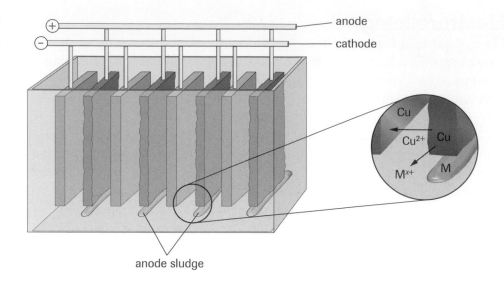

Figure 7
The photo shows an electrolytic tank house, where hundreds of copper sheets are electrolyzed at a time.

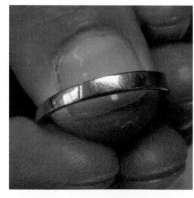

Figure 8
Gold plating can be used to manufacture attractive gold jewellery at a fraction of the cost of using solid 14-karat or 18-karat gold.

electroplating depositing a layer of metal onto another object at the cathode of an electrolytic cell

Electroplating

Metals such as silver, gold, and chromium are popular because of their corrosion resistance and attractive finishes. Making jewellery entirely from these metals, however, is very expensive and, in some cases, undesirable. Pure gold, for example, is too soft for jewellery. An attractive but inexpensive ring can be made by **electroplating** a thin coating of gold onto a steel or aluminum ring (**Figure 8**). Electroplating is a multi-billion dollar industry in Canada.

In an electroplating cell, the object to be coated is the cathode. The solution in the cell contains ions of the metal to be plated, such as gold. As the cell in **Figure 9** operates, for example, the gold ions gain three electrons each, forming a thin layer of solid gold on the ring. Gold ions that are released from the anode replace those that are used.

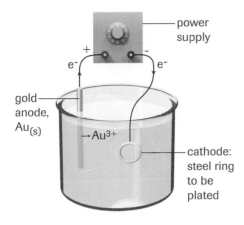

power supply

e⁻ ← → e⁻

gold anode, Au₍ₛ₎

→Au³⁺

cathode: steel ring to be plated

Figure 9
In gold electroplating, gold ions from the solution are reduced to metallic gold on the object. The gold anode re-supplies the solution with more ions.

Many common objects are electroplated. The Canadian "loonie," for example, is made of a nickel core electroplated with a thin coating of bronze (**Figure 10**).

Other common examples of electroplated metals include

- "tin" cans (tin plated onto steel cans);
- costume jewellery (gold or silver plated onto nickel or steel);
- silver-plated cutlery (silver plated onto steel);
- chrome plating (chromium plated onto steel).

Figure 10
Since 1998, the loonie has been 91.5% nickel with 8.5% bronze plating (a mixture of tin and copper).

The Environmental Impact of Electrolysis

Every industrial process has an impact on the environment. Industries that use electrolysis are no different. Two of the most serious effects result from the amount of power they consume and the types of chemicals they employ.

Power Consumption and Acid Rain

Aluminum production, electrorefining, and electroplating all consume large quantities of electricity. In Ontario, approximately one-third of the electricity that is generated comes from the burning of fossil fuels. Some fossil fuels contain sulfur impurities. These impurities are oxidized to sulfur dioxide

Figure 11
Chromium(VI) compounds are used for the chrome finish on cars and motorcycles.

when the fossil fuels are burned. Therefore, the consumption of electricity by metal plating and refining industries contributes indirectly to acid rain.

Chemical Hazards

Industrial electroplating often involves hazardous chemicals and working conditions. Here are two examples:

- Some of the best electroplating finishes are produced using metal cyanide compounds. These compounds are extremely poisonous if swallowed. Furthermore, if there is any acidity in the plating tank, deadly hydrogen cyanide gas is released. There has been considerable research into eliminating cyanides from the plating industry.

- Many of the metals that are used are either toxic or carcinogenic (cancer-causing). For example, chromium(VI) compounds (**Figure 11**) are known carcinogens. It is not surprising that there are strict rules governing the transport, handling, and disposal of the chemicals that are used in this industry. These rules have partly come about because of serious industrial accidents or chemical spills that have happened in the past.

▶ Section 5.18 Questions

Understanding Concepts

1. Describe two reasons why aluminum is an extremely versatile metal.

2. What physical property makes aluminum oxide extremely difficult to electrolyze?

3. Which two products are produced during the Hall-Héroult process? At which electrode is each product formed?

4. Why is Canada a major world producer of aluminum despite not having any bauxite—the mineral from which aluminum is extracted?

5. What environmental problem could be eliminated if the carbon anodes that are used to manufacture aluminum could be made from one piece of carbon?

6. (a) What is electrorefining?
   (b) Why is the electrorefining of copper necessary?

7. Why is the anode mud in an electrorefining tank processed?

8. (a) Sketch and label a cell that could be used to electrorefine nickel. Suggest a possible electrolyte.
   (b) At which electrode does the pure metal form?
   (c) At which electrode does oxidation occur?

9. Galvanized steel can be produced by electroplating steel with zinc. Sketch a diagram of a cell that could be used to galvanize steel. Label the anode, cathode, and electrolyte.

10. Why is a solid gold anode always used in a gold-plating cell?

11. Identify two chemical hazards that might be encountered by employees in an electroplating plant.

12. What is the relationship between industrial electrolytic processes (such as electroplating and electrorefining metals) and acid rain?

▶ *Section 5.18* **Questions** *continued*

Making Connections

13. Auto manufacturers are using more parts that contain aluminum and magnesium than ever before.
    (a) What effects do magnesium and aluminum parts have on the mass and fuel efficiency of a car?
    (b) Discuss the impact on the environment of using aluminum parts in vehicles.

14. Why should we recycle as much aluminum as possible?

15. Search the Internet for the Material Safety Data Sheet (MSDS) for chromium(VI) oxide, commonly called chromium trioxide. You will need this document to answer the following questions.
    (a) Imagine that you are a new employee in a chrome electroplating plant. You need to become familiar with the properties of chromium trioxide. Identify three possible health effects of exposure to chromium trioxide that are given on the MSDS.
    (b) The toxicity of a chemical is expressed by an LD_{50} rating found on its MSDS. An LD_{50} rating is the quantity of a substance that researchers estimate would kill 50% of a particular species, such as rats, if exposed to this quantity of the substance. What is the LD_{50} rating for chromium trioxide?
    (c) Rubbing alcohol is poisonous, having an LD_{50} rating of 5800 mg/kg. Some Canadians have either died or suffered serious injuries from accidentally drinking rubbing alcohol. Chromium trioxide, however, is even more toxic. Calculate approximately how much more toxic chromium trioxide is, compared with rubbing alcohol.

     www.science.nelson.com

Extension

16. (a) Vapour deposition is a new technique of coating one metal with another. Research how vapour deposition works.
    (b) Identify some consumer products that are plated using vapour deposition.

    GO www.science.nelson.com

Key Understandings

5.1 Oxidation–Reduction Reactions

- Understand the meanings of the following terms: oxidation, reduction, redox reaction.
- Identify the reactant that is oxidized and the reactant that is reduced in a redox reaction.
- Research information about the career of a film-processing technician.

5.2 Redox Reactions of Nonmetals

- Use oxidation numbers to identify the reactant that is oxidized and the reactant that is reduced in a redox reaction.
- Use oxidation numbers to establish whether or not a reaction is a redox reaction.

5.3 Activity: Developing an Activity Series of Metals

- Interpret your observations from experiments to organize several metals in order of reactivity.
- Safely handle laboratory equipment and materials.

5.4 The Activity Series of Metals

- Understand that metals have different reactivities and that they can be arranged in an activity series.
- Use the activity series to predict the occurrence of single displacement reactions.

5.5 Investigation: Testing the Activity Series

- Use the activity series to predict which metals will react spontaneously with compounds of other metals in solution.
- Safely experiment to test your predictions.

5.6 Investigation: Conductors and Nonconductors

- Categorize a variety of substances as metals, acids, bases, salt solutions, and molecular substances.
- Collect experimental evidence to classify the various categories of matter as conductors or nonconductors of electricity.

5.7 Galvanic Cells

- Name the parts of a galvanic cell, describe what each part does, and explain where oxidation and reduction take place.

- Describe the operation of a galvanic cell in terms of half-cell reactions, direction of electron flow, and direction of ion migration.
- Describe how a galvanic cell produces electricity.

5.8 Activity: Designer Cells

- Build a galvanic cell, and measure its cell potential.
- Determine the advantages and disadvantages of your galvanic cell: for example, power output, safety, portability, rechargeability, and voltage.

5.9 Consumer Cells and Batteries

- Understand that commercial batteries are simply well-designed collections of galvanic cells.
- Understand that any electricity-producing cell contains two different electrodes and at least one electrolyte.

5.10 Explore an Issue: Should Ni–Cd Batteries Be Banned?

- Understand the importance of properly disposing of batteries that contain cadmium.
- Identify available alternatives to Ni–Cd batteries.

5.11 Tech Connect: Hydrogen Fuel Cells

- Explain how a hydrogen fuel cell works.
- Describe the benefits and drawbacks of fuel cells over other forms of power.

5.12 Corrosion

- Describe the chemical reactions that are involved in corrosion.
- Describe how a corroding metal is considered to be a galvanic cell.
- Describe the effects of road salt and acid rain on the process of corrosion.

5.13 Case Study: Piercing Problems

- Become aware of how the reactivity of various metals can cause health problems.
- Identify ways to reduce the hazards of body piercing.

5.14 Investigation: Factors That Affect the Rate of Corrosion

- Design your own mini-experiments to discover how four factors affect the rate of corrosion.

- Safely carry out your procedures, collect and analyze your evidence, and make a statement about the four factors that affect the rate of corrosion.

5.15 Preventing Corrosion

- Identify and explain various techniques that are used to prevent corrosion of metals.
- Suggest ways to reduce the corrosive effects of road salt and acid rain.
- Research the career of a materials engineering technologist.

5.16 Activity: Impressed Current

- Perform an activity to demonstrate how an impressed current can prevent corrosion.

5.17 Electrolytic Cells

- Name the parts of an electrolytic cell, describe what each part does, and explain where oxidation and reduction take place.
- Compare the similarities and differences between galvanic and electrolytic cells.

5.18 Applications of Electrolysis

- Explain the production of aluminum.
- Outline some of the environmental impacts that are associated with aluminum production.
- Explain how electrolytic processes are used in metal refining and in the electroplating industry.
- Evaluate the impact of electrolytic processes on the environment: for example, acid rain and high energy consumption.

▶ MAKE a summary

1. Draw a simple concept map that shows how the following terms are related: oxidation, reduction, oxidation number, electrons.
2. (a) Draw and label one diagram of a simple galvanic cell and another diagram of an electrolytic cell. On your diagrams, indicate the names of the electrodes and the direction of electron flow and ion flow. Identify where oxidation and reduction take place.
   (b) Beneath the galvanic cell, list at least two applications of galvanic cells.
   (c) Beneath the electrolytic cell, list at least three applications of electrolysis. For each application, list the anode, cathode, and electrolyte involved.

Problems You Can Solve

5.1

- Identify the reactant that is oxidized and the reactant that is reduced in a redox reaction.

5.2

- Assign oxidation numbers, and use them to identify a redox reaction.

5.4

- Use the activity series to predict whether a particular redox reaction will occur.

5.7

- Write cell reactions for both halves of a galvanic cell.

Key Terms

5.1
oxidation
reduction
redox reaction

5.2
oxidation number

5.4
activity series of metals

5.7
galvanic cell
spontaneous reaction
half-cell
electrode
anode
cathode
salt bridge
cell potential (voltage)

5.9
battery

primary cell
secondary cell

5.11
fuel cell

5.12
corrosion

5.15
galvanizing
cathodic protection

sacrificial anode
impressed current

5.17
nonspontaneous reaction
electrolysis
electrolytic cell

5.18
electrorefining
electroplating

PERFORMANCE TASK

Building an Electrochemical Invention

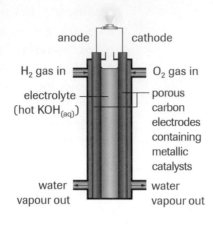

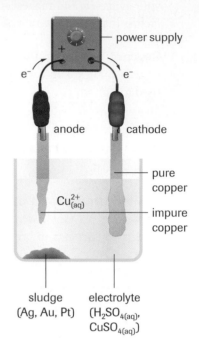

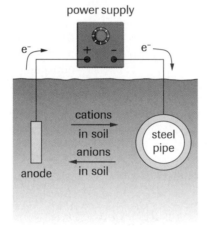

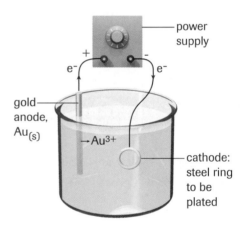

Figure 1

In this unit, you have developed an activity series of metals, tested substances for conductivity, built galvanic cells, investigated corrosion, and electrolyzed solutions. As well, you have learned about many useful electrochemical inventions and processes (**Figure 1**). In this project, you are going to build your own electrochemical invention.

This project can be divided into the following three steps: planning and proposal, development, and marketing.

Planning and Proposal

(a) Preparation

Research and plan an invention that uses some of the electrochemical processes you have studied in this unit. Some possible inventions are

- an unusual battery that can be used to power a light, buzzer, or motor;
- an electroplating procedure;
- a new rust-prevention product or process;
- a solar cell that can be used to power an electroplating process.

Using a variety of resources, research inventions or processes that are similar to your idea for an invention.

(b) Proposal

To refine your idea, prepare a one-page summary that includes

- a diagram of your invention;
- a list of materials required;
- a one- or two-paragraph summary of how your invention will work.

Development

(c) Procedure

Write detailed steps outlining how you will build your invention and how you will test its effectiveness. Include all the necessary safety precautions. Ask your teacher to approve your procedure before you continue.

(d) Building and Testing

Build, test, and refine your invention. Record your observations.

Marketing

(e) Develop and present an advertisement or television infomercial to sell your invention. Your infomercial (or ad) should

- demonstrate your understanding of how the invention works;
- state the benefits of using your invention;
- demonstrate the correct use of grammar, scientific terms, symbols, chemical equations, and SI units.

Understanding Concepts

1. Copy and complete each of the following sentences. (Do not write in this textbook.)
   (a) The gain of electrons is called
   _____.
   (b) The loss of electrons is called
   _____.
   (c) An element's oxidation number _____ when the element is oxidized.
   (d) An element's oxidation number _____ when the element is reduced. (5.1, 5.2)

2. Identify the reactant oxidized and the reactant reduced in each of the following reaction equations:
   (a) $Zn_{(s)} + Cl_{2(g)} \rightarrow ZnCl_{2(s)}$
   (b) $2\,Ca_{(s)} + O_{2(g)} \rightarrow 2\,CaO_{(s)}$
   (c) $Zn_{(s)} + Cu^{2+}_{(aq)} \rightarrow Cu_{(s)} + Zn^{2+}_{(aq)}$
   (d) $2\,AgNO_{3(aq)} + Zn_{(s)} \rightarrow Zn(NO_3)_{2(aq)} + 2\,Ag_{(s)}$
   (e) $Fe_2O_{3(s)} + 2\,Al_{(s)} \rightarrow 2\,Fe_{(s)} + Al_2O_{3(s)}$
   (f) $2\,H_{2(g)} + O_{2(g)} \rightarrow 2\,H_2O_{(g)}$
   (g) $2\,Mg_{(s)} + CO_{2(g)} \rightarrow 2\,MgO_{(s)} + C_{(s)}$
   (h) $Cl_{2(g)} + 2\,KI_{(aq)} \rightarrow I_{2(aq)} + 2KCl_{(aq)}$ (5.1)

3. Determine the charge or oxidation number of the underlined element in each of the following compounds or ions:
   (a) $\underline{Ca}Cl_2$ (e) $\underline{Fe}(NO_3)_2$
   (b) $H_2\underline{S}$ (f) $\underline{Fe}_2O_3$
   (c) $\underline{Pb}O_2$ (g) $\underline{S}O_3^{2-}$
   (d) $Zn\underline{S}O_4$ (h) $\underline{C}_2O_4^{2-}$ (5.2)

4. Is it possible for oxidation to occur without reduction? Justify your answer. (5.1)

5. Use oxidation numbers to determine whether or not each of the following reactions is a redox reaction:
   (a) $2\,H_2S_{(g)} + O_{2(g)} \rightarrow 2\,S_{(s)} + 2\,H_2O_{(g)}$
   (b) $CH_{4(g)} + 2\,O_{2(g)} \rightarrow CO_{2(g)} + 2\,H_2O_{(g)}$
   (c) $ZnO_{(s)} + 2\,HCl_{(aq)} \rightarrow ZnCl_{2(aq)} + H_2O_{(l)}$
   (d) $2\,PbO_{(s)} + PbS_{(s)} \rightarrow 3\,Pb_{(s)} + SO_{2(g)}$ (5.2)

6. Use the activity series to predict whether or not each of the following reactions occurs. Write a balanced chemical equation if a reaction does occur. Write "no reaction" if a reaction does not occur.
   (a) $Mg_{(s)} + Cu(NO_3)_{2(aq)} \rightarrow$
   (b) $Fe_{(s)} + Na_2SO_{4(aq)} \rightarrow$
   (c) $Ni_{(s)} + AgNO_{3(aq)} \rightarrow$
   (d) $Ca_{(s)} + 2\,HCl_{(aq)} \rightarrow$ (5.4)

7. Use the activity series to predict what happens if a steel (iron) spoon is used to stir a solution of copper(II) sulfate, $CuSO_{4(aq)}$. Assume that $Fe^{2+}_{(aq)}$ ions are formed. (5.4)

8. Because aluminum is high in the activity series, you know that it is a reactive metal. Many gas barbecues are made out of aluminum, however (**Figure 1**). Explain why an aluminum barbecue can remain outside all year with little evidence of corrosion. (5.4)

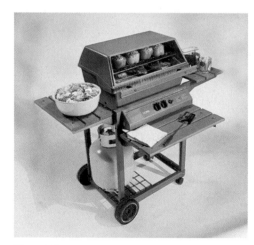

Figure 1
If aluminum is so reactive, why is it used to manufacture barbecues?

9. (a) Why is it necessary to separate the reactants in a galvanic cell?
   (b) Describe the function of the salt bridge in a galvanic cell.
   (c) What happens to the cell potential if the salt bridge is removed? Why? (5.7)

10. Given each pair of half-reactions, write the chemical equation for the overall cell reaction of the galvanic cell.

    Anode half-reaction: **Cathode half-reaction:**

    (a) $Zn_{(s)} \rightarrow Zn^{2+}_{(aq)} + 2e^-$ $Cu^{2+}_{(aq)} + 2e^- \rightarrow Cu_{(s)}$

    (b) $Ni_{(s)} \rightarrow Ni^{2+}_{(aq)} + 2e^-$ $Ag^+_{(aq)} + e^- \rightarrow Ag_{(s)}$

                                                      (5.7)

11. Why is it necessary to use a porous fabric in an alkaline dry cell? (5.9)

12. How are a fuel cell and an alkaline dry cell similar? How are they different? (5.11)

13. (a) What determines the voltage of a battery or cell?

    (b) What determines the amount of electrical energy that a battery or cell can produce? (5.9)

14. (a) Write the anode and cathode half-reactions for the rusting of iron.

    (b) Use your answer for (a) to explain why steel objects rust very slowly in desert climates.

    (c) Why is a puddle of salt water on rusting steel similar to the salt bridge in a galvanic cell?

    (d) Why is galvanizing a more effective form of rust protection than painting or oiling? (5.12)

15. (a) Based on the activity series, which metal is more reactive: iron or aluminum? Why?

    (b) Which of these metals experiences a major corrosion problem? Explain why. (5.12)

16. (a) How does cathodic protection protect against corrosion?

    (b) Name two types of cathodic protection, and describe how they differ. (5.15)

17. List two ways in which a galvanic cell and an electrolytic cell differ. (5.17)

18. Sketch a diagram of an electrolytic cell that can be used to produce nickel metal from a solution of nickel sulfate, $NiSO_{4(aq)}$. Label the anode, cathode, direction of electron flow, and direction of ion flow. (5.17)

19. (a) What is electrorefining?

    (b) Electrorefining can be used to produce extremely pure samples of gold (**Figure 2**). In the process, sheets of impure and very

pure gold are suspended in a tank that contains gold(III) chloride and acid. Sketch a diagram of an electrolytic cell that is used for electrorefining gold. Label the anode and the cathode in your cell. (5.18)

Figure 2
Because gold is so expensive, a small amount of impurity would significantly affect its value.

20. Titanium is an extremely hard, lightweight metal that is used to make the striking surface on golf clubs. One step in the production of titanium is

    $TiCl_{4(g)} + 2\,Mg_{(l)} \rightarrow Ti_{(l)} + 2\,MgCl_{2(l)}$

    (a) Given this equation, which metal is more reactive: titanium or magnesium?

    (b) Predict whether titanium would be above or below magnesium in the activity series. Justify your prediction. (5.4)

21. Galvanic cells are constructed using the following half-cells. (Assume that all solution concentrations are identical.)

    - Cell 1: $Mg_{(s)}$ in $Mg(NO_3)_{2(aq)}$ and $Cu_{(s)}$ in $Cu(NO_3)_{2(aq)}$
    - Cell 2: $Zn_{(s)}$ in $Zn(NO_3)_{2(aq)}$ and $Cu_{(s)}$ in $Cu(NO_3)_{2(aq)}$
    - Cell 3: $Fe_{(s)}$ in $Fe(NO_3)_{2(aq)}$ and $Cu_{(s)}$ in $Cu(NO_3)_{2(aq)}$

    (a) Use the activity series to rank the metal electrodes in order from *least* reactive to *most* reactive.

    (b) Rank the galvanic cells in order from *smallest* to *largest* cell potential. Justify your answer. (5.7)

22. The water temperature and the amount of dissolved oxygen both decrease the farther you descend into an ocean. Why might the rate of corrosion of a steel object differ at depths of 100 m and 1000 m? (Assume that the salt content of the water is the same at both depths.)
(5.12)

23. **Figure 3** shows the formation of tin metal on an electrode surface during the electrolysis of a tin(II) chloride solution, $SnCl_{2(aq)}$.
    (a) Identify the ions that were oxidized and the ions that were reduced during the operation of this cell.
    (b) Write chemical equations for the anode and cathode half-reactions.
    (c) Write the chemical equation for the overall reaction in this cell. (5.17)

Figure 3
Tin metal, $Sn_{(s)}$, and chlorine gas, $Cl_{2(g)}$, are produced during the electrolysis of $SnCl_{2(aq)}$.

Applying Inquiry Skills

24. Small pieces of silver, copper, and magnesium metals are placed in solutions that contain the following ions: $Ag^+_{(aq)}$, $Cu^{2+}_{(aq)}$, and $Mg^{2+}_{(aq)}$.
    (a) Copy **Table 1** and complete it by predicting which metal–ion combinations result in spontaneous chemical reactions.
    (b) Write net ionic equations for the reactions that occur spontaneously. (5.4)

Table 1 Predicted Reactions

Metal \ Ion	$Ag^+_{(aq)}$	$Cu^{2+}_{(aq)}$	$Mg^{2+}_{(aq)}$
$Ag_{(s)}$			
$Cu_{(s)}$			
$Mg_{(s)}$			

25. The following reaction takes place in a test tube:
$$Mg_{(s)} + 2\,HCl_{(aq)} \rightarrow H_{2(g)} + MgCl_{2(aq)}$$
    (a) Describe what changes you would observe on the surface of the magnesium.
    (b) How would the rate of bubbling differ if magnesium were replaced with zinc? Why?
    (c) How would the rate of bubbling differ if magnesium were replaced with copper? Why? (5.4)

26. (a) Draw a diagram of a galvanic cell that can be constructed using the following materials: 2 beakers; connecting wires; glass U-tube; cotton batting; distilled water; strip of magnesium, $Mg_{(s)}$; solid magnesium nitrate, $Mg(NO_3)_{2(s)}$; solid sodium nitrate, $NaNO_{3(s)}$; strip of copper, $Cu_{(s)}$; solid copper(II) sulfate, $CuSO_{4(s)}$.
    (b) Identify the anode and the cathode in the cell.
    (c) Indicate the directions of electron flow and ion flow in the cell.
    (d) Describe why the cell would not be a practical source of electricity. (5.7)

27. Identical nails are placed in test tubes that contain equal volumes of the following 0.1-mol/L solutions: sodium hydroxide, $NaOH_{(aq)}$; sodium chloride, $NaCl_{(aq)}$; hydrochloric acid, $HCl_{(aq)}$. Compare the amount of corrosion you would expect to see in the three test tubes. (5.12, 5.14)

28. (a) What is a "sacrificial anode"?
    (b) Design an experiment to test the prediction that magnesium can act as a sacrificial anode. Assume that the following materials are available: 2 beakers; salt water; 2 identical nails; magnesium ribbon.
    (c) What observations would you expect from your experiment? (5.15)

29. The bases of some expensive cooking pots are made of copper. Sketch and label a diagram of a cell that could be used to electroplate copper onto the base of a steel pot. (5.18)

Making Connections

30. (a) Describe why using salt on roads promotes the rusting of automobiles.
    (b) Describe some environmental problems related to the use of road salt.
    (c) Research some more environmentally friendly alternatives to salt. (5.12)

 www.science.nelson.com

31. (a) Describe the link between acid rain and metal refining.
    (b) Describe the link between acid rain and corrosion.
    (c) State two other ways in which the electrolytic production of metals can affect the environment.
    (d) Suggest some actions that the general public could take to minimize these environmental effects. (5.18)

32. Consider the reaction of magnesium with dry ice (solid carbon dioxide):

$$2\,Mg_{(s)} + CO_{2(g)} \rightarrow 2\,MgO_{(s)} + C_{(s)}$$

    (a) Identify the reactant oxidized and the reactant reduced in this reaction.
    (b) Find and observe a demonstration of this reaction on the Internet. Record your observations.
    (c) Use your observations to explain why compressed carbon dioxide fire extinguishers cannot be used to extinguish a magnesium fire.
    (d) A magnesium fire is an example of what firefighters refer to as a Class D fire. Research the recommended way to put out a Class D fire. (5.1, 5.2)

 www.science.nelson.com

Extension

33. (a) Research what substances are present in different grades of gold, besides gold itself.
    (b) Why are other substances added to gold?
    (c) What kinds of gold would you recommend that people with nickel allergies should avoid? Explain your reasoning. (5.13)

 www.science.nelson.com

34. Being able to produce oxygen while submerged allows submarines to stay underwater for weeks at a time without having to surface (**Figure 4**). Oxygen can be produced by electrolyzing water:

$$2\,H_2O_{(l)} + energy \rightarrow 2\,H_{2(g)} + O_{2(g)}$$

    (a) Draw and label an electrolytic cell that could be used to produce oxygen. (*Hint:* Use graphite electrodes.)
    (b) What is a likely source of energy for this reaction?
    (c) Suggest some of the hazards of this reaction. What precautions should be taken to avoid them? (5.17)

Figure 4
Electrolysis of water provides the oxygen that is needed for this submarine to remain underwater for long periods of time.

35. The Martian atmosphere has far too little oxygen to support life. Future settlers on Mars might be able to use redox reactions to extract oxygen from compounds in the ground, just as the early metallurgists on Earth did to extract iron. Suggest a method that could be used to extract oxygen. Illustrate your method with a labelled diagram. Outline some of the design challenges that might be encountered. (5.17)

Appendices

A1 Scientific Inquiry

Planning an Investigation

In our attempts to further our understanding of the natural world, we encounter questions, mysteries, or events that are not easily explained. We can use controlled experiments or observational studies to help us look for answers or explanations. The methods used in scientific inquiry depend, to a large degree, on the purpose of the inquiry.

Controlled Experiments

Controlled experiments are performed when the purpose of the inquiry is to create or test a scientific concept. In a controlled experiment, an independent variable is purposefully and steadily changed to determine its effect on a second, dependent variable. All other variables are controlled or kept constant.

The common components of controlled experiments are outlined in the flow chart below. *Even though the sequence is presented as linear, there are normally many cycles through the steps during an actual experiment.*

Process Description

Choose a topic that interests you. Determine whether you are going to carry out a given procedure or develop a new experimental design. Indicate your decision in a statement of the purpose.

Your Question forms the basis for your investigation. Controlled experiments are about relationships, so the Question could be about the effects on variable A when variable B is changed. The Question could be about what causes the change in variable A. In this case, you might speculate about possible variables and determine which variable causes the change.

A hypothesis is a tentative explanation. You must be able to test your hypothesis, which can range in certainty from an educated guess to a concept that is widely accepted in the scientific community. A prediction is based on a hypothesis or a more established scientific explanation, such as a theory. In the prediction, you state what outcome you expect from your investigation.

The design of a controlled experiment identifies how you plan to manipulate the independent variable, measure the response of the dependent variable, and control all the other variables.

Stating the Purpose → **Asking the Question** → **Hypothesizing/ Predicting** → **Designing the Investigation**

Example: A Test of the Collision Model

The purpose of this investigation is to provide some concrete evidence to support or refute the collision model.

How does changing the concentration of hydrochloric acid affect the time required for the acid to react completely with a fixed quantity of zinc?

According to the collision model, if the concentration of hydrochloric acid is increased, then the time required for the reaction with zinc will decrease. The following reasoning supports this prediction: A higher concentration of $HCl_{(aq)}$ produces more collisions per second between the aqueous ions in hydrochloric acid and the zinc atoms. More collisions per second produces more reactions per second and, therefore, a shorter time is required to consume the zinc.

The same amount of zinc metal is made to react with different known concentrations of excess hydrochloric acid. The time for the zinc to react completely is measured for each concentration of acid solution. The independent variable is the concentration of hydrochloric acid. The dependent variable is the time for the zinc to be consumed. The temperature of the solution, the quantity of zinc, the surface area of the zinc in contact with the acid, and the volume of the acid are all controlled variables.

There are many ways to gather and record observations during an investigation. It is helpful to plan ahead and think about what data you will need and how best to record them. This helps you clarify your thinking about the Question posed at the beginning, the variables, the number of trials, the procedure, and your skills. It will also help you organize your evidence for easier analysis later.

After thoroughly analyzing your observations, you may have sufficient and appropriate evidence to answer the Question posed at the beginning of the investigation.

At this stage of the investigation, you will evaluate the processes that you followed to plan and perform the investigation. Evaluating the processes includes reviewing the design and the procedure. You will also evaluate the outcome of the investigation, which involves assessing the evidence—whether it supports the hypothesis or not—and the hypothesis itself.

In preparing your report, your aim should be to describe your design and procedure accurately, and to report your observations accurately and honestly.

Gathering, recording, and organizing observations

Analyzing the observations

Evaluating the evidence and the hypothesis

Reporting on the investigation

Time to completion for the reaction will be measured using a stopwatch and recorded in a table like **Table 1**.

The observations will be presented in graphical format, with time on the horizontal axis and concentration of $HCl_{(aq)}$ on the vertical axis. In this format, any trends or patterns will be easier to see.

For a sample evaluation, see the sample lab report in Appendix A4.

For the format of a typical lab report, see the sample lab report in Appendix A4.

Table 1 Reaction Time for Zinc with $HCl_{(aq)}$

Concentration of $HCl_{(aq)}$ (mol/L)	Time for reaction (s)
2.5	
2.0	
1.5	
1.0	
0.5	

Observational Studies

Often the purpose of inquiry is simply to study a natural phenomenon with the intention of gaining scientifically significant information to answer a question. Observational studies involve observing a subject or phenomenon in an unobtrusive or unstructured manner, often with no specific hypothesis. A hypothesis to describe or explain the observations may, however, be generated after repeated observations, and modified as new information is collected over time.

The flow chart below summarizes the stages and processes of scientific inquiry through an observational study. *Even though the sequence is presented as linear, there are normally many cycles through the steps during an actual study.*

Process Description

Choose a topic that interests you. Determine whether you are going to replicate or revise a previous study, or create a new study. Indicate your decision in a statement of the purpose.

In planning an observational study, it is important to pose a general question about the natural world. You may or may not follow the Question with the creation of a hypothesis.

A hypothesis is a tentative explanation. In an observational study, a hypothesis can be formed after observations have been made and information has been gathered on a topic. A hypothesis may be created in the analysis.

The design of an observational study describes how you will make observations that are relevant to the Question.

Stating the Purpose	→	Asking the Question	→	Hypothesizing/Predicting	→	Designing the Investigation

Example: Water Quality in Public Swimming Area

Although the quality of public swimming areas is normally tested by the municipal or provincial Department of Health, no chemical analysis is done unless a problem arises. The purpose of this study is to carry out an environmental assessment to determine the chemical quality of the local public swimming area.

What common chemicals are found in the local public swimming area, and in what concentrations are they present?

At this point, we have no indication of which chemicals are present in the area, what their concentrations may be, and whether there are any threats to swimmers. We have no hypothesis and can make no predictions.

We will take a sample of water from five different locations within the local swimming area each week for a month. Note will be made of other significant conditions (such as heavy rain or wind) that are present during the course of the study. The water samples will be tested for organic and inorganic chemicals that may pose a health hazard. The testing facilities at the Department of Health will be used to determine the presence and concentrations of chemicals.

Table 2   Presence and Concentration of Chemicals in Public Swimming Area

Chemicals	Week 1 Sample area					Week 2 Sample area					Week 3 Sample area					Week 4 Sample area				
	1	2	3	4	5	1	2	3	4	5	1	2	3	4	5	1	2	3	4	5

There are many ways to gather and record observations during an investigation. During your observational study, you should quantify your observations where possible. All observations should be objective and unambiguous. Consider ways to organize your information for easier analysis.

After thoroughly analyzing your observations, you may have sufficient and appropriate evidence to answer the Question posed at the beginning of the study. You may also have enough observations and information to form a hypothesis.

At this stage of the study, you will evaluate the processes used to plan and perform the study. Evaluating the processes includes evaluating the materials, the design, the procedure, and your skills. Often the results of such studies suggest further studies, perhaps correlational studies or controlled experiments to explore tentative hypotheses you may have developed.

In preparing your report, your aim should be to describe your design and procedure accurately, and to report your observations accurately and honestly.

Gathering, recording, and organizing observations → **Analyzing the observations** → **Evaluating the evidence and the hypothesis** → **Reporting on the investigation**

The data will be recorded in a table like **Table 2**. The chemicals to be tested for include the following:
— lead
— mercury
— cadmium
— nitrates/nitrites
— volatile organic compounds such as benzene, toluene, and carbon tetrachloride
— petroleum products
— chlorine

The concentrations of the chemicals found in the swimming area are determined.

We must determine if our sampling and testing procedures are appropriate. Is the number of samples sufficient? Were they taken at the proper sites? Was the testing of the samples carried out with care and precision?

The presence of chemicals in concentrations higher than the acceptable levels will alert us to potential problems with the swimming area. This might suggest further studies to determine the possible source(s) of the chemical(s).

For the format of a typical lab report, see the sample lab report in Appendix A4.

A2 Decision Making

Modern life is filled with environmental and social issues that have scientific and technological dimensions. An issue is defined as a problem that has at least two possible solutions rather than a single solution. There can be many positions, generally determined by the values that an individual or a society holds, on a single issue. Which solution is "best" is a matter of opinion; ideally, the solution that is put into practice is the one that is most appropriate for society as a whole.

The common elements of the decision-making process are outlined in the flow chart below. *Even though the sequence is presented as linear, you may go through several cycles before deciding that you are ready to defend a decision.*

Process Description

The first step in understanding an issue is to explain why it is an issue, describe the problems associated with the issue, and identify the individuals or groups (called stakeholders) involved in the issue. You could brainstorm the following questions to research the issue: Who? What? Where? When? Why? How? Develop background information on the issue by clarifying facts and concepts, and identifying relevant attributes, features, or characteristics of the problem.

Examine the issue and think of as many alternative solutions as you can. At this point, it does not matter if the solutions seem unrealistic. To analyze the alternatives, you should examine the issue from a variety of perspectives. Stakeholders may bring different viewpoints to an issue, and these viewpoints may influence their position on the issue. Brainstorm or hypothesize how various stakeholders would feel about your alternatives.

Formulate a research question that helps to limit, narrow, or define the issue. Then develop a plan to find reliable and relevant sources of information. Outline the stages of your information search: gathering, sorting, evaluating, selecting, and integrating relevant information. You may consider using a flow chart, a concept map, or another graphic organizer to outline the stages of your information search. Gather information from many sources, including newspapers, magazines, scientific journals, the Internet, and the library.

Defining the issue → **Identifying alternatives/positions** → **Researching the issue**

Example: The Issue of Pesticide Use

In recent years, the use of pesticides (herbicides, insecticides, and fungicides) on lawns has increased despite reports of health and environmental risks. Several attempts are being made to deal with the increased use, including publicity campaigns by various groups, and attempts to ban or limit the use of pesticides at municipal and other government levels. A list of possible stakeholders in this issue is started in **Table 1** (p. 456).

Develop background information on the issue by clarifying information and concepts, and identifying relevant attributes, features, or characteristics of the problem. For example:

- While more research is needed on the health risks, many lawn chemicals currently in use are known carcinogens. As well, there are numerous other less serious symptoms (such as headaches, nausea, fever, and breathing difficulties) associated with pesticide poisoning.
- Manufacturers point out that the pesticides they manufacture have been approved for use by the federal government. Pesticides considered unsafe, such as DDT and fenitrothion, have been banned.

One possible solution for people concerned about pesticide use is to ban the production of pesticides. A solution for government might be to enforce stricter regulations governing pesticide use.

Think about how different stakeholders might feel about the alternatives. For example, citizens may be affected by the use of pesticides in their neighbourhood. What would be their perspective? What would be the perspective of a parent of small children? A farmer? A pest-control business owner? A chemist? A gardener? Employees and owners of a company that produces pesticides? An environmentalist? (See **Table 2** for a start on this process.) Remember that one person can have more than one perspective. It is also possible that two people, looking at an issue from the same perspective, might disagree about the best solution or even the available information. For example, scientists might disagree about the degree of risk associated with pesticide use.

Begin your search for reliable and relevant sources of information about the issue with a question such as, "What does the research say about the risks associated with pesticide use?" or "What are the established positions of various groups on the issue?"

There are five steps that must be completed to analyze the issue effectively:

1. Establish criteria for determining the relevance and significance of the data you have gathered.

2. Evaluate the sources of information.

3. Identify and determine what assumptions have been made. Challenge unsupported evidence.

4. Determine any relationships associated with the issue.

5. Evaluate the alternative solutions, possibly by conducting a risk–benefit analysis.

After analyzing your information, you can answer your research question and take an informed position on the issue. You should be able to defend your solution in an appropriate format—debate, class discussion, speech, position paper, multimedia presentation (such as a computer slide show), brochure, poster, or video.

Your position on the issue must be justified using supporting information that you have researched. You should be able to defend your position to people with different perspectives. Ask yourself the following questions:

- Do I have supporting evidence from a variety of sources?

- Can I state my position clearly?

- Can I show why this issue is relevant and important to society?

- Do I have solid arguments (with solid evidence) supporting my position?

- Have I considered arguments against my position, and identified their faults?

- Have I analyzed the strong and weak points of each perspective?

The final phase of decision making includes evaluating the decision itself and the process used to reach the decision. After you have made a decision, carefully examine the thinking that led to your decision.

Some questions to guide your evaluation:

- What was my initial perspective on the issue? How has my perspective changed since I first began to explore the issue?

- How did we make our decision? What process did we use? What steps did we follow?

- In what ways does our decision resolve the issue?

- What are the likely short- and long-term effects of the decision?

- To what extent am I satisfied with the final decision?

- What reasons would I give to explain our decision?

- If we had to make this decision again, what would I do differently?

Analyzing the issue	**Defending the decision**	**Evaluating the process**

After reviewing government, chemical industry, and university studies, and reading newspaper articles and papers by environmental groups, we concluded that research seems to indicate that the active ingredients in many common pesticides are carcinogenic and therefore pose a significant risk to health.

There are reports that contradict our view, and domestic pesticides have been approved for use by federal government agencies. There are many jobs, some of them based in our town, that rely on continued use of pesticides.

After performing a risk–benefit analysis of the various alternative solutions, we decided that we should attempt to reduce or eliminate the use of pesticides on lawns.

Table 3 (p. 457) shows a risk–benefit analysis of allowing pesticide use on lawns.

In our defence at the town hall meeting, we will concentrate on our evidence that there are alternative methods of pest control that are effective and safe. By concentrating on this, and on reasonable doubt about the safety of pesticide use, we hope to be able to counter arguments by opponents.

We tried to obtain information from a variety of reputable sources; however, some of the research is highly technical, and it is possible that we misunderstood its main points or misjudged its relevance.

In the town hall meeting, we created a bylaw to eliminate pesticide use on town property and to limit pesticide use on private property to exceptional circumstances. We realize that this decision will not satisfy all stakeholders, but we believe it is the best solution given the evidence at our disposal.

This decision may cause painful changes for industries that produce pesticides and for several service industries. We believe that after a transition period, these industries will survive to produce and market safe alternatives to conventional pesticides.

Table 1 Potential Stakeholders in the Pesticide Debate

Stakeholders	Viewpoint (perspectives)
parent	Children are more susceptible to pesticide poisoning than adults and should not be put at risk. (social)
scientists	1. Active ingredients in many pesticides are known carcinogens. 2. Levels of the active ingredient in pesticides pose no risk (or a risk) to humans with short-term exposure. (scientific)
doctor	Environmental factors that pose any risk to human health should be eliminated or severely restricted. (ecological/legal)
environmentalist	Pesticides from lawns are percolating into rivers, streams, and ground water and are affecting wildlife. (ecological)
pest-control business owner	Used properly, pesticides pose no risk to humans. Only trained persons should be allowed to use pesticides. The pest-control industry is a valuable contributor to the economy. (scientific/technological/legal/economic)
owners of chemical company	Pesticides have been tested and approved by the federal government. (legal) Jobs will be lost if these pesticides are banned. (economic/social)

Table 2 Perspectives on an Issue

Perspective	Focus of the perspective
cultural	customs and practices of a particular group
ecological	interactions among organisms and their natural habitat
economic	the production, distribution, and consumption of wealth
educational	the effects on learning
emotional	feelings and emotions
environmental	the effects on physical surroundings
esthetic	artistic, tasteful, beautiful
moral/ethical	what is good/bad, right/wrong
legal	the rights and responsibilities of individuals and groups
spiritual	the effects on personal beliefs
political	the effects on the aims of a political group or party
scientific	logical or research information based
social	the effects on human relationships, the community, or society
technological	machines and industrial processes

A Risk–Benefit Analysis Model

Risk–benefit analysis is a tool used to organize and analyze information gathered in research. A thorough analysis of the risks and benefits associated with each alternative solution can help you decide on the best alternative.

- Research as many aspects of the proposal as possible. Look at it from different perspectives.
- Collect as much evidence as you can, including reasonable projections of likely outcomes if the proposal is adopted.
- Classify each potential result as being either a benefit or a risk.
- Quantify the size of the potential benefit or risk (perhaps as a dollar figure, as a number of lives affected, or on a scale of 1 to 5).
- Estimate the probability (percentage) of that event occurring.
- By multiplying the size of a benefit (or risk) by the probability of its happening, you can calculate a probability value for each potential result.

- Total the probability values of all the potential risks and all the potential benefits.
- Compare the sums to help you decide whether to accept the proposed action.

Table 3 shows an incomplete risk–benefit analysis of one option in the lawn pesticide issue—making no changes in regulations. Note that although you should try to be objective in your assessment, the beliefs of the person making the risk–benefit analysis will have an effect on the final sums. The possible outcomes considered for analysis, the assessment of the relative importance of a cost or benefit, and the probability of the cost or benefit actually arising will vary according to who does the analysis. For example, would you agree completely with the values placed in the "Cost" and "Benefit" columns of the analysis in **Table 3**?

Table 3   Risk–Benefit Analysis of Continuing Use of Pesticides on Lawns

Risks				Benefits			
Possible result	Cost of result (scale of 1 to 5)	Probability of result occurring (%)	Cost × probability	Possible result	Benefit of result (scale of 1 to 5)	Probability of result occurring (%)	Benefit × probability
Pesticide use on lawns presents human health risks.	very serious 5	inconclusive research (60%)	300	Pesticides eliminate pests, which also present health risk.	high 4	somewhat likely (60%)	240
Pesticide use on lawns affects other species.	serious 4	likely (80%)	320	Lawn-care business is a valuable part of local economy.	high 4	certain (100%)	400
Health-care costs will increase.	very serious 5	likely (80%)	400	Well-kept lawn increases property value.	medium 3	likely (80%)	240
Total risk value			**1020**	**Total benefit value**			**880**

A3 Technological Problem Solving

There is a difference between science and technology. The goal of science is to understand the natural world. The goal of technological problem solving is to develop or revise a product or a process in response to a human need. The product or process must fulfill its function but, in contrast with scientific problem solving, it is not essential to understand why or how it works.

Technological solutions are evaluated based on such criteria as simplicity, reliability, efficiency, cost, and ecological and political consequences.

Even though the sequence presented in the flow chart below is linear, there are normally many cycles through the steps in any problem-solving attempt.

Process Description

This process involves recognizing and identifying the need for a technological solution. You need to state clearly the question(s) that you want to investigate to solve the problem and the criteria you will use to plan and evaluate your solution. In any design, some criteria may be more important than others. For example, if a tool measures accurately and is economical, but is not safe, then it is clearly unacceptable.

Use your prior knowledge, experience, and creativity to propose possible solutions.

During brainstorming, the goal is to generate many ideas without judging them. They can be evaluated and accepted or rejected later.

To visualize the possible solutions, it is helpful to draw sketches. Sketches are often better than verbal descriptions for communicating an idea.

Planning is the heart of the entire process. Your plan will outline your processes, identify potential sources of information and materials, define your resource parameters, and establish evaluation criteria.

Seven types of resources are generally used when developing technological solutions to problems: people, information, materials, tools, energy, capital, and time.

Defining the problem → **Identifying possible solutions** → **Planning**

Example: Inventing a pH Meter

We often need to solve technological problems before we can conduct a scientific investigation. For example, imagine that you are asked to conduct an investigation in which you cannot, for safety reasons, use the traditional, commercial pH indicators. Your task, then, is to design a safe chemical indicator for pH that will be as effective as those available commercially.

If you are not given criteria for the solution to the problem (criteria are often given in technological problem solving), you can establish your own by asking some basic questions about the situation and the function of the device.

In this case, you are asked to design and produce a chemical pH indicator to meet the following criteria:

- must be able to measure pH to at least the nearest 0.5 on the pH scale
- must be safe enough to pose no health hazard if spilled on the skin or ingested
- must have a shelf life of at least one month
- must be at least as economical as a comparable commercial product
- must be produced from readily available materials

Design 1: Poison Primrose Flower Extract
This extract is made by mixing dried and ground poison primrose flower petals, boiling the resulting powder in isopropyl alcohol, and filtering the resulting mixture to isolate the extract.

Design 2: Red Cabbage Extract
This extract is made by mixing chopped red cabbage with water in a blender and straining the juice (extract) from the resulting mush.

People: The human resources required to solve this problem include you and your partner.
Information: You already understand the concepts of acidity and basicity. You will need to understand fully the pH scale. You may also need to find out about naturally occurring substances that react with acids or bases to produce different (such as visible) effects.
Materials: Within the limitations imposed by your proposed solution, cost, availability, safety, and time, you can use whatever materials you deem necessary.
Tools: Your design should not require any specialized tools or machines that are not immediately available.
Energy: The solution to this problem should not require any external source of energy.
Capital: The dollar cost must be low. Your solution must cost less to build than a comparable commercial product would cost to buy.
Time: Because of the time limit on the scientific investigation, there is an even shorter time limit on the production of the indicator. You should be able to produce your indicator within 60 min. (This does not include designing and testing.)

The solution will be evaluated on how well it meets the design criteria established earlier.

In this phase, you will construct and test your prototype using trial and error. Try to change only one variable at a time. Use failures to inform the decisions you make before your next trial. You may also complete a cost–benefit analysis on the prototype.

To help you decide on the best solution, you can rate each potential solution based on the design criteria. Use a five-point rating scale, with 1 being poor, 2 fair, 3 good, 4 very good, and 5 excellent. You can then compare your proposed solutions by totalling the scores.

Once you have made the choice among the possible solutions, you need to produce and test a prototype. While making the prototype, you may need to experiment with the characteristics of different components. A model, on a smaller scale, might help you decide whether the product will be functional. The test of your prototype should answer three basic questions:

- Does the prototype solve the problem?
- Does it satisfy the design criteria?
- Are there any unanticipated problems with the design?

If these questions cannot be answered satisfactorily, you may have to modify the design or select another potential solution.

In presenting your solution, you will communicate your solution, identify potential applications, and put your solution to use.

Once the prototype has been produced and tested, the best presentation of the solution is a demonstration of its use—a test under actual conditions. This demonstration can serve as a further test of the design, as well. Any feedback should be considered for future redesign. Remember that no solution should be considered the absolute final solution.

Evaluation is not restricted to the final step. However, it is important to evaluate the final product using the criteria established earlier and to evaluate the processes used while arriving at the solution. Consider the following questions:

- To what degree does the final product meet the design criteria?
- Did you have to make any compromises in the design? If so, are there ways to minimize the effects of the compromises?
- Did you exceed any of the resource parameters?
- Are there other possible solutions that deserve future consideration?
- How did your group work as a team?

Constructing/testing solutions → **Presenting the preferred solution** → **Evaluating the solution and process**

Table 1 illustrates the rating for two different designs. Note that although Design 1 came out with the highest rating, there is one factor (safety) that suggests we should go with Design 2. This is what is referred to as a tradeoff. By reviewing or evaluating product and processes, we may be able to modify Design 2 to optimize its performance on the other criteria.

Table 1 Design Analysis

Criterion	Design 1	Design 2
accuracy	5	3
safety	3	5
shelf life	3	4
economy	4	4
materials	5	3
Total score	**20**	**19**

The chosen design was presented to the chemistry class. A set of 12 test tubes was set up. Each test tube contained a colourless solution of known pH, ranging from pH 1 to pH 14. Ten drops of red cabbage extract was added to each test tube. Students observed the colour changes and rated the results using a rating scale similar to the one used in the product testing stages. Students were given a list of the criteria used in the design and production stages, and were asked to provide comments regarding the indicator's performance.

Our chosen product, red cabbage extract, meets most of the established criteria. Feedback from the chemistry class demonstration was positive. Unfortunately, the indicator does not change colour in the intervals pH 3 to 4 and pH 9 to 10, which limits how it can be used.

Kept in the refrigerator, red cabbage extract is still acting as a good pH indicator one month after it was prepared. Red cabbage is readily available. Its extract is safe for all uses and cheap to prepare.

In general, our group worked well as a team. However, some individuals pitched in a little more than others, especially in areas where they felt they they were more skilled. We have agreed that in future projects, every member of the team will work on something that is new to them, which will give each of us a chance to learn, and everyone will do their fair share in the final cleanup.

A4 Lab Reports

When carrying out investigations, it is important that scientists keep records of their plans and results, and share their findings. In order to have their investigations repeated (replicated) and accepted by the scientific community, scientists generally share their work by publishing papers in which they provide details of their design, materials, procedure, evidence, analysis, and evaluation.

Lab reports are prepared after an investigation is completed. To ensure that you can accurately describe the investigation, it is important to keep thorough and accurate records of your activities as you carry out the investigation.

Investigators use a similar format in their final reports or lab books, although the headings and order may vary. Your lab book or report should reflect the type of scientific inquiry that you used in the investigation and should be based on the following headings, as appropriate. (See **Figure 1**, on pages 462–463, for a sample lab report.)

Title

At the beginning of your report, write the section number and title of your investigation. In this course, the title is usually given, but if you are designing your own investigation, create a title that suggests what the investigation is about. Include the date the investigation was conducted and the names of all lab partners (if you worked as a team).

Purpose

State the purpose of the investigation. Why are you doing the investigation?

Question

This is the Question that you attempted to answer in the investigation. If it is appropriate to do so, state the Question in terms of independent and dependent variables.

Hypothesis/Prediction

Based on your reasoning or on a concept that you have studied, formulate an explanation of what should happen (a hypothesis). From your hypothesis you may make a prediction (a statement of what you expect to observe) before carrying out the investigation. Depending on the nature of your investigation, you may or may not have a hypothesis or a prediction.

Experimental Design

This is a brief general overview (one to three sentences) of what was done. If your investigation involved independent, dependent, and controlled variables, list them. Identify any control or control group that was used in the investigation.

Materials

This is a detailed list of all materials used, including sizes and quantities where appropriate. Be sure to include safety equipment such as goggles, lab apron, prospective gloves, and tongs, where needed. Draw a diagram to show any complicated setup of apparatus.

Procedure

Describe, in detailed, numbered steps, the procedure you followed to carry out your investigation. Include steps to clean up and dispose of waste.

Observations

This includes all qualitative and quantitative observations you made. Be as precise as possible when describing quantitative observations, include any unexpected observations, and present your information in a form that is easily understood. If you have only a few observations, this can be a list. For controlled experiments and for many observations, a table will be more appropriate.

Analysis

Interpret your observations and present the evidence in the form of tables, graphs, or illustrations, each with a title. Include any calculations, and show the results of your calculations in a table. Make statements about any patterns or trends you observed. Conclude the analysis with a statement based only on the evidence you have gathered, answering the question that initiated the investigation.

Evaluation

The evaluation is your judgment about the quality of evidence obtained and about the validity of the prediction and hypothesis (if present). This section can be divided into two parts: evaluation of the investigation and evaluation of the prediction (and hypothesis). The following questions and suggestions should help you in each part of the process.

Evaluation of the Investigation

- Did the design enable you to answer the question?
- As far as you know, is the design the best available or are there flaws that could be corrected?
- Were the steps in the investigation in the correct order and adequate to gather sufficient evidence?
- What steps, if done incorrectly, could have significantly affected the results?
- What improvements could be made to the procedure?

Sum up your conclusions about the procedure in a statement that begins like this: "The procedure is judged to be adequate/inadequate because…"

- What specialized skills (such as measuring) might have an effect on the results?
- Was the evidence from repeated trials reasonably similar?
- Can the measurements be made more precise?

Sum up your conclusions about the required skills in a statement that begins like this: "The skills are judged to be adequate/inadequate because…"

- What are the sources of uncertainty and error in my investigation?
- Based on any uncertainties and errors you have identified, do you have enough confidence in your results to proceed with the evaluation of the prediction and hypothesis?

State your confidence level in a statement like this: "Based on my evaluation of the investigation, I am certain/I am moderately certain/I am very certain of my results."

Evaluation of the Prediction (and Hypothesis)

- Does the predicted answer clearly agree with the answer in your analysis?
- Can any difference be accounted for by the sources of uncertainty or error listed earlier in the evaluation?

Sum up your evaluation of the prediction in a statement that begins like this: "The prediction is judged to be verified/inconclusive/falsified because…"

- Is the hypothesis supported by the evidence?
- Is there a need to revise the hypothesis or to replace it with a new hypothesis?

If the prediction was verified, the hypothesis behind it is supported. If the results were inconclusive or the prediction is falsified, then the hypothesis is questionable. Sum up your evaluation of the hypothesis in a statement that begins like this: "The hypothesis being tested is judged to be acceptable/unacceptable because…"

Investigation 2.5: The Effect of Concentration on Reaction Time

Conducted December 15, 2009

By Barry L. and Lakshmi B.

Purpose

The purpose of this investigation is to test one of the ideas of the collision model.

Question

How does changing the concentration of hydrochloric acid affect the time required for the reaction of hydrochloric acid with a fixed quantity of zinc?

Hypothesis/Prediction

According to the collision model, if the concentration of hydrochloric acid is increased, then the time required for the reaction with zinc will decrease. The following reasoning supports this hypothesis: A higher concentration produces more collisions per second between the hydrochloric acid particles and the zinc atoms. More collisions per second produces more reactions per second and, therefore, a shorter time is required to consume the zinc.

Experimental Design

Different known concentrations of excess hydrochloric acid react with zinc metal. The time for the zinc to react completely is measured for each concentration of acid solution. The independent variable is the concentration of hydrochloric acid. The dependent variable is the time for the zinc to be consumed. The temperature of the solution, the quantity of zinc, the surface area of the zinc in contact with the acid, and the volume of the acid are all controlled variables.

Materials

lab apron	safety glasses
four 10-mL graduated cylinders	four 18-mm by 150-mm test tubes and a test-tube rack
clock or watch (precise to nearest second)	four pieces of zinc metal strip, 5 mm by 5 mm
stock solutions of $HCl_{(aq)}$: 2.0 mol/L, 1.5 mol/L, 1.0 mol/L, 0.5 mol/L	solution of a weak base (baking soda)

Procedure

1. 15 mL of 2.0 mol/L $HCl_{(aq)}$ was transferred into an 18-mm by 150-mm test tube.

2. A piece of $Zn_{(s)}$ was carefully placed into the hydrochloric acid solution. The starting time of the reaction was noted.

3. The time required for all of the zinc to react was measured and recorded.

4. Steps 2 to 4 were repeated using 1.5 mol/L, 1.0 mol/L, and 0.5 mol/L $HCl_{(aq)}$.

5. Any acid remaining in the solutions was neutralized with a solution of the weak base and then poured down the sink with large amounts of water.

Figure 1
Sample lab report

Observations

The Effect of Concentration on Reaction Time

Concentration of $HCl_{(aq)}$ (mol/L)	Time for reaction (s)
2.0	70
1.5	80
1.0	144
0.5	258

Analysis

The evidence is plotted on a graph of time versus concentration of $HCl_{(aq)}$ (below). The graph tends to level off at the two highest concentrations. From this trend, we can predict, both from the graph and from common sense, that if you keep increasing the concentration, the reaction time will never reach zero. We might also predict that as the concentration gets very low, the time required for all of the zinc to react will become very long.

Based on the evidence gathered in this investigation, increasing the concentration of hydrochloric acid decreases the time required for the reaction of hydrochloric acid with a fixed quantity of zinc.

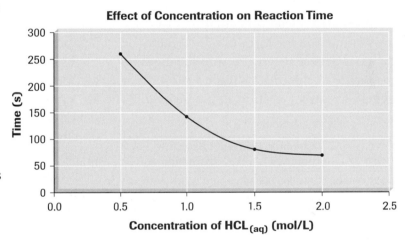

Effect of Concentration on Reaction Time

Evaluation

The design, materials, and skills used in this investigation are adequate because this experiment produced the type of evidence needed to answer the question with a high degree of certainty. The variables were easy to measure, manipulate, and control.

The procedure is also considered to be adequate since the steps are simple and straightforward. We could have improved the procedure by extending the range of concentrations, by stirring, and by performing more than one trial for each concentration.

Sources of uncertainty in this investigation include the purity of the zinc metal strip, the concentration of the stock acid, and the determination of when the last bit of zinc had reacted.

The hypothesis is supported by the evidence, which clearly shows that the reaction time decreased as the concentration increased. Based on the evidence, the collision model is also acceptable.

Synthesis

Other investigations using one of the controlled variables (such as temperature of the acid or surface area of the zinc) as the independent variable could be carried out to determine their effect on the reaction rate. Additional investigations studying the effect of concentration on reaction rate using different reactants and reaction types could be conducted.

A5 Math Skills

Scientific Notation

It is difficult to work with very large or very small numbers when they are written in common decimal notation. Usually it is possible to accommodate such numbers by changing the SI prefix so that the number falls between 0.1 and 1000. For example, 237 000 000 mm can be expressed as 237 km, and 0.000 000 895 kg can be expressed as 0.895 mg. However, this prefix change is not always possible, either because an appropriate prefix does not exist or because a particular unit of measurement must be used. In these cases, the best method of dealing with very large and very small numbers is to write them using scientific notation. Scientific notation expresses a number by writing it in the form $a \times 10^n$, where the coefficient a is between 1 and 10, and the digits in the coefficient a are all significant. **Table 1** shows situations where scientific notation would be used.

Table 1   Examples of Scientific Notation

Expression	Common decimal notation	Scientific notation
124.5 million kilometres	124 500 000 km	1.245×10^8 km
6 billion nanometres	6 000 000 000 nm	6×10^9 nm
5 gigabytes	5 000 000 000 bytes	5×10^9 bytes

To multiply numbers in scientific notation, multiply the coefficients and add the exponents; the answer is expressed in scientific notation. Note that when writing a number in scientific notation, the coefficient should be between 1 and 10 and should be rounded to the same certainty (number of significant digits) as the measurement with the least certainty (fewest number of significant digits). Look at the following examples:

$$(4.73 \times 10^5 \text{ m})(5.82 \times 10^7 \text{ m}) = 27.5 \times 10^{12} \text{ m}^2 = 2.75 \times 10^{13} \text{ m}^2$$

$$(3.9 \times 10^4 \text{ N})(5.3 \times 10^{-3} \text{ m}) = 21 \times 10^1 \text{ N·m} = 2.1 \times 10^2 \text{ N·m}$$

On many calculators, scientific notation is entered using a special key, labelled EXP or EE. This key includes "$\times 10$" from the scientific notation, so you need to enter only the exponent. For example, to enter

7.5×10^4	press	7.5 EXP 4
3.6×10^{-3}	press	3.6 EXP +/−3

Uncertainty in Measurements

There are two types of quantities that are used in science: exact values and measurements. Exact values include defined quantities (1 m = 100 cm) and counted values (5 cars in a parking lot). Measurements, however, are not exact because there is some uncertainty or error associated with every measurement.

There are two types of measurement error. **Random error** results when an estimate is made to obtain the last significant figure for any measurement. The size of the random error is determined by the precision of the measuring instrument. For example, when measuring length, it is necessary to estimate between the marks on the measuring tape. If these marks are 1 cm apart, the random error will be greater and the precision will be less than if the marks are 1 mm apart.

Systematic error is associated with an inherent problem with the measuring system, such as the presence of an interfering substance, incorrect calibration, or room conditions. For example, if the balance is not zeroed at the beginning or if the metre stick is slightly worn, all the measurements will have a systematic error.

The precision of measurements depends on the gradations of the measuring device. **Precision** is the place value of the last measurable digit. For example, a measurement of 12.74 cm is more precise than a measurement of 12.7 cm because the first value was measured to hundredths of a centimetre, whereas the latter value was measured to tenths of a centimetre.

When adding or subtracting measurements of different precision, the answer is rounded to the same precision as the least precise measurement. For example, using a calculator, add

$$11.7 \text{ cm} + 3.29 \text{ cm} + 0.542 \text{ cm} = 15.532 \text{ cm}$$

The answer must be rounded to 15.5 cm because the first measurement limits the precision to a tenth of a centimetre.

No matter how precise a measurement is, it still may not be accurate. **Accuracy** refers to how close a value is to its true value. The comparison of the two

(a)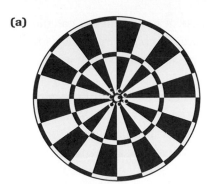

(b)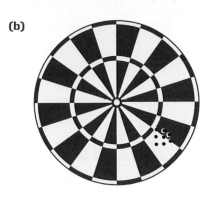

(c)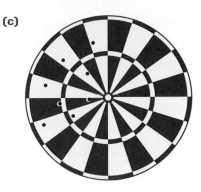

Figure 1
The positions of the darts in these diagrams are analogous to measured or calculated results in a laboratory setting. In **(a)** the results are precise and accurate, in **(b)** they are precise but not accurate, and in **(c)** they are neither precise nor accurate.

values can be expressed as a percentage difference. The percentage difference is calculated as follows:

$$\% \text{ difference} = \frac{|\text{experimental value} - \text{predicted value}|}{\text{predicted value}} \times 100\%$$

Figure 1 shows an analogy between precision and accuracy, and the positions of darts thrown at a dartboard.

How certain you are about a measurement depends on two factors: the precision of the instrument used and the size of the measured quantity. More precise instruments give more certain values. For example, a mass measurement of 13 g is less precise than a measurement of 12.76 g; you are more certain about the second measurement than the first. Certainty also depends on the measurement. For example, consider the measurements 0.4 cm and 15.9 cm. Both measurements have the same precision. If the measuring instrument is precise to ± 0.1 cm, however, the first measurement is 0.4 ± 0.1 cm (0.3 cm or 0.5 cm) for an error of 25%, whereas the second measurement is 15.9 ± 0.1 cm (15.8 cm or 16.0 cm) for an error of 0.6%. For both factors—the precision of the instrument used and the value of the measured quantity—the more digits there are in a measurement, the more certain you are about the measurement.

Significant Digits

The certainty of any measurement is communicated by the number of significant digits in the measurement. In a measured or calculated value, significant digits are the digits that are certain plus one estimated (uncertain) digit. Significant digits include all digits correctly reported from a measurement.

Follow these rules to decide if a digit is significant:

1. If a decimal point is present, zeros to the left of the first non-zero digit (leading zeros) are not significant.

2. If a decimal point is not present, zeros to the right of the last non-zero digit (trailing zeros) are not significant.

3. All other digits are significant.

4. When a measurement is written in scientific notation, all digits in the coefficient are significant.

5. Counted and defined values have infinite significant digits.

Table 2 shows some examples of significant digits.

Table 2   Significant Digits

Measurement	Number of significant digits
32.07 m	4
0.0041 g	2
5×10^5 kg	1
6400 s	2
100 people (counted)	infinite

An answer obtained by multiplying and/or dividing measurements is rounded to the same number of significant digits as the measurement with the fewest number of significant digits. For example, you could use a calculator to solve the following equation:

77.8 km/h $\times$ 0.8967 h = 69.763 26 km

However, the certainty of the answer is limited to three significant digits, so the answer is rounded up to 69.8 km.

Rounding Off

The following rules should be used when rounding answers to calculations.

1. When the first digit discarded is less than 5, the last digit retained should not be changed.

   3.141 326 rounded to four digits is 3.141.

2. When the first digit discarded is greater than 5, or if it is a 5 followed by at least one digit other than zero, the last digit retained is increased by 1 unit.

   2.221 672 rounded to four digits is 2.222.

   4.168 501 rounded to four digits is 4.169.

3. When the first digit discarded is 5 followed by only zeros, the last digit retained is increased by 1 if it is odd, but not changed if it is even.

   2.35 rounded to two digits is 2.4.

   2.45 rounded to two digits is 2.4.

   -6.35 rounded to two digits is -6.4.

Measuring and Estimating

Many people believe that all measurements are *reliable* (consistent over many trials), *precise* (to as many decimal places as possible), and *accurate* (representing the actual value). There are many things that can go wrong when measuring, however.

- There may be limitations that make the instrument or its use unreliable (inconsistent).
- The investigator may make a mistake or fail to follow the correct techniques when reading the measurement to the available precision (number of decimal places).
- The instrument may be faulty or inaccurate. A similar instrument may give different readings.

For example, when measuring the temperature of a liquid, it is important to keep the thermometer at the proper depth and the bulb of the thermometer away from the bottom and sides of the container. If you sit a thermometer with its bulb at the bottom of a liquid-filled container, you will be measuring the temperature of the bottom of the container and not the temperature of the liquid. There are similar concerns with other measurements.

To be sure that you have measured correctly, you should repeat your measurements at least three times.

If your measurements appear to be reliable, calculate the mean and use this value. To be more certain about the accuracy, repeat the measurements with a different instrument.

Every measurement is a best estimate of the actual value. The measuring instrument and the skill of the investigator determine the certainty and the precision of the measurement. The usual rule is to make a measurement that estimates between the smallest divisions on the scale of the instrument.

Logarithms

Any positive number N can be expressed as a power of some base b where $b > 1$. Some obvious examples are

$$16 = 2^4 \qquad \text{base 2, exponent 4}$$
$$25 = 5^2 \qquad \text{base 5, exponent 2}$$
$$27 = 3^3 \qquad \text{base 3, exponent 3}$$
$$0.001 = 10^{-3} \qquad \text{base 10, exponent } -3$$

In each of these examples, the exponent is an integer. Exponents may be any real number, however, not just an integer. If you use the x^y button on your calculator, you can experiment to obtain a better understanding of this concept.

The most common base is base 10. Some examples for base 10 are

$$10^{0.5} = 3.162$$
$$10^{1.3} = 19.95$$
$$10^{-2.7} = 0.001\ 995$$

By definition, the exponent to which a base b must be raised to produce a given number N is called the **logarithm** of N to base b (abbreviated as $\log_b$). When the value of the base is not written, it is assumed to be base 10. Logarithms to base 10 are called **common logarithms**. We can express the previous examples as logarithms:

$$\log 3.162 = 0.5$$
$$\log 19.95 = 1.3$$
$$\log 0.001\ 995 = -2.7$$

Most measurement scales you have encountered are linear in nature. For example, a speed of 80 km/h is twice as fast as a speed of 40 km/h and four times as fast as a speed of 20 km/h. However, there are several examples in science where the range of values of the variable being measured is so great that it is more convenient to use a logarithmic scale to base 10. One example of this is the scale for measuring the acidity of a solution (the pH scale). For example, a solution with a pH of 3 is 10 times more acidic than a solution with a pH of 4 and 100 times (10^2) more acidic than a solution with a pH of 5. Other situations that use logarithmic scales are sound intensity (the dB scale) and the intensity of earthquakes (the Richter scale).

Tables and Graphs

Both tables and graphs are used to summarize information and to illustrate patterns or relationships. Preparing tables and graphs requires some knowledge of accepted practice and some skill in designing the table or graph to best describe the information.

Tables

1. Write a title that describes the contents or the relationship among the entries in the table.

2. The rows or columns with the controlled variables and independent variable usually precede the row or column with the dependent variable.

3. Give each row or column a heading, including unit symbols in parentheses where necessary. Units are not usually written in the main body of the table (**Table 3**).

Table 3 The Effect of Concentration on Reaction Time

Concentration of HCl$_{(aq)}$ (mol/L)	Time for Reaction (s)
2.0	70
1.5	80
1.0	144
0.5	258

Graphs

1. Write a title and label the axes (**Figure 2**).
   (a) The title should be at the top of the graph. A statement of the two variables is often used as a title: for example, "Solubility versus Temperature for Sodium Chloride."
   (b) Label the horizontal (x) axis with the name of the independent variable and the

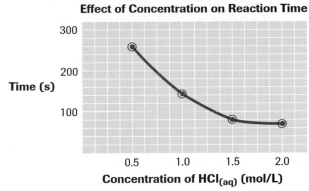

Figure 2

vertical (y) axis with the name of the dependent variable.
   (c) Include the unit symbols in parentheses on each axis label: for example, "Time (s)."

2. Assign numbers to the scale on each axis.
   (a) As a general rule, the points should be spread out so that at least one-half of the graph paper is used.
   (b) Choose a scale that is easy to read and has equal divisions. Each division (or square) must represent a small simple number of units of the variable: for example, 0.1, 0.2, 0.5, or 1.0.
   (c) It is not necessary to have the same scale on each axis or to start a scale at zero.
   (d) Do not label every division line on the axis. Scales on graphs are labelled in a way similar to the way scales on rulers are labelled.

3. Plot the points.
   (a) Locate each point by making a small dot in pencil. When you have drawn and checked all the points, draw an X over each point or circle each point in ink. The size of the circle can be used to indicate the precision of the measurement.
   (b) Be suspicious of a point that is obviously not part of the pattern. Double check the location of such a point, but do not eliminate the point from the graph just because it does not align with the rest.

4. Draw the best fitting curve.
   (a) Using a sharp pencil, draw a line that best represents the trend shown by the collection of points. Do not force the line to go through

each point. Imprecision of experimental measurements may cause some of the points to be misaligned.

(b) If the collection of points appears to fall in a straight line, use a ruler to draw the line. Otherwise draw a smooth curve that best represents the pattern of the points.

(c) Since the points are ink and the line is pencil, it is easy to change the position of the line if your first curve does not fit the points to your satisfaction.

Problem Solving in Chemistry: The Factor-Label Method

Solving problems is a basic aspect of working in all sciences, including chemistry. One of the characteristics of good chemists is their ability to solve problems. Although there are several different methods for solving mathematical problems in chemistry, the factor-label method (also known as dimensional analysis) is one of the most useful methods.

The Factor-Label Method

The factor-label method was developed as a logical and consistent way of converting a quantity in one unit into the equivalent quantity in another unit. For example, if you are asked to determine how many seconds there are in 2 min, you might quickly answer, "There are 120 s in 2 min." If asked to explain your answer, you might say, "Since there are 60 s in 1 min, there are 2×60 s in 2 min, or 120 s." Notice that the solution to the problem is based on the *equality* 1 min = 60 s. Mathematically,

1 min = 60 s
2(1 min) = 2(60 s)
2 min = 120 s

To solve the same problem using the factor-label method, you determine the required value (the value you are asked to find) by multiplying the given value by a *conversion factor*.

required value = given value $\times$ conversion factor

The conversion factor is an equality that relates the units of the required value (e.g., seconds) to the units of the given value (e.g., minutes). For the equality 1 min = 60 s, the conversion factor is obtained by stating the equality in the form of a fraction equal to 1. In this case,

$$\frac{1 \text{ min}}{60 \text{ s}} = 1 \quad \text{or} \quad \frac{60 \text{ s}}{1 \text{ min}} = 1$$

These fractions are equal to 1 because, in both cases, the numerators and denominators are of equal value: 60 s is the same length of time as 1 min, and vice versa. All conversion factors equal 1. The only difference is that one fraction is inverted when compared to the other fraction.

The conversion factor you use in the solution to a problem depends on the units of the given value. *Choose the form of the conversion factor whose denominator has the same units as the given value.* Since multiplying by a conversion factor is like multiplying by 1, only the units change.

Using the factor-label method, how many seconds are in a time (t) of 2 min?

$$t \text{ s in 2 min} = 2.0 \text{ min} \times \frac{60 \text{ s}}{1 \text{ min}}$$

We chose the conversion factor $\frac{60 \text{ s}}{1 \text{ min}}$ because the unit in the denominator of the conversion factor (min) is the same as the unit of the given value (min). By performing the numerical calculation, and cancelling like units in the numerator and denominator (min), the required value is

$$t = 2.0 \text{ m\!i\!n} \times \frac{60 \text{ s}}{1 \text{ m\!i\!n}}$$
$$t = 120 \text{ s}$$

> ▶ **SAMPLE** problem

One tablet of a popular antacid medication contains 0.25 g of calcium carbonate, $CaCO_{3(s)}$. What mass of calcium carbonate is in 48 tablets?

Step 1: List Given Values

number of tablets = 48
$m_{CaCO_{3(s)}} = ?$

Step 2: State Problem in Form: required value = given value × conversion factor

$m_{CaCO_{3(s)}} = 48$ tablets × conversion factor

Step 3: Identify Equality and Two Possible Forms of Conversion Factor

Equality: 1 tablet = 0.25 g $CaCO_{3(s)}$

Possible conversion factors:

$$\frac{1 \text{ tablet}}{0.25 \text{ g } CaCO_{3(s)}} \quad \text{or} \quad \frac{0.25 \text{ g } CaCO_{3(s)}}{1 \text{ tablet}}$$

Step 4: Substitute Appropriate Conversion Factor into Equation and Solve

In this case, we choose $\dfrac{0.25 \text{ g } CaCO_{3(s)}}{1 \text{ tablet}}$ as the conversion factor because the unit "tablet" in the denominator cancels the unit "tablet" in the given value.

$m_{CaCO_{3(s)}} = 48$ tablets × conversion factor

$$= 48 \text{ tablets} \times \frac{0.25 \text{ g } CaCO_{3(s)}}{1 \text{ tablet}}$$

$m_{CaCO_{3(s)}} = 12$ g

Therefore, 12 g of calcium carbonate is needed to make 48 tablets of the antacid.

Example

One iron nail contains 2.6 g of iron, $Fe_{(s)}$. How many nails can be made from 143 g of iron?

Solution

$m_{Fe_{(s)}} = 143$ g

number of nails = ?

$$\text{number of nails} = 143 \text{ g } Fe_{(s)} \times \frac{1 \text{ nail}}{2.6 \text{ g } Fe_{(s)}}$$

number of nails = 55 nails

Therefore, 55 nails can be made from 143 g of iron.

▶ **Practice**

1. A 2003 Canadian $1 (loonie) coin contains 6.4 g of nickel. How many loonies can be made with 460.8 g of nickel?
2. A single compact disc (CD) has a mass of 23 g. How many discs are in a 575-g package of CDs? (Disregard the packaging material.)
3. A 250-mL cup of corn flakes contains 2.0 g of protein. What mass of protein is contained in 850 mL of corn flakes?

Answers

1. 72 loonies
2. 25 CDs
3. 6.8 g

A6 Laboratory Skills and Techniques

Using a Bunsen Burner

Practise and memorize the following. Note the safety caution. You are responsible for your safety and the safety of others near you.

1. Turn the air and gas adjustments to the off position (**Figure 1**).

2. Connect the burner hose to the gas outlet on the bench.

3. Turn the bench gas valve to the fully on position.

4. If you suspect that there may be a gas leak, replace the burner. (Give the leaky burner to your teacher.)

5. While holding a lit match above and to one side of the barrel, open the burner gas valve until a small yellow flame results (**Figure 2**). If a striker is used instead of matches, generate sparks over the top of the barrel (**Figure 3**).

6. Adjust the air flow to obtain a pale blue flame with a dual cone (**Figure 4**). For most Bunsen burners, rotating the barrel adjusts the air intake. Rotate the barrel slowly. If too much air is added, the flame may go out. If this happens, immediately turn off the gas flow and relight the burner as outlined in step 5.

7. Adjust the gas valve on the burner to increase or decrease the height of the blue flame. The hottest part of the flame is the tip of the inner blue cone. Usually a 5 to 10 cm flame, which just about touches the object heated, is used.

Figure 2
A yellow flame is relatively cool and easier to obtain on lighting.

Figure 3
To generate a spark with a striker, pull the side of the handle containing the flint up and across.

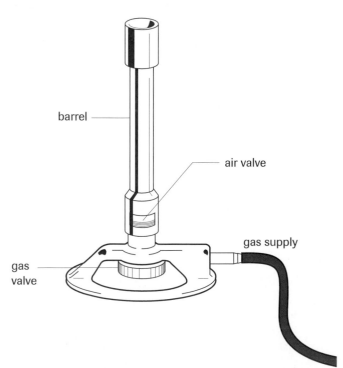

Figure 1
The parts of a common Bunsen burner

barrel

air valve

gas supply

gas valve

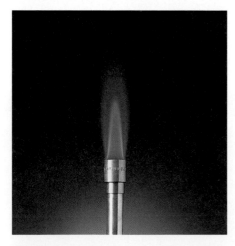

Figure 4
A pale blue-violet flame is much hotter than a yellow flame. The hottest point is at the tip of the inner blue cone.

8. Bunsen burners, when lit, should not be left unattended. If the burner is on but not being used, adjust the air and gas intakes to obtain a small yellow flame. This flame is more visible and therefore less likely to cause problems.

Using a Laboratory Balance

A balance is a sensitive instrument that is used to measure the mass of an object. There are two types of balances: electronic (**Figure 5**) and mechanical (**Figure 6**).

Figure 5
An electronic balance

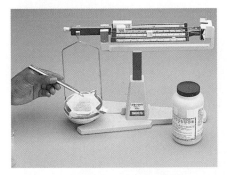

Figure 6
On this type of mechanical balance, the sample is balanced by moving masses on several beams.

Below are some general rules that you should follow when using a balance:

- All balances must be handled carefully and kept clean.
- Always place chemicals into a container, such as a beaker or plastic boat, to avoid contamination and corrosion of the balance pan.

- To avoid error due to convection currents in the air, allow hot or cold samples to return to room temperature before placing them on the balance.
- Always record masses showing the correct precision. On a centigram balance, mass is measured to the nearest hundredth of a gram (0.01 g).
- When you need to move a balance, hold the instrument by the base and steady the beam. Never lift a balance by the beams or pans.
- To avoid contaminating a whole bottle of reagent, a scoop should not be placed in the original container of a chemical. A quantity of the chemical should be poured out of the original reagent bottle into a clean, dry beaker or bottle, from which samples can be taken. Another acceptable technique for dispensing a small quantity of chemical is to rotate or tap the chemical bottle.

Using an Electronic Balance

Electronic balances are sensitive instruments, requiring care in their use. Be gentle when placing objects on the pan, and remove the pan when cleaning it. Since an electronic balance is sensitive to small movements and changes in level, do not lean on the lab bench.

To use an electronic balance, follow the steps below:

1. Place a container or weighing paper on the balance.
2. Reset (tare) the balance so the mass of the container registers as zero.
3. Add chemical until the desired mass of chemical is displayed. The last digit may not be constant, indicating uncertainty due to air currents or the high sensitivity of the balance.
4. Remove the container and sample.

Using a Mechanical Balance

There are different kinds of mechanical balance. This general procedure applies to most kinds:

1. Clean and zero the balance. (Turn the zero adjustment screw so that the beam is balanced when the instrument is set to read 0 g and no load is on the pan.)

2. Place the container on the pan.

3. Move the largest beam mass one notch at a time until the beam drops. Then move the mass back one notch.

4. Repeat this process with the next smaller mass, and continue until all the masses have been moved and the beam is balanced. (If you are using a dial-type balance, the final step will be to turn the dial until the beam balances.)

5. Record the mass of the container.

6. If you need a specific mass of a substance, set the masses on the beams to correspond to the total mass of the container plus the desired sample.

7. Add the chemical until the beam is once again balanced.

8. Remove the sample from the pan, and return all beam masses to the zero position. (For a dial-type balance, return the dial to the zero position.)

Using a Pipette

A pipette is a specially designed glass tube used to measure precise volumes of liquids. There are two types of pipettes and a variety of sizes for each type. A volumetric pipette (**Figure 7(a)**) transfers a fixed volume, such as 10.00 mL or 25.00 mL, accurate to within 0.04 mL, for example. A graduated pipette (**Figure 7(b)**) measures a range of volumes within the limit of the scale, just as a graduated cylinder does. A 10-mL graduated pipette delivers volumes accurate to within 0.1 mL.

To use a pipette, follow the steps below:

1. Rinse the pipette with small volumes of distilled water using a wash bottle, and then with the sample solution. A clean pipette has no visible residue or liquid drops clinging to the inside wall. Rinsing with aqueous ammonia and scrubbing with a pipe cleaner might be necessary to clean the pipette.

2. Hold the pipette with your thumb and fingers near the top. Leave your index finger free.

3. Place the pipette in the sample solution, resting the tip on the bottom of the container if possible. Be careful that the tip does not hit the sides of the container.

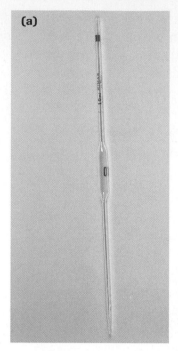
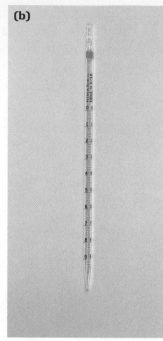

Figure 7
(a) A volumetric pipette delivers the volume printed on the label if the temperature is near room temperature.
(b) To use a graduated pipette, you must be able to start and stop the flow of the liquid.

4. Squeeze the bulb into the palm of your hand, and place the bulb firmly and squarely on the end of the pipette (**Figure 8**) with your thumb across the top of the bulb.

Figure 8
Release the bulb slowly. Pressing down with your thumb placed across the top of the bulb maintains a good seal. Setting the pipette tip on the bottom slows the rise or fall of the liquid.

5. Release your grip on the bulb until the liquid has risen above the calibration line. (This may require bringing the level up in stages: remove the bulb, put your finger on the pipette, squeeze the air out of the bulb, replace the bulb, and continue the procedure.)

6. Remove the bulb, placing your index finger over the top.

7. Wipe all solution from the outside of the pipette using a paper towel.

8. While touching the tip of the pipette to the inside of a waste beaker, gently roll your index finger (or rotate the pipette between your thumb and fingers) to allow the liquid level to drop until the bottom of the meniscus reaches the calibration line (**Figure 9**). To avoid parallax errors, set the meniscus at eye level. Stop the flow when the bottom of the meniscus is on the calibration line. Use the bulb to raise the level of the liquid again if necessary.

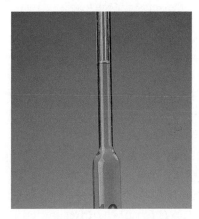

Figure 9
To allow the liquid to drop slowly to the calibration line, it is necessary for your finger and the pipette top to be dry. Also keep the tip on the bottom to slow down the flow.

9. While holding the pipette vertically, touch the pipette tip to the inside wall of a clean receiving container. Remove your finger or adjust the valve, and allow the liquid to drain freely until the solution stops flowing.

10. Finish by touching the pipette tip to the inside of the container at about a 45° angle (**Figure 10**). Do not shake the pipette. The delivery pipette is calibrated to leave a small volume in the tip.

Figure 10
A vertical volumetric pipette is drained by gravity, and then the tip is placed against the inside wall of the container. A small volume is expected to remain in the tip.

Crystallization

Crystallization is used to separate a solid from a solution by evaporating the solvent or lowering the temperature. Evaporating the solvent is useful for quantitative analysis of a solution; lowering the temperature is commonly used to purify and separate a solid whose solubility is temperature-sensitive. Chemicals that have a low boiling point or decompose on heating cannot be separated by crystallization using a heat source. Fractional distillation is an alternative design for the separation of a mixture of liquids.

1. Measure the mass of a clean beaker or evaporating dish.

2. Place a precisely measured volume of the solution in the container.

3. Set the container aside to evaporate the solution slowly, or warm the container gently on a hot plate or with a Bunsen burner.

4. When the contents appear dry, measure the mass of the container and solid.

5. Heat the solid with a hot plate or burner, cool it, and measure the mass again.

6. Repeat step 5 until the final mass remains constant. (Constant mass indicates that all of the solvent has evaporated.)

Filtration

In filtration, solid is separated from a mixture using a porous filter paper. The more porous papers are called qualitative filter papers. Quantitative filter papers allow only invisibly small particles through the pores of the paper.

1. Set up a filtration apparatus (**Figure 11**): stand, funnel holder, filter funnel, waste beaker, wash bottle, and a stirring rod with a flat plastic or rubber end for scraping.

Figure 11
The tip of the funnel should touch the inside wall of the collecting beaker.

2. Fold the filter paper along its diameter, and then fold it again to form a cone. A better seal of the filter paper on the funnel is obtained if a small piece of the outside corner of the filter paper is torn off (**Figure 12**).

3. Measure and record the mass of the filter paper after removing the corner.

4. While holding the open filter paper in the funnel, wet the entire paper and seal the top edge firmly against the funnel with the tip of the cone centred in the bottom of the funnel.

5. With the stirring rod touching the spout of the beaker, decant most of the solution into the funnel (**Figure 13**). Transferring the solid too soon clogs the pores of the filter paper. Keep the level of liquid about two-thirds up the height of the filter paper. The stirring rod should be rinsed each time it is removed.

Figure 13
The separation technique of pouring off clear liquid is called decanting. Pouring along the stirring rod prevents drops of liquid from going down the outside of the beaker when you stop pouring.

6. When most of the solution has been filtered, pour the remaining solid and solution into the funnel. Use the wash bottle and the flat end of the stirring rod to clean any remaining solid from the beaker.

7. Use the wash bottle to rinse the stirring rod and the beaker.

8. Wash the solid two or three times to ensure that no solution is left in the filter paper. Direct a gentle stream of water around the top of the filter paper.

9. When the filtrate has stopped dripping from the funnel, remove the filter paper. Press your thumb against the thick (three-fold) side of the filter

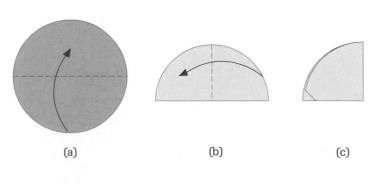

(a)   (b)   (c)   (d)

Figure 12
To prepare a filter paper, fold it in half twice and then remove the outside corner as shown.

paper, and slide the paper up the inside of the funnel.

10. Transfer the filter paper from the funnel onto a labelled watch glass, and unfold the paper to let the precipitate dry.

11. Determine the mass of the filter paper and dry precipitate.

Preparation of Standard Solutions

Laboratory procedures often call for the use of a solution of specific, precise concentration. The apparatus that is used to prepare such a solution is a volumetric flask. A meniscus finder is useful in setting the bottom of the meniscus on the calibration line (**Figure 14**).

Figure 14
Raise the meniscus finder along the back of the neck of the volumetric flask until the meniscus is outlined as a sharp, black line against a white background.

Preparing a Standard Solution from a Solid Reagent

1. Calculate the required mass of solute from the volume and concentration of the solution.

2. Obtain the required mass of solute in a clean, dry beaker or weighing boat. (Refer to "Using a Laboratory Balance" earlier in this section.)

3. Dissolve the solid in pure water using less than one-half of the final solution volume.

4. Transfer the solution and all water used to rinse the equipment into a clean volumetric flask. (The beaker and any other equipment should be rinsed two or three times with pure water.)

5. Add pure water, using a medicine dropper for the final few millilitres while using a meniscus finder to set the bottom of the meniscus on the calibration line.

6. Stopper the flask, and mix the solution by slowly inverting the flask several times.

Preparing a Standard Solution by Dilution

1. Calculate the volume of concentrated reagent required.

2. Add approximately one-half of the final volume of pure water to the volumetric flask.

3. Measure the required volume of stock solution using a pipette. (Refer to "Using a Pipette" earlier in this section.)

4. Transfer the stock solution slowly into the volumetric flask while mixing.

5. Add pure water. Use a medicine dropper and a meniscus finder to set the bottom of the meniscus on the calibration line.

6. Stopper the flask, and mix the solution by slowly inverting the flask several times.

Titration

Titration is used in the volumetric analysis of a solution of an unknown concentration. Titration involves adding a solution (the titrant) from a burette to another solution (the sample) in an Erlenmeyer flask until a recognizable endpoint, such as a colour change, occurs.

1. Rinse the burette with small volumes of distilled water using a wash bottle. Using a burette funnel, rinse with small volumes of the titrant (**Figure 15**, on the next page). (If liquid droplets remain on the sides of the burette after rinsing, scrub the burette with a burette brush. If the tip of the burette is chipped or broken, replace the tip or the whole burette.)

2. Using a small burette funnel, pour the solution into the burette until the level is near the top. Open the stopcock for maximum flow to clear any air bubbles from the tip and to bring the liquid level down to the scale.

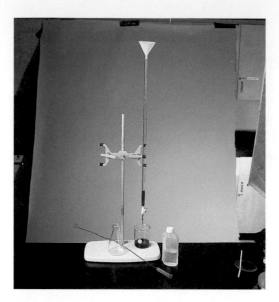

Figure 15
A burette should be rinsed with water and then the titrant before use. Use a burette brush only if necessary.

3 Record the initial burette reading to the nearest 0.01 mL. (Estimate the second decimal place.) Avoid parallax errors by reading volumes at eye level with the aid of a meniscus finder.

4. Pipette a sample of the solution of unknown concentration into a clean Erlenmeyer flask. Place a white piece of paper beneath the Erlenmeyer flask to make it easier to detect colour changes.

5. Add an indicator if one is required. Add the smallest quantity necessary (usually one to two drops) to produce a noticeable colour change in your sample.

6. Add the solution from the burette quickly at first, and then slowly, drop by drop, near the endpoint (**Figure 16**). Stop as soon as a drop of the titrant produces a permanent colour change in the sample solution. A permanent colour change is considered to be a noticeable change that lasts for 10 s after swirling.

7. Record the final burette reading to the nearest 0.01 mL.

8. The final burette reading for one trial becomes the initial burette reading for the next trial. Three trials with results within 0.02 mL are normally required for a reliable analysis of a solution of unknown concentration.

Figure 16
Near the endpoint, continuous gentle swirling of the solution is particularly important.

9. Drain and rinse the burette with pure water. Store the burette upside down with the stopcock open.

Diagnostic Tests

The tests described in **Table 1** are commonly used to detect the presence of a specific substance. Thousands more are possible. All diagnostic tests include a brief procedure, some expected evidence, and an interpretation of the evidence obtained. This is conveniently communicated using the format "If [procedure] and [evidence], then [analysis]."

Diagnostic tests can be designed using any characteristic property of a substance. For example, diagnostic tests for acids, bases, and neutral substances can be specified in terms of the pH values of the solutions. For specific chemical reactions, properties of the products that the reactants do not have (such as the insolubility of a precipitate, the production of a gas, or the colour of ions in aqueous solutions) can be used to design diagnostic tests.

If possible, you should use a control to illustrate that the test does not give the same results with other substances. For example, in the test for oxygen, inserting a glowing splint into a test tube that contains only air is used to compare the effect of air on the splint with a test tube in which you expect oxygen has been collected.

For a test to be valid, it usually has to be conducted both before and after a chemical change. Consider this control when planning your designs and procedures.

Table 1 Some Standard Diagnostic Tests

Substance Tested	Diagnostic Test
water	If cobalt(II) chloride paper is exposed to a liquid or vapour, and the paper turns from blue to pink, then water is likely present.
oxygen	If a glowing splint is inserted into the test tube, and the splint glows brighter or relights, then oxygen gas is likely present.
hydrogen	If a flame is inserted into the test tube, and a squeal or pop is heard, then hydrogen is likely present.
carbon dioxide	If the gas is bubbled into a limewater solution, and the limewater turns cloudy, then carbon dioxide is likely present.
halogens	If a few millilitres of chlorinated hydrocarbon solvent is added, with shaking, to a solution in a test tube, and the colour of the solvent appears to be • light yellow-green, then chlorine is likely present; • orange, then bromine is likely present; • purple, then iodine is likely present.
acid	If strips of blue and red litmus paper are dipped into the solution, and the blue litmus turns red, then an acid is present.
base	If strips of blue and red litmus paper are dipped into the solution, and the red litmus turns blue, then a base is present.
neutral solution	If strips of blue and red litmus paper are dipped into the solution, and neither changes colour, then only neutral substances are likely present.
neutral ionic solution	If a neutral substance is tested for conductivity with a voltmeter or multimeter, and the solution conducts a current, then a neutral ionic substance is likely present.
neutral molecular solution	If a neutral solution is tested and does not conduct a current, then a neutral molecular substance is likely present.

B1 Safety Conventions and Symbols

Although every effort is undertaken to make the science experience a safe one, there are inherent risks associated with some scientific investigations. These risks are generally associated with the materials and equipment used and the disregard of safety instructions that accompany investigations. However, there may also be risks associated with the location of the investigation, whether in the science laboratory, at home, or outdoors. Most of these risks pose no more danger than one would normally experience in everyday life. With an awareness of the possible hazards, knowledge of the rules, appropriate behaviour, and a little common sense, these risks can be practically eliminated.

Remember that you share the responsibility for not only your own safety but also the safety of those around you. Always alert the teacher in case of an accident.

In this textbook, chemicals, equipment, and procedures that are hazardous are highlighted in red and are preceded by the appropriate Workplace Hazardous Materials Information System (WHMIS) symbol or by ✋ .

WHMIS Symbols and HHPSs

The Workplace Hazardous Materials Information System (WHMIS) provides workers and students with complete and accurate information regarding hazardous products. All chemical products that are supplied to schools, businesses, and industries must contain standardized labels and be accompanied by Material Safety Data Sheets (MSDSs) providing detailed information about the product. Clear and standardized labelling is an important component of WHMIS (**Table 1**). The labels must be present on the product's original container or be added to other containers if the product is transferred.

The Canadian Hazardous Products Act requires manufacturers of consumer products containing chemicals to include a symbol that specifies both the nature of the primary hazard and the degree of this hazard. In addition, any secondary hazards, first-aid treatment, storage, and disposal must be noted.

Household Hazardous Product Symbols (HHPSs) are used to show the hazard and the degree of the hazard by the type of border surrounding the illustration (**Figure 1**).

	Corrosive
	This material can burn your skin and eyes. If you swallow it, it will damage your throat and stomach.
	Flammable
	This product or the gas (or vapour) from it can catch fire quickly. Keep this product away from heat, flames, and sparks.
	Explosive
	Container will explode if it is heated or if a hole is punched in it. Metal or plastic can fly out and hurt your eyes and other parts of your body.
	Poison
	If you swallow or lick this product, you could become very sick or die. Some products with this symbol on the label can hurt you even if you breathe (or inhale) them.

Danger

Warning

Caution

Figure 1
Hazardous household product symbols

Table 1 The Workplace Hazardous Materials Information System (WHMIS)

Class and type of compounds	WHMIS symbol	Risks	Precautions
Class A: *Compressed Gas* Material that is normally gaseous and kept in a pressurized container		• could explode due to pressure • could explode if heated or dropped • may be possible hazard from both the force of explosion and the release of contents	• ensure container is always secured • store in designated areas • do not drop or allow to fall
Class B: *Flammable and Combustible Materials* Materials that will continue to burn after being exposed to a flame or other ignition source		• may ignite spontaneously • may release flammable products if allowed to degrade or if exposed to water	• store in designated areas • work in well-ventilated areas • avoid heating • avoid sparks and flames • ensure that electrical sources are safe
Class C: *Oxidizing Materials* Materials that can cause other materials to burn or support combustion		• can cause skin or eye burns • increase fire and explosion hazards • may cause combustibles to explode or react violently	• store away from combustibles • wear body, hand, face, and eye protection • store in containers that will not rust or oxidize
Class D: *Toxic Materials Immediate and Severe* Poisons and potentially fatal materials that cause immediate and severe harm		• may be fatal if ingested or inhaled • may be absorbed through the skin • small volumes have a toxic effect	• avoid breathing dust or vapours • avoid contact with skin or eyes • wear protective clothing, and face and eye protection • work in well-ventilated areas, and wear breathing protection
Class D: *Toxic Materials Long Term Concealed* Materials that have a harmful effect after repeated exposures or over a long period		• may cause death or permanent injury • may cause birth defects or sterility • may cause cancer • may be sensitizers causing allergies	• wear appropriate personal protection • work in a well-ventilated area • store in appropriate designated areas • avoid direct contact • use hand, body, face, and eye protection • ensure that respiratory and body protection is appropriate for the specific hazard
Class D: *Biohazardous Infectious Materials* Infectious agents or a biological toxin causing a serious disease or death		• may cause anaphylactic shock • include viruses, yeasts, moulds, bacteria, and parasites that affect humans • include fluids containing toxic products • include cellular components	• take special training required to handle materials • work in designated biological areas with appropriate engineering controls • avoid forming aerosols • avoid breathing vapours • avoid contamination of people and/or area • store in special designated areas
Class E: *Corrosive Materials* Materials that react with metals and living tissue		• may irritate eyes and skin on exposure • may cause severe burns/tissue damage on longer exposure • may damage lungs if inhaled • may cause blindness if contact eyes • may cause environmental damage as a result of fumes	• wear body, hand, face, and eye protection • use breathing apparatus • ensure protective equipment is appropriate • work in a well-ventilated area • avoid all direct body contact • use appropriate storage containers and ensure nonventing closures
Class F: *Dangerously Reactive Materials* Materials that may have unexpected reactions		• may react with water • may be chemically unstable • may explode if exposed to shock or heat • may release toxic or flammable vapours • may vigorously polymerize • may burn unexpectedly	• handle with care, avoiding vibrations, shocks, and sudden temperature changes • store in appropriate containers • ensure storage containers are sealed • store and work in designated areas

B2 Safety in the Laboratory

General Safety Rules

Safety in the laboratory is an attitude and a habit more than a set of rules. It is easier to prevent accidents than to deal with the consequences of an accident. Most of the following rules are common sense.

- Do not enter a laboratory or prep room unless a teacher or another supervisor is present, or you have permission to do so.
- Familiarize yourself with your school's safety regulations.
- Make your teacher aware of any allergies or other health problems you may have.
- Listen carefully to any instructions given by your teacher, and follow them closely.
- Wear eye protection, lab aprons or coats, and protective gloves when appropriate.
- Wear closed shoes (not sandals) when working in the laboratory.
- Place your books and bags away from the work area. Keep your work area clear of all materials except those that you will use in the investigation.
- Do not chew gum, eat, or drink in the laboratory. Food should not be stored in refrigerators in laboratories.
- Know the location of MSDS information, exits, and all safety equipment, such as the fire blanket, fire extinguisher, and eyewash station.
- Use stands, clamps, and holders to secure any potentially dangerous or fragile equipment that could be tipped over.
- Avoid sudden or rapid motion in the laboratory that may interfere with someone carrying or working with chemicals or using sharp instruments.
- Never engage in horseplay or practical jokes in the laboratory.
- Ask for assistance when you are not sure how to do a procedural step.

- Never attempt unauthorized experiments.
- Never work in a crowded area or alone in the laboratory.
- Report all accidents.
- Clean up all spills, even spills of water, immediately.
- Always wash your hands with soap and water before or immediately after you leave the laboratory. Wash your hands before you touch any food.
- Do not forget safety procedures when you leave the laboratory. Accidents can also occur outdoors, at home, or at work.

Eye and Face Safety

- Wear approved eye protection in a laboratory, no matter how simple or safe the task appears to be. Keep the eye protection over your eyes, not on top of your head. For certain experiments, full face protection (safety goggles or a face shield) may be necessary.
- Never look directly into the opening of flasks or test tubes.
- If, in spite of all precautions, you get a chemical in your eye, quickly use the eyewash or nearest cold running water. Continue to rinse the eye with water for at least 15 min. This is a very long time—have someone time you. Have another student inform your teacher of the accident. The injured eye should be examined by a doctor.
- If you must wear contact lenses in the laboratory, be extra careful. Whether or not you wear contact lenses, do not touch your eyes without first washing your hands. If you do wear contact lenses, make sure that your teacher is aware of it. Carry your lens case and a pair of glasses with you.
- If a piece of glass or another foreign object enters your eye, seek immediate medical attention.

- Do not stare directly at any bright source of light (such as a piece of burning magnesium ribbon, lasers, or the Sun). You will not feel any pain if your retina is being damaged by intense radiation. You cannot rely on the sensation of pain to protect you.
- When working with lasers, be aware that a reflected laser beam can act like a direct beam on the eye.

Handling Glassware Safely

- Never use glassware that is cracked or chipped. Give such glassware to your teacher or dispose of it as directed. Do not put the item back into circulation.
- Never pick up broken glassware with your fingers. Use a broom and dustpan.
- Do not put broken glassware into garbage containers. Dispose of glass fragments in special containers marked "Broken Glass."
- Heat glassware only if it is approved for heating. Check with your teacher before heating any glassware.
- If you cut yourself, inform your teacher immediately. Embedded glass or continued bleeding requires medical attention.
- If you need to insert glass tubing or a thermometer into a rubber stopper, get a cork borer of a suitable size. Insert the borer in the hole of the rubber stopper, starting from the small end of the stopper. Once the borer is pushed all the way through the hole, insert the tubing or thermometer through the borer. Ease the borer out of the hole, leaving the tubing or thermometer inside. To remove the tubing or thermometer from the stopper, push the borer from the small end through the stopper until it shows from the other end. Ease the tubing or thermometer out of the borer.
- Protect your hands with heavy gloves or several layers of cloth before inserting glass into rubber stoppers.

- Be very careful while cleaning glassware. There is an increased risk of breakage from dropping when the glassware is wet and slippery.

Using Sharp Instruments Safely

- Make sure that your instruments are sharp. Surprisingly, one of the main causes of accidents with cutting instruments is the use of dull instruments. Dull cutting instruments require more pressure than sharp instruments and are therefore much more likely to slip.
- Select the appropriate instrument for the task. Never use a knife when scissors would work better.
- Always cut away from yourself and others.
- If you cut yourself, inform your teacher immediately and get appropriate first aid.
- Be careful when working with wire cutters or wood saws. Use a cutting board when needed.

Heat and Fire Safety

- In a laboratory where burners or hot plates are being used, never pick up a glass object without first checking the temperature by lightly and quickly touching the item, or by placing your hand near, but not on, the item. Glass items that have been heated stay hot for a long time, but do not appear to be hot. Metal items such as ring stands and hot plates can also cause burns; take care when touching them.
- Do not use a Bunsen burner near wooden shelves, flammable liquids, or any other item that is combustible.
- Before using a Bunsen burner, make sure that long hair is tied back. Do not wear loose clothing (wide long sleeves should be tied back or rolled up).
- Never look down the barrel of a Bunsen burner.
- Always pick up a burner by the base, never by the barrel.
- Never leave a lighted Bunsen burner unattended.

- If you burn yourself, *immediately* run cold water gently over the burned area or immerse the burned area in cold water and inform your teacher.
- Make sure that heating equipment, such as a burner, hot plate, or electrical equipment, is secure on the bench and clamped in place when necessary.
- Always assume that hot plates and electric heaters are hot, and use protective gloves when handling them.
- Keep a clear workplace when performing experiments with heat.
- When heating a test tube over a Bunsen burner, use a test-tube holder and a spurt cap. Holding the test tube at an angle, facing away from you and others, gently move the test tube backward and forward through the flame.
- Remember to include a "cooling" time in your experiment plan. Do not put away hot equipment.
- Very small fires in a container may be extinguished by covering the container with a wet paper towel or ceramic square.
- For larger fires, inform your teacher and follow your teacher's instructions for using fire extinguishers, blankets, and alarms, and for evacuation. Do not attempt to deal with a fire by yourself.
- If anyone's clothes or hair catch fire, tell the person to drop to the floor and roll. Then use a fire blanket to help smother the flames.

Electrical Safety

- Water or wet hands should never be used near electrical equipment.
- Do not operate electrical equipment near running water or any large containers of water.
- Check the condition of electrical equipment. Do not use if wires or plugs are damaged, or if the ground pin has been removed.
- Make sure that electrical cords are not placed where someone could trip over them.

- When unplugging equipment, remove the plug gently from the socket. Do not pull on the cord.
- When using variable power supplies, start at low voltage and increase slowly.

Handling Chemicals Safely

Many chemicals are hazardous to some degree. When using chemicals, operate under the following principles:

1. Never underestimate the risks associated with chemicals. Assume that any unknown chemicals are hazardous.
2. If you can substitute, use a less hazardous chemical wherever possible.
3. Reduce exposure to chemicals to the absolute minimum. Avoid direct skin contact if possible.
4. When using chemicals, ensure that there is adequate ventilation.

The following guidelines do not address every possible situation but, used with common sense, are appropriate for situations in the high-school laboratory.

- Consult the MSDS before you use a chemical.
- Wear appropriate eye protection at all times where chemicals are used or stored. Wear a lab coat and/or other protective clothing (such as an apron and gloves).
- When carrying chemicals, hold containers carefully using two hands, one around the container and one underneath.
- Read all labels to ensure that the chemicals you have selected are the intended ones. Never use the contents of a container that has no label or has an illegible label. Give any such containers to your teacher.
- Label all chemical containers correctly to avoid confusion about contents.
- Never pipette or start a siphon by mouth. Use a pipette bulb or similar device.
- Pour liquid chemicals carefully (down the side of the receiving container or down a stirring rod) to ensure that they do not splash. Always pour from

the side opposite the label. If everyone follows this rule, drips will always form on the same side, away from your hand.

- Always pour volatile chemicals in a fume hood or in a well-ventilated area.
- Never smell or taste chemicals.
- Return chemicals to their correct storage place. Chemicals are stored by hazard class.
- If you spill a chemical, use a chemical spill kit to clean up.
- Do not return surplus chemicals to stock bottles. Dispose of excess chemicals in the appropriate manner.
- Clean up your work area, the fume hood, and any other area where chemicals were used.
- Wash your hands immediately after handling chemicals and before leaving the lab, even if you wore gloves.

Waste Disposal

Waste disposal at school, at home, or at work is a social and environmental issue. To protect the environment, federal and provincial governments have regulations to control wastes, especially chemical wastes. For example, the WHMIS program applies to controlled products that are being handled. (When being transported, they are regulated under the *Transport of Dangerous Goods Act*, and for disposal they are subject to federal, provincial, and municipal regulations.) Most laboratory waste can be washed down the drain or, if it is in solid form, placed in ordinary garbage containers. Some waste, however, must be treated more carefully. It is your responsibility to follow procedures and dispose of waste in the safest possible manner according to your teacher's instructions.

Flammable Substances

Flammable liquids should not be washed down the drain. Special fire-resistant containers are used to store flammable liquid waste. Waste solids that pose a fire hazard should be stored in fireproof containers. Care must be taken not to allow flammable waste to come into contact with any sparks, flames, other ignition sources, or oxidizing materials. The method of disposal depends on the nature of the substance.

Corrosive Solutions

Solutions that are corrosive but not toxic, such as acids and bases, can usually be washed down the drain, but care should be taken to ensure that they are first either neutralized or diluted to low concentration. While disposing of such a substance, use large quantities of water and continue to pour water down the drain for a few minutes after all the substance has been washed away.

Heavy Metal Solutions

Heavy metal compounds (for example, lead, mercury, and cadmium compounds) should not be flushed down the drain. These substances are cumulative poisons and should be kept out of the environment. Pour any heavy metal waste into the special container marked "Heavy Metal Waste." Remember that paper towels used to wipe up solutions of heavy metals, as well as filter papers with heavy metal compounds embedded in them, should be treated as solid toxic waste.

Disposal of heavy metal solutions is usually accomplished by precipitating the metal ion (for example, as lead(II) silicate) and disposing of the solid. Heavy metal compounds should not be placed in school garbage containers. Usually, waste disposal companies collect materials that require special disposal and dispose of them as required by law.

Toxic Substances

Toxic chemicals and solutions of toxic substances should not be poured down the drain. They should be retained for disposal by a licensed waste disposal company.

Organic Material

Remains of plants and animals can generally be disposed of in the normal school garbage containers. Animal dissection specimens should be rinsed thoroughly to rid them of any excess preservative and sealed in plastic bags.

Fungi and bacterial cultures should be autoclaved or treated with a fungicide or antibacterial soap before disposal.

First Aid

The following guidelines apply if an injury, such as a burn, cut, chemical spill, ingestion, inhalation, or splash in eyes, happens to you or to one of your classmates.

- If an injury occurs, inform your teacher immediately. If the injury appears serious, call for emergency assistance immediately.
- Know the location of the first-aid kit, fire blanket, eyewash station, and shower, and be familiar with the contents/operation.
- If the injury is the result of a chemical, drench the affected area with a continuous flow of water for 30 min. Clothing should be removed as necessary. Inform your teacher. Retrieve the Material Safety Data Sheet (MSDS) for the chemical; this sheet provides information about the first-aid requirements for the chemical. If the chemical is splashed in your eyes, have another student assist you in getting to the eyewash station immediately. Rinse with the eyes open for at least 15 min.
- If you have ingested or inhaled a hazardous substance, inform your teacher immediately. The MSDS will give information about the first-aid requirements for the substance in question. Contact the Poison Control Centre in your area.
- If the injury is from a burn, immediately immerse the affected area in cold water. This will reduce the temperature and prevent further tissue damage.
- In the event of electrical shock, do not touch the affected person or the equipment the person was using. Break contact by switching off the source of electricity or by removing the plug.
- If a classmate's injury has rendered him or her unconscious, notify your teacher immediately. Your teacher will perform CPR if necessary. Do not administer CPR unless under specific instructions from your teacher. You can assist by keeping the person warm and reassured.

C1 Units, Symbols, and Prefixes

Throughout *Nelson Chemistry 12: College Preparation,* including this reference section, we have attempted to be consistent in the presentation and usage of quantities, units, and their symbols. As far as possible, the Système international d'unités (SI) has been used. Some other units have also been included, however, because of their practical importance, wide usage, or use in specialized fields. In our interpretations and usage, *Nelson Chemistry 12: College Preparation* has followed the most recent *Canadian Metric Practice Guide* (CAN/CSA–Z234.1–89), published in 1989 and reaffirmed in 1995 by the Canadian Standards Association.

Numerical Prefixes

Prefix	Power	Symbol
deca-	10^1	da
hecto-	10^2	h
kilo-	10^3	k*
mega-	10^6	M*
giga-	10^9	G*
tera-	10^{12}	T
peta-	10^{15}	P
exa-	10^{18}	E
deci-	10^{-1}	d
centi-	10^{-2}	c*
milli-	10^{-3}	m*
micro-	10^{-6}	μ*
nano-	10^{-9}	n*
pico-	10^{-12}	p
femto-	10^{-15}	f
atto-	10^{-18}	a

* commonly used

Examples of Prefix Use

0.0034 mol $= 3.4 \times 10^{-3}$ mol $=$ 3.4 **milli**moles or 3.4 mmol

1530 L $= 1.53 \times 10^3$ L $=$ 1.53 **kilo**litres or 1.53 kL

SI Base Units

Quantity	Symbol	Unit name	Symbol
amount of substance	n	mole	mol
electric current	I	ampere	A
length	L, l, h, d, w	metre	m
luminous intensity	I_v	candela	cd
mass	m	kilogram	kg
temperature	T	kelvin	K
time	t	second	s

Defined (Exact) Quantities

1 mL	$=$	1 cm^3
1 kL	$=$	1 m^3
1000 kg	$=$	1 t
1 Mg	$=$	1 t
1 atm	$=$	101.325 kPa
0°C	$=$	273.15 K
STP	$=$	0°C and 101.325 kPa
SATP	$=$	25°C and 100 kPa

Common Multiples

Multiple	Prefix
1	mono–
2	bi–, di–
3	tri–
4	tetra–
5	penta–
6	hexa–
7	hepta–
8	octa–
9	nona–
10	deca–

C2 Common Chemicals

You live in a chemical world. As one bumper sticker asks, "What in the world isn't chemistry?" Every natural and technologically produced substance around you is composed of chemicals. Many of these chemicals are used to make your life easier or safer, and some of them have life-saving properties. Following is a list of selected common chemicals.

Common name	Recommended name	Formula	Common use/source
acetic acid	ethanoic acid	$HC_2H_3O_{2(aq)}$; $CH_3COOH_{(aq)}$	vinegar
acetone	propanone	$(CH_3)_2CO_{(l)}$	nail polish remover
acetylene	ethyne	$C_2H_{2(g)}$	cutting/welding torch
ASA (Aspirin®)	acetylsalicylic acid	$HC_9H_7O_{4(s)}$	pain-relief medication
baking soda	sodium hydrogen carbonate	$NaHCO_{3(s)}$	leavening agent
battery acid	sulfuric acid	$H_2SO_{4(aq)}$	car batteries
bleach	sodium hypochlorite	$NaClO_{(s)}$	bleach for clothing
bluestone	copper(II) sulfate pentahydrate	$CuSO_4{\cdot}5\,H_2O_{(s)}$	algicide, fungicide
brine	aqueous sodium chloride	$NaCl_{(aq)}$	water-softening agent
CFC	chlorofluorocarbon	$C_xCl_yF_{z(l)}$; e.g., $C_2Cl_2F_{4(l)}$	refrigerant
charcoal/graphite	carbon	$C_{(s)}$	fuel, lead pencils
citric acid	2-hydroxy-1,2,3-propanetricarboxylic acid	$H_3C_8H_5O_{7(s)}$	in fruit and beverages
carbon dioxide	carbon dioxide	$CO_{2(g)}$	dry ice, carbonated beverages
ethylene	ethene	$C_2H_{4(g)}$	for polymerization
ethylene glycol	1,2-ethanediol	$C_2H_4(OH)_{2(l)}$	radiator antifreeze
freon-12	dichlorodifluoromethane	$CCl_2F_{2(l)}$	refrigerant
Glauber's salt	sodium sulfate decahydrate	$Na_2SO_4{\cdot}10\,H_2O_{(s)}$	solar heat storage
glucose	D-glucose, dextrose	$C_6H_{12}O_{6(s)}$	in plants and blood
grain alcohol	ethanol (ethyl alcohol)	$C_2H_5OH_{(l)}$	beverage alcohol
gypsum	calcium sulfate dihydrate	$CaSO_4{\cdot}2\,H_2O_{(s)}$	wallboard
lime (quicklime)	calcium oxide	$CaO_{(s)}$	masonry
limestone	calcium carbonate	$CaCO_{3(s)}$	chalk, building materials
lye (caustic soda)	sodium hydroxide	$NaOH_{(s)}$	oven/drain cleaner
malachite	copper(II) hydroxide carbonate	$Cu(OH)_2{\cdot}CuCO_{3(s)}$	copper mineral
methyl hydrate	methanol (methyl alcohol)	$CH_3OH_{(l)}$	gas line antifreeze
milk of magnesia	magnesium hydroxide	$Mg(OH)_{2(s)}$	antacid (for indigestion)
MSG	monosodium glutamate	$NaC_5H_8NO_{4(s)}$	flavour enhancer
muriatic acid	hydrochloric acid	$HCl_{(aq)}$	concrete etching
natural gas	methane	$CH_{4(g)}$	fuel
PCB	polychlorinated biphenyl	$(C_6H_xCl_y)_2$; e.g., $(C_6H_4Cl_2)_{2(l)}$	in transformers
potash	potassium chloride	$KCl_{(s)}$	fertilizer
road salt	calcium chloride or sodium chloride	$CaCl_{2(s)}$ or $NaCl_{(s)}$	for melting ice
rotten-egg gas	hydrogen sulfide	$H_2S_{(g)}$	in natural gas
rubbing alcohol	2-propanol (also isopropanol)	$CH_3CHOHCH_{3(l)}$	for massage
sand (silica)	silicon dioxide	$SiO_{2(s)}$	in glassmaking
slaked lime	calcium hydroxide	$Ca(OH)_{2(s)}$	limewater
soda ash	sodium carbonate	$Na_2CO_{3(s)}$	in laundry detergents
sugar	sucrose	$C_{12}H_{22}O_{11(s)}$	sweetener
table salt	sodium chloride	$NaCl_{(s)}$	seasoning
vitamin C	ascorbic acid	$H_2C_6H_6O_{6(s)}$	vitamin supplement
washing soda	sodium carbonate decahydrate	$Na_2CO_3{\cdot}10\,H_2O_{(s)}$	water softener

C3 The Elements

Element	Symbol	Atomic number	Ionization energy (kJ/mol)	Electro-negativity	Electron affinity (kJ/mol)	Ionic radius (pm)	Common ion charge
actinium	Ac	89	509	1.1		111	3+
aluminum	Al	13	578	1.5	42.5	50	3+
americium	Am	95	578	1.3		97.5	3+
antimony	Sb	51	834	1.9	100.9	76	3+
argon	Ar	18	1521				
arsenic	As	33	947	2.0	78	222	
astatine	At	85		2.2	[270]	227	1–
barium	Ba	56	503	0.89	[14]	135	2+
berkelium	Bk	97	601	1.3		98	3+
beryllium	Be	4	899	1.5		31	2+
bismuth	Bi	83	703	1.9	91.3	96	3+
boron	B	5	801	2.04	26.7		
bromine	Br	35	1140	2.8	324.54	196	1–
cadmium	Cd	48	868	1.69		97	2+
calcium	Ca	20	590	1.00	1.78	99	2+
californium	Cf	98	608	1.3		95	3+
carbon	C	6	1086	2.55	121.85		
cerium	Ce	58	528	1.12		102	3+
cesium	Cs	55	376	0.7	45.50	169	1+
chlorine	Cl	17	1251	3.0	348.57	181	1–
chromium	Cr	24	653	1.6	64.3	64	3+
cobalt	Co	27	758	1.8	63.9	74.5	2+
copper	Cu	29	745	1.90	119.2	72	2+
curium	Cm	96	581	1.3		97	3+
dysprosium	Dy	66	572	1.22		91.2	3+
einsteinium	Es	99	619	1.3		98	3+
erbium	Er	68	589	1.24		89.0	3+
europium	Eu	63	547	1.2		94.7	3+
fermium	Fm	100	627	1.3		97	3+
fluorine	F	9	1681	4.0	328.16	136	1–
francium	Fr	87		0.7	[44]	180	1+
gadolinium	Gd	64	592	1.1		93.8	3+
gallium	Ga	31	579	1.6	29	62.0	3+
germanium	Ge	32	762	1.8	119.0	53.0	4+
gold	Au	79	890	2.4	222.75	91	3+
hafnium	Hf	72	680	1.3	[≈0]	78	4+
helium	He	2	2372				
holmium	Ho	67	581	1.23		90.1	3+
hydrogen	H	1	1312	2.1	72.55	10^{-3}/154	1+/1–
indium	In	49	558	1.7	29	81	3+
iodine	I	53	1008	2.5	295.15	216	1–
iridium	Ir	77	880	2.2	151.0	64	4+
iron	Fe	26	759	1.8	14.6	64.5	3+
krypton	Kr	36	1351				
lanthanum	La	57	538	1.10	[48]	106	3+
lawrencium	Lr	103				94	3+
lead	Pb	82	716	1.8	35.1	120	2+
lithium	Li	3	520	0.98	59.63	68	1+
lutetium	Lu	71	524	1.2		86.1	3+
magnesium	Mg	12	738	1.2		65	2+

Element	Symbol	Atomic number	Ionization energy (kJ/mol)	Electro-negativity	Electron affinity (kJ/mol)	Ionic radius (pm)	Common ion charge
manganese	Mn	25	717	1.5		80	2+
mendelevium	Md	101	635	1.3		114	2+
mercury	Hg	80	1007	1.9		110	2+
molybdenum	Mo	42	685	1.8	72.2	62	6+
neodymium	Nd	60	530	1.2		98.3	3+
neon	Ne	10	2081				
neptunium	Np	93	605	1.3		75	5+
nickel	Ni	28	737	1.8	111.5	72	2+
niobium	Nb	41	664	1.6	86.2	72	5+
nitrogen	N	7	1402	3.0			
nobelium	No	102	642	1.3		110	2+
osmium	Os	76	840	2.2	[19]	65	4+
oxygen	O	8	1314	3.50	140.98	140	
palladium	Pd	46	805	2.2	54.2	86	2+
phosphorus	P	15	1012	2.1	72.03	212	
platinum	Pt	78	870	2.2	205.3	70	4+
plutonium	Pu	94	585	1.3		86	4+
polonium	Po	84	812	2.0	[183]	65	4+
potassium	K	19	419	0.8	48.38	138	1+
praseodymium	Pr	59	523	1.13		99	3+
promethium	Pm	61	535	1.2		97	3+
protactinium	Pa	91	568	1.5		78	5+
radium	Ra	88	509	0.9		148	2+
radon	Rn	86	1037				
rhenium	Re	75	760	1.9	[14]	60	7+
rhodium	Rh	45	720	2.2		75	3+
rubidium	Rb	37	403	0.82	46.88	148	1+
ruthenium	Ru	44	711	2.2	[101]	77	3+
samarium	Sm	62	543	1.17		95.8	3+
scandium	Sc	21	631	1.3	18.1	81	3+
selenium	Se	34	941	2.4	194.96	198	
silicon	Si	14	786	1.8			
silver	Ag	47	731	1.93	125.6	126	1+
sodium	Na	11	496	0.93	52.87	95	1+
strontium	Sr	38	549	0.95	4.6	113	2+
sulfur	S	16	1000	2.5	200.41	184	
tantalum	Ta	73	761	1.5	31.1	68	5+
technetium	Tc	43	702	1.9	[53]	58	
tellurium	Te	52	869	2.1	190.15	221	2-
terbium	Tb	65	564	1.2	(-48)	92.3	3+
thallium	Tl	81	589	1.8	-9	144	1+
thorium	Th	90	587	1.3		94	4+
thulium	Tm	69	596	1.25		88.0	3+
tin	Sn	50	709	1.8	107.3	71	4+
titanium	Ti	22	658	1.54	7.6	68	4+
tungsten	W	74	770	1.7	78.6	65	6+
uranium	U	92	598	1.7		73	6+
vanadium	V	23	650	1.63	50.7	59	5+
xenon	Xe	54	1170				
ytterbium	Yb	70	603	1.1		86.8	3+
yttrium	Y	39	616	1.3	29.6	93	3+
zinc	Zn	30	906	1.65		74.0	2+
zirconium	Zr	40	660	1.4	41.1	79	4+

Bracketed values are calculated.
Values in parentheses are estimated.
Values in this table are taken from Lange's *Handbook of Chemistry*.

C4 Line Spectra of Elements

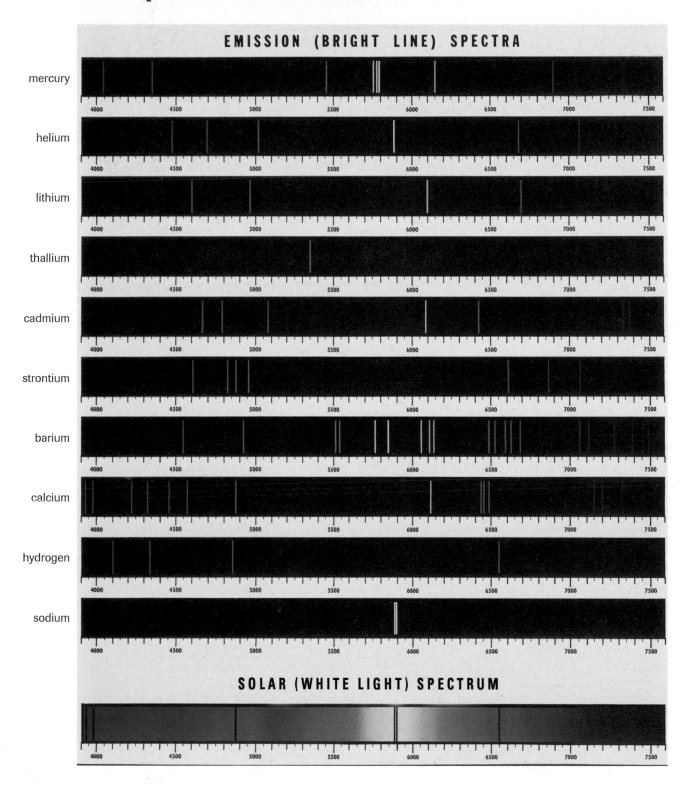

C5 Cations and Anions

Common Cations

Ion	Name
H^+	hydrogen
Li^+	lithium
Na^+	sodium
K^+	potassium
Cs^+	cesium
Be^{2+}	beryllium
Mg^{2+}	magnesium
Ca^{2+}	calcium
Ba^{2+}	barium
Al^{3+}	aluminum
Ag^+	silver

Ion	Flame
Li^+	bright red
Na^+	yellow
K^+	violet
Ca^{2+}	yellow-red
Sr^{2+}	bright red

Common Anions

Ion	Name
H^-	hydride
F^-	fluoride
Cl^-	chloride
Br^-	bromide
I^-	iodide
O^{2-}	oxide
S^{2-}	sulfide
N^{3-}	nitride
P^{3-}	phosphide

Ion	Flame
Ba^{2+}	yellow-green
Cu^{2+}	blue (halides) green (others)
Pb^{2+}	light blue-grey
Zn^{2+}	whitish green

Ion Colours

Ion	Solution colour
Groups 1, 2, 17	colourless
$Cr^{2+}_{(aq)}$	blue
$Cr^{3+}_{(aq)}$	green
$Co^{2+}_{(aq)}$	pink
$Cu^+_{(aq)}$	green
$Cu^{2+}_{(aq)}$	blue
$Fe^{2+}_{(aq)}$	pale green
$Fe^{3+}_{(aq)}$	yellow-brown
$Mn^{2+}_{(aq)}$	pale pink
$Ni^{2+}_{(aq)}$	green
$CrO_4^{2-}{}_{(aq)}$	yellow
$Cr_2O_7^{2-}{}_{(aq)}$	orange
$MnO_4^-{}_{(aq)}$	purple

Solubility Rules for Ionic Compounds in Water

Anions	+	Cations	→	Solubility of compounds
most		alkali ions (Li^+, Na^+, K^+, Rb^+, Cs^+, Fr^+)		soluble
most		hydrogen ion, $H^+_{(aq)}$		soluble
most		ammonium ion, NH_4^+		soluble
nitrate, NO_3^-		most		soluble
acetate, $C_2H_3O_2^-$		Ag^+		low solubility
		most others		soluble
chloride, Cl^- bromide, Br^- iodide, I^-		Ag^+, Pb^{2+}, Hg_2^{2+}, Cu^+, Tl^+		low solubility
		all others		soluble
sulfate, SO_4^{2-}		Ca^{2+}, Sr^{2+}, Ba^{2+}, Pb^{2+}, Ra^{2+}		low solubility
		all others		soluble
sulfide, S^{2-}		alkali ions, $H^+_{(aq)}$, NH_4^+, Be^{2+}, Mg^{2+}, Ca^{2+}, Sr^{2+}, Ba^{2+}, Ra^{2+}		soluble
		all others		low solubility
hydroxide, OH^-		alkali ions, $H^+_{(aq)}$, NH_4^+, Sr^{2+}, Ba^{2+}, Ra^{2+}, Tl^+		soluble
		all others		low solubility
phosphate, PO_4^{3-} carbonate, CO_3^{2-} sulfite, SO_3^{2-}		alkali ions, $H^+_{(aq)}$, NH_4^+		soluble
		all others		low solubility

C6 Names and Formulas of Inorganic Compounds

Prior to the late 1700s, there was no systematic method of naming compounds. Today, we are aware of a very large number of compounds, and there is every possibility that more will be discovered in the future. We need a system that is easy to use and provides information on the composition of every compound. As a result, scientists now use the system of *chemical nomenclature* chosen by the International Union of Pure and Applied Chemistry (IUPAC), sometimes called the IUPAC system, for naming compounds. Just as many immigrants to Canada have learned a new language, you will soon become familiar with the language of chemical nomenclature. Chemical nomenclature provides a systematic means of both naming and identifying compounds.

When describing a chemical compound, chemists may use its name or its formula. The formula of an ionic compound tells you the ratio of elements present. The formula of a molecular compound indicates the number of atoms of each element present in a molecule of the compound. Every chemical compound has a unique name. Knowing the names of common chemicals and being able to write their formulas correctly are useful skills in chemistry.

Naming Binary Ionic Compounds

The simplest compounds are called *binary compounds*. Binary ionic compounds consist of two types of monatomic ions (ions with of one charged atom). In the formula of a binary ionic compound, the metal cation is always written first, followed by the nonmetal anion. (This order reflects the periodic table: metals to the left and nonmetals to the right.) The name of the metal is stated in full and the name of the nonmetal ion has an *-ide* suffix. For example, $NaCl_{(s)}$ is sodium chloride, and $LiBr_{(s)}$ is lithium bromide. (See **Table 1**.) Binary ionic compounds can be made up of more than two ions, provided there are only two kinds of atoms. For example, $Al_2O_{3(s)}$ is aluminum oxide.

Table 1 Names and Ionic Charges of Some Nonmetals

Name of element	Symbol	Ionic charge	Name of compound
fluorine	F	1–	fluoride
chlorine	Cl	1–	chloride
bromine	Br	1–	bromide
iodine	I	1–	iodide
oxygen	O	2–	oxyde
sulfur	S	2–	sulphide
nitrogen	N	3–	nitride
phosphorous	P	3–	phosphide

If you know what ions make up a compound, you can often predict the compound's formula. First, you determine the charges on each type of ion making up the compound. The charge on an ion is sometimes called the *valence*. You then balance the charges to determine the simplest ratio in which they combine. You can predict the charge on the most common ion formed by each representative element by counting the number of electrons that would have to be gained or lost to obtain a stable octet. (Ionic charges for the most common ions of elements are shown in the periodic table at the back of this textbook.) For example, the compound magnesium chloride is composed of the ions Mg^{2+} and Cl^-. For a net charge of zero, the ratio of magnesium to chlorine ions must be 1 : 2. The formula of magnesium chloride is therefore $MgCl_{2(s)}$.

Example 1
Predict the formula for magnesium oxide—a source of dietary magnesium when used as a food additive.

Solution
Charge on magnesium ion: 2+
Charge on oxygen ion: 2–
The ratio of magnesium ions to oxide ions that produces a net charge of zero is 1 : 1.
The formula is therefore MgO.

You may be familiar with a method known as the crisscross rule for predicting the formula of an ionic compound. The crisscross rule works as follows:

1. Write the symbol of each of the elements in the order in which they appear in the name of the compound.

2. Write the valence number (electrons lost or gained in forming an element's most stable ion) above the symbol of each of the elements.

3. Crisscross the numbers written above the symbols such that the valence number of one element becomes a subscript on the other.

4. Divide each subscript by the highest common factor. The resulting subscripts indicate the ratio of ions present in the compound.

5. Omit any subscript equal to 1 from the formula.

For example, if you were to use the crisscross rule to predict the formula for magnesium chloride, you would go through these steps.

1. Write symbols in order.

   Mg   Cl

2. Write valences above symbols.

3. Crisscross the valences, making them subscripts.

   Mg_1 Cl_2

4. Divide subscripts by highest common factor (= 1).

   Mg_1 Cl_2

5. Remove 1 subscripts.

   $MgCl_2$

Example 2
Use the crisscross rule to predict the formula for barium sulfide.

Solution

The formula for barium sulfide is BaS.

Most transition metals and some representative metals can form more than one kind of ion. Metals that can have more than one valence, or charge, are classified as *multivalent*. For example, iron can form an Fe^{2+} ion or an Fe^{3+} ion, although Fe^{3+} is more common. The periodic table at the end of this textbook shows the most common ion of each element first, with one alternative ion charge below. It does not list all of the possible ions of the element.

Copper is a multivalent element. It is capable of bonding with chlorine in two different ratios to form two different chloride compounds: $CuCl$ and $CuCl_2$. How are the names of these compounds different? The IUPAC system of naming compounds that contain multivalent ions is very simple. The name of the metal ion includes the charge on the ion, indicated by Roman numerals in brackets. Consequently, $CuCl_{(s)}$ (in which copper has a charge of 1+) is copper(I) chloride, and $CuCl_{2(s)}$ (in which copper has a charge of 2+) is copper(II) chloride. This system of naming is sometimes referred to as the Stock system.

To determine the chemical name of a compound that contains a multivalent metal ion, you have to figure out the necessary charge on this ion to yield a net charge of zero. If you are given the formula, you simply have to calculate the equivalent negative charge. The metal's charge is then written, in Roman numerals, after the name of the metal. For example, if you were asked to name the compound MnO_2 using the IUPAC system, you would first look at the charge on the nonmetal ions. In this case, the charge on each O is 2–, so the total negative charge is 4–. The charge on the Mn ion must be 4+. Consequently, the IUPAC name for MnO_2 is manganese(IV) oxide.

If the ion of a multivalent metal is not specified in a name, it is assumed that the charge on the ion is the most common one.

Table 2 Classical and IUPAC Names of Common Multivalent Metal Ions

Metal	Ion	Classical name	IUPAC name
iron	Fe^{2+}	ferrous	iron(II)
	Fe^{3+}	ferric	iron(III)
copper	Cu^+	cuprous	copper(I)
	Cu^{2+}	cupric	copper(II)
tin	Sn^{2+}	stannous	tin(II)
	Sn^{4+}	stannic	tin(IV)
lead	Pb^{2+}	plumbous	lead(II)
	Pb^{4+}	plumbic	lead(IV)
antimony	Sb^{3+}	stibnous	antimony(III)
	Sb^{5+}	stibnic	antimony(V)
cobalt	Co^{2+}	cobaltous	cobalt(II)
	Co^{3+}	cobaltic	cobalt(III)
gold	Au^+	aurous	gold(I)
	Au^{2+}	auric	gold(II)
mercury	Hg^+	mercurous	mercury(I)
	Hg^{2+}	mercuric	mercury(II)

Example 3
The formula of a compound is found to be $SnCl_2$. What is its IUPAC name?

Solution
Charge on each Cl ion: 1–
Total negative charge: 2–
Charge on Sn ion: 2+
The IUPAC name for $SnCl_2$ is tin(II) chloride.

Example 4
The formula of a compound is found to be Fe_2O_3. What is its IUPAC name?

Solution
Total negative charge: 6–
Total positive charge: 6+
Charge on each Fe ion: 3+
The IUPAC name of Fe_2O_3 is iron(III) oxide.

The classical nomenclature system has, in the past, been used for naming compounds that contain multivalent metals with no more than two possible charges. In this system, the Latin name for the element along with the suffix *-ic* was applied to the larger charge, and the suffix *-ous* was applied to the smaller charge. The compounds formed by copper and chlorine were therefore known as cuprous chloride ($CuCl_{(s)}$) and cupric chloride ($CuCl_{2(s)}$).

In many industries, the classical system is still used extensively. **Table 2** shows a comparison of the classical and IUPAC names of multivalent metal ions.

> ▶ **Practice**
>
> **1.** Write the IUPAC names for the following binary ionic compounds:
> (a) $Na_2O_{(s)}$ (g) $Ni_2O_{3(s)}$
> (b) $SnCl_{4(s)}$ (h) $Ag_2S_{(s)}$
> (c) $ZnI_{2(s)}$ (i) $FeCl_{2(s)}$
> (d) $SrCl_{2(s)}$ (j) $KBr_{(s)}$
> (e) $AlBr_{3(s)}$ (k) $CuI_{2(s)}$
> (f) $PbCl_{4(s)}$ (l) $NiS_{(s)}$
>
> (See Appendix E for answers.)

Naming Compounds with Polyatomic Ions

Many familiar compounds (such as sulfuric acid, $H_2SO_{4(aq)}$, used in car batteries, and sodium phosphate, $Na_3PO_{4(aq)}$, a food additive typically found in processed cheese) are composed of three different elements. Compounds of this type are classified as *tertiary compounds.* (Many compounds that contain polyatomic ions consist of more than three different elements, but you will not be dealing with these compounds at this stage.) Tertiary ionic compounds are composed of a metal ion and a polyatomic ion (a covalently bonded group of atoms, possessing a net charge). You treat polyatomic ions much like regular monatomic ions when you write them in formulas or chemical equations.

Polyatomic ions that include oxygen are called *oxyanions.* One example is the nitrate ion, NO_3^-. In determining the name of a compound that contains an oxyanion, the first part of the name is easy. It is the name of the metal cation. The second part requires more thought. You have to consider the three parts of the ion indicated in **Figure 1**, on the next page.

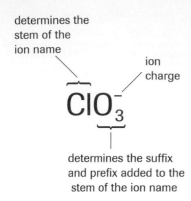

determines the stem of the ion name

ion charge

$$ClO_3^-$$

determines the suffix and prefix added to the stem of the ion name

Figure 1
The chlorate anion

There are four polyatomic ions formed from combinations of chlorine and oxygen. Note that all of these oxyanions have the same charge, despite the fact that their formulas are different.

- ClO^- is the hypochlorite ion.
- ClO_2^- is the chlorite ion.
- ClO_3^- is the chlorate ion.
- ClO_4^- is the perchlorate ion.

Note that in each name the stem is -chlor-. The suffixes and prefixes vary according to the number of oxygen atoms in the ion, as described below:

- The *per–ate* oxyanion has one more oxygen atom than does the *-ate* oxyanion.
- The *-ite* oxyanion has one fewer oxygen atom than does the *-ate* oxyanion.
- The *hypo–ite* oxyanion has one fewer oxygen atom than the *-ite* oxyanion.

Table 3 indicates some polyatomic ions that are commonly found in compounds.

You can use the crisscross rule to predict the formulas of ionic compounds that contain polyatomic ions. From the ion charges (whether for a single ion or for a polyatomic ion), determine the number of each ion necessary to yield a net charge of zero.

Table 3 IUPAC Names and Formulas of Some Common Polyatomic Ions

Ion	Name	Ion	Name
$C_2H_3O_2^-$	acetate	CO_3^{2-}	carbonate
ClO_3^-	chlorate*	CrO_4^{2-}	chromate
ClO_2^-	chlorite*	$Cr_2O_7^{2-}$	dichromate
CN^-	cyanide	HPO_4^{2-}	hydrogen phosphate
$H_2PO_4^-$	dihydrogen phosphate	$C_2O_4^{2-}$	oxalate
HCO_3^-	hydrogen carbonate (bicarbonate)	O_2^{2-}	peroxide
HSO_4^-	hydrogen sulfate (bisulfate)	SiO_3^{2-}	silicate
HS^-	hydrogen sulfide (bisulfide)	SO_4^{2-}	sulfate
HSO_3^-	hydrogen sulfite (bisulfite)	SO_3^{2-}	sulfite
ClO^-, OCl^-	hypochlorite*	$S_2O_3^{2-}$	thiosulfate
OH^-	hydroxide	BO_3^{3-}	borate
NO_2^-	nitrite	PO_4^{3-}	phosphate
NO_3^-	nitrate	$P_3O_{10}^{5-}$	tripolyphosphate
ClO_4^-	perchlorate*	NH_4^+	ammonium
MnO_4^-	permanganate	H_3O^+	hydronium
SCN^-	thiocyanate	Hg_2^{2+}	mercury(I)

*There are also corresponding ions that contains Br and I instead of Cl.

Example 5
Write the formula of copper(II) nitrate. (Use **Table 3** to find the charge of a nitrate ion.)

Solution
Charge on each Cu ion: 2+
Charge on each NO_3 ion: 1−

$$\overset{2}{Cu}\diagdown\diagup\overset{1}{NO_3}$$

Cu₁   (NO₃)₂

The formula of copper(II) nitrate is $Cu(NO_3)_2$.

Notice how brackets are used in the formula if there is more than one of the polyatomic ions. Brackets are not required with one polyatomic ion or with simple compounds.

Example 6
Use IUPAC nomenclature to name $CaCO_3$, which is commonly added to breakfast cereals as a source of calcium.

Solution
$CaCO_3$ is calcium carbonate.

Naming Hydrates

Many tertiary ionic compounds form crystals that contain molecules of water within the crystal structure. Such compounds are referred to as *hydrates*. When heat is applied to a hydrate, it decomposes to produce water vapour and an associated ionic compound, indicating that the water is loosely held to the ionic compound. The water molecules are assumed to be electrically neutral in the compound. When this water, called *water of hydration*, is removed, the product is referred to as *anhydrous*. Bluestone, hydrated copper(II) sulfate, is an example of a hydrate. Its formula is written $CuSO_4 \cdot 5\ H_2O_{(s)}$. Notice how the chemical formula

includes both the formula of the compound and the formula of water. This is true of the chemical formulas of all hydrated compounds. The formula of bluestone indicates the association of five water molecules with each unit of copper(II) sulfate (**Figure 2**). The IUPAC names of ionic hydrates indicate the number of water molecules by a Greek prefix (**Table 4**), so bluestone, or $CuSO_4 \cdot 5\ H_2O_{(s)}$, is called copper(II) sulfate pentahydrate.

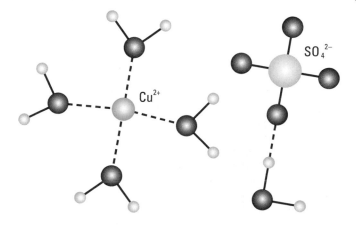

Figure 2
In a model of the compound copper(II) sulfate pentahydrate, the copper(II) ions are surrounded by four water molecules. The fifth water molecule is hydrogen-bonded to the sulfate ion.

Table 4   Prefixes Used when Naming Hydrated Compounds

Number of water molecules in chemical formula	Prefix in chemical nomenclature
1	mono-
2	di-
3	tri-
4	tetra-
5	penta-
6	hexa-
7	hepta-
8	octa-
9	nona-
10	deca-

2. Use IUPAC chemical nomenclature to name each of the following ionic compounds containing polyatomic ions:

(a) $LiClO_{3(s)}$

(b) $BaSO_{4(s)}$

(c) $Hg_2CO_{3(s)}$

(d) $Mg(NO_3)_{2(s)}$

(e) $Fe(BrO_3)_{3(s)}$

(f) $Na_3PO_{4(s)}$

(g) $NH_4IO_{3(s)}$

(h) $AuC_2H_3O_{2(s)}$

(i) $Zn_3(PO_4)_{2(s)}$

(j) $Sb(ClO_3)_{5(s)}$

(k) $MnSO_{3(s)}$

(l) $KBrO_{(s)}$

(m) $AlPO_{5(s)}$

(n) $Ag_2SO_{4(s)}$

(o) $Hg(BrO_3)_{2(s)}$

(p) $Fe_2(CO_3)_{3(s)}$

(q) $NH_4ClO_{(s)}$

(r) $Au(NO_3)_{3(s)}$

(s) $Mg(BrO_3)_{2(s)}$

(t) $NaIO_{(s)}$

(u) $Zn(ClO_2)_{2(s)}$

(v) $SnCO_{3(s)}$

(w) $SrSO_{3(s)}$

(x) $NiPO_{4(s)}$

(y) $Cu(C_2H_3O_2)_{2(s)}$

(z) $Ba_3(PO_5)_{2(s)}$

3. Name each of the following hydrated ionic compounds:

(a) $Na_2SO_4 \cdot 10\ H_2O_{(s)}$

(b) $MgSO_4 \cdot 7\ H_2O_{(s)}$

(See Appendix E for answers.)

Naming Molecular Compounds

If a binary compound is formed from two nonmetals, it is classified as a molecular compound. To name compounds that are formed from two nonmetals, you attach a Greek prefix to the name of each element in the binary compound, indicating the number of atoms of this element in the molecule. The common prefixes and their numerical equivalences are shown in **Table 5**. If there is only one of the first type of atom, you leave out the prefix *mono*.

Suppose that you are asked to write the IUPAC name of the chemical compound represented by the formula N_2O. Looking at the first element, you can see that the subscript after the nitrogen is 2, so the prefix for nitrogen is *di*. Looking at the second element, you can see that there is only one oxygen atom, so the prefix for oxygen is *mono*. Therefore, the formula's IUPAC name is dinitrogen monoxide.

Table 5 Prefixes Used when Naming Binary Covalent Compounds

Subscript in chemical formula	Prefix in chemical nomenclature
1	mono-
2	di-
3	tri-
4	tetra-
5	penta-
6	hexa-
7	hepta-
8	octa-
9	nona-
10	deca-

Example 7

What is the IUPAC name for the chemical compound CF_4?

Solution

C: carbon (not monocarbon)

F: tetrafluoride

The IUPAC name for CF_4 is carbon tetrafluoride.

Once again, hydrogen is an exception to this rule. The common practice is not to use the prefix system for hydrogen. For example, H_2S it is not called dihydrogen sulfide, but simply hydrogen sulfide.

4. Name the compound indicated by each of the following formulas:

(a) $SF_{6(g)}$

(b) $N_2O_{3(g)}$

(c) $NO_{2(g)}$

(d) $PCl_{3(l)}$

(e) $PCl_{5(s)}$

(f) $IF_{7(g)}$

(g) $BF_{3(g)}$

(h) $P_2S_{5(s)}$

(i) $P_2O_{5(s)}$

(See Appendix E for answers.)

Naming Acids

Acids are well-known, long-established chemicals. They were originally named decades or even centuries ago, and the use of classical names persists.

Binary acids are named by using the prefix *hydro-* with the stem of the name of the most electronegative element and the ending *-ic*. The name "hydrogen" does not appear. Instead, the word "acid" is added after the *hydro-stem-ic* combination, as indicated in **Table 6**. Consequently, the name of $HCl_{(aq)}$ is hydrochloric acid. Similarly, $HBr_{(aq)}$ is hydrobromic acid.

Table 6 Naming Binary Acids

Formula	Name
$HF_{(aq)}$	hydrofluoric acid
$HCl_{(aq)}$	hydrochloric acid
$HBr_{(aq)}$	hydrobromic acid
$HI_{(aq)}$	hydroiodic acid
$H_2S_{(aq)}$	hydrosulfuric acid

Note that the difference between the solution and the pure binary compound is indicated by the presence or absence of the subscript (aq) in the formula.

A second group of acids is named in the same way as binary acids. In this group, the IUPAC names of the polyatomic ions end in *-ide* (for example, the cyanide ion, CN^-). Looking at **Table 6**, you can see that the name of the acidic solution $HCN_{(aq)}$ is hydrocyanic acid.

A third group of acids is formed from various combinations of oxyanions (negative polyatomic ions consisting of a nonmetal plus oxygen) with hydrogen. Perhaps the best-known example is $H_2SO_{4(aq)}$, or sulfuric acid, which is one of the most widely produced industrial chemicals in the world. Phosphoric acid, $H_3PO_{4(aq)}$, is another example of an acid formed from an oxyanion and hydrogen. These acids are classified as *oxyacids* because they incorporate oxyanions. **Table 7** lists the names of some oxyacids.

The names of oxyacids can be derived according to the simple rules in **Table 8**.

Table 7 Oxyacids

Acid	Name
$HNO_{3(aq)}$	nitric acid
$HNO_{2(aq)}$	nitrous acid
$H_2SO_{4(aq)}$	sulfuric acid
$H_2SO_{3(aq)}$	sulfurous acid
$H_3PO_{4(aq)}$	phosphoric acid
$HC_2H_3O_{2(aq)}$	acetic acid
$HClO_{4(aq)}$	perchloric acid
$HBrO_{4(aq)}$	perbromic acid
$HIO_{4(aq)}$	periodic acid
$HClO_{3(aq)}$	chloric acid
$HBrO_{3(aq)}$	bromic acid
$HIO_{3(aq)}$	iodic acid
$HClO_{2(aq)}$	chlorous acid
$HClO_{(aq)}$	hypochlorous acid
$HBrO_{(aq)}$	hypobromous acid
$HIO_{(aq)}$	hypoiodous acid
$HFO_{(aq)}$	hypofluorous acid

When naming oxyacids, you omit the word "hydrogen" and add the word "acid." For example, to name the acidic solution with the formula $HNO_{2(aq)}$, you would first consider the IUPAC name: hydrogen nitrite (from **Table 3**). You change "nitrite" to "nitrous," drop the "hydrogen" from the front of the name, and add "acid" to the end. Thus, $HNO_{2(aq)}$ is called nitrous acid.

If you were asked to write the formula for an acid, you would first have to figure out the names of the ions involved, then their symbols or formulas, and then their ratio. For example, what is the formula for phosphoric acid? The *-ic* ending indicates the presence of the *-ate* oxyanion of phosphorus: phosphate. The phosphate oxyanion is PO_4^{3-} with a charge of 3–. The cation in oxyacids is always

Table 8 Rules for Naming Acids and Oxyanions

Name of oxyanion	Example	Formula	Name of acid	Example
per–ate	persulfate	SO_5^{2-}	*per–ic* acid	persulfuric acid
–ate	sulfate	SO_4^{2-}	*–ic* acid	sulfuric acid
–ite	sulfite	SO_3^{2-}	*–ous* acid	sulfurous acid
hypo–ite	hyposulfite	SO_2^{2-}	*hypo–ous* acid	hyposulfurous acid

hydrogen, which has a charge of 1+. To find the ratio of the ions, use the crisscross rule:

This method gives the subscripts for each ion.

$$H_3 \qquad (PO_4)_1$$

Divide each subscript by the highest common factor—in this case, 1.

$$H_3 \qquad (PO_4)_1$$

The hydrogen ions and the phosphate oxyanions combine in a ratio of 3:1. Therefore, the correct formula is $H_3PO_{4(aq)}$.

Example 8
What is the formula for the oxyacid sulfurous acid?

Solution
Sulfurous indicates the sulfite ion: SO_3^{2-}

The formula for sulfurous acid is $H_2SO_{3(aq)}$.

Example 9
What is the formula for the oxyacid hypochlorous acid?

Solution
Hypochlorous indicates one fewer oxygen atom than a chlorite ion: ClO^-

The formula for hypochlorous acid is $HClO_{(aq)}$.

Naming Bases

Chemists have discovered that all aqueous solutions of ionic hydroxides are bases. Aqueous ionic hydroxides, such as $NaOH_{(aq)}$ and $Ba(OH)_{2(aq)}$, are formed from a combination of a metal cation with one or more hydroxide anions. The name of the base is the name of the ionic hydroxide: in this case, aqueous sodium hydroxide and aqueous barium hydroxide.

▶ **Practice**

5. Name each of the following compounds:
   (a) $H_2SO_{3(aq)}$ (e) $H_2S_{(aq)}$
   (b) $H_3PO_{4(aq)}$ (f) $HCl_{(aq)}$
   (c) $HCN_{(aq)}$ (g) $H_2SO_{4(aq)}$
   (d) $H_2CO_{3(aq)}$ (h) $H_3PO_{4(aq)}$
6. Write the name of each the following bases:
   (a) $KOH_{(aq)}$ (b) $Ca(OH)_{2(aq)}$

(See Appendix E for answers.)

C7 Acids and Bases

Concentrated Reagents•

Reagent	Formula	Molar mass (g/mol)	Concentration (mol/L)	Concentration (mass %)
acetic acid	$HC_2H_3O_{2(aq)}$	60.05	17.45	99.8
carbonic acid	$H_2CO_{3(aq)}$	62.03	0.039	0.17
formic acid	$HCOOH_{(aq)}$	46.03	23.6	90.5
hydrobromic acid	$HBr_{(aq)}$	80.91	8.84	48.0
hydrochloric acid	$HCl_{(aq)}$	36.46	12.1	37.2
hydrofluoric acid	$HFl_{(aq)}$	20.01	28.9	49.0
nitric acid	$HNO_{3(aq)}$	63.02	15.9	70.4
perchloric acid	$HClO_{4(aq)}$	100.46	11.7	70.5
phosphoric acid	$H_3PO_{4(aq)}$	98.00	14.8	85.5
sulfurous acid	$H_2SO_{3(aq)}$	82.08	0.73	6.0
sulfuric acid	$H_2SO_{4(aq)}$	98.08	18.0	96.0
ammonia	$NH_{3(aq)}$	17.04	14.8	28.0
potassium hydroxide	$KOH_{(aq)}$	56.11	11.7	45.0
sodium hydroxide	$NaOH_{(aq)}$	40.00	19.4	50.5

• typical concentrations of commercial concentrated reagents

Acid–Base Indicators

Common name	Colour of $HIn_{(aq)}$	pH range	Colour of $In^-_{(aq)}$	Common name	Colour of $HIn_{(aq)}$	pH range	Colour of $In^-_{(aq)}$
methyl violet	yellow	0.0–1.6	blue	p-nitrophenol	colourless	5.3–7.6	yellow
cresol red (acid range)	red	0.2–1.8	yellow	litmus	red	6.0–8.0	blue
cresol purple (acid range)	red	1.2–2.8	yellow	bromothymol blue	yellow	6.2–7.6	blue
thymol blue (acid range)	red	1.2–2.8	yellow	neutral red	red	6.8–8.0	yellow
tropeolin oo	red	1.3–3.2	yellow	phenol red	yellow	6.4–8.0	red
orange iv	red	1.4–2.8	yellow	m-nitrophenol	colourless	6.4–8.8	yellow
benzopurpurine-4B	violet	2.2–4.2	red	cresol red	yellow	7.2–8.8	red
2,6-dinotrophenol	colourless	2.4–4.0	yellow	m-cresol purple	yellow	7.6–9.2	purple
2,4-dinotrophenol	colourless	2.5–4.3	yellow	thymol blue	yellow	8.0–9.6	blue
methyl yellow	red	2.9–4.0	yellow	phenolphthalein	colourless	8.0–10.0	red
congo red	blue	3.0–5.0	red	α-naphtholbenzein	yellow	9.0–11.0	blue
methyl orange	red	3.1–4.4	yellow	thymolphthalein	colourless	9.4–10.6	blue
bromophenol blue	yellow	3.0–4.6	blue-violet	alizarin yellow r	yellow	10.0–12.0	violet
bromocresol green	yellow	4.0–5.6	blue	tropeolin o	yellow	11.0–13.0	orange-brown
methyl red	red	4.4–6.2	yellow	nitramine	colourless	10.8–13.0	orange-brown
chlorophenol red	yellow	5.4–6.8	red	indigo carmine	blue	11.4–13.0	yellow
bromocresol purple	yellow	5.2–6.8	purple	1,3,5-trinitrobenzene	colourless	12.0–14.0	orange
bromophenol red	yellow	5.2–6.8	red				

D1 Matter and Chemical Bonding

SUMMARY

(a)

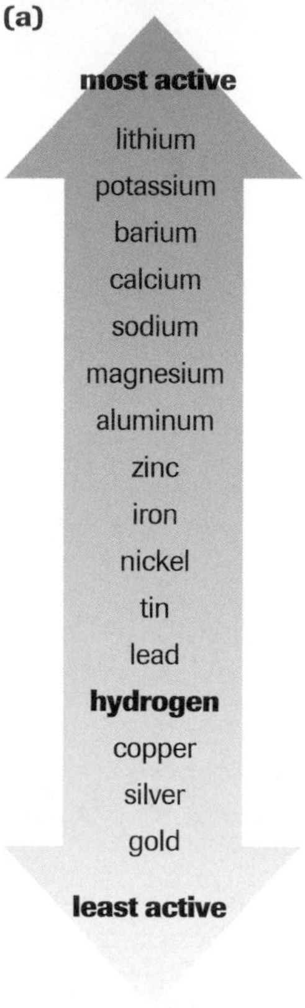

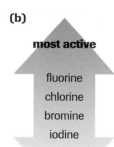

most active

lithium
potassium
barium
calcium
sodium
magnesium
aluminum
zinc
iron
nickel
tin
lead
hydrogen
copper
silver
gold

least active

(b)

most active

fluorine
chlorine
bromine
iodine

least active

Table 1 Summary of Bonding Characteristics

Intramolecular force	Bonding model
ionic bond	• involves an electron transfer, resulting in the formation of cations and anions • cations and anions attract each other
polar covalent bond	• involves unequal sharing of pairs of electrons by atoms of two different elements • can involve one, two, or three pairs of electrons: single (weakest), double, or triple (strongest) bonds
nonpolar covalent bond	• involves equal sharing of pairs of electrons • can involve one, two, or three pairs of electrons: single (weakest), double, or triple (strongest) bonds

Table 2 Summary of Reaction Types

Reaction type	Reactants	Products
combustion	metal + oxygen	metal oxide
	nonmetal + oxygen	nonmetal oxide
	fossil fuel + oxygen	carbon dioxide + water
synthesis	element + element	compound
	element + compound	more complex compound
	compound + compound	more complex compound
decomposition	binary compound	element + element
	complex compound	simpler compound + simpler compound or simpler compound + element(s)
single displacement	A + BC	B + AC
double displacement	AB + CD	AD + CB

Figure 1
(a) In the activity series of metals, each metal will displace any metal listed below it. Hydrogen is usually included in the activity series, even though it is not a metal, because hydrogen can form positive ions, just like the metals.
(b) The halogens can also be ordered in an activity series.

D2 Balancing Chemical Equations

Before you can analyze a sample quantitatively, you have to know the quantitative relationship between reactants and products. Writing a balanced equation is almost always the first step in a quantitative analysis. When balancing a chemical equation, always obey the *law of conservation of mass*. This law states that, in a chemical reaction, the total mass of all reactants equals the total mass of all products. It means that there must be as many atoms of each element on the right-hand side of the equation as there are on the left-hand side of the equation.

You will balance equations by inspection, which means that you count the number of atoms of each element on the left-hand side and right-hand side of the equation (before and after the reaction), and adjust the coefficients to make sure that there are equal numbers of each element on each side of the equation. Remember that you cannot change the subscripts in a chemical formula because the ratio of atoms in a compound is fixed. You can only change the coefficients (the number of entities). The following summary outlines a systematic approach that helps to reduce the time spent in a random trial-and-error method.

Balancing Chemical Equations by Inspection

Before beginning the balancing procedure, make sure that all molecular formulas are written correctly.

> **LEARNING TIP**
>
> **Reactants and Products**
> Recall that "left-hand side of the equation" refers to the reactant side of the equation and "right-hand side of the equation" refers to the product side.

> ▶ **SAMPLE** problem
>
> ### Balancing Equations by Inspection
>
> **(a) Balance the following equation by inspection:**
>
> $Ca(OH)_{2(aq)} + HCl_{(aq)} \rightarrow CaCl_{2(aq)} + H_2O_{(l)}$
>
> **Step 1: Write Unbalanced Equation**
> Since you are given the equation in the question, this step is already completed.
>
> **Step 2: Balance Atoms other than Hydrogen and Oxygen**
> Balance calcium first, and then chlorine. (Although the number 1 is not normally shown as a coefficient in a balanced equation, it is convenient to show the 1 when balancing and remove it at the end.) Fix the number of calcium atoms on both sides of the equation by placing a 1 in front of $Ca(OH)_{2(aq)}$ and $CaCl_{2(aq)}$. (*Fixing* means that you have placed a coefficient in front of the formula and "fixed" the number of atoms.)
>
> $1\ Ca(OH)_{2(aq)} + HCl_{(aq)} \rightarrow 1\ CaCl_{2(aq)} + H_2O_{(l)}$
>
> There are two chlorine atoms on the right-hand side of the equation and only one chlorine atom on the left-hand side. Balance chlorine by placing the number 2 in front of $HCl_{(aq)}$.
>
> $1\ Ca(OH)_{2(aq)} + 2\ HCl_{(aq)} \rightarrow 1\ CaCl_{2(aq)} + H_2O_{(l)}$

Step 3: Balance Oxygen

There are two atoms of oxygen on the left-hand side of the equation and one on the right-hand side. Place the number 2 in front of $H_2O_{(l)}$ to balance oxygen.

$$1\,Ca(OH)_{2(aq)} + 2\,HCl_{(aq)} \rightarrow 1\,CaCl_{2(aq)} + 2\,H_2O_{(l)}$$

Step 4: Balance Hydrogen

There are four atoms of hydrogen on the left-hand side of the equation $(2 + 2 = 4)$ and four on the right-hand side $(2 \times 2 = 4)$. The number of hydrogen atoms is already balanced, meaning that the entire equation should be balanced.

Step 5: Perform Final Check, and Remove the Coefficient 1

The number of atoms on the left-hand side of the equation equals the number of atoms on the right-hand side. The equation is balanced.

$$Ca(OH)_{2(aq)} + 2\,HCl_{(aq)} \rightarrow CaCl_{2(aq)} + 2\,H_2O_{(l)}$$

(b) Balance the equation for the complete combustion of ethyne, $C_2H_{2(g)}$.

Step 1: Write Unbalanced Equation

The complete combustion of hydrocarbons always involves a reaction of the hydrocarbon with oxygen. Water and carbon dioxide are the only products. Write the unbalanced equation as follows:

$$C_2H_{2(g)} + O_{2(g)} \rightarrow CO_{2(g)} + H_2O_{(l)}$$

Step 2: Balance Atoms Other than Hydrogen and Oxygen

Balance carbon.

$$1\,C_2H_{2(g)} + O_{2(g)} \rightarrow 2\,CO_{2(g)} + H_2O_{(l)}$$

Step 3: Balance Oxygen

When you balanced carbon, you did not fix the number of oxygen atoms on either side of the equation. Do not attempt to balance oxygen at this time. When oxygen is not fixed on one side of the equation, reverse the order of steps 3 and 4 by balancing hydrogen first and then oxygen.

Revised Step 3: Balance Hydrogen

$$1\,C_2H_{2(g)} + O_{2(g)} \rightarrow 2\,CO_{2(g)} + 1\,H_2O_{(l)}$$

Step 4: Balance Oxygen

There are five atoms of oxygen on the right-hand side of the equation $(4 + 1 = 5)$ and two atoms of oxygen on the left-hand side. Since there is no whole number that can be multiplied by 2 to give 5, temporarily place a fractional coefficient of $\dfrac{5}{2}$ in front of $O_{2(g)}$, because $\dfrac{5}{2} \times 2 = 5$, to balance the oxygen atoms on both sides of the equation.

$$1\,C_2H_{2(g)} + \frac{5}{2}\,O_{2(g)} \rightarrow 2\,CO_{2(g)} + 1\,H_2O_{(l)}$$

A fractional coefficient is a temporary solution because $\dfrac{5}{2} = 2.5$, which means that there are two oxygen molecules and one oxygen atom on the left-hand side of the equation. Since combustion is a reaction with oxygen molecules, not oxygen atoms, you need to remove the fractional coefficient in front of $O_{2(g)}$ by multiplying all entities in the equation by the denominator of the fractional coefficient, 2.

$$2\left(1\ C_2H_{2(g)} + \frac{5}{2}\ O_{2(g)} \to 2\ CO_{2(g)} + 1\ H_2O_{(l)}\right)$$
$$2\ C_2H_{2(g)} + 5\ O_{2(g)} \to 4\ CO_{2(g)} + 2\ H_2O_{(l)}$$

Step 5: Perform Final Check, and Remove the Coefficient 1

The number of all the atoms on the left-hand side of the equation equals the number of all the atoms on the right-hand side. There are no coefficients of 1. The equation is balanced.

$$2\ C_2H_{2(g)} + 5\ O_{2(g)} \to 4\ CO_{2(g)} + 2\ H_2O_{(l)}$$

Example

Balance the following equation by inspection:

$$Na_3PO_{4(aq)} + CaCl_{2(aq)} \to Ca_3(PO_4)_{2(s)} + NaCl_{(aq)}$$

Solution

$$2\ Na_3PO_{4(aq)} + 3\ CaCl_{2(aq)} \to Ca_3(PO_4)_{2(s)} + 6\ NaCl_{(aq)}$$

▶ Practice

Balance each of the following equations by inspection:

(a) $Na_{(s)} + H_2O_{(l)} \to NaOH_{(aq)} + H_{2(g)}$

(b) $CaCO_{3(s)} + HCl_{(aq)} \to CaCl_{2(s)} + H_2O_{(l)} + CO_{2(g)}$

(c) $MnO_{2(s)} + HCl_{(aq)} \to MnCl_{2(aq)} + Cl_{2(g)} + H_2O_{(l)}$

(d) $Pb(NO_3)_{2(aq)} + K_2S_{(aq)} \to PbS_{(s)} + KNO_{3(aq)}$

(e) the complete combustion of 2-pentene, $C_5H_{10(l)}$

(f) $(NH_4)_2CO_{3(s)} \to NH_{3(g)} + CO_{2(g)} + H_2O_{(l)}$

(g) the complete combustion of benzene, $C_6H_{6(l)}$, with oxygen to produce water and carbon dioxide

(See Appendix E for answers.)

D3 Quantities in Chemical Reactions

SUMMARY

Table 1 Stoichiometry: Symbols and Units

Symbol	Quantity	Unit
n	amount	mol
m	mass	mg, g, kg
M	molar mass	g/mol
N	number of entities	atoms, ions, formula units, molecules
N_A	Avogadro's constant, 6.02×10^{23} entities	–

D4 Solutions and Solubility

SUMMARY

Molar Concentration (mol/L)

$$\text{molar concentration} = \frac{\text{amount of solute (in moles)}}{\text{volume of solution (in litres)}}$$

$$c = \frac{n}{v} \quad \text{or} \quad n = vc \quad \text{or} \quad v = \frac{n}{c}$$

Preparing Standard Solution by Diluting Stock Solution

$$v_i c_i = v_f c_f$$

where

v_i = initial volume (volume of stock solution used)

c_i = initial concentration (concentration of stock solution used)

v_f = final volume (volume of dilute solution)

c_f = final concentration (concentration of dilute solution)

Hydrogen Ion Concentration and pH

pH is the negative power of 10 of the hydrogen ion concentration.

$$\text{pH} = -\log[H^+_{(aq)}] \quad \text{or} \quad [H^+_{(aq)}] = 10^{-\text{pH}}$$

Solution:	acidic	neutral	basic
$[H^+_{(aq)}]$:	$>10^{-7}$	10^{-7}	$<10^{-7}$
pH:	<7	7	>7

Note the inverse relationship between $[H^+_{(aq)}]$ and pH. The higher the hydrogen ion molar concentration, the lower the pH.

D5 **Hydrocarbons**

SUMMARY

Hydrocarbons

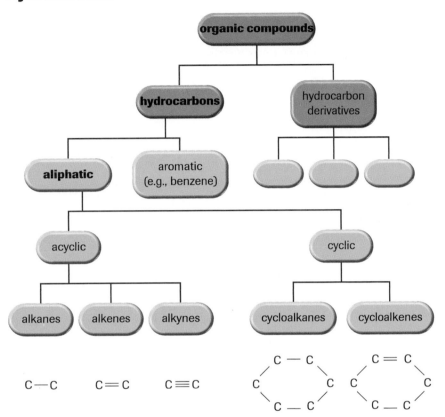

Table 1 Prefixes for Naming Alkanes, Alkenes, and Alkynes

Prefix	Number of carbon atoms
meth-	1
eth-	2
prop-	3
but-	4
pent-	5
hex-	6
hept-	7
oct-	8
non-	9
dec-	10

Figure 1
This classification system helps scientists organize their knowledge of organic compounds.

Isomers

Structural isomers are chemicals with the same molecular formula, but with different structures and different names.

Unit 2 Quantities in Chemistry

Are You Ready?, p. 76–77

5. (a) $2 Na_{(s)} + F_{2(g)} \rightarrow 2 NaF_{(s)}$

(b) $16 Al_{(s)} + 3 S_{8(s)} \rightarrow 8 Al_2S_{3(s)}$

(c) $7 CO_{(g)} + 14 H_{2(g)} \rightarrow C_7H_{14(l)} + 7 H_2O_{(l)}$

6. (a) $Ca_{(s)} + 2 H_2O_{(l)} \rightarrow Ca(OH)_{2(aq)} + H_{2(g)}$

(b) $Pb(NO_3)_{2(aq)} + 2 KI_{(aq)} \rightarrow PbI_{2(s)} + 2 KNO_{3(aq)}$

8. (a) 317.26 g

9.

Decimal	Scientific
0.010 m	1.0×10^{-2} m
401 mL	4.01×10^2 mL
385.5 g	3.855×10^2 g

10. (a) $y = 6$

(b) 60.1%

(c) 7.38×10^7

Section 2.1, p. 92

8. 4439 dimes

9. 4.81×10^{-1} mol

10. (b) 12.01 g/mol

(c) 65.38 g

(d) gold

Section 2.2, p. 106

1. 58.33 g/mol

2. 48.00 g/mol

3. 538 g

4. 342.34 g

5. (a) 6.02×10^{22} atoms

(b) 2.1×10^{24} formula units

6. (a) 14.6 mol

(b) 1.95 mol

(c) 0.793 mol

(d) 1.53×10^{-3} mol

(e) 4.0 mol

7. (a) 45.5 g

(b) 38.6 g

(c) 355 g

(d) 608 g

(e) 62 g

8. (a) 1.5×10^{24} molecules

(b) 8.8×10^{22} molecules

(c) 4.1×10^{22} molecules

9. (a) 4.4 g

(b) 18 g

(c) 3.2 g

10. 1.6×10^{24} molecules

11. 3.1×10^{20} molecules

12. water = 25 mol; sugar = 0.3 mol; vinegar = 0.040 mol; salt = 0.03 mol

Section 2.3, p. 120

4. (a) CH_2O

(b) NH_3

(c) CH

5. $K_2Cr_2O_7$

6. (a) $C_{11}H_{12}N_2O_5Cl_2$

(b) $C_6H_8N_2O_2S$

7. (a) 11.21% H
88.79% O

(b) 54.09% Ca
43.18% O
2.73% H

8. $Na_2C_8H_4O_4$

9. $C_7H_6OS_2$

Section 2.5, p. 137

1. 7200 g

2. 0.38 mg

3. 8.6 g

4. 1.64 mol/L

5. (a) 0.341 mol/L

(b) 70.0 mL

6. 10 mL

Section 2.9, p. 148

3. reactants = 3:1
products = 2:1

5. 4.01×10^{23} molecules

Section 2.10, p. 153

4. (a) 19 g

(b) 11 g

5. 42.5 g

6. 381.3 g

Section 2.12, p. 159

3. (a) $AgNO_{3(aq)} + NaBr_{(aq)} \rightarrow AgBr_{(s)} + NaNO_{3(aq)}$

(b) 5.53 g

(c) 5.03 g

(d) 91.0%

4. 96.4%

Unit 2 Review, pp. 169–173

1. (a) 4, 3, 2

2. (a) $12; 6.02 \times 10^{23}$

(b) 4×10^{25} g

3. (a) 6.02×10^{23} atoms

(b) 200.59 g

5. (a) 8.3×10^{-3} mol;
5.0×10^{21} atoms

(b) 0.0330 mol;
1.99×10^{22} atoms

6. (a) 166.14 g/mol

(b) 0.23 g

(c) 0.016 mol

(d) 1.0×10^{22} atoms

9. (a) CH_3O

(b) CH_3O or $C_2H_6O_2$

10. $A = C_{14}H_{28}O_4$
$B = CH_5N$

13. 0.58 mg

14. (a) 0.12 mol/L

(b) 0.121 mol/L

(c) 1.7×10^{-2} mol/L

15. (a) 0.1 mol/L; 0.4% W/V

(b) 0.02 mol/L;
0.08% W/V

16. (a) $3 Fe + 4 H_2O \rightarrow Fe_3O_4 + 4 H_2$

(b) $H_2SO_4 + 2 NaOH \rightarrow 2 H_2O + Na_2SO_4$

(c) $4 Cu + O_2 \rightarrow 2 Cu_2O$

(d) $Fe_2(SO_4)_3 + 12 KSCN \rightarrow 2 K_3Fe(SCN)_6 + 3 K_2SO_4$

17. (a) $C_2H_6O_{(l)} + 3 O_{2(g)} \rightarrow 2 CO_{2(g)} + 3 H_2O_{(l)}$

(b) $C_3H_8O_3 + 3 HNO_3 \rightarrow C_3H_5N_3O_9 + 3 H_2O$

18. (a) 3.76 mol; 2.10×10^2 g

(b) 84.8%

19. 0.139 mol; 4.45 g

20. (a) 255.1 g

(b) 2.22×10^2 g

21. (a) rainwater = 2:1
tap water = 2:1

Unit 4 Chemistry in the Environment

Section 4.3, p. 284

5. (a) 5.5 mg

Section 4.7, p. 304

8. (a) 10^{-7} mol/L

(b) 10^{-11} mol/L

(c) 10^{-2} mol/L

(d) 10^{-4} mol/L

(e) 10^{-14} mol/L

9. (a) pH = 3

(b) pH = 5

(c) pH = 7

(d) pH = 10

10. pH = 7

11. 100

12. (a) $[H^+_{(aq)}] = [Br^-_{(aq)}] = 0.15$ mol/L

(b) $[K^+_{(aq)}] = 0.15$ mol/L;
$[OH^-_{(aq)}] = 0.15$ mol/L

(c) $[NH_4^+_{(aq)}] = 0.15$ mol/L
$[Cl^-_{(aq)}] = 0.15$ mol/L

13. 3-ppm solution

Section 4.11, p. 323

4. 0.2778 mol/L

5. 20.49 mL

Section 4.14, p. 341–342

4. (a) 25°C = 298 K

(b) 30°C = 303 K

(c) −35°C = 238 K

(d) 312 K = 39°C

(e) 208 K = −65°C

5. (a) 62 L

(b) 2.3

7. 500 kPa

Unit 4 Review, pp. 363–367

7. 1.55 mg

31. (b) 0.140 mol/L

Appendix C6 Practice Questions, pp. 493–498

1. (a) sodium oxide
   (b) tin(IV) chloride
   (c) zinc iodide
   (d) strontium chloride
   (e) aluminum bromide
   (f) lead(IV) chloride
   (g) nickel(III) oxide
   (h) silver sulfide
   (i) iron(II) chloride
   (j) potassium bromide
   (k) copper(II) iodide
   (l) nickel(II) sulfide

2. (a) lithium chlorate
   (b) barium sulfate
   (c) mercury(I) carbonate
   (d) magnesium nitrate
   (e) iron(III) bromate
   (f) sodium phosphate
   (g) ammonium iodate

   (h) gold(I) acetate
   (i) zinc phosphate
   (j) antimony(V) chlorate
   (k) manganese(II) sulfite
   (l) potassium hypobromite
   (m) aluminum perphosphate
   (n) silver sulfate
   (o) mercury(II) bromate
   (p) iron(III) carbonate
   (q) ammonium hypochlorate
   (r) gold(III) nitrate
   (s) magnesium bromate
   (t) sodium iodate
   (u) zinc chlorite
   (v) tin(II) carbonate
   (w) strontium sulfite
   (x) nickel(III) phosphate
   (y) copper(II) acetate
   (z) barium perphosphate

3. (a) sodium sulfate decahydrate

   (b) magnesium sulfate heptahydrate

4. (a) sulfur hexafluoride
   (b) dinitrogen trioxide
   (c) nitrogen dioxide
   (d) phosphorus trichloride
   (e) phosphorus pentachloride
   (f) iodine heptafluoride
   (g) boron trifluoride
   (h) diphosphorus pentasulfide
   (i) diphosphorus pentoxide

5. (a) sulfurous acid
   (b) phosphoric acid
   (c) hydrocyanic acid
   (d) carbonic acid
   (e) hydrosulfuric acid
   (f) hydrochloric acid
   (g) sulfuric acid
   (h) phosphoric acid

6. (a) potassium hydroxide
   (b) calcium hydroxide

Appendix D2 Practice Questions, p. 503

(a) $2\,Na_{(s)} + 2\,H_2O_{(l)} \rightarrow 2\,NaOH_{(aq)} + H_{2(g)}$

(b) $CaCO_{3(s)} + 2\,HCl_{(aq)} \rightarrow CaCl_{2(s)} + H_2O_{(l)} + CO_{2(g)}$

(c) $MnO_{2(s)} + 4\,HCl_{(aq)} \rightarrow MnCl_{2(aq)} + Cl_{2(g)} + 2\,H_2O_{(l)}$

(d) $Pb(NO_3)_{2(aq)} + K_2S_{(aq)} \rightarrow PbS_{(s)} + 2\,KNO_{3(aq)}$

(e) $2\,C_5H_{10(l)} + 15\,O_{2(g)} \rightarrow 10\,CO_{2(g)} + 10\,H_2O_{(l)}$

(f) $(NH_4)_2CO_{3(s)} \rightarrow 2\,NH_{3(g)} + CO_{2(g)} + H_2O_{(l)}$

(g) $2\,C_6H_{6(l)} + 15\,O_{2(g)} \rightarrow 12\,CO_{2(g)} + 6\,H_2O_{(l)}$

Glossary

A

absolute zero 0 K or –273°C; believed to be the lowest possible temperature

acid a substance that neutralizes bases, conducts electricity in solution, and has a pH that is less than 7 (neutral) in solution

acid deposition acid-forming pollutants that fall to Earth as wet deposition (such as rain, hail, drizzle, fog, and snow) and as dry deposition (such as dust and other particulate matter)

acid precipitation any form of natural precipitation that has an unusually high acidity (pH less than 5.6)

activity series of metals a list of metals arranged in order of their reactivity, based on observations gathered from single displacement reactions

actual yield the quantity of product that is actually produced in a chemical reaction

addition polymer a polymer that is formed when monomer units are linked through addition reactions; all atoms in the monomer are retained in the polymer

addition reaction a reaction of an alkene or alkyne in which a molecule, such as hydrogen or a halogen, is added to a double or triple bond

aeration the process in which agitated water absorbs oxygen from the air

alcohol an organic compound that is characterized by the presence of a hydroxyl functional group; general formula R–OH

aldehyde an organic compound that is characterized by a terminal carbonyl functional group; that is, the carbon atom of a carbonyl group is bonded to at least one hydrogen atom

alkali metal a Group I metal

alkane a hydrocarbon that has only single bonds between carbon atoms; general formula C_nH_{2n+2}

alkene a hydrocarbon that contains at least one C=C double bond; general formula C_nH_{2n}

alkyl group a substitution group or branch derived from a hydrocarbon; general formula R

alkyne a hydrocarbon that contains at least one C≡C triple bond; general formula C_nH_{2n-2}

amide an organic compound in which the carbon atom of a carbonyl group (C=O) is bonded to a nitrogen atom

amine an ammonia molecule in which one or more hydrogen atoms are replaced by alkyl groups

amino acid a compound in which an amino group and a carboxyl group are attached to the same carbon atom

anion an atom that carries an overall negative charge because it has more electrons than protons

anode the electrode at which oxidation occurs

aqueous solution a solution in which water is the solvent; denoted by the subscript (aq)

aquifer an underground formation of porous rock that collects or holds ground water

atmospheric pressure the force per unit area that is exerted by air on all objects within the atmosphere

atom the smallest particle of an element that has all the properties of this element

atomic mass the mass of one atom of an element, expressed in atomic mass units, u

Avogadro's constant, N_A the number of entities in one mole; 6.02×10^{23}

B

base a substance that neutralizes acids, conducts electricity in solution, and has a pH that is greater than 7 (neutral) in solution

battery a group of two or more galvanic cells connected in series

Boyle's law As the pressure on a gas increases, the volume of the gas decreases proportionally, provided that the temperature and amount of gas remain constant.

C

carbohydrate a compound of carbon, hydrogen, and oxygen; general formula $C_x(H_2O)_y$

carbonyl group a functional group that contains a carbon atom joined to an oxygen atom with a double covalent bond; general formula C=O

carboxyl group a functional group that consists of a hydroxyl group attached to the carbon atom of a carbonyl group; general formula –COOH

carboxylic acid an organic compound that is characterized by the presence of a carboxyl group; general formula R–COOH

cathode the electrode at which reduction occurs

cathodic protection a form of corrosion prevention in which the metal being protected is forced to be the cathode of a cell, using either impressed current or a sacrificial anode

cation an atom that carries an overall positive charge because it has fewer electrons than protons

cell potential (voltage) a measure of the potential difference across the electrodes

centrifuge a piece of laboratory equipment that spins solutions at very high speeds, to separate the different particles from each other based on their densities

Charles's law As the temperature of a gas increases, the volume increases proportionately, provided that the pressure and the amount of gas remain constant.

chemical entity a chemical unit, such as an atom, an ion, or a molecule

chemical reaction a change that forms one or more new substances with different physical and chemical properties

combustible liquid a liquid that will ignite and burn at temperatures higher than normal working temperatures

combustion analyzer laboratory instrument that is used to determine the percentages of carbon, hydrogen, oxygen, and nitrogen in a compound

combustion reaction a chemical reaction that occurs when a substance reacts rapidly with oxygen, releasing energy

compound a molecule that contains two or more atoms of different elements, or a combination of oppositely charged ions, in fixed proportions

concentrated a relatively large amount of solute per unit volume of solution

concentration a ratio of the quantity of solute to the quantity of solution; symbol c

condensation polymer a polymer that is formed when monomer units are linked through condensation reactions

conductivity the ability of a substance to conduct electricity; a physical property of matter

continuous spectrum an uninterrupted pattern of colours that is observed when a narrow beam of white light passes through a prism

corrosion the deterioration of a metal as a result of slow oxidation

covalent bond a bond that arises when two atoms share one or more pairs of electrons between them. The shared electron pairs are attracted to the nuclei of both atoms.

cracking the process in which large straight-chain hydrocarbon molecules are converted into smaller branched-chain hydrocarbon molecules, usually by catalytic heating

crosslinks covalent bonds that form between polymer chains

D

Dalton's law of partial pressures The total pressure of a mixture of nonreacting gases is equal to the sum of the partial pressures of the individual gases.

decomposition reaction a chemical reaction in which a molecule or ionic compound is broken down into simpler entities

dilute a relatively small amount of solute per unit volume of solution

dilution the process of decreasing the concentration of a solution by adding more solvent

dipole–dipole force (DDF) an intermolecular force of attraction that forms between the slightly positive end of one polar molecule and the slightly negative end of an adjacent polar molecule

dissociate separate into positive and negative ions

double displacement reaction a chemical reaction in which two compounds in aqueous solution react to form two new compounds

E

electrode a solid electrical conductor where, in a galvanic cell, the electron transfers occur

electrolysis the process of supplying electrical energy to force a nonspontaneous redox reaction to occur

electrolyte a compound that, when dissolved in water, produces a solution that conducts electricity

electrolytic cell a cell that consists of a combination of two electrodes, an electrolyte solution, and an external battery or power source

electromagnetic energy light energy that travels in the form of waves

electron a negatively charged subatomic particle

electronegativity a measure of an atom's ability to attract a shared pair of electrons within a covalent bond

electroplating depositing a layer of metal onto another object at the cathode of an electrolytic cell

electrorefining the process of increasing the purity of a metal using an electrolytic cell

empirical formula a formula that gives the lowest ratio of atoms in a compound

empirical knowledge knowledge coming directly from observations

endpoint the point in a titration at which a sharp change in property, such as a colour change, occurs

enzyme a molecule, found within a biological system, that increases the speed of a chemical reaction

ester an organic compound that is characterized by the presence of a carbonyl group bonded to an oxygen atom

esterification a condensation reaction in which a carboxylic acid and an alcohol combine to produce an ester and water

ether an organic compound that has two alkyl groups (the same or different) attached to an oxygen atom; general formula R–O–R

excess reagent the reactant that is present in more than the required amount for a complete reaction to occur

excited state an electron's state when it absorbs energy and jumps to a higher energy level

F

flame emission spectroscopy a qualitative analysis technique that is used to determine the presence of a substance by exciting the substance's electrons using a flame and then detecting the electromagnetic energy emitted when the excited electrons return to their ground state

flammable liquid a liquid that will readily ignite and burn at room temperature

flashpoint the lowest temperature at which a flammable or combustible liquid will vaporize sufficiently to form a burnable mixture with air

formula unit the smallest amount of a substance having the composition given by its chemical formula

formula unit mass the mass of one formula unit of an ionic compound, expressed in atomic mass units, u

fractional distillation the separation of components of petroleum by distillation, using differences in boiling points

frequency the number of cycles per second

fuel cell a galvanic cell that produces electricity by a continually supplied fuel, such as hydrogen or methanol

functional group a particular combination of atoms that contributes to the physical and chemical characteristics of a substance

G

galvanic cell a device that converts chemical energy from redox reactions into electrical energy

galvanizing the process of coating iron or steel with a thin layer of zinc

global warming the increase in the average temperature of Earth's atmosphere

greenhouse effect a theory describing how heat is trapped near Earth's surface by carbon dioxide, atmospheric water vapour, and other gases

greenhouse gases gases in the atmosphere that trap heat close to Earth's surface, contributing to the greenhouse effect

ground state the energy level in which an electron is most stable; the electron does not release any electromagnetic radiation in this state

ground water water from precipitation that seeps underground and collects in aquifers

H

half-cell one of the two compartments of a galvanic cell, composed of an electrode and an electrolyte solution

hard water water that contains dissolved calcium, magnesium, and iron ions

homogeneous mixture a mixture that has uniform chemical and physical properties throughout because the components are uniformly distributed

hydrocarbon an organic compound that contains only carbon and hydrogen atoms in its molecular structure

hydrogen bond an intermolecular attraction between –OH and –NH groups in different molecules

hydrolysis a reaction in which a bond is broken by the addition of the components of water, resulting in the formation of two or more products

hydronium ion a hydrogen ion covalently bonded to a water molecule, represented as H_3O^+

I

impressed current a form of cathodic corrosion protection in which the metal object to be protected is attached to the negative terminal of a power source, making the object the cathode in a cell

indicator a chemical that changes colour when the acidity of a solution (its pH) changes

inference a judgment or opinion that is based on observations and/or conclusions from testing

intermolecular bonds bonds between molecules; forces of attraction that form between a molecule and its neighbouring molecules

ion an atom (or group of atoms) that has lost or gained one or more electrons

ionic bond the bond that results from the electrostatic force of attraction that holds positive and negative ions together

ionic compound a compound that consists of cations and anions held together by ionic bonds

ionic crystal a solid that consists of large numbers of cations and anions arranged in a repeating three-dimensional pattern

ionization a reaction in which electrically neutral molecules (or atoms) produce ions

isotope an atom of an element that has the same number of protons as the element, but different numbers of neutrons

isotopic abundance the relative quantities of isotopes in a natural sample of an element, expressed as percentages

IUPAC International Union of Pure and Applied Chemistry, the organization that establishes the conventions used by chemists

K

Kelvin temperature scale a temperature scale that has 0 K at absolute zero and degrees with the same magnitude as the degrees on the Celsius temperature scale

ketone an organic compound in which the carbon atom of a carbonyl group is bonded to two carbon atoms

kinetic molecular theory All substances contain particles that are in constant, random motion.

L

law of constant composition A compound contains elements in certain fixed proportions (ratios), regardless of how the compound is prepared, or where it is found in nature.

Lewis structure a representation of covalent bonding based on Lewis symbols, with shared electron pairs shown as lines and lone pairs shown as dots

Lewis symbol a diagram composed of a chemical symbol and dots, depicting the valence electrons of an atom or ion

limiting reagent the reactant that is completely consumed in a chemical reaction

line spectrum a discontinuous spectrum that is produced when light emitted by an element is directed through a prism or a diffraction grating; unique to an element

London dispersion force (LDF) an intermolecular force of attraction that forms between atoms of neighbouring molecules as a result of a temporary imbalance in the position of the atoms' electrons; forms between all molecules, polar and nonpolar

lone pair a pair of valence electrons that is not involved in bonding

M

macroscopic large enough to be seen with unaided eyes

mass spectrometer laboratory instrument that is used to measure the molar mass of a compound

model a representation of a theoretical concept

molar concentration the concentration of a solution expressed as moles of solute per unit volume of solution (mol/L)

molar mass the mass, in grams, of one mole of a chemical entity; symbol M

mole 6.02×10^{23} entities

mole ratio the ratio of the amount, in moles, of reactants and products in a chemical reaction

molecular element a molecule that contains two or more atoms of one type of element

molecular formula a formula that indicates the actual numbers of atoms in one molecule of a compound

molecular mass the mass of one molecule, expressed in atomic mass units, u

molecule two or more atoms that are joined by covalent bonds

monatomic ion an ion that is composed of only one atom

monomer a molecule that is linked with other similar molecules to form a polymer

N

nanometre 10^{-9} m; unit nm

net ionic equation an equation that depicts only the ions that are involved in a chemical reaction

neutralization a reaction between an acid and a base that yields a salt and water

neutron an uncharged subatomic particle found in the nucleus of an atom

nonelectrolyte a compound that, when dissolved in water, does not produce a solution that conducts electricity

nonpolar covalent bond a bond in which an electron pair is shared equally between a pair of atoms having the same electronegativity

nonpolar molecule a molecule that has no charged ends

nonspontaneous reaction a reaction that proceeds only with outside assistance

nucleus the positively charged centre of an atom

O

observation a statement that is based on what you see, hear, taste, touch, and smell

octet rule Atoms gain or lose electrons in their outermost shells in order to attain a noble gas configuration.

organic family a group of organic compounds with common structural features that impart characteristic physical and chemical properties

oxidation a process in which chemical entities lose electrons

oxidation number the apparent charge an atom would have if it gained or lost its bonding electrons

P

partial pressure the pressure, p, that a gas in a mixture would exert if it were the only gas in the same volume, at the same temperature

parts per million concentration unit that is used for very low concentrations; one part solute per million parts of solution; unit ppm

percentage composition the percentage, by mass, of each element in a compound

percentage yield the ratio, expressed as a percentage, of the actual or experimental quantity of product obtained (actual yield) to the maximum possible quantity of product (theoretical yield) derived from a stoichiometry calculation

petroleum a mixture of gases and liquids, composed of hydrocarbon molecules up to 40 carbon atoms long

pH a numerical scale that is used to measure how acidic or basic a solution is; negative of the exponent to the base 10 of the hydrogen ion concentration

plastic a synthetic substance that can be moulded (often under heat and pressure) and retains the shape it is moulded into

polar covalent bond a bond in which an electron pair is shared unequally between a pair of atoms that have different electronegativities

polar molecule a molecule that has a slightly positive charge on one end and a slightly negative charge on the other end

polyamide a polymer that is formed by condensation reactions resulting in amide linkages between monomers

polyatomic ion an ion that is composed of two or more atoms

polyester a polymer that is formed by condensation reactions resulting in ester linkages between monomers

polymer a molecule that consists of many repeating subunits

precipitate a solid formed in a reaction that takes place in aqueous solution

primary cell a cell that cannot be recharged

proton a positively charged subatomic particle found in the nucleus of an atom

Q

quantitative analysis measurement of the quantity of a chemical entity

quantized possessing a specific value or amount (quantity)

R

redox reaction a reaction in which one reactant is oxidized and the other reactant is reduced

reduction a process in which chemical entities gain electrons

S

sacrificial anode a form of cathodic corrosion protection in which a metal that is more easily oxidized than iron (or some other structural metal) is electrically connected to an iron object

salt bridge a container of electrolyte solution that connects the two solutions of a galvanic cell

saturated containing only single C–C bonds

secondary cell a rechargeable cell

single displacement reaction the reaction of an element and a compound to produce a new element and a new compound

soda-lime process a water-softening process in which sodium carbonate and calcium hydroxide cause calcium carbonate and magnesium carbonate to precipitate

solubility a measure of the extent to which a solute dissolves in a solvent at a given temperature and pressure

solute a pure substance in a solution that is dissolved by a solvent; usually the substance in lesser quantity

solution a homogeneous mixture of two or more pure substances

solvent the pure substance in a solution that dissolves other components; usually the substance in greater quantity

spectator ion an ion that is present during a chemical reaction but does not participate in the reaction

spectroscope an optical instrument that separates light energy into its component wavelengths; used in qualitative analysis

spontaneous combustion igniting and burning by itself, without an outside ignition source

spontaneous reaction a reaction that proceeds on its own without outside assistance

standard solution a solution of precisely and accurately known concentration

stoichiometry mathematical procedures for calculating the quantities of reactants and products involved in chemical reactions

strong acid an acid that completely ionizes in water to form ions and, therefore, is a good conductor of electricity

structural isomer a compound that has the same molecular formula as another compound but a different molecular structure

supernate the part of a centrifuged solution that does not settle to the bottom of the centrifuge tube

surface water water in lakes, ponds, rivers, and streams

synthesis reaction a chemical reaction in which two or more simple substances combine to form a more complex substance

T

theoretical knowledge knowledge based on ideas that are created to explain observations

theoretical yield the quantity of product calculated from a balanced equation

theory an explanation of a large number of related observations

titrant the solution in a burette during a titration

titration a laboratory procedure that involves the carefully measured and controlled addition of a solution, usually from a burette, into a measured volume of the sample being analyzed

total ionic equation a chemical equation that illustrates all soluble ionic compounds in their ionic form

U

unified atomic mass unit (u) the mass of one-twelfth of a carbon-12 atom

unsaturated containing at least one double or triple bond between carbon atoms

V

valence electrons electrons that are found in the outermost shell of an atom

van der Waals forces forces of attraction between molecules, such as the dipole–dipole force and the London dispersion force

visible spectrum the region of the electromagnetic spectrum that the human eye can detect

W

wavelength the distance between successive crests or troughs in a wave

weak acid an acid that partially ionizes in water to form ions and, therefore, is a poor conductor of electricity

Y

yield the quantity of product produced in a chemical reaction

Glossary

▸ Index

Excited state, 21
Extrusion, 241
Exxon Valdez, 271

F

Factor-label method, 87, 468–69
Fats, 189
Fertilizers, 161–62, 281, 282
Fever, and infrared energy, 17
Film-processing technicians, 374
Filtration, 474–75
Fingerprints, 56
Firefighters, 217
Fireworks, 1
Fish, 228
Fixing, 502
Flame emission spectroscopy, 25
Flame tests, 23–24
Flammable liquids, 215
Flashpoint, 215
Fluorine, 40
Forensic science, 8, 48, 75
Formic acid, 218, 219
Formula unit, 34
Formula unit mass, 81
Fractional distillation, 194–95
Freon, 202
Frequency, 16
Frogs, 327
Fuel cells, 411
Functional groups, 199–201

G

Gallstones, 206
Galvani, Luigi, 400
Galvanic cells, 394–99. *See also* Batteries;
    Cells
    constructing, 401–2
    and electrolytic cells, 430–31
Galvanic corrosion, 415
Galvanized steel, 413, 415
Galvanizing, 422–23
Gas chromatograph, 357
Gas chromatograph–mass spectrometer, 79
Gases
    greenhouse. *See* Greenhouse gases
    identifying, 19–20
    noble, 30
    properties of, 330
    solubility in water, 278–79
Gas laws, 330–41
Gasoline, 195, 215
Gas pressure, 331–32
    and solubility, 340–41
    and volume, 332–33
Geological technicians, 25
Global warming, 346
Glycerine, 206

Glycerol, 206
Glyptal, 245
Gold, in jewellery, 94, 169, 388, 417, 418, 436
Gold plating, 436, 437
Grapes, 205
Graphs, 467–68
Gravimetric analysis, 138
Great Lakes, 266, 271–72, 280, 281, 288
    and acid rain, 326
    environmental protection of, 290–92
Great Lakes Action Plan, 290
Greenfreeze, 202–3
Greenhouse effect, 345
Greenhouse gases, 344–48, 434. *See also* Air
    pollution; Carbon dioxide
Ground-level ozone, 350–52, 354
Ground state, 22
Ground water, 270
Guar gum, 244
Guest, Kelly, 142–43

H

Haber, Fritz, 161–62
Haber process, 161–62
Half-cells, 395
Half-reactions, 396
Hall, Charles Martin, 433
Hall-Héroult process, 433–34
Halogens, 202
Hamilton Harbour, 291–92
Hard water, 275–76, 361–62
    softening, 277–78
Hazardous household product symbols
    (HHPSs), 282
Hazardous household wastes, disposal of,
    283, 292, 305–7
HCFCs, 202–3
Heat capacity, of water, 271–72
HFCs. *See* Hydrofluorocarbons
Helium, 30, 95
Henry, William, 340
Henry's law, 340–41
Héroult, Paul, 433
HHPSs. *See* Household Hazardous Product
    Symbols
High-pressure liquid chromatography, 79
Hoffman apparatus, 171
Homogeneous mixture, 50
Homogenization, 79
Household cleaning products, analyzing, 63
Household Hazardous Product Symbols
    (HHPSs), 478
Hydrates, naming, 495–96
Hydrocarbons, 49
    addition reactions, 188–90
    combustion reactions of, 187–88
    defined, 180
    examples, 181
    naming, 181–86
    physical properties of, 187

summary, 505
uses, 194
Hydrochloric acid, 295–96
Hydrofluorocarbons (HFCs), 202–3, 346,
    347
Hydrofluoroether (HFE), 347
Hydrogen, and Haber process, 161–62
Hydrogen atom, Bohr model of, 21–22
Hydrogen balloons, 335
Hydrogen bonds, 200, 272
Hydrogen chloride, 41, 43, 294
Hydrogen cyanide, 213, 438
Hydrogen fuel cells, 411
Hydrogen ion concentrations, 297–301
Hydrogen peroxide, 50, 52
Hydrogen protons, and MRI, 28
Hydrolysis, 224
Hydronium ion, 294
Hydroxide ion, 293
    concentrations, 297–301
Hydroxyl group, 205, 218

I

Impressed current, 424, 427–28
Indicator, 26
Inference, 10–11
Infrared energy, 6, 17
Infrared light waves, 16
Ink, 223
Inspection, balancing chemical equations by,
    501
Intermolecular bonds, 42–44
International Union of Pure and Applied
    Chemistry (IUPAC), 35, 491
Intravenous solution, 124
Investigations, planning, 450–53
Ion colours, 490
Ionic bonds, 33
Ionic compounds, 33–34, 36
    binary, naming, 491–93
    molar mass of, 85
    solubility rules for, 54, 490
    tertiary, 493
Ionic crystals, 33
Ionization, 294
Ionization equations, 295
Ions
    determining presence of, in solution, 64
    formation of, 30–31
    monatomic, 33
    polyatomic, 33
    positive/negative, 398
    in water, testing for, 285–87
    word origin, 397
Iron, 272
    rusting of, 413–14
Isotope, 15
Isotopic abundance, 80
IUPAC. *See* International Union of Pure and
    Applied Chemistry

Index

Photo Credits

Unit Opening Photos: Unit 1: pp. iv and viii, Al Harvey; Unit 2: pp. iv and 74, Samuel Ashfield/Science Photo Library; Unit 3: pp. v and 174, Dawn Goss/First Light; Unit 4: pp. vi and 260, Bettmann/CORBIS; Unit 5: pp. vi and 368, Reuters New Media/CORBIS.

Unit 1: p. 2, David Wrobel/Visuals Unlimited; p. 6 (Fig 1), 1751L/NASA; p. 6 (Fig 2), Copyright © 1995,1996,1997,1998 by Terran Technologies, Inc. All rights reserved. *http://mineral.galleries.com*; p. 6 (Fig 3), EL2000-00308/NASA; p. 10, John Marshall Mantel/CORBIS; p. 12, Dave Starrett; p. 18, Jerry DiMarco/Montana State University; p. 24, Lester V. Bergman/CORBIS; p. 25, Ivy Images; p. 26, Richard Megna/Fundamental Photographers Inc.; p. 27 (top), Dick Hemingway; p. 27 (bottom), Bank of Canada; p. 28, Roger Ressmeyer/CORBIS; p. 33, Andrew Syred/Tony Stone Images; p. 49, CORBIS/Magma; p. 51, Richard Siemens; p. 52, Dave Starrett; p. 55, Jeremy Jones; p. 70 (Fig 3a), Richard Siemens; p. 70 (Fig 3b), Galen Rowell/CORBIS; p. 72, Robert Estall/CORBIS/Magma.

Unit 2: p. 78, Sheila Terry/Science Photo Library; p. 79, Rosenfeld Images Ltd./SPL; p. 82 (Fig 1a), PhotoEdit/Jeff Greenberg; p. 82 (Fig 1b), Colin Cuthbert/Science Photo Library; p. 83 (Fig 2), Science Photo Library; p. 83 (Fig 3), Richard Siemens; p. 94, Hans Georg/CORBIS; p. 95, Dick Hemingway; p. 97, Andrew Lambert Photography/Science Photo Library; p. 100, Royalty Free/CORBIS; p. 107 (Fig 1), Science Photo Library; p. 107 (Fig 2), Lawrence Livermore National Laboratory; p. 123 (Fig 1), Michael S. Yamashita/CORBIS; p. 123 (Fig 2), Peter Byron/PhotoEdit; p. 142, CP Photo/Andrew Vaughan; p. 148, Geographical Visual Aids; p. 150 (Fig 4a), Science Photo Library; p. 155, Dave Starrett; p. 158, Bob Krist/CORBIS; p. 161, CORBIS/Magma; p. 162, Potash & Phosphate Institute; p. 169 (Fig 1), Chris Collins/CORBIS; p. 169 (Fig 2), Visuals Unlimited; p. 171, Ivy Images; p. 172 (Fig 5a), Lester V. Bergman/CORBIS; p. 172 (Fig 5b), John Sohldon/Visuals Unlimited; p. 173, Dave Starrett.

Unit 3: p. 178, Zefa Visual Media, Germany; p. 179, Dave Starrett; p. 180, Charles & Josette Lenars/CORBIS/Magma; p. 182 (top), Rodney Hyett/Elizabeth Whiting & Associates/CORBIS/Magma; p. 182 (bottom), Lary Stepanowicz/Visuals Unlimited; p. 187, Michael Coyne/The Image Bank; p. 194, CORBIS/Magma; p. 195, Hamish Robertson; p. 199, Paul A Sauders/CORBIS/Magma; p. 202, Ivy Images; p. 203, NASA/Science Photo Library; p. 205, Ron Watts/CORBIS/Magma; p. 206, David M. Martin, M.D.; p. 207, Science Photo Library; p. 212, Joe MacDonald/CORBIS/Magma; p. 213 (Fig 2a), Michelle Garrett/CORBIS/Magma; p. 213 (Fig 2b), CORBIS/Magma; p. 216, Ivy Images; p. 218, ATP/CORBIS/Magma; p. 219 (Fig 2), Michael Freeman/CORBIS/Magma; p. 219 (Fig 3), EyeSquared; p. 220, Catherine Karnow/CORBIS/Magma; p. 221, Michelle Garrett/CORBIS/Magma; p. 223 (Fig 1), CORBIS/Magma; p. 223 (bottom) Dallas and John Heaton/CORBIS/Magma; p. 225, Elio Ciol/CORBIS/Magma; p. 228, Sinclair Stammers/Science Photo Library; p. 229, Custom Medical Stock Photo; p. 236, Bob Krist/CORBIS/Magma; p. 238 (Fig 1), Gregg Epperson/MaXx Images Inc.; p. 238 (Fig 3), Foodpix; p. 244 (Fig 1), Ivy Images; p. 244 (Fig 2), Ivy Images; p. 247, CORBIS/Magma; p. 254, Gail Mooney/CORBIS/Magma.

Unit 4: p. 266, Worldsat International Inc.; p. 268, Premium Stock/First Light; p. 274, Associated Press/CP Picture Archive; p. 276, Francoise Sauze/Science Photo Library; p. 277, WaterSoft DS-32P Softener. Courtesy of Shelson Crawford, Water Softening and Purification Specialist; p. 280 (Fig 1), Kevin Frayer/CP Picture Archive; p. 281, Gurmanlen/Morina/Visuals Unlimited; p. 288 (Fig 2), Courtesy of Pure Water Corporation; p. 288 (Fig 3), Courtesy of Trojan Technologies; p. 291 (Fig 2), Ron Pozzer/CP Picture Archive; p. 291 (Fig 3), Courtesy of the City of Hamilton, Ontario; p. 292, Courtesy of Metropolitan Council, St. Paul, MN; p. 305, Dave Starrett; p. 314, Dave Starrett; p. 316, EyeSquared; p. 317, Jack Chiang/CP Picture Archive; p. 318, Richard Siemens; p. 325, Ivy Images; p. 326, Environment Canada; p. 328, V. Wilkinson/Valan Photos; p. 330, ISS001_E5011/NASA; p. 331, CORBIS/Magma; p. 342, CORBIS/Magma; p. 347, Keith Gosse MBR/CP Picture Archive; p. 352 (Fig 1), Ken Straiton/First Light; p. 352 (Fig 2), Ontario Ministry of the Environment; p. 357, Ontario Ministry of the Environment; p. 364, Mark E. Gibson/Visuals Unlimited; p. 365, Robert Landau/CORBIS/Magma; p. 366 (Fig 5), Martin/Visuals Unlimited; p. 366 (Fig 6), R. Gerth/Masterfile.

Unit 5: p. 372 (Fig 1a), Peter Johnson/CORBIS; p. 372 (Fig 1b), Reuters New Media/CORBIS; p. 372 (Fig 2), Bettmann/CORBIS; p. 372 (Fig 3), A. Gragera/Latin Stock/Science Photo Library; p. 373, Dick Hemingway; p. 375, Dan Farrall/Photodisc; p. 378 (Fig 6), Minnesota Historical Society/CORBIS; p. 379, Hugh Rose/Visuals Unlimited; p. 380 (top), Royalty Free/CORBIS; p. 380 (bottom), Tom Brakefield/CORBIS; p. 382 (Fig 3), D.R. Degginger; p. 382 (Fig 4), Martin Bond/Science Photo Library; p. 384 (Fig 5), CORBIS/Magma; p. 384 (Fig 6), Ivy Images; p. 385 (Fig 1), John Sohldon/Visuals Unlimited; p. 386, Richard Siemens; p. 387, Richard Megna/Fundamental Photographs; p. 388, Hamish Robertson; p. 390, Crown Copyright/Health & Safety Laboratory/Science Photo Library; p. 391, Andrew Lambert Photography/Science Photo Library; p. 394 (Fig 1), Al Harvey; p. 394 (Fig 2), Courtesy of Professor Jean-Claude Nicole, *Europa Star* 3/2000; p. 403 (Fig 2), J.A. Wilkinson/Valan Photos; p. 404, Lester V. Bergman/CORBIS; p. 406, Ken Lucas/Visuals Unlimited; p. 410 (Fig 4), Courtesy of Makita Corporation; p. 412 (Fig 1), The Mariners' Museum/CORBIS; p. 412 (Fig 2), Ralph White/CORBIS; p. 412 (Fig 3), Charles O'Rear/CORBIS; p. 413 (Fig 4), Valan Photos; p. 413 (Fig 5), CORBIS/Magma; p. 414 (Fig 7), Eye Ubiquitous/CORBIS/Magma; p. 414 (Fig 9), Ivy Images; p. 415 (Fig 11), Ivy Images; p. 417 (Fig 1), Dick Hemingway; p. 417 (top), Professor Jean-Claude Nicole, *Europa Star* 3/2000; p. 418 (Fig 2), Professor Jean-Claude Nicole, *Europa Star* 3/2000; p. 418 (Fig 3), Alan Jakubek/CORBIS; p. 419 (top), Dick Hemingway; p. 419 (bottom), Matthias Kulka/CORBIS; p. 422 (Fig 1), Inga Spence/Visuals Unlimited; p. 423 (Fig 3), Kevin and Betty Collins/Visuals Unlimited; p. 432 (Fig 4), Dick Hemingway; p. 433 (Fig 1), Rosenfeld Images Ltd./SPL; p. 433 (Fig 2), Charles O'Rear/CORBIS; p. 435, George B. Diebold/CORBIS; p. 436 (Fig 7), Al Harvey; p. 436 (Fig 8), Dick Hemingway; p. 437 (Fig 10 left), Courtesy of The Westaim Corporation; p. 437 (Fig 10 right), Dick Hemingway; p. 437 (top), Dick Hemingway; p. 437 (bottom), © Rich Lasalle/agefotostock/First Light; p. 438, Agefotostock/First Light; p. 444, Richard Siemens; p. 445, Imagestate/First Light; p. 446, SynapsChemTools; p. 447, Robin Adshead/The Military Picture Library/CORBIS.

Appendices: vii and 448, Phil Schermeister/CORBIS/Magma; Appendix 6A (Fig 2, 3, 4, 5, 6, 8, 9, 10, 11, 13, 14, 15, 16), Richard Siemens; Appendix 6A (Fig 7a, b), Dave Starrett.